Arno Simons, Adrian Wüthrich, Michael Zichert, Gerd Graßhoff (eds.)
Understanding Science with Large Language Models?

Science Studies

Arno Simons is senior researcher at the Institute of History and Philosophy of Science, Technology, and Literature at Technische Universität Berlin and a member of the Network Epistemology in Practice (NEPI) project. Trained as an interdisciplinary scholar, he has contributed theoretical and methodological work across policy and environmental science, translational biomedicine, science and technology studies, media studies, and computer science. His expertise includes quantitative and qualitative social science methods, computational linguistics, and natural language processing, which he regularly combines in interdisciplinary, multi-method research.
Adrian Wüthrich is a guest professor for the History and Philosophy of Modern Science at Technische Universität Berlin. He leads the project Network Epistemology in Practice (NEPI), funded by a European Research Council Consolidator Grant, which investigates collective knowledge generation in large collaborations. His earlier work addressed interpretational debates in quantum mechanics, the role of images in science, causal reasoning in discovery processes, and the history of concepts in physics. His research often combines historical, philosophical, and computational approaches.
Michael Zichert is a PhD candidate in the project Network Epistemology in Practice (NEPI) at Technische Universität Berlin, where he studies how scientific knowledge is produced in large-scale collaborations. He holds a BA in History of Science from Technische Universität Berlin and an MA in Digital Humanities from Universität Leipzig. His research combines computational methods with the history of science, focusing on diachronic semantic change, social network analysis, and the application of large language models in the field.
Gerd Graßhoff is a professor for the History of Ancient Science at Humboldt-Universität zu Berlin and former director of the Excellence Cluster Topoi. He pioneered computational approaches to the history of science, applying machine learning and causal reasoning to ancient astronomical texts and scientific discovery processes. A Max Planck Fellow and member of Deutsche Akademie der Naturforscher Leopoldina, his research spans from Babylonian astronomy to digital humanities and AI methods for historical research.

Arno Simons, Adrian Wüthrich, Michael Zichert, Gerd Graßhoff (eds.)

Understanding Science with Large Language Models?

Potentials for the History, Philosophy, and Sociology of Science

[transcript]

This work was supported by the European Union under an ERC Consolidator Grant (Project No. 101044932, "Network Epistemology in Practice (NEPI)"). Views and opinions expressed are however those of the author only and do not necessarily reflect those of the European Union or the European Research Council. Neither the European Union nor the granting authority can be held responsible for them.

Bibliographic information published by the Deutsche Nationalbibliothek
The Deutsche Nationalbibliothek lists this publication in the Deutsche Nationalbibliografie; detailed bibliographic data are available online at https://dnb.dnb.de

transcript Verlag | Hermannstraße 26 | D-33602 Bielefeld | live@transcript-verlag.de

Cover design: Kordula Röckenhaus
Cover illustration: Adrian Wüthrich and Magdalena González (AI-assisted)
Printing: Elanders Waiblingen GmbH, Waiblingen
https://doi.org/10.14361/9783839447529
Print-ISBN: 978-3-8376-7994-6 | PDF-ISBN: 978-3-8394-4752-9
ISSN of series: 2703-1543 | eISSN of series: 2703-1551

Printed on permanent acid-free text paper.

Contents

Part 2: Historicizing LLMs

Part 3: Linguistic change, markers, and differences

Part 4: Knowledge graphs and topic modeling

Part 5: Retrieval-augmented generation (RAG)

Part 6: Citation context analysis

Foreword

This volume grew out of the international workshop "Large Language Models for the History, Philosophy, and Sociology of Science" held at Technische Universität Berlin, April 2–4, 2025, and organized by us, the Editors of this volume.[1] The workshop featured eighteen contributed short talks and two keynote presentations—one by Iryna Gurevych on Natural Language Processing for the analysis of intertextual relationships and the other jointly by Nina Tahmasebi and Pierluigi Cassotti on large-scale text analysis for the study of cultural and societal change. All contributions focused on the potential of large language models (LLMs) to address long-standing challenges in the history, philosophy, and sociology of science (HPSS). Participants explored concrete use cases and more general proposals for integrating LLMs into HPSS research. Video recordings of most presentations are available online.[2]

The workshop had two seeds. One was the ERC-project "Network Epistemology in Practice (NEPI)", led by Adrian Wüthrich. As part of NEPI, Arno Simons was among the first to pre-train a BERT model exclusively on texts from physics[3] and suggested that this kind of work should be discussed in a broader HPSS setting. It did not take much to convince Michael Zichert, also a member of the NEPI team and likewise working with LLMs to analyse conceptual developments in physics, that this was a worthwhile topic for a focused meeting. The second seed was Gerd Graßhoff, who has long advocated the use of computational methods in the history and philosophy of science, especially for the analysis of discovery processes. As a cooperation partner of the NEPI project, he had been considering a workshop on novel AI-assisted methods in history and philosophy of science. The Berlin event and this volume resulted from our decision to join forces.

The workshop centered on two key themes. First, the participants discussed the specific needs and desiderata of HPSS researchers and what they hoped LLMs can contribute to their work. Second, they critically assessed the current state of LLM development and

1 https://www.tu.berlin/hps-mod-sci/llms-for-hpss/workshop-llms-for-hpss
2 YouTube playlist "Large Language Models for the History, Philosophy, and Sociology of Science": https://youtube.com/playlist?list=PL5rAX6ywmP7O_nT99Osd74uino78BJMVT&si=4W52BV4pJot3XuyF
3 https://huggingface.co/arnosimons/astro-hep-bert

asked what extent their research goals are already becoming attainable with these systems.

The use of LLMs in HPSS connects to recent trends in these disciplines. In the history of science, computational approaches are in the process of establishing themselves among the standard repertoire of tools, with remarkable successes already achieved, for example through network analyses. Subfields of sociology of science have long relied on quantitative methods such as bibliometrics and scientometrics. More recently, the philosophy of science has experienced a shift towards more empirical work, including large-scale, algorithmic analyses of scientific or methodological concepts. Across these areas, computational tools can not only help reduce the workload in traditional research but, more importantly, also open up new avenues which to explore would otherwise be out of reach.

Analyses of co-occurrences and word frequencies, as well as more advanced techniques such as topic modeling, have made it possible to move beyond purely structural features of scientific activity and begin to touch the semantics of scientific language. However, a deeper understanding of scientific concepts, the structure of scientific arguments, and the process through which knowledge is transformed and circulates have remained formidable challenges for computational approaches in these fields.

The main motivation for the workshop and for this volume was that, with the advent of LLMs, the sketched situation might be changing. Natural language processing and machine learning have made rapid progress in capturing and analyzing meaning and grammatical structures of texts. This creates the prospect that LLMs can help HPSS researchers address some of the challenges just mentioned. At the same time, beyond general issues such as opacity, bias and interpretability, the use of LLMs for HPSS raises specific obstacles that stem from the specialized nature of scientific language and from the particular perspectives and aims of HPSS. A central goal of the workshop was therefore to ask how, given these obstacles, recent advances in LLM development might help tackle long-standing challenges in HPSS, and where their limits lie.

The invitation to contribute to this volume was circulated among workshop participants and attendees, with additional co-authors explicitly welcomed. Not all workshop speakers appear in this book, and some authors joined the project later, but all chapters bear the imprint of the discussions in Berlin. Arno Simons took the lead in developing the focus of the volume, coordinating the collective writing process with all contributors and managing the interaction with the publisher. All four of us—Simons, Wüthrich, Zichert and Graßhoff—were involved in the reviewing process for the volume. Julia Kim supported the organisation of the workshop in administrative matters, and Svenja Goetz and Lea Stengel helped with various organisational tasks during the workshop days and with the recording of talks, which was set up by Oliver Ziegler's uniCam team. We thank them all for their valuable support. Of course, neither the workshop nor this volume would have been possible without the many contributions of workshop presenters, the audience and the chapter authors. We thank all of you.

The workshop and this publication were funded by the European Union through the NEPI project (Grant no. 101044932). Views and opinions expressed are, however, those of the organisers and editors only and do not necessarily reflect those of the European

Union or the European Research Council. Neither the European Union nor the granting authority can be held responsible for them.

The Editors

Doing science studies with large language models
An introduction

Arno Simons, Adrian Wüthrich, and Michael Zichert

Introduction

Large language models (LLMs) are rapidly becoming part of the everyday infrastructure of scholarly work. They can read, summarise, search across, and transform texts at a speed and scale that changes the way researchers find and engage with literature and sources. As with any powerful new technique, their uptake raises questions about reliability, transparency and appropriate use, and these questions are intensified by the opacity and commercial governance of many current systems. For researchers in the history, philosophy and sociology of science (HPSS), this creates a double task: to explore how LLMs can be integrated into their own research practices, and to critically assess and guide that integration. HPSS is well positioned to do both, given its reflexive stance and close engagement with science and technology, but this double role also creates a tension between optimistic exploration and cautionary, critical reflection. The chapters in this volume are written from within this tension.

The volume documents how the HPSS community thought and felt about the arrival of LLMs as research tools in mid 2025. It captures the ideas, doubts, hopes, and arguments sparked by the ERC-funded international workshop "Large Language Models for the History, Philosophy, and Sociology of Science", held in Berlin in April 2025[1], and develops them further through continued reflection and exchange. All contributions were drafted and revised in a shared online document, where authors could read, comment on and respond to each other's chapters as they wrote. Cross-references and explicit dialogue across chapters were thus part of the design rather than an afterthought. This framing invited openness, experimentation, and personal voice, and encouraged authors to build on their own styles, interests, and positions within or adjacent to HPSS. In doing so, the volume carried the workshop's lively, collaborative and intellectually diverse atmosphere into written form. While many authors had presented their work at the workshop,

1 See https://www.tu.berlin/hps-mod-sci/llms-for-hpss/workshop-llms-for-hpss for general information about the workshop and https://www.youtube.com/playlist?list=PL5rAX6ywmP7O_nT99Osd74uino78BJMVT for video recordings of the talks.

not all speakers are represented in this volume, and some contributors joined the project later without having taken part in the Berlin meeting.

Building on our survey paper (Simons et al., 2026), this volume is situated within a growing literature on LLMs as research tools in fields related to HPSS. Programmatic publications and reviews have appeared in scientometrics (Zhang et al., 2025a), (digital) humanities (Karjus, 2025; Lozić and Štular, 2023; Zhong et al., 2025), quantitative (Thapa et al., 2025; Ziems et al., 2024) and qualitative (Friese and Morgan, 2026) social sciences (see also Christou, 2025; Davidson and Karrell, 2025), and related fields (Varnum et al., 2024). Computer scientists are also increasingly calling for interpretative and humanistic perspectives in the development and evaluation of LLMs and explicitly invite collaboration with the humanities (Ehrlich and Hazzan, 2025; Hemment and Kommers, 2025). In parallel, work on LLMs for scientific research more generally is expanding (Binz et al., 2025; Luo et al., 2025; Ren et al., 2025), including efforts to adapt models to scientific languages and tasks (Ho et al., 2024; Zhang et al., 2024) and to support literature reviews and research synthesis (Han et al., 2024; Scherbakov et al., 2025; Zhuang et al., 2025).

In the remainder of this chapter we briefly sketch what current LLMs are, how they are trained and in which broad ways they can be used. We then outline major opportunities and challenges that LLMs pose for research in HPSS. The chapter closes with an overview of the structure of the volume and a short introduction to each contribution.

What are LLMs, how are they trained, and how can we use them?

By LLMs we mean neural language models based on the transformer neural network architecture (Vaswani et al., 2017).[2] Transformers combine ideas that had been developed in earlier neural language models, such as contextualised word embeddings, encoder and decoder modules, and attention mechanisms, in a way that scales to very large datasets and model sizes. Although the original transformer was introduced in machine translation, it has since become the standard architecture for many kinds of language and multimodal models (Yin et al., 2024).

In everyday use, people often equate LLMs with chat oriented, generative systems such as ChatGPT. Technically, these are one branch of transformer models, built around the decoder part of the architecture and trained with a simple next-token prediction objective. Given a sequence of tokens, the model learns to predict the next token in the sequence, where a token can be a word, a subword, or sometimes just a single character (Radford et al., 2018; 2019). During training and generation the model is only allowed to use its representation of the tokens to the left of the prediction point. This left to right

2 The term "large" in large language models is a moving target. In some discussions it is reserved for frontier-scale generative models with tens or hundreds of billions of parameters, roughly from the size of GPT-3 onward. In this introduction we use the term LLM in a broader sense to refer to all transformer-based language models, including encoder-only models like the original BERT whose parameter counts are "only" in the hundreds of millions, since these models also play an important role in HPSS workflows.

setup is what makes these models generative. At inference time they can be asked to continue a text, answer a question, or transform an input, one token at a time.

In contrast, encoder-based transformer models are trained differently. The paradigmatic example is BERT (Devlin et al., 2018). Instead of predicting the next token, BERT-style models learn to fill in masked tokens in a sentence or paragraph. During training, a certain proportion of tokens in the input is replaced by a special mask symbol and the model is asked to recover the original tokens. Crucially, it can use both left and right context to do so. This bidirectional attention lets encoder models take into account the entire surrounding sentence or passage when deciding which token fits best.[3]

Both kinds of models, decoder-based and encoder-based[4], learn internal vector representations for each token in context, often called contextualised word embeddings (CWEs). Roughly speaking, these are numerical encodings of how a word is used in a particular sentence, such that uses with similar meanings end up with similar vectors (Ethayarajh, 2019; Jawahar, 2019). In decoder models, these representations are geared towards predicting the next token, using only information from the past in a sentence. In encoder models, they are shaped by information from both sides of a token and are therefore particularly informative about local meaning and sentence level structure.

This distinction matters because it gives rise to different application types, also within HPSS (Simons et al., 2026). Decoder-based, generative LLMs are extremely flexible text generators and can also be used for structured extraction. The introduction of GPT-2 (Radford et al., 2019) and GPT-3 (Brown et al., 2020) showed that training larger networks on larger datasets led to the emergence of unanticipated skills, such as few- and zero-shot learning. In response, the AI industry scaled up decoder models at high speed and added further post training steps, such as instruction tuning and reinforcement learning from human feedback (RLHF), to improve their ability to follow prompts (Ouyang et al., 2022). More recently, specialised prompting and training techniques for step-by-step, so-called chain-of-thought (CoT) "reasoning" (Wei et al., 2022) have further increased their performance on tasks that involve mathematical proofs, logic puzzles or multi-step argumentation. These efforts resulted in today's proprietary, energy intensive generative frontier models and their striking ability to solve many tasks through natural language prompting, including many tasks directly relevant to HPSS. While this success has concentrated attention and investment on very large decoder models, it has not made encoder models obsolete.

---

3 After BERT, researchers improved this encoder-style "masked token" approach in two main ways. RoBERTa (Liu et al., 2019) largely kept the same masked-word objective but showed that changes to the training recipe can make this kind of learning work better in practice. ELECTRA (Clark et al., 2020) introduced an alternative objective in which the model learns to detect tokens that have been replaced, offering a different and often more efficient route to learning strong bidirectional representations. DeBERTaV3 (He et al., 2022) builds on these ideas by pairing a strong encoder architecture with ELECTRA-style training, illustrating that post-BERT progress also came from rethinking the pretraining objective itself.

4 Some transformers are encoder–decoder hybrids, such as BART (Lewis et al., 2019) and T5 (Raffel et al., 2020), combining bidirectional encoding with left-to-right generation. This blurs the split, but the roles still matter as components that can appear alone or together.

Encoder-based models with their context-rich CWEs still play a central role for HPSS, especially for tasks that concern word and sentence meaning, such as semantic similarity, topic exploration, or tracking conceptual change (Simons et al., 2026; in this volume, see Ahmadi, 2026; Simons, 2026a). They are often adapted in different ways. A common strategy is task specific fine-tuning for classification, sequence labelling or token tagging on relatively small labelled datasets. Another is to fine-tune them to produce sentence embeddings (Reimers and Gurevych, 2019), that is, single vectors that represent whole sentences or paragraphs and can be compared with similarity measures. Such embeddings underpin state-of-the-art text-clustering and topic-exploration tools (Grootendorst, 2022). They are also widely used in many retrieval-augmented generation (RAG) setups to index and search large corpora, so that an encoder model first retrieves the most relevant passages which are then handed to a generative model that composes an answer on top of them (Gao et al., 2024). In other words, BERT-like models quietly power the search and filtering stages that make many GPT-style systems look well informed.[5]

For many HPSS-related tasks it therefore makes sense to ask not only which single model type fits best, but how differently specialised models can be combined in task-specific pipelines. In many current systems, these pipelines are wrapped into so called "agents", in which a generative LLM acts as a controller that breaks a task into steps, calls external tools such as search engines, databases, code interpreters or encoder models, and integrates their outputs into a final answer (Huang et al., 2024; Ren et al., 2025; Schmidgall et al., 2025). Emerging multi-agent setups go one step further by coordinating several such agents in different roles, for example as planners, critics or domain specialists (Guo et al., 2024; Tran et al., 2025). LLMs are also increasingly used as judges, for instance to rate or compare candidate answers from different prompts or models (Li et al., 2025). Beyond model choice, HPSS scholars can experiment with prompting strategies, including CoT and prompt chaining (Sun et al., 2024), to decompose complex research questions into smaller steps that LLMs can handle more reliably. Many chapters in this volume, such as Danilova et al. (2026), Scharnhorst et al. (2026), and Schlattmann et al. (2026), illustrate such a division of labor, discussing the combination of encoder and decoder models or multiple prompt stages within a single workflow.

Opportunities

Natural language text is one of the most important data sources in HPSS research, which makes LLMs, with the improved context-sensitive semantic capacities and broad task performance sketched above, natural candidates for many kinds of text-related work.

5 On the encoder side, development continues beyond the original BERT generation. Some BERT-style models have been pushed to much larger parameter counts, including DeBERTa (He et al., 2022; 2021) and multilingual encoders such as XLM-R (Goyal et al., 2021). In addition, prompt-based methods like pattern-exploiting training (PET) show that masked language models can support few-shot classification by reframing labels as fill-in-the-blank decisions, offering an alternative to full task-specific fine-tuning (Clavié et al., 2025; Schick and Schütze, 2021; He et al., 2022). These newer encoder-focused lines of work are worth following to see which yield robust, meaning-sensitive tools that could become useful for HPSS.

HPSS scholars routinely engage with a wide range of genres, from scientific publications across virtually all fields and subfields, through transdisciplinary materials such as industry reports, policy papers and media coverage, to field notes, archival documents and interview transcripts. Modern LLMs can also process structured material, for example tables, databases or code, and increasingly multimodal inputs such as images, figures and video (Yin et al., 2024). Taken together, these capacities open up a broad landscape of possible applications, from large-scale corpus analysis to interactive exploratory reading. At the same time, they lead HPSS into largely uncharted territory, where questions about methodological fit, expertise, trust, and control over the tools become newly pressing. In this section we highlight some of the main opportunities that LLMs create for HPSS, before turning to the challenges and risks that accompany them in the next section.

A first opportunity is domain and task adaptability. Frontier generative models are strong multitask and in-context learners, so they can often be applied to very HPSS-specific problems simply by careful prompt design, for example to generate alternative readings of citation contexts (Simons, 2026b). At the same time, both decoder and encoder models can be adapted more deeply to particular domains by continued pretraining on targeted corpora, such as physics articles for tracing shifting meanings of key terms like "Planck" or "virtual" over time (Zichert et al., 2025; Zichert and Simons, 2026; Simons, 2026a). This kind of domain adaptation sharpens a model's sensitivity to local vocabularies, genres and histories of use, which is crucial for many HPSS questions. Together, prompt-based task adaptation and corpus-based domain adaptation make it possible to build task-specific pipelines in which different models do what they are best at. For example, Danilova et al. (2026) first use small decoder LLMs to generate structured historical texts, then use encoder models to check how diverse and historically plausible these texts are and to map out the main topics they cover, and finally treat this vetted synthetic corpus as training data for a future genre classifier, which could itself be an encoder-based model.

A second opportunity is the methodological tunability of LLMs. They can support multiple epistemic roles in HPSS, with adjustable levels of delegation, commitment, and evidential responsibility. Depending on how they are configured and validated in context, LLMs can be used to delegate well-specified tasks such as document classification, first-pass coding, or information extraction. In this mode, they return relatively stable outputs that can serve as inputs to later analysis or as deliverables in their own right. They thus scale up familiar close reading practices to large corpora (e.g., Alves et al., 2026; Boulanger, 2026; D'Alessandro, 2026). But the same architectures can also be positioned in lower-commitment, supportive roles, for example as conversational partners in the exploratory reading of particular sources (e.g., Hill, 2026) or entire archives (e.g., Scharnhorst et al., 2026), where outputs function as suggestions, contrasts, or "what if" scenarios. Importantly, the dimensions do not always align: models can be used to automate exploratory work (for example by generating candidate codes or hypotheses at scale), or to support more evidential tasks (for example by producing checkable extractions with quotes and pointers that the researcher verifies). Across projects, and sometimes across stages of a single project, this tunability lets researchers choose how much agency to delegate, how strongly to treat outputs as results, and how explicitly evidential the model's contributions are supposed to be (cf. Simons et al., 2026).

A third opportunity lies in the ability of LLM-based methods to act as translation layers that improve interoperability across scales, methods, and heterogeneous materials. Because they can operate at different textual scales, from passages to full documents and large corpora, LLM-based methods facilitate the design of multi-level analyses that move between abstracts, sections, and full texts (cf. Malaterre and Lareau, 2026). In addition, they enable movement between fine-grained textual interpretation and aggregate analysis, allowing researchers to relate local readings and categories to broader distributions and patterns, including in mixed-method settings (e.g., Schlattmann et al., 2026). This interoperability also shows up in the ease with which LLM components can be inserted into otherwise familiar pipelines, for example embedding-based semantic expansion of key terms to broaden or refine term sets in conceptual and terminological studies (e.g., Aguilar-Valdez et al., 2026; Malaterre and Lareau, 2026), or the generation of synthetic or semi-synthetic training data to bootstrap genre classifiers or coding models when annotated examples are scarce (e.g., Danilova et al., 2026). Overall, the opportunity lies in using LLMs as mediators that make different scales of text, heterogeneous sources, and qualitative and quantitative modes of analysis interoperable, with modular insertion points that let researchers build those connections where a project most needs them. Finally, as multimodal systems mature, this kind of mediation can extend beyond text to figures, tables, diagrams, and other visual or audiovisual materials that are central to scientific communication and archival work (see Graßhoff, 2026).

A fourth opportunity concerns documentation, reproducibility, and shared infrastructure, especially for interpretative work. LLM-based analyses are often implemented through prompts, scripts, and configuration files that spell out, in a concrete, machine-readable way, what was asked of the model, which data were used, and how intermediate steps were wired together. Such files can be run, modified, and inspected by others. They also make prompts and pipelines visible as concrete objects of study in their own right. For earlier computational methods in HPSS, such as topic models or citation networks, this kind of scripted reproducibility was already possible, but it mostly applied to pattern-oriented analysis. LLMs extend this possibility into domains of interpretative work, for example qualitative coding, summarising or reconstructing arguments, that previously lived almost entirely in tacit practices and prose descriptions (e.g., Hitch, 2024; Morgan, 2023; Zhang et al., 2025b; but see Nguyen and Welch, 2025, for a sharp critique of applying LLMs to qualitative research). If their pipelines are carefully documented, and if the models are clearly version-pinned and run with fixed, deterministic decoding settings (for example with temperature set to zero), then even interpretative LLM-based studies might come much closer to the reproducibility standards of other computational methods. With proprietary frontier models, strict repeatability of outputs is harder to achieve (see the fourth challenge below), yet publishing prompts, code, and logs could still improve procedural transparency. We see an opportunity in turning such workflows into reusable models, shared datasets or shared prompt libraries tailored to particular sources, thereby creating a common infrastructure that others can adopt, adapt and extend, and that helps robust methods spread across projects, institutions and subfields of HPSS and beyond (cf. Boulanger, 2026; Scharnhorst et al., 2026; Simons, 2026b).

A fifth opportunity is the way LLMs push HPSS toward a more reflexive articulation of concepts and reasoning. To obtain useful behaviour from a model, researchers must specify tasks, categories, examples, and evaluation criteria with an unusual degree of explicitness. Writing and revising prompts, programming pipelines, curating training or few-shot examples and testing how sensitive outputs are to small changes all tend to surface tacit assumptions about conceptualization, context or relevance (e.g., Simons, 2026b; Liesegang and Gläser, 2026; Zichert and Simons, 2026). In this way, model steering and evaluation can double as methodological self-study, where the frictions, failures, and ambiguities encountered with the system help clarify where our own concepts are ill-suited, overloaded or contested, and where they are robust.

Challenges

The opportunities just outlined do not come for free. They are intertwined with a series of challenges and open questions that HPSS needs to confront if LLMs are to become responsible and intellectually productive parts of its research practices.

A first challenge concerns the trade-off between accessibility and literacy (Simons et al., 2026; see also Marx, 2024). Frontier models in polished interfaces make it very easy to run impressive-looking analyses without much technical knowledge, which lowers the barrier to entry but makes it tempting to rely on off-the-shelf behaviour for genuinely difficult HPSS tasks. Many of these tasks, however, require careful prompt design, task-specific evaluation and often some form of corpus adaptation or fine-tuning, which presupposes at least basic coding skills, familiarity with the tools and interfaces used to run and adapt models, and an understanding of how these systems fail. This creates a tension between accessible tools that many can use and more accurate, tailored workflows that only a subset of researchers can build and diagnose, and raises the question of how HPSS as a field can develop and share LLM literacy rather than leaving it to individual enthusiasts (cf. Meding and Daugh, 2026b; Wagner and Hermes, 2026; Vogl et al., 2026).

A second challenge is the limited temporal and contextual sensitivity of current models (see Part 2 of this volume, "Historicizing LLMs", which includes chapters by Büttner, 2026; Olival et al., 2026; Danilova et al., 2026; and Wolf, 2026). Most LLMs are not explicitly trained to track historical change in language, concepts or genres, yet this is central to the history of science and to historically-informed strands of philosophy and sociology of science. Today's models tend to blur or collapse historical distance, for example by projecting contemporary meanings into older texts or by smoothing over shifts in terminology that are precisely what HPSS wants to study. Addressing this may require not only new training regimes and datasets, but also architectural innovations such as explicit time encodings or attention mechanisms that respect temporal structure (cf. Büttner, 2026). HPSS scholars are well placed to formulate what genuine time awareness would mean in practice and should therefore take part in defining and testing such developments rather than treating them as purely technical questions left to others.

A third challenge concerns interpretative work and the absence of fixed ground truths. Much HPSS research deals with shifting categories, multiple plausible readings and situated judgments that cannot easily be reduced to single correct labels (cf. Alves et al., 2026, who show how disputed labels and shifting boundaries complicate "ground truth" in discipline classification). This sits uneasily with standard machine-learning notions of training data and evaluation, which assume clear targets against which model performance can be scored. HPSS therefore needs to reflect on what it means to use LLMs in domains where disagreement and ambiguity are not bugs but core features of the object of study, and to experiment with evaluation practices that foreground qualitative appraisal, comparative reading and argumentation rather than simple accuracy metrics (e.g., Meding and Daugs, 2026a). In doing so, the field can connect to neighbouring interpretative disciplines in the humanities and social sciences that face similar issues (Friese and Morgan, 2026; Hitch, 2024; Morgan, 2023; Zhang et al., 2025b), and respond to emerging calls for a "qualitative turn" in contemporary AI (Hemment and Kommers, 2025). Such a turn treats model outputs, especially from LLMs, as cultural artefacts that are subject to the kinds of judgments and curatorial practices familiar from literary studies, art criticism or ethnography. At a time when some interpretative communities are highly skeptical of (Nguyen and Welch, 2025) or calling for bans on generative AI in their own domains (Jowsey et al., 2025) while major social-science and computer-science preprint servers are narrowing the space for interdisciplinary debates about LLMs from both ends (arXiv, 2025; SocOpen, 2025; see also Brainard, 2025), it becomes all the more important for HPSS to claim room within its own institutions for sustained, technically-informed discussion of these tools (cf. Simons et al., 2026).

A fourth, more general challenge follows from these first two points. If serious HPSS work with LLMs requires literacy, temporal sensitivity, and a nuanced view of interpretative judgment, then the field needs to co-shape its own tools (cf. Eberle, 2026). This includes defining benchmarks and task formulations that reflect HPSS-specific questions (e.g., Boulanger, 2026), building and curating domain-specific corpora, and finding ways to deal with "ground truth" scarcity in areas where categories are contested or annotations are expensive. It also involves organising prompt, code, and model sharing in ways that make successful workflows visible and reusable, in line with the fourth opportunity above, for example through shared repositories, community guidelines or small purpose-built platforms. A key part of this agenda is to work toward open-science solutions, wherever possible using or co-developing open models and datasets so that HPSS research does not depend too much on closed, evolving systems controlled by a few large providers (Valleriani, 2025). This will often require collaboration beyond HPSS, for instance co-developing general-purpose open models with other scientific communities, or working with neighbouring fields that face similar challenges, such as historical linguistics or digital humanities for historical languages, while focusing HPSS-internal efforts on more specialised models and datasets. Without such collective efforts, HPSS risks either importing ill-fitting tools and evaluation practices from other fields or fragmenting into isolated projects that cannot easily learn from each other.

A fifth challenge is how to engage with LLMs critically, neither demonising them nor using them naively (e.g., Meding and Daugs, 2026b). Beyond technical literacy (see the first challenge above), HPSS needs reflective frameworks for integrating these systems into established methodologies without letting them quietly deflect or erode those methods (e.g., Khutsishvili, 2026). Models can offer seemingly convincing shortcuts, for example quick summaries, instant codings or plausible-sounding explanations, that risk nudging researchers away from slow reading, archival work or theory-driven interpretation. They can also reconfigure qualitative workflows in ways that shift or endanger the epistemic authority of researchers themselves, raising questions about who is recognised as the primary knower and interpreter in a project (Khutsishvili, 2026; Nguyen and Welch, 2025). At the same time, LLM-driven automation intervenes in very material struggles over scarce academic resources and workloads, fuelling justified concerns about the outsourcing or elimination of research and teaching tasks, especially for early-career and precariously employed scholars. The field will have to decide in what roles LLMs can serve as co-analysts, pattern-finders or sparring-partners, how to disclose and justify their use in publications, and how to resist the pressure to align research questions with what is easy for the models rather than what is substantively important.

Finally, HPSS faces the general challenges that come with LLMs, which are not specific to this field but still shape its choices (cf. Bender et al, 2021; Guest et al., 2025; Lang, 2026; Khutsishvili, 2026; Strubell et al., 2019). On the epistemic side, there is the black-box character of large models and their tendency to hallucinate or fabricate details in ways that are hard to detect and correct. On the structural side, there is dependence on a small number of large technology companies, their changing business models, and their entanglement with political and economic agendas that HPSS itself often studies. On the ethical side, there are issues of energy use and climate impact, the exploitation of human annotation labour, opaque data collection and privacy risks. Engaging with LLMs in HPSS therefore also means taking a position on these wider conditions, or at least making them visible as part of the cost and context of any research that relies on such systems.

Outline of the book

The volume is organised into six parts.

Part 1, "General challenges and limitations", gathers six chapters that set the conceptual and critical scene. **Sarah Lang**'s opening chapter maps key ethical and epistemic risks of bringing proprietary LLMs into humanities workflows, from opaque and biased training data and "open-washing" to exploitative data labour, environmental costs, and threats to research integrity such as hallucinations and paper mills. She critiques explainable AI as insufficient on its own and foregrounds dataset documentation and auditing, care-based and solidaristic approaches, and stronger institutional and funding requirements for accountable AI. **Kristina Khutsishvili** examines how LLMs fracture epistemic authority by shifting scientific authorship from the act of judgement toward the act of approval. Drawing on themes of recognition and "imagined communities", she argues that AI-enabled writing and synthesis can blur boundaries of belonging and reframe scientists as curators of outputs whose genesis is opaque. For HPSS, this both threatens interpretive identity through mimicry and increases its relevance as a source of epistemological orientation about trust, responsibility, and what counts as inquiry. **Oliver Eberle**, writing from both ML and HPSS perspectives, argues that reliable use of frontier and "reasoning" models in HPSS hinges on better evaluation and interpretability. He surveys approaches from behavioral benchmarks to mechanistic explanation methods, highlights the lack of HPSS-specific datasets and ground-truthed evaluations, and calls for curated materials, culturally and temporally grounded benchmarks, and stronger HPSS participation in AI research and policy to shape trustworthy systems. **Holle Meding and Aurel Daugs** map five recurrent limitations of LLMs for historical scholarship, from hallucinations and chrono-insensitivity to informational presentism and Anglocentrism, alignment effects, and black-box opacity. They argue for a cautious, task-specific use where concepts, prompts, parameters, and evaluation criteria are made explicit, and they illustrate this with an evaluated named entity recognition (NER) workflow on German memory-discourse data that treats LLM outputs as checkable components rather than as autonomous historical narration. **Andreas Wagner and Jürgen Hermes** stage a dialogue about "encoded humanities" that pushes back against the one-size-fits-all, chat-centric drive toward generative systems. They argue that many humanities tasks are better served by encoder-based pipelines that support traceability, minimal computing, and reproducible workflows, while reserving generative models for carefully framed exploratory roles like hypothesis generation or cultural analytics under conditions of openness and critical verification. Finally, **William D'Alessandro** reports a case study in philosophy of mathematical practice, using Gemini to mine thousands of arXiv papers and assemble a large annotated dataset of mathematicians' explanatory aims. He finds that current LLMs can assist impressively with long-context retrieval and disciplined analytical comparison of existing views, yet tend to fail at producing genuinely novel philosophical theories. He argues that this gap is largely a training and evaluation problem: unlike math or coding, philosophical progress offers fewer repeatable, reinforcement-learning-friendly patterns and little consensus ground truth.

Part 2, "Historicizing LLMs", includes four chapters that turn explicitly to temporality. **Jochen Büttner** argues that most LLMs treat texts from many centuries as if they belonged to one undifferentiated "present", which blurs historical change and pushes outputs toward modern language and assumptions, making anachronism a structural risk for historical work. He surveys strategies for temporal conditioning, from period-focused pretraining and time-slicing to time tokens, time embeddings, temporal attention, and temporal RAG, and concludes that time grounding will likely be most feasible and useful in targeted tasks given the scale and metadata hurdles of building time-conditioned foundation models. **Fernanda Olival, Helena Freire Cameron, António Branco, and Renata Vieira** draw on their work with the 18th-century Portuguese Memórias Paroquiais to show how generative LLMs stumble on pluricentric language variation and uneven digital representation across Portuguese varieties, historical spelling and transcription/normalization constraints, and formally structured tasks like named entity recognition, where a Portuguese BERT-like full-context model outperforms generative models. They argue for tailored historical corpora, domain fine-tuning, and hybrid workflows that enforce formal constraints, with close human and multidisciplinary oversight. **Vera Danilova, Julia Reed, Andrew Burchell, Gijs Aangenendt, and Ylva Söderfeldt** test zero-shot synthetic genre generation for 20th-century patient-organisation medical periodicals in four European languages using small open-weight models. Combining historian evaluation with diversity metrics, they find reasonable genre mimicry but weak historical accuracy and strong year and digit preference biases, highlighting a diversity–hallucination trade-off and motivating a cautious view of "historical plausibility" and downstream reuse. **Jeffrey C. Wolf** closes the part with a reflection on where LLMs actually fit in multilingual, longue-durée digital history. Rather than chat-style generation, he argues that encoder embedding models are the workhorse for cross-lingual semantic matching and text reuse at scale, but only if they are evaluated and adapted to four hard constraints: strong multilingual coverage, OCR noise, modern out-of-domain bias, and semantic shift across centuries. He outlines the trade-offs in temporal modeling, motivates corpus-specific fine-tuning and preprocessing, and stresses interdisciplinary collaboration and transparent workflow design.

Part 3, "Linguistic change, markers, and differences", brings together six chapters that all use distributional approaches to study scientific language and its evolution. **Christophe Malaterre and Francis Lareau** propose "epistemic framings" as patterned ways disciplines articulate what counts as knowledge and how it is justified, and show how embeddings and LLMs can help detect and map these framings at scale. They argue for hybrid workflows that combine curated epistemic markers with embedding-based expansion, syntactic pattern mining for multiword expressions, and selective LLM annotation or labeling where full-corpus LLM use is still too costly. **Michael Zichert and Arno Simons** offer a programmatic and methodological chapter on computational conceptual history for HPSS. They situate LLM-based methods within a longer trajectory of computational attempts to model conceptual change and use this perspective to highlight the specific opportunities and challenges that LLMs introduce for this undertaking. They argue that contextualized embeddings and hybrid workflows expand how polysemy and change can be modelled, but that corpus construction, concept operationalization, evaluation, and interpretive responsibility remain the decisive bottlenecks, now sharpened by model choice, training data, and infrastructure. **Arno Simons**' chapter then provides a detailed case study. Focusing on the multi-sense term "Planck" in astrophysics and high-energy physics, he compares several general and domain-adapted BERT models, including his own Astro-HEP-BERT, showing how CWEs can disambiguate senses, induce sense clusters and track diachronic shifts in their relative prominence. The chapter demonstrates what highly specialised, domain-adapted models can do for HPSS-style conceptual history and offers practical guidance on model choice, domain adaptation, and validation. Using similar CWE-based methods, **Elaheh Sadat Ahmadi** turns to "codification", understood as the degree to which a field stabilises its terminology. Drawing on sociological debates about codification, she proposes a semantic uniformity score derived from contextual embeddings and applies it to astrophysics and sociology, comparing how consistently key terms are used in each field. The chapter not only reports clear differences in semantic uniformity but also reflects on limits and validation strategies, thus linking classic codification theory to contemporary distributional methods. **Diego Alves, Sergei Bagdasarov, Badr Mohammed Abdullah, and Stefania Degaetano-Ortlieb** use a fine-tuned Llama-3 model to classify Royal Society Corpus articles across more than three centuries (1665–1996) into a small, explicitly defined set of disciplines, treating disciplines as evolving linguistic registers. Through manual annotation, error analysis, and a close look at "out-of-list" labels, they show both strong practical performance and how misclassifications often stem from diachronic lexical and semantic change, shifting genre conventions, and genuinely fuzzy boundaries between disciplinary sublanguages. Using embeddings of

LLM-generated TLDR summaries, they visualise overlaps between disciplinary language varieties as a way to probe interdisciplinarity, and argue that discipline taxonomies are historically loaded instruments that may require period-sensitive refinement. Finally, **Sofía Aguilar-Valdez, Bách Phan-Tất, Dirk Speelman, Dirk Geeraerts, and Stefania Degaetano-Ortlieb** offer a deliberate counterpoint by not using LLMs for their core analysis, arguing that detecting signals of the late 18th-century shift from phlogiston to oxygen requires evidence traceable to a defined corpus. They instead combine LDA topic modeling and information-theoretic change measures with dependency-based slot–filler patterns around "air" and its modifiers, and then contrast this approach with ChatGPT and Grok to clarify what LLMs can supply as macro-narrative versus what they miss when the aim is inspectable, reproducible linguistic markers.

Part 4, "Knowledge graphs and topic modeling", consists of four chapters and moves from word- and phrase-level analysis to structured representations of scientific knowledge and discourse. **Christian Boulanger** shows how building a socio-legal knowledge graph for disciplinary history runs aground on missing and messy citation data in non-English, footnote-heavy law and humanities texts. To tackle this, he develops an open, TEI-annotated gold-standard corpus and sets up an LLM-based extraction and evaluation workflow using LLamore, benchmarking it against Grobid and highlighting trade-offs in accuracy, speed, and compute. He argues that opening bibliometrically neglected domains will require open evaluation data, shared standards for interchange, and collaborative infrastructure building. **Raphael Schlattmann, Aleksandra Kaye, and Malte Vogl** trace the development of a two-stage, human-in-the-loop pipeline that extracts and validates open triples from biographical lexicons and then turns them into an ontology-guided knowledge graph with linking and constraint checks. Using an interview format plus LLM "sparring partner" critiques, they make the iterative toolbuilding process, trade-offs between nuance and operationalisation, and the structuralist commitments of graph-based history unusually explicit. **Francis Lareau and Christophe Malaterre** compare topic modeling on titles, abstracts, and full texts in astrobiology journals, contrasting LDA with embedding-based BERTopic and evaluating models with diversity, coherence, recall, and topic-balance measures. They find that title-only models are generally too information-poor, that abstracts often hit a "Goldilocks" sweet spot with high coherence at far lower cost, and that full texts can improve recall while sometimes reducing coherence due to methodological noise. The chapter argues for goal-driven, and sometimes multi-level, use of different text structures rather than assuming that more text is always better. Finally, **Malte Vogl, Alexander von Schwerin, and Sabrina Kirschke** reflect on an interdisciplinary project that combines BERTopic's embedding-based topic modelling with a generative LLM used to produce concise, meaningful topic labels for qualitative narrative analysis of the German genetic engineering discourse. They show how LLM-assisted labeling plus interactive visualisations can help researchers orient themselves in a large, time-stamped corpus, while also surfacing heavy translation work between developer and users and an 'extra-dark' black box problem around the lack of best-practice norms, missing training, and reproducibility.

Part 5, "Retrieval-augmented generation (RAG)", gathers five chapters that focus on concrete LLM workflows in qualitative and archival research built around RAG architectures. In their second chapter, **Holle Meding and Aurel Daugs** present and critically evaluate a modular RAG pipeline built on a 100,000+ article corpus from DER SPIEGEL, a major German weekly news magazine, showing how chunking, embeddings, and retrieval settings shape both semantic discovery and source-traceable responses. The pipeline was developed and piloted in a collaborative project with students, and it is evaluated via a pragmatic "silver standard", foregrounding retrieval quality as the key bottleneck, advocating prompts that force uncertainty when evidence is missing, and arguing that RAG only becomes reliable when paired with historian-led evaluation and source criticism. **Miira Hill** introduces the "data interview" as a reflexive, RAG-based approach to qualitative content analysis grounded in the Empirical Theory of Science (EToS), which treats knowledge as socially produced and stabilized through communicative practices, institutions, and infrastructures. Accordingly, LLM outputs are handled as situated, dialogical artifacts that require interpretation and validation rather than being taken as results. In a case study of German Facebook comments on migration and history, she shows how multi-interviewer prompting, corpus-grounding, and team-based coding can scale interpretation while surfacing persistent risks like misclassification, repetitive evidence, and normative "false balance" in sensitive content. **Andrea Scharnhorst, Han Yang, Jetze Touber, Kim Ferguson, Philipp Mayr, and Vyacheslav Tykhonov** situate RAG in the context of cultural-heritage and research-data infrastructures. Tracing an engineering journey from "archives for everyone" to "chatting with collections" on the Dataverse platform and in the MuseIT project, they show how local RAG chatbots that combine LLMs with knowledge-organisation systems can empower curators and users of cultural data, provided they are co-designed with galleries, libraries, archives, and museums and embedded in open, interoperable and commons-oriented infrastructures. **Gerd Graßhoff** introduces the AI-Reporter pipeline, which takes slides and recordings of academic talks and turns them into structured, publication-ready chapters. Using the opening lecture of our Berlin workshop as an example, he demonstrates how multimodal LLM pipelines can create a new genre of scientific communication that bridges live presentation and written text, while still relying on human editorial control for curation and quality. **Jeffrey C. Wolf**'s second contribution is an experiment in recursive reflection. Adapting an open-source NotebookLM RAG variant to run locally, he generates extended scholarly dialogues about his own workshop talk and uses them to explore what happens when we use LLMs to discuss LLMs in HPSS. The resulting conversations are technically impressive yet shallow, and he argues that such computational reflexivity, while playful and revealing, still falls short of the deeper, situated reflexivity that human scholars practice.

Part 6, "Citation context analysis (CCA)", includes three chapters that turn to an established method at the intersection of scientometrics and the sociology of science and explores how LLMs challenge and transform it. **Lydia Liesegang and Jochen Gläser** argue that LLM-based automation turns CCA itself back into an object of inquiry, because LLM support forces explicit rules that expose CCA's tacit assumptions. They challenge three assumptions: that nearby text reliably says something about the cited work, that coding schemes can ignore missing values, and that fixed sentence or paragraph windows match the real unit of analysis, the argument. They also warn about circular validation: if the "gold standard" annotations were produced using the same flawed rules, an LLM can score well by copying them while still misreading what the citation is doing. They conclude that useful LLM support needs theory-driven CCA designs and evaluation that checks argumentative meaning, not just agreement with legacy labels. **Arno Simons, Hiba Arnaout, and Iryna Gurevych** propose a roadmap for "reconstructive" CCA that treats citation meaning as plural and inter-contextual, emerging across citing passages, cited sources, and patterns of uptake over time. They translate these aims into LLM-supported workflows from context detection and classification to citation–entity and citation–argument mining, similarity and clustering of citation framings, and chained, time-aware interpretative readings that surface alternatives while keeping human judgment, evidence visibility, and reproducible settings central. In doing so, they also sketch several forward-looking tasks that are still rare or largely unexplored in computational CCA, especially those aimed at tracing evolving meanings and argumentative linkages rather than refining evaluative metrics. Overall, they argue that LLMs can widen the scale and nuance of citation interpretation only when used reflexively as tools for inquiry rather than as mere substitutes for the hermeneutic work of researchers. In the final chapter, **Arno Simons** argues for "scaling in" thick, case-based citation interpretation, testing it on a canonical hard case from the CCA literature. He develops a two-stage GPT-5 workflow that pairs a stable surface label with cross-text checking and multiple, text-grounded interpretative hypotheses. A 2×3 prompt-variation study then shows how scaffolding and framing steer which plausible readings and vocabularies the model foregrounds, reinforcing that LLMs work best as inspectable co-analysts whose outputs remain curated and judged by humans.

In sum, the contributions in this volume probe and stretch traditional HPSS methodologies and invite readers to engage critically with a new item in their toolkit: LLMs. Taken together, they exemplify the openness and reflexivity toward LLM use that we hope this volume will foster.

Use of LLMs in the production of this volume

Parts of this introduction, as well as many of the chapters in this volume, were written with support from LLMs, such as OpenAI's ChatGPT. Any text produced with LLM assistance was reviewed, edited and, where necessary, substantially rewritten by the authors, who bear full responsibility for the published versions. Where particular uses of LLMs are methodologically relevant, they are documented in the respective chapters.

References

Aguilar-Valdez S, Phan-Tất B, Speelman D, et al. (2026) Discursive parallels of the chemical revolution. Topic modelling and distributional analysis. In: Simons A, Wüthrich A, Zichert M, et al. (eds) *Understanding Science with Large Language Models? Potentials for the History, Philosophy, and Sociology of Science*. Bielefeld: transcript, part-3.

Ahmadi ES (2026) Exploring disciplinary differences in semantic uniformity. A computational approach to codification. In: Simons A, Wüthrich A, Zichert M, et al. (eds)

Understanding Science with Large Language Models? Potentials for the History, Philosophy, and Sociology of Science. Bielefeld: transcript, part-3.

Alves D, Bagdasarov S, Abdullah BM, et al. (2026) Use of large language models in the classification of scientific texts into disciplines. In: Simons A, Wüthrich A, Zichert M, et al. (eds) *Understanding Science with Large Language Models? Potentials for the History, Philosophy, and Sociology of Science*. Bielefeld: transcript, part-3.

arXiv (2025) Attention Authors: Updated Practice for Review Articles and Position Papers in arXiv CS Category – arXiv blog. Available at: https://blog.arxiv.org/2025/10/31/attention-authors-updated-practice-for-review-articles-and-position-papers-in-arxiv-cs-category/.

Bender EM, Gebru T, McMillan-Major A, et al. (2021) On the Dangers of Stochastic Parrots: Can Language Models Be Too Big? 🦜. In: *Proceedings of the 2021 ACM conference on fairness, accountability, and transparency*, 2021, pp. 610–623.

Binz M, Alaniz S, Roskies A, et al. (2025) How should the advancement of large language models affect the practice of science? *Proceedings of the National Academy of Sciences* 122(5).

Boulanger C (2026) The potential of LLMs for constructing a socio-legal knowledge graph. In: Simons A, Wüthrich A, Zichert M, et al. (eds) *Understanding Science with Large Language Models? Potentials for the History, Philosophy, and Sociology of Science*. Bielefeld: transcript, part-4.

Brainard J (2025) As scientists explore AI-written text, journals hammer out policies. *Science*, 10 December. Available at: https://www.science.org/content/article/new-preprint-server-welcomes-papers-written-and-reviewed-ai.

Brown TB, Mann B, Ryder N, et al. (2020) Language Models are Few-Shot Learners. http://arxiv.org/abs/2005.14165.

Büttner J (2026) Why pursue temporally-grounded AI for historical disciplines, and what makes it so challenging? In: Simons A, Wüthrich A, Zichert M, et al. (eds) *Understanding Science with Large Language Models? Potentials for the History, Philosophy, and Sociology of Science*. Bielefeld: transcript, part-2.

Christou PA (2025) *Artificial Intelligence (AI) in Social Research*. CABI.

Clark K, Luong M-T, Le QV, et al. (2020) Electra: Pre-training text encoders as discriminators rather than generators. http://arxiv.org/abs/2003.10555.

D'Alessandro W (2026) LLMs as philosophers: what can they do, and why aren't they better? In: Simons A, Wüthrich A, Zichert M, et al. (eds) *Understanding Science with Large Language Models? Potentials for the History, Philosophy, and Sociology of Science*. Bielefeld: transcript, part-1.

Danilova VV, Reed J, Burchell A, et al. (2026) Zero-shot generation of synthetic historical data with LLMs. In: Simons A, Wüthrich A, Zichert M, et al. (eds) *Understanding Science with Large Language Models? Potentials for the History, Philosophy, and Sociology of Science*. Bielefeld: transcript, part-2.

Davidson T and Karell D (2025) Integrating Generative Artificial Intelligence into Social Science Research: Measurement, Prompting, and Simulation. *Sociological Methods & Research*.

Devlin J, Chang M-W, Lee K, et al. (2018) BERT: Pre-training of Deep Bidirectional Transformers for Language Understanding. http://arxiv.org/abs/1810.04805.

Eberle O (2026) Grounding AI in humanistic inquiry. Interdisciplinary challenges for evaluation and interpretability. In: Simons A, Wüthrich A, Zichert M, et al. (eds) *Understanding Science with Large Language Models? Potentials for the History, Philosophy, and Sociology of Science*. Bielefeld: transcript, part-1.

Ehrlich N and Hazzan O (2025) The Converging Paths of Computer Science and the Humanities in the Age of GenAI – Communications of the ACM. In: *Artificial Intelligence and Machine Learning*. Available at: https://cacm.acm.org/blogcacm/the-converging-paths-of-computer-science-and-the-humanities-in-the-age-of-genai/.

Ethayarajh K (2019) How Contextual are Contextualized Word Representations? Comparing the Geometry of BERT, ELMo, and GPT-2 Embeddings. http://arxiv.org/abs/1909.00512.

Graßhoff G (2026) AI-Reporter: a path to a new genre of scientific communication. From presentation to publication through agentic LLMs. In: Simons A, Wüthrich A, Zichert M, et al. (eds) *Understanding Science with Large Language Models? Potentials for the History, Philosophy, and Sociology of Science*. Bielefeld: transcript, part-5.

Guest, O., Suarez, M., Müller, B., et al. (2025). Against the Uncritical Adoption of "AI" Technologies in Academia. https://doi.org/10.5281/zenodo.17065099.

Jawahar G, Sagot B and Seddah D (2019). What does BERT learn about the structure of language? *ACL 2019–57th Annual Meeting of the Association for Computational Linguistics*.

Friese S and Morgan D (eds) (2026) *Qualitative Data Analysis with Artificial Intelligence: Theory, Methods and Practice*. SAGE.

Gao Y, Xiong Y, Gao X, et al. (2024) Retrieval-Augmented Generation for Large Language Models: A Survey. arXiv:2312.10997. http://arxiv.org/abs/2312.10997.

Grootendorst M (2022) BERTopic: Neural topic modeling with a class-based TF-IDF procedure. http://arxiv.org/abs/2203.05794.

Guo T, Chen X, Wang Y, et al. (2024) Large Language Model based Multi-Agents: A Survey of Progress and Challenges. http://arxiv.org/abs/2402.01680.

Han B, Susnjak T and Mathrani A (2024) Automating systematic literature reviews with retrieval-augmented generation: a comprehensive overview. *Applied Sciences* 14(19).

He P, Liu X, Gao J, et al. (2021) DeBERTa: Decoding-enhanced BERT with Disentangled Attention. http://arxiv.org/abs/2006.03654.

He P, Gao J and Chen W (2022) DeBERTaV3: Improving DeBERTa using ELECTRA-Style Pre-Training with Gradient-Disentangled Embedding Sharing. In: *The Eleventh International Conference on Learning Representations*, 29 September 2022. Available at: https://openreview.net/forum?id=sE7-XhLxHA.

Hemment D and Kommers C (2025) Doing AI Differently. Rethinking the foundations of AI via the humanities. The Alan Turing Institute, London, UK. Available at: https://www.turing.ac.uk/news/publications/doing-ai-differently.

Hill M (2026) The data interview. Reflexive integration of large language models in qualitative content analysis. In: Simons A, Wüthrich A, Zichert M, et al. (eds) *Understanding Science with Large Language Models? Potentials for the History, Philosophy, and Sociology of Science*. Bielefeld: transcript, part-5.

Hitch D (2024) Artificial Intelligence Augmented Qualitative Analysis: The Way of the Future? *Qualitative Health Research* 34(7): 595–606.

Lund BD, Wang T, Mannuru NR, et al. (2023) ChatGPT and a new academic reality: Artificial Intelligence-written research papers and the ethics of the large language models in scholarly publishing. *Journal of the Association for Information Science and Technology* 74(5): 570–581.

Ho X, Nguyen AKD, Dao AT, et al. (2024) A Survey of Pre-trained Language Models for Processing Scientific Text. http://arxiv.org/abs/2401.17824.

Huang X, Liu W, Chen X, et al. (2024) Understanding the planning of LLM agents: A survey. http://arxiv.org/abs/2402.02716.

Jowsey T, Braun V, Clarke V, et al. (2025) We reject the use of generative artificial intelligence for reflexive qualitative research, SSRN Scholarly Paper. Available at: https://papers.ssrn.com/abstract=5676462.

Khutsishvili K (2026) AI and the scientist. On the fracture of epistemic authority. In: Simons A, Wüthrich A, Zichert M, et al. (eds) *Understanding Science with Large Language Models? Potentials for the History, Philosophy, and Sociology of Science*. Bielefeld: transcript, part-1.

Lang S (2026) Critical concerns for using LLMs in the (computational) humanities and beyond. In: Simons A, Wüthrich A, Zichert M, et al. (eds) *Understanding Science with Large Language Models? Potentials for the History, Philosophy, and Sociology of Science*. Bielefeld: transcript, part-1.

Lewis M, Liu Y, Goyal N, et al. (2019) BART: Denoising Sequence-to-Sequence Pre-training for Natural Language Generation, Translation, and Comprehension. http://arxiv.org/abs/1910.13461.

Li D, Jiang B, Huang L, et al. (2025) From Generation to Judgment: Opportunities and Challenges of LLM-as-a-judge. In: *Proceedings of the 2025 Conference on Empirical Methods in Natural Language Processing* (eds C Christodoulopoulos, T Chakraborty, C Rose, et al.), Suzhou, China, November 2025, pp. 2757–2791. Association for Computational Linguistics. Available at: https://aclanthology.org/2025.emnlp-main.138/.

Liesegang L and Gläser J (2026) Supporting citation context analysis with large language models raises questions that should have been asked 40 years ago. In: Simons A, Wüthrich A, Zichert M, et al. (eds) *Understanding Science with Large Language Models? Potentials for the History, Philosophy, and Sociology of Science*. Bielefeld: transcript, part-6.

Liu Y, Ott M, Goyal N, et al. (2019) RoBERTa: A Robustly Optimized BERT Pretraining Approach. http://arxiv.org/abs/1907.11692.

Lozić E and Štular B (2023) Fluent but Not Factual: A Comparative Analysis of ChatGPT and Other AI Chatbots' Proficiency and Originality in Scientific Writing for Humanities. *Future Internet* 15(10).

Luo Z, Yang Z, Xu Z, et al. (2025) LLM4SR: A Survey on Large Language Models for Scientific Research. http://arxiv.org/abs/2501.04306.

Malaterre C and Lareau F (2026) Epistemic framings in science. Charting scientific knowledge with embeddings and LLMs. In: Simons A, Wüthrich A, Zichert M, et al. (eds) *Understanding Science with Large Language Models? Potentials for the History, Philosophy, and Sociology of Science*. Bielefeld: transcript, part-3.

Marx V (2024) Quest for AI literacy. *Nature Methods* 21(8): 1412–1415.

Meding H and Daugs A (2026a) From RAGs to rich responses. Enhancing LLM reliability through retrieval-augmented generation. In: Simons A, Wüthrich A, Zichert M, et al. (eds) *Understanding Science with Large Language Models? Potentials for the History, Philosophy, and Sociology of Science*. Bielefeld: transcript, part-5.

Meding H and Daugs A (2026b) On the use and limitations of large language models in historical scholarship. In: Simons A, Wüthrich A, Zichert M, et al. (eds) *Understanding Science with Large Language Models? Potentials for the History, Philosophy, and Sociology of Science*. Bielefeld: transcript, part-1.

Morgan DL (2023) Exploring the Use of Artificial Intelligence for Qualitative Data Analysis: The Case of ChatGPT. *International Journal of Qualitative Methods* 22: 16094069231211248.

Nguyen DC and Welch C (2025) Generative Artificial Intelligence in Qualitative Data Analysis: Analyzing—Or Just Chatting? *Organizational Research Methods*: 10944281251377154.

Olival F, Cameron HF, Branco A, et al. (2026) Generative LLMs and history research. Limitations for languages, periods, and tasks. In: Simons A, Wüthrich A, Zichert M, et al. (eds) *Understanding Science with Large Language Models? Potentials for the History, Philosophy, and Sociology of Science*. Bielefeld: transcript, part-2.

Radford A, Narasimhan K, Salimans T, et al. (2018) Improving Language Understanding by Generative Pre-Training. OpenAI. Available at: https://cdn.openai.com/research-covers/language-unsupervised/language_understanding_paper.pdf.

Radford A, Wu J, Child R, et al. (2019) Language models are unsupervised multitask learners. OpenAI. Available at: https://cdn.openai.com/better-language-models/language_models_are_unsupervised_multitask_learners.pdf.

Reimers N and Gurevych I (2019) Sentence-BERT: Sentence Embeddings using Siamese BERT-Networks. http://arxiv.org/abs/1908.10084.

Ren S, Jian P, Ren Z, et al. (2025) Towards Scientific Intelligence: A Survey of LLM-based Scientific Agents. http://arxiv.org/abs/2503.24047.

Scharnhorst A, Yang H, Touber J, et al. (2026) Co-creation of AI technology, empowering curators of cultural heritage information and guarding research commons. In: Simons A, Wüthrich A, Zichert M, et al. (eds) *Understanding Science with Large Language Models? Potentials for the History, Philosophy, and Sociology of Science*. Bielefeld: transcript, part-5.

Scherbakov D, Hubig N, Jansari V, et al. (2024) The emergence of Large Language Models (LLM) as a tool in literature reviews: an LLM automated systematic review. http://arxiv.org/abs/2409.04600.

Schlattmann R, Kaye A and Vogl M (2026) From source to structure. Extracting knowledge graphs with LLMs. In: Simons A, Wüthrich A, Zichert M, et al. (eds) *Understanding Science with Large Language Models? Potentials for the History, Philosophy, and Sociology of Science*. Bielefeld: transcript, part-4.

Schmidgall S, Su Y, Wang Z, et al. (2025) Agent Laboratory: Using LLM Agents as Research Assistants. http://arxiv.org/abs/2501.04227.

Simons A (2026a) Meaning at the Planck scale? Contextualized word embeddings for doing history, philosophy, and sociology of science. In: Simons A, Wüthrich A, Zichert M, et al. (eds) *Understanding Science with Large Language Models? Potentials for the History, Philosophy, and Sociology of Science*. Bielefeld: transcript, part-3.

Simons A (2026b) Scaling In, Not Up? Testing Thick Citation Context Analysis with GPT-5 and Fragile Prompts. In: Simons A, Wüthrich A, Zichert M, et al. (eds) *Understanding Science with Large Language Models? Potentials for the History, Philosophy, and Sociology of Science*. Bielefeld: transcript, part-6.

Simons A, Arnaout H and Gurevych I (2026) Reconstructive citation context analysis using large language models. A roadmap. In: Simons A, Wüthrich A, Zichert M, et al. (eds) *Understanding Science with Large Language Models? Potentials for the History, Philosophy, and Sociology of Science*. Bielefeld: transcript, part-6.

Simons A, Zichert M and Wüthrich A (2026) Large language models for history, philosophy, and sociology of science: Interpretive uses, methodological challenges, and critical perspectives. *Studies in History and Philosophy of Science* 117: 102151. https://doi.org/10.1016/j.shpsa.2026.102151.

SocOpen (2025) Moderation Policy. In: *SocOpen: Home of SocArXiv*. Available at: https://socopen.org/moderation-policy/.

Strubell E, Ganesh A and McCallum A (2019) Energy and Policy Considerations for Deep Learning in NLP. In: *Proceedings of the 57th Annual Meeting of the Association for Computational Linguistics* (eds A Korhonen, D Traum, and L Màrquez), Florence, Italy, July 2019, pp. 3645–3650.

Thapa S, Shiwakoti S, Shah SB, et al. (2025) Large language models (LLM) in computational social science: prospects, current state, and challenges. *Social Network Analysis and Mining* 15(1): 4.

Tran K-T, Dao D, Nguyen M-D, et al. (2025) Multi-Agent Collaboration Mechanisms: A Survey of LLMs. http://arxiv.org/abs/2501.06322.

Valleriani M (2025) Large language models that power AI should be publicly owned. *The Guardian*, 26 May. Available at: https://www.theguardian.com/technology/2025/may/26/large-language-models-that-power-ai-should-be-publicly-owned.

Varnum MEW, Baumard N, Atari M, et al. (2024) Large Language Models based on historical text could offer informative tools for behavioral science. *Proceedings of the National Academy of Sciences* 121(42): e2407639121.

Vaswani A, Shazeer N, Parmar N, et al. (2017) Attention is all you need. *Advances in neural information processing systems* 30.

Vogl M, von Schwerin A and Kirschke S (2026) Large language models in interdisciplinary research settings. A reflection. In: Simons A, Wüthrich A, Zichert M, et al. (eds) *Understanding Science with Large Language Models? Potentials for the History, Philosophy, and Sociology of Science*. Bielefeld: transcript, part-4.

Wagner A and Hermes J (2026) Encoded humanities, or: not everything has to be generative. A dialogue on AI tasks and roles. In: Simons A, Wüthrich A, Zichert M, et al. (eds) *Understanding Science with Large Language Models? Potentials for the History, Philosophy, and Sociology of Science*. Bielefeld: transcript, part-1.

Wei J, Wang X, Schuurmans D, et al. (2022) Chain-of-Thought Prompting Elicits Reasoning in Large Language Models. *Advances in Neural Information Processing Systems* 35: 24824–24837.

Wolf JC (2026) LLMs and multilingual historical corpora in a digital history project. Reflections from the Berlin workshop. In: Simons A, Wüthrich A, Zichert M, et al. (eds)

Understanding Science with Large Language Models? Potentials for the History, Philosophy, and Sociology of Science. Bielefeld: transcript, part-2.

Yin S, Fu C, Zhao S, et al. (2024) A Survey on Multimodal Large Language Models. http://arxiv.org/abs/2306.13549.

Zhang H, Wu C, Xie J, et al. (2025b) Redefining Qualitative Analysis in the AI Era: Utilizing ChatGPT for Efficient Thematic Analysis. *Computers in Human Behavior: Artificial Humans* 4: 100144.

Zhang Y, Chen X, Jin B, et al. (2024) A Comprehensive Survey of Scientific Large Language Models and Their Applications in Scientific Discovery. http://arxiv.org/abs/2406.10833.

Zhang Y, Zhang C and Kousha K (2025a) Editorial: artificial intelligence for scientometrics (Part I). *Scientometrics* 130(10): 5281–5284.

Zhong T, Yang Z, Liu Z, et al. (2025) Opportunities and Challenges of Large Language Models for Low-Resource Languages in Humanities Research. http://arxiv.org/abs/2412.04497.

Zhuang Z, Chen J, Xu H, et al. (2025) Large language models for automated scholarly paper review: A survey. *Information Fusion*: 103332.

Zichert M and Simons A (2026) From early digital methods to LLMs. Computational conceptual history of scientific concepts. In: Simons A, Wüthrich A, Zichert M, et al. (eds) *Understanding Science with Large Language Models? Potentials for the History, Philosophy, and Sociology of Science*. Bielefeld: transcript, part-3.

Zichert M, Simons A and Wüthrich A (2025) Expanding conceptual histories: using contextualized word embeddings for the history and philosophy of the virtual particle concept. *Computational Humanities Research* 1: e16.

Ziems C, Held W, Shaikh O, et al. (2024) Can large language models transform computational social science? *Computational Linguistics* 50(1). MIT Press: 237–291.

Part 1: General challenges and limitations

Critical concerns for using LLMs in the (computational) humanities and beyond

Sarah Lang

1. Introduction

When using large language models (LLMs) and other AI tools in their research, humanities scholars must engage with the ethical concerns these systems raise. Yet standard AI ethics remain largely shaped by legal and technical frameworks that prioritise compliance and explainability. These approaches fail to account for the complexities of working with historical data. This contribution outlines key ethical and epistemic concerns surrounding the use of LLMs within computational humanities and related domains, with a particular focus on the implications for historical and humanities data.[1] While the field of AI ethics has become increasingly preoccupied with frameworks such as explainable AI (XAI), this paper draws attention to the limitations of such approaches when adopted in isolation, and it foregrounds the importance of critical, humanities-informed methodologies, including feminist praxis, activist scholarship, and data-centric critique attentive to labour and power.

Explainable AI has gained traction as a technical means of addressing opacity in algorithmic systems.[2] Within the context of AI ethics (Liao, 2020b), its popularity is understandable: providing a post hoc explanation of a model's outputs often appears to enhance transparency and accountability. However, scholars working from critical theory, feminist technology studies, and the digital humanities have questioned the efficacy of such explanations. These critiques argue that XAI can function more as a technical patch than a substantive response to the systemic issues that underlie model behaviour (Rudin, 2019; Klein and D'Ignazio, 2024). Explanations may create the impression of positive change, thereby allowing development to continue without addressing the more fundamental problems embedded in the data or model architecture. Nevertheless, this contri-

1 In this edited collection, there are a number of papers which go in somewhat similar directions. For instance, Meding and Daugs (2026) discuss the reliability versus added value of LLMs for historians. Khutsishvili (2026) is a short paper on epistemic authority. The introduction by the volume editors gives an excellent overview of related issues (cf. Simons et al., 2026).

2 See contribution by Oliver Eberle (2026).

bution does not reject explainability altogether. When used critically, XAI may serve as a valuable diagnostic tool, capable of exposing underlying issues that demand structural correction. The emphasis here is not on discarding technical solutions, but rather on reframing their use within a broader critical practice attentive to root causes rather than surface symptoms.

Given the limitations of explainability, there is growing scholarly attention to training data as the root site of many ethical problems. The flawed nature of most available datasets (Paullada et al., 2021), particularly those involving historical or cultural material, creates challenges for those working in the computational humanities (Lang and Suárez Cronauer, 2026). Concerns range from unclear provenance and embedded power relations to epistemic opacity and skewed representations. These are especially salient in domains like the history, philosophy, and sociology of science (HPSS), where the interpretation and contextualisation of source material are central to scholarly practice.

Other pressing concerns addressed in this paper include the redistribution and devaluation of labour (particularly in the form of ghost work, Gray and Suri, 2019), challenges to research integrity such as hallucinated citations and AI-generated misinformation, and the environmental cost of large-scale model training. These issues are contextualised through references to relevant scholarship, such as Kate Crawford's critique of AI's material footprint (Crawford, 2021) or recent work on "open-washing" LLMs, where transparency is often claimed but rarely truly delivered (Liesenfeld and Dingemanse, 2024). The paper also discusses how methodologies from fields such as critical archival studies, critical data studies, and the critical digital humanities can offer tools such as auditing (Vecchione et al., 2021), which is a particularly effective form of intervention for analyzing and curating datasets. These approaches can help articulate the epistemological stakes of working with LLMs and offer tools to mitigate their ethical shortcomings.

When examining the use of large language models from a critical and ethical standpoint within the history, philosophy, and sociology of science, a number of distinct issues arise. This overview serves as an initial orientation for those wishing to engage with these concerns, whether to reflect more critically on their own use of LLMs or to locate key literature and frameworks. While some suggestions for mitigating specific risks are included, the format does not allow for a comprehensive treatment of each topic.[3]

Importantly, this is not a call for refusal, though refusal remains a valid stance within some feminist and activist circles when technological systems reproduce harm or fail to meet the ethical standards of the communities they purport to serve. Rather, the aim here is to offer a constructive intervention that bridges critical perspectives with technical practice. By doing so, the article addresses an audience that includes technically proficient researchers in the history, philosophy or sociology of science and knowledge engaging in LLM-assisted scholarship or computational humanists who may not be trained in ethics but are seeking accessible and informed frameworks for evaluating and improving their work.

3 More in-depth analyses are currently in preparation by the author, and readers are encouraged to consult forthcoming work for more details: Lang and Suárez Cronauer (2026; forthcoming-2026) and Lang et al. (2025).

The conclusion offers a set of future-oriented reflections. It suggests that fields such as the digital humanities, history of science, and science and technology studies can provide essential contributions to the ethical development of AI, not least through their longstanding engagement with questions of knowledge production, power, and historical context. In particular, it advocates for participatory models of development grounded in solidarity, the feminist ethics of care (Gray and Witt, 2021), and the CARE principles (Collective benefit, Authority to control, Responsibility, and Ethics: GIDA, 2021). These frameworks provide both a conceptual and practical foundation for guiding the use of LLMs in a manner that is responsive to the specific needs and values of the humanities. While perfection may remain impossible to achieve in the end, meaningful ethical progress is still both necessary and possible. However, ethical engagement with LLMs in the context of HPSS and the digital humanities must move beyond mere regulatory compliance and technical mitigation. It requires an expanded ethical toolkit, attentive to the specific epistemic, historical, and social stakes of the field. This contribution offers a preliminary guide to that endeavour.

2. AI ethics and the issue of explanation

The ethics of LLMs, particularly as applied in digital humanities and HPSS contexts, remains a developing field.[4] This is not to imply that ethical questions have gone unaddressed; rather, the available literature often appears either overly specialised or not immediately applicable. Much work on AI ethics exists in adjacent disciplines, such as law and medicine where long-standing traditions and regulatory structures have shaped the debate. In these fields, concerns about privacy, personal data, safety, and legal compliance dominate the discourse (e.g., Hildebrandt, 2021). The EU AI Act and similar guidelines are frequently cited as important milestones. However, such documents are primarily oriented towards preventing legal violations or safeguarding individuals from harm, rather than offering guidance for the responsible use of AI in humanities contexts. Their focus, in short, tends to be on preventing illegal wrongdoing, rather than promoting ethical engagement.

Broadly speaking, three disciplinary traditions can be identified within AI ethics: one rooted in computer science (Kuipers, 2020), another in philosophy (Gunkel, 2020), and a third informed by intersectional feminism, critical theory, and activist scholarship (e.g., O'Neil and Gunn, 2020; Gray and Witt, 2021), including areas such as critical digital humanities, critical code and data studies, and archival theory. Each of these traditions brings distinct assumptions and goals. For those unfamiliar with their respective vocabularies and commitments, the differences can be challenging to navigate.

From a computer science perspective, explainable AI (XAI) has emerged as a prominent response to concerns about algorithmic opacity. The aim is to render the internal

4 That is not to say that there is no literature on AI ethics. This would be far from the truth, as the literature is abundant. However, AI ethics is by no means straightforward, and there is even an ethics of AI ethics (Powers and Ganascia, 2020). For an introduction, the following handbooks are an excellent starting point: Dubber et al., 2020; Liao, 2020a; Floridi, 2023.

workings of models more transparent, thereby fostering trust and enabling oversight. Yet such technical solutions are often criticised within critical humanities approaches for failing to address the structural origins of harm (D'Ignazio and Klein, 2020; Klein and D'Ignazio, 2024). As scholars informed by feminist praxis and critical theory have argued, explanation alone may provide a sense of closure or accountability without actually disrupting the logics that produce bias or injustice. In this sense, explainability risks becoming a superficial remedy – a "band-aid" that leaves the core problems intact. In addition, scholars have questioned whether transparency, which is often heralded as an ideal in the context of black box algorithms, is actually the ideal solution many seem to take it to be (Ananny and Crawford, 2016). Nevertheless, explanation remains a necessary, if insufficient, component of ethical AI. It is especially vital for philosophical approaches to machine ethics, which require insight into the reasoning processes of AI systems in order to evaluate moral acts. In the digital humanities, ethical discussions have been less institutionalised and to this day often remain at a surface level, engaging solely with legal issues such as access or privacy. As Berry notes, the field has yet to establish a professional ethics (Berry, 2022a) that could address its specific entanglements with AI and computational infrastructures. This gap is particularly evident given that we use historical data with major concerns about provenance, representation and archival bias.

Humanities scholars dealing with historical datasets simply have a different focus than other disciplines when it comes to ethics (Lang and Suárez Cronauer, 2026). What they seek is not a compliance framework, but a deeper understanding of how to use AI well: how to engage responsibly with tools that hold powerful potential, yet are methodologically alien to traditional humanities practices. In this respect, the existing AI ethics literature does not usually speak to the key concerns of researchers working at the intersection of computational methods and historical inquiry.

Technical fixes, even those motivated by ethical considerations, rarely account for the interpretive complexities and disciplinary norms that shape humanities scholarship. Instead, more fruitful models may be found in audit-based approaches informed by critical theory (Berry, 2022b, 2023). These include dataset audits (Vecchione et al., 2021), critical metadata practices, and frameworks developed within critical archival studies and data feminism (D'Ignazio and Klein, 2020). Such methods aim to uncover the epistemological and political dimensions of data and model construction, moving beyond mere explanation towards structural critique and transformation. While no solution is perfect, especially given the inherently flawed nature of most available datasets, these approaches offer tools for more critical engagement with the problematic legacy of historical data.

3. Auditing to tackle bias at its root

Another central concern in the critical evaluation of large language models within the history, philosophy, and sociology of science pertains to the political orientations and implicit commitments of the actors and institutions that create data and develop AI technologies. The values embedded in AI systems, shaped by the political, economic, or ideological contexts of their creators (Crawford, 2021), profoundly influence the outputs these models produce. Since LLMs are trained on massive, opaque, and uncurated datasets that incorporate some of the most discriminatory and toxic content available online, their core behaviour reflects these origins (Paullada et al., 2021). Particularly concerning is the use of superficial interventions to manage the outputs of LLMs, such as post hoc filtering mechanisms intended to remove racist or otherwise harmful content. While these filters may suppress problematic outputs at the interface level, the underlying model architecture remains unchanged, and therefore the biases embedded in the training data persist. Data workers involved in this post hoc filtering are constantly exposed to such material, resulting in psychological harm and exploitation, leading scholars to be complicit in ethically problematic industry practices, potentially without being aware of those concerns (Yang et al., 2025). Yet, the fragility of superficial fixes becomes particularly evident when safeguards are bypassed or relaxed, revealing that the models still generate outputs shaped by racist, sexist, or otherwise exclusionary patterns. Feminist scholars have critiqued this form of sanitisation as inadequate (Klein and D'Ignazio, 2024) and potentially precarious in the context of shifting political climates where such filters can be modified or removed. In a critique of discriminatory search engine results, Noble (2018) documented how Google's response to harmful content often involved redirecting queries rather than addressing structural causes – an approach that resonates with current responses to problematic LLM outputs. As in the case of Google's search algorithms, filtering in LLMs does not correct the underlying patterns; it merely reroutes them, allowing the problem to reappear under slightly different conditions. These examples reinforce a broader point: even where technical interventions can serve as useful starting points, they must be accompanied by critical reflection on the historical, political, and epistemological conditions in which both data and technologies are embedded. Without this, even the most advanced systems risk reproducing old exclusions under the guise of innovation. All in all, these dynamics highlight the long-lasting consequences of training LLMs – or any AI for that matter – on datasets whose exact contents are unknown or undisclosed. Their opacity makes rigorous critique or assessment of the data, and by extension the model, exceptionally difficult. Adding to this, current AI development depends on precarious, outsourced, and exploitative labour practices (Gray and Suri, 2019).

For this reason, feminist and critical approaches consistently call attention to training data as the root site of ethical concern. Addressing systemic bias in AI systems requires far more than technical adjustments or user-facing filters. It demands sustained engagement with the datasets that shape model behaviour. As a response, many scholars advocate for dataset (Lang and Suárez Cronauer, forthcoming-2026) and algorithm audits (Vecchione et al., 2021) to expose and interrogate the foundations on which these systems are built. Beyond dataset documentation (e.g., Gebru et al., 2021), algorithm audits have also revealed embedded biases in model outputs (Vecchione et al., 2021). For example, prompting an image-generating model to visualise an inclusive bathroom revealed a stark disability bias: the model was unable to produce a usable representation of a disabled-accessible facility (Manzoor et al., 2025). This not only points to a lack of spatial reasoning capacity within the model architecture but also underscores how bias in training data results in the erasure of marginalised needs and experiences. Such audits highlight the importance of testing model performance in ways that reflect diverse, real-world scenarios, particularly those not centred in dominant cultural narratives. While such interventions are likely the most effective, it must also be acknowledged that not all researchers may have the capacity to undertake them extensively: Time, expertise, and financial constraints, such as within the framework of grant-funded projects, make it difficult to incorporate comprehensive ethical review or auditing processes unless they are explicitly required and supported by funding bodies.

Scholars frequently rely on inherited or repurposed datasets for practical reasons, but these usually carry the imprint of earlier exclusions and biases. This is especially true when dealing with historical data. The process of digitisation itself frequently amplifies already-dominant voices, as materials that have survived previous curatorial selections tend to be those already privileged by institutional and historical forces. As a result, many early digital humanities projects have ended up reproducing familiar canons, such as works by Goethe or Shakespeare, despite the fact that these topics were arguably already over-researched and of diminishing interest even in traditional, non-digital scholarship. However, audits can also be applied to datasets assembled for humanities research to mitigate the effects of these sometimes unavoidable circumstances (Lang and Suárez Cronauer, forthcoming-2026). If we cannot fundamentally change the datasets we have, we can at least make visible what they contain and what is missing.

4. AI's unsustainability

Environmental concerns further complicate the ethical landscape of LLMs. The energy demands associated with training and deploying large models are considerable and projected to rise sharply in the absence of regulation (Strubell et al., 2019; Bender et al., 2021; Crawford, 2021: chapter "Earth", 2024; Luccioni et al., 2024; Lang et al., 2025). While some have argued that the environmental impact of occasional personal use is negligible (Masley, 2025), particularly when compared to high-emission activities regularly practiced by many scholars such as air travel, this should not justify indiscriminate use. In many humanities research contexts, smaller and more specialised machine learning models may be more appropriate: smaller, non-LLM models are not only easier to finetune and more stable, but also significantly less resource-intensive.

For instance, the Data Science Lab at the Staatsbibliothek zu Berlin prefers traditional models over large-scale foundation models (cf. Neudecker, 2023). These choices reflect not only a commitment to environmental, financial and technical sustainability but also frequently provide better results and are more efficient to implement for specialized Humanities use cases. Additionally, large models require high-performance computing infrastructure that many institutions cannot maintain. In contrast, using only what is necessary aligns with both budgetary and ecological responsibility.

Energy consumption is often the primary concern in discussions of sustainability, though its calculation is not straightforward (Lang et al., 2025). While some tools now exist for estimating the energy usage of custom-trained models, many consumer-facing AI services such as ChatGPT do not provide transparent information. Institutions like the German GWDG, for instance, offer high-performance computing resources to researchers, but these, too, still lack tools for estimating or monitoring energy consumption.

As Kate Crawford has pointed out, water usage is another significant but under-discussed issue when it comes to AI's environmental impacts (Crawford, 2024). Many server farms are located in desert regions and require substantial water resources for cooling that might otherwise meet basic human needs. This trade-off raises serious ethical questions about the prioritisation of computational infrastructure over the well-being of local populations.

Electronic waste (e-waste) represents another environmental cost (Strubell et al., 2019). As AI systems evolve, so too must the hardware that supports them. This leads to frequent upgrades and the disposal of equipment that has either become outdated or reached the end of its serviceable lifecycle. Much of this e-waste is exported to the Global South, where it is often dumped without proper recycling or regulatory oversight. The environmental burden of high-end computing thus extends far beyond the institutions that use it. However, even technologically unrelated issues like, quite simply, long-term research data preservation or the digital preservation of digitized cultural heritage collections, are equally affected by this issue.

Researchers must therefore be mindful of the full material footprint of AI usage. Running LLMs at scale, particularly when not necessary, contributes to global ecological degradation. Yet, efficient prompting and careful API use can reduce both financial costs and environmental impact. Since energy seems to correlate with the volume of input and output tokens, optimising these variables may have a tangible ecological benefit in larger scale usage scenarios. Some programming libraries already provide tools for tracking environmental impacts of one's token usage during inference, which may become increasingly important in large-scale projects conducted by libraries or digital humanities centres.[5]

5. Redistributing labour and research integrity

While explainable AI (XAI) should not be overestimated in its capacity to bring about structural improvements to artificial intelligence systems, it remains a valuable diagnostic instrument. It can illuminate how algorithms operate, expose systemic gaps, and support broader efforts in algorithm auditing. For example, recent work employing feature embeddings has revealed that the large language model Claude, when tasked with solving mathematical problems, does not actually perform the mathematical operations it is supposed to perform but rather, manages to get to a correct answer somehow through hardly understandable next-token prediction (Lindsey et al., 2025). More significantly, although Claude produces textbook-style explanations, it does not actually arrive at its results through the application of mathematical rules it could have learned from the maths textbooks in its training data. Instead, it relies on statistical next-token prediction. This discrepancy between the model's generated explanation and its underlying mechanism underscores a critical issue: large language models can

5 However, these usually do not account for costs during training, which are tracked in other scores, such as the AI Energy Score. For more information and a concrete example, see Lang et al. (2025).

fabricate plausible yet misleading accounts of their reasoning processes, thereby raising important concerns about the reliability of self-explanation in machine systems.

Hallucinations (Xiao and Wang, 2021; Singh and Singh, 2025), now widely discussed in both academic and public discourse, pose a threat to research quality and scholarly integrity.[6] While not illegal, the use of inaccurate or misleading AI-generated content in academic work can reinforce harmful ideologies or perpetuate misinformation. As these technologies become more integrated into scholarly workflows, vigilance is required – not only to correct the outputs but to question the infrastructures that produce them.

Transparency remains a major issue in this context. Many AI systems are presented as "open" or accessible while the datasets used to train them remain closed (Liesenfeld and Dingemanse, 2024). This "open-washing" obscures the fact that models are often trained on massive corpora containing unconsented, pirated and harmful content. Kate Crawford's work has shown that this problem predates LLMs: even earlier machine learning systems already relied on uncurated datasets drawn from surveillance, military, or corporate contexts (Crawford, 2021: chapter "Labour"). Once released, such datasets often persist in circulation and continue to shape new models.[7] Even when the flaws of a dataset are well known, the lack of viable alternatives and the sunk cost of development mean that the same corpora continue to be used, and the harms embedded in them are further perpetuated.

A result of this is that even when researchers themselves are not directly working with such datasets, the tools or algorithms they – perhaps even inadvertently – use may be

6 For a more in-depth treatment of hallucinations in the context of this edited volume, see the contribution Meding and Daugs (2026) or Singh and Singh (2025). However, the term *hallucination* has been criticised in various contexts as problematic. First, it psychologises AI, further humanising it and obscuring the fact that this is a technical system. What is called a "hallucination" is not a pathological, psychological delusion but an expected technical behaviour. On a technical level, the term is also misleading, as it assumes that producing incorrect information is a mistake. In reality, these systems are not "making errors" in that sense. All outputs of large language models are the result of mathematical operations and computational programming functions. The model is doing precisely what it is designed to do: predict the most probable next token, given the current context, with a certain degree of randomness. There is no perception involved; it is simply statistics. Anthropomorphic framing not only humanises, psychologises, and pathologises this technical functionality but also implies that it is a malfunction when in fact, this is exactly how the system is intended to operate. The so-called "hallucination" is not a bug; it is an inherent feature of the model's architecture. The term makes it appear to us as though the model should not be doing this, but in reality, next-token prediction is the core behaviour. Similarly, the model's tendency to generalise and omit details is not a malfunction but a built-in characteristic. LLMs are designed to produce knowledge that is as generalisable as possible, which inherently leads to forgetting specific details and favouring central, mainstream knowledge over less standard, more diverse knowledge from the margins. This has the built-in side effect of erasing diversity and nuance, precisely the kinds of details in historical sources that are of greatest interest to historians. An example criticizing the term "hallucination" is: Jason Bell on LinkedIn: https://www.linkedin.com/feed/update/urn:li:activity:7353693695822438400

7 There is a range of publications addressing the lifecycle of AI datasets, including their often flawed creation circumstances and their out-of-control persistence and continued use in AI even after they have been publicly revoked because of their blatant ethical flaws: e.g., Luccioni et al., 2022; Orr and Crawford, 2023; Luccioni and Crawford, 2024; Orr and Crawford, 2024.

shaped by these legacies. Many of these datasets originated in institutional contexts with little accountability or ethical oversight, and their continued use embeds these values into current technologies. For scholars in the humanities especially these entanglements cannot be overlooked.

LLMs tend to produce generalised summaries that obscure or omit specific details, and their architecture favours information from the mainstream rather than from the margins (Peterson, 2025), whose nuance and diversity historians are usually more interested in than in blatant overgeneralisations. Although skilled prompting can help mitigate some of these tendencies, the problem is rooted in the design and training of the models. As such, outputs must be thoroughly verified, especially in scholarly work, and scholars may find that models simply do not produce adequate results in their specialised areas of expertise if these are not well represented in the data the algorithms were trained on.

On top of all that, contrary to promises of labour-saving automation, LLMs often shift the labour of research rather than reducing it. Rather than freeing scholars from repetitive tasks, these systems require them to function as human fact-checkers for unreliable machine outputs. While some tasks may indeed be accelerated, the overall result is frequently an increase in cognitive and editorial labour.

And while it is not illegal to produce substandard academic writing, the proliferation of fraudulent or low-quality papers generated through paper mills represents a serious threat to scholarly integrity and to public trust in science more broadly. While early examples were often easily identified, contemporary high-performing tools make detection significantly more difficult (Else and Van Noorden, 2021; Liverpool, 2023). Academic publishers now face an ongoing and escalating challenge, effectively engaged in an arms race against those producing fake submissions. This dynamic not only undermines scientific credibility but also places additional labour demands on journal editors, reviewers, and the publishing infrastructure as a whole.

The increasing use of AI-assisted writing tools contributes further to the system that generates publication pressure in the first place: While AI tools can support writing workflows in meaningful and ethically acceptable ways (Abernethy, 2024), they also risk accelerating publication rates in ways that reinforce the already problematic "publish or perish" culture in academia (Rawat and Meena, 2014; Johnson et al., 2024; Kendall and Teixeira da Silva, 2024; Hegde, 2025). As researchers are able to produce more text more quickly, expectations around productivity may rise, potentially diminishing the time available for reflection, critical engagement, and scholarly rigour. These developments are not necessarily unique to history, philosophy and sociology of science and knowledge research but should nonetheless be considered within the broader context in which such research now unfolds.

Regulatory and funding institutions must take a more active role in mandating ethical interventions, including dataset audits and representational analyses, as part of project deliverables (cf. Lang and Suárez Cronauer, 2026). Encouragingly, some funding agencies are beginning to request AI ethics statements. However, if such statements remain vague, requiring only that researchers affirm the absence of ethical concerns, then their practical value is minimal. There is also a real risk of "ethics-washing," wherein researchers are incentivised to downplay or overlook issues to streamline approval (Powers and Ganascia, 2020; Lambrecht and Moreno, 2024). Moreover, these initiatives, through their reliance on virtue ethics, usually assume that good people will intuitively know how to act ethically, downplaying the substantial skill and expertise involved in thoughtfully evaluating the ethics of data practices. As the authors of *Data Feminism* have pointed out, those who have not experienced structural discrimination themselves are often unaware of how their work may negatively affect others – a phenomenon they refer to as the "privilege hazard" (D'Ignazio and Klein, 2020: 28–29; Lang and Suárez Cronauer, 2026). Without the deliberate inclusion of diverse perspectives and affected communities, researchers cannot be expected to reliably detect harms their technologies reproduce or exacerbate. To act more ethically, scholars need tools and strategies. Methods such as Explainable AI (XAI), although critiqued earlier in this article for its limited capacity to address systemic bias, can nevertheless offer valuable starting points for identifying and understanding model behaviour. In the absence of transparency from developers, third-party audits have become increasingly important. Resources informed by critical archival studies – such as data sheets, data papers, and related documentation practices (Gebru et al., 2021) – offer structured ways to reflect the situatedness of datasets and clarify their limitations. These tools provide a means of acknowledging the epistemological and ethical boundaries of a dataset, but they have yet to be adopted at scale within the development of large foundation models.

6. Humanity, solidarity and care for better AI

Amid the challenges outlined in this article, there are also grounds for optimism. Frameworks like data-centric AI present new opportunities. The use of smaller and more efficient architectures in particular may be better suited to many humanities tasks. These models are often easier to fine-tune, more stable, and more transparent than large-scale foundation models, and can be retrained on domain-specific corpora. Although such retraining is not always feasible in humanities contexts due to limited data or computational resources, the possibility remains significant. It is also encouraging to note the increasing visibility and impact of scholars engaged in critical AI ethics, whose work continues to inform more accountable and reflective approaches.

A key aim of this paper has been to introduce readers to these resources and to encourage engagement with existing work in the field. Many useful tools and theoretical frameworks are already available, and increased uptake would strengthen both our practice and our discourse. What is needed is not only more incentive for such engagement, but also greater self-awareness across the field. The digital humanities, for example, already contain a wealth of literature on the ethical implications of data, algorithms, and infrastructure. The challenge now is to mobilise this work effectively and to make it part of standard practice.

Central to this endeavour is the practice of dataset documentation and auditing. Whether applied to one's own data or to third-party datasets, audits can help illuminate representational imbalances and ethical risks. Data papers and dataset metadata provide crucial context about the origins, composition, and limitations of datasets, thus enabling more responsible reuse. For instance, knowing that 50–90% of authors in a given dataset are male allows researchers to at least state that the dataset will likely reflect a male gaze. While this does not resolve the issue, it creates a baseline for critical reflection and makes visible biases that may otherwise remain implicit. Truly addressing such biases will require filling data gaps, where possible, yet this is especially challenging in historical research where records may be incomplete or unrecoverable. Nevertheless, efforts can be made to mitigate these problems whether through more inclusive digitisation, redesigning research projects to account for diverse perspectives from the outset, or even initiating new collections that avoid replicating past exclusions. In the long term, these steps might even support retraining foundation models on less problematic grounds.

Such efforts cannot be undertaken by individual researchers alone. As Valleriani (2025) has argued, state institutions should take a more active role in developing AI infrastructures, including foundation models. Currently, these models are largely built by private corporations with opaque motivations, and their training data and objectives are often misaligned with scholarly or public values. If large language models are to become core infrastructures for knowledge production, as they already are in many fields, it is imperative that their development is subjected to ethical scrutiny and democratic oversight. The humanities, with their critical and contextualising capacities, are well positioned to contribute to this process not only by critiquing existing systems, but by actively shaping the values and materials that future models are built upon. Importantly, researchers need not be AI specialists to participate in this work. Dataset audits, for example, can be conducted by anyone familiar with the material, and they offer an accessible entry point into AI ethics. By critically examining the datasets they use or create, scholars can contribute meaningfully to the development of more equitable and transparent AI systems. The cumulative effect of such individual actions may help shift the field as a whole.

It is no longer sufficient to ask how we might better *use* LLMs. We must also ask how *we* might contribute to making them better. This includes contributing the intellectual and cultural materials, methodological insights, and ethical reflections that only the humanities can provide. Regulatory institutions must support this process through clearer requirements, enforceable standards, and publicly accountable infrastructures. Creating foundation models whose content is transparent and whose construction is participatory would ensure both community control and alignment with shared ethical aims. Such an approach would represent a decisive alternative to systems designed to serve narrow political or economic interests and would contribute to the development of more equitable and responsible technologies.

A significant yet often overlooked factor in the development of artificial intelligence, computing, and adjacent fields is the absence of solidarity, particularly in the historical and ongoing extractive practices of the tech industry. Many of the ethical and epistemological challenges currently facing these domains can be traced to systemic undervaluation and invisibilisation of labour. For instance, large-scale projects such as Google Books were made possible through the labour of underpaid and insufficiently acknowledged book scanners. Similarly, the infrastructure supporting contemporary AI relies on 'ghost workers' and underpaid contributors from platforms such as Amazon Mechanical Turk (Gray and Suri, 2019).

These patterns are not unique to industry alone. Within the digital humanities and the historical sciences, tasks dismissed as "menial" such as metadata collection or the work of librarians have historically been marginalised, both in terms of compensation and attribution (Lang and Suárez Cronauer, 2026). This systemic disregard for certain forms of labour has long-term consequences. One tangible outcome is the difficulty scholars now face in tracing the origins of specific data or annotations, due to the absence of proper documentation and attribution. The lack of recognition for this invisible labour has contributed to a structural opacity in digital infrastructures, which in turn affects how knowledge is produced, evaluated, and utilised. This demands a renewed commitment to solidarity within both scholarly and technical communities. Valuing diverse forms of expertise and properly crediting all contributors regardless of the perceived status of their labour are essential steps in reshaping the ethics and culture of AI development. While individual efforts may appear limited, collective action across the field holds the potential for meaningful change. A non-extractive, solidaristic model of knowledge production is both necessary and achievable.

Contemporary AI development has largely been driven by unregulated and extractive practices, frequently marked by non-consensual data collection, lack of transparency, and motivations aligned with economic and military interests (Crawford, 2021: chapter "Data"). These practices stand in stark contrast to the ethical commitments we stand for within scholarship and, in particular, the humanities. A more ethical framework would centre solidarity, aiming for collective benefit and co-liberation rather than economic gain for the few. Such an approach resonates with implementing the CARE principles (Collective benefit, Authority to control, Responsibility, and Ethics: GIDA, 2021) in the context of AI and a (feminist) ethics of care (Gray and Witt, 2021), which may offer a more constructive ethical foundation than current emphases on explainability. The latter is often ideologically tethered to ideals of rigour and quantification that implicitly devalue humanities knowledge as "soft" or secondary (Riley, 2017). Yet it is precisely this form of expertise that is capable of exposing the limitations of technical solutions. Humanities scholars are well positioned to advance this critical perspective. The foundation of future AI must not be left to unaccountable actors. It is time to build models, literally and ethically, that we can collectively stand behind.[8]

References

Abernethy NJ (2024) Let stochastic parrots squawk: Why academic journals should allow large language models to coauthor articles. *AI Ethics*. DOI: 10.1007/s43681-024-00575-7.

8 This chapter was written with support from large language models (LLMs). All model-generated text was reviewed and, where necessary, rewritten by the authors, who remain fully responsible for the final version. For details on the use of LLMs in this volume, see the statement in the volume's introduction.

Ananny M and Crawford K (2016) Seeing without knowing: Limitations of the transparency ideal and its application to algorithmic accountability. *New Media & Society*. DOI: 10.1177/1461444816676645.

Bender EM, Gebru T, McMillan-Major A and Shmitchell S (2021) On the dangers of stochastic parrots: Can language models be too big? In: *Proceedings of the 2021 ACM Conference on Fairness, Accountability, and Transparency (FAccT '21)*, 610–623. DOI: 10.1145/3442188.3445922.

Berry DM (2022a) AI, ethics, and digital humanities. In: O'Sullivan J (ed.) *The Bloomsbury Handbook to the Digital Humanities*. London: Bloomsbury Academic, pp.445–458. DOI: 10.5040/9781350232143.ch-42.

Berry DM (2022b) Critical digital humanities. In: O'Sullivan J (ed.) *The Bloomsbury Handbook to the Digital Humanities*. London: Bloomsbury Academic, pp.125–136. DOI: 10.5040/9781350232143.ch-12.

Berry DM (2023) Tracing 'toxicity' through code: Towards a method of explainability and interpretability in software. *Digital Humanities Quarterly* 17(2). Available at: https://dhq.digitalhumanities.org/vol/17/2/000706/000706.html (accessed 2025-11-25).

Crawford K (2021) *Atlas of AI: Power, Politics, and the Planetary Costs of Artificial Intelligence*. New Haven: Yale University Press.

Crawford K (2024) Generative AI's environmental costs are soaring – and mostly secret. *Nature* 626(8000): 693. DOI: 10.1038/d41586-024-00478-x.

D'Ignazio C and Klein LF (2020) *Data Feminism*. Cambridge, MA: MIT Press.

Dubber MD, Pasquale F and Das S (eds) (2020) *The Oxford Handbook of Ethics of AI*. Oxford: Oxford University Press. DOI: 10.1093/oxfordhb/9780190067397.001.0001.

Eberle O (2026) Grounding AI in humanistic inquiry. Interdisciplinary challenges for evaluation and interpretability. In: Simons A, Wüthrich A, Zichert M, et al. (eds) *Understanding Science with Large Language Models? Potentials for the History, Philosophy, and Sociology of Science*. Bielefeld: transcript, part-1.

Else H and Van Noorden R (2021) The fight against fake-paper factories that churn out sham science. *Nature* 591: 516–519. DOI: 10.1038/d41586-021-00733-5.

Floridi L (2023) *The Ethics of Artificial Intelligence – Principles, Challenges, and Opportunities*. Oxford: Oxford University Press.

Gebru T, Morgenstern J, Vecchione B, Vaughan JW, Wallach H, Daumé III H and Crawford K (2021) Datasheets for datasets. *Communications of the ACM* 64(12): 86–92. DOI: 10.1145/3458723.

GIDA (Global Indigenous Data Alliance) (2021) CARE principles of Indigenous data governance. Available at: https://www.gida-global.org/care/ (accessed 2025-11-25).

Gray J and Witt A (2021) A feminist data ethics of care for machine learning: The what, why, who and how. *First Monday* 26(12). DOI: 10.5210/fm.v26i12.11833.

Gray ML and Suri S (2019) *Ghost Work: How to Stop Silicon Valley from Building a New Global Underclass*. Boston: Eamon Dolan Books.

Gunkel DJ (2020) Perspectives on ethics of AI: Philosophy. In: Dubber MD, Pasquale F and Das S (eds) *The Oxford Handbook of Ethics of AI*. Oxford: Oxford University Press, pp.538–553. DOI: 10.1093/oxfordhb/9780190067397.013.35.

Hegde SN (2025) Publish and cherish, not 'publish or perish.' *APIK Journal of Internal Medicine* 13(3): 159–160. DOI: 10.4103/ajim.ajim_39_25.

Hildebrandt M (2021) The issue of bias: The framing powers of machine learning. In: Pelillo M and Scantamburlo T (eds) *Machines We Trust: Perspectives on Dependable AI*. Cambridge, MA: MIT Press. DOI: 10.7551/mitpress/12186.003.0009.

Johnson TF, Simmons BI, Millard J, Strydom T, Danet A, Sweeny AR and Evans LC (2024) Pressure to publish introduces large-language model risks. *Methods in Ecology and Evolution* 15(10): 1771–1773. DOI: 10.1111/2041-210X.14397.

Kendall G and Teixeira da Silva JA (2024) Risks of abuse of large language models, like ChatGPT, in scientific publishing: Authorship, predatory publishing, and paper mills. *Learned Publishing* 37(1): 55–62. DOI: 10.1002/leap.1578.

Khutsishvili K (2026) AI and the scientist. On the fracture of epistemic authority. In: Simons A, Wüthrich A, Zichert M, et al. (eds) *Understanding Science with Large Language Models? Potentials for the History, Philosophy, and Sociology of Science*. Bielefeld: transcript, part-1.

Klein L and D'Ignazio C (2024) Data feminism for AI. In: *Proceedings of the 2024 ACM Conference on Fairness, Accountability, and Transparency (FAccT '24)*, 100–112. New York: Association for Computing Machinery. DOI: 10.1145/3630106.3658543.

Kuipers B (2020) Perspectives on ethics of AI: Computer science. In: Dubber MD, Pasquale F and Das S (eds) *The Oxford Handbook of Ethics of AI*. Oxford: Oxford University Press, pp.420–441. DOI: 10.1093/oxfordhb/9780190067397.013.27.

Lambrecht F and Moreno M (2024) What is AI ethics? *American Philosophical Quarterly* 61(4): 387–401. DOI: 10.5406/21521123.61.4.07.

Lang S and Suárez Cronauer E (forthcoming-2026) Dataset audits for mitigating data gaps. *Computational Humanities Research Journal*.

Lang S and Suárez Cronauer E (2026) Beyond data feminism: Toward ethical data work in the (digital) humanities. *Zeitschrift für digitale Geisteswissenschaften (ZfdG)*. DOI: 10.17175/wp_2026.

Lang S, Pitawanik W, Belouin P, Sevink E, Olszynko-Gryn J, Freeborn A and Benson E (2025) *Quantifying the environmental footprint of curating datasets with LLMs*. Working paper. DOI 10.5281/zenodo.17902821.

Liao SM (ed.) (2020a) *Ethics of Artificial Intelligence*. Oxford: Oxford University Press. DOI: 10.1093/oso/9780190905033.001.0001.

Liao SM (2020b) A short introduction to the ethics of artificial intelligence. In: Liao SM (ed.) *Ethics of Artificial Intelligence*. Oxford: Oxford University Press, pp.1–42. DOI: 10.1093/oso/9780190905033.003.0001.

Liesenfeld A and Dingemanse M (2024) Rethinking open source generative AI: Open-washing and the EU AI Act. In: *Proceedings of the 2024 ACM Conference on Fairness, Accountability, and Transparency (FAccT '24)*, 1774–1787. DOI: 10.1145/3630106.3659005.

Lindsey J, Gurnee W, Ameisen E, Chen B, Pearce A, Turner NL, Citro C *et al.* (2025) On the biology of a large language model. *Transformer Circuits Thread*. Available at: https://transformer-circuits.pub/2025/attribution-graphs/biology.html (accessed 2025–11–25).

Liverpool L (2023) AI intensifies fight against 'paper mills' that churn out fake research. *Nature* 618: 222–223. DOI: 10.1038/d41586-023-01780-w.

Luccioni AS, Corry F, Sridharan H, Ananny M, Schultz J and Crawford K (2022) A framework for deprecating datasets: Standardizing documentation, identification, and communication. In: *Proceedings of the 2022 ACM Conference on Fairness, Accountability, and Transparency*.

Luccioni AS and Crawford K (2024) The nine lives of ImageNet: A sociotechnical retrospective of a foundation dataset and the limits of automated essentialism. *Journal of Data-Centric Machine Learning Research*. Available at: https://data.mlr.press/assets/pdf/v01-4.pdf (accessed 2025–11-25).

Luccioni AS, Jernite Y and Strubell E (2024) Power hungry processing: Watts driving the cost of AI deployment? In: *Proceedings of the 2024 ACM Conference on Fairness, Accountability, and Transparency (FAccT '24)*, 85–99. DOI: 10.1145/3630106.3658542.

Manzoor R, Hussain W and Anjum ML (2025) Out of dataset, out of algorithm, out of mind: A critical evaluation of AI bias against disabled people. *AI & Society* 40: 3941–3951. DOI: 10.1007/s00146-024-02168-8.

Masley A (2025) Why using ChatGPT is not bad for the environment – A cheat sheet. *Substack* (blog), April 2025. Available at: https://andymasley.substack.com/p/a-cheat-sheet-for-conversations-about (accessed 2025–11-25).

Meding H and Daugs A (2026) On the use and limitations of large language models in historical scholarship. In: Simons A, Wüthrich A, Zichert M, et al. (eds) *Understanding Science with Large Language Models? Potentials for the History, Philosophy, and Sociology of Science*. Bielefeld: transcript, part-1.

Neudecker C (2023) Digital curation and artificial intelligence – Opportunities and risks for cultural heritage institutions. In: Thiel S and Bernhardt J (eds) *AI in Museums: Reflections, Perspectives and Applications*. Bielefeld: transcript Verlag, pp.149–162.

Noble SU (2018) *Algorithms of Oppression: How Search Engines Reinforce Racism*. New York: NYU Press.

O'Neil C and Gunn H (2020) Near-term artificial intelligence and the ethical matrix. In: Liao SM (ed.) *Ethics of Artificial Intelligence*. Oxford: Oxford University Press, pp.237–270. DOI: 10.1093/oso/9780190905033.003.0009.

Orr W and Crawford K (2023) The social construction of datasets. *New Media & Society*.

Orr W and Crawford K (2024) Building better datasets: Seven recommendations for responsible design from dataset creators. *Journal of Data-Centric Machine Learning Research*. Available at: https://openreview.net/forum?id=6bd8BrRKTW (accessed 2025–11-25).

Paullada A, Raji ID, Bender EM, Denton E and Hanna A (2021) Data and its (dis)contents: A survey of dataset development and use in machine learning research. *Patterns* 2(11): 100336. DOI: 10.1016/j.patter.2021.100336.

Peterson AJ (2025) AI and the problem of knowledge collapse. *AI & Society* 40: 3249–3269. DOI: 10.1007/s00146-024-02173-x.

Powers TM and Ganascia J-G (2020) The ethics of the ethics of AI. In: Dubber MD, Pasquale F and Das S (eds) *The Oxford Handbook of Ethics of AI*. Oxford: Oxford University Press, pp.26–51. DOI: 10.1093/oxfordhb/9780190067397.013.2.

Rawat S and Meena S (2014) Publish or perish: Where are we heading? *Journal of Research in Medical Sciences* 19(2): 87–89.

Riley D (2017) Rigor/us: Building boundaries and disciplining diversity with standards of merit. *Engineering Studies* 9(3): 249–265. DOI: 10.1080/19378629.2017.1408631.

Rudin C (2019) Stop explaining black box machine learning models for high stakes decisions and use interpretable models instead. *Nature Machine Intelligence* 1(5): 206–215.

Singh D and Singh A (2025) Ubiquity of LLM hallucinations across critical domains: A survey. In: Yuan S, Malliaros F and Zheng X (eds) *Trends and Applications in Knowledge Discovery and Data Mining*. Lecture Notes in Computer Science, vol.15835. Singapore: Springer, pp.115–132. DOI: 10.1007/978-981-96-8197-6_9.

Simons A, Zichert M and Wüthrich A (2026) Large language models for history, philosophy, and sociology of science: Interpretive uses, methodological challenges, and critical perspectives. *Studies in History and Philosophy of Science* 117: 102151. https://doi.org/10.1016/j.shpsa.2026.102151.

Strubell E, Ganesh A and McCallum A (2019) Energy and policy considerations for deep learning in NLP. In: *Proceedings of the 57th Annual Meeting of the Association for Computational Linguistics*, Florence, Italy, pp.3645–3650. Florence: Association for Computational Linguistics. Available at: https://aclanthology.org/P19-1355 (accessed 2025–11-25).

Valleriani M (2025) Large language models that power AI should be publicly owned. *The Guardian*, 26 May. Available at: https://www.theguardian.com/technology/2025/may/26/large-language-models-that-power-ai-should-be-publicly-owned (accessed 2025–11-25).

Vecchione B, Barocas S and Levy K (2021) Algorithmic auditing and social justice: Lessons from the history of audit studies. In: *Equity and Access in Algorithms, Mechanisms, and Optimization (EAAMO '21)*. New York: ACM. DOI: 10.1145/3465416.3483294.

Xiao Y and Wang WY (2021) On hallucination and predictive uncertainty in conditional language generation. In: *Proceedings of the 16th Conference of the European Chapter of the Association for Computational Linguistics: Main Volume*, Online, pp.2734–2744. Association for Computational Linguistics. Available at: https://aclanthology.org/2021.eacl-main.236 (accessed 2025–11-25).

Yang T, Strippel C, Keiner A, Baker D, Chávez A, Kauffman K, Pohl M, Sinders C and Miceli M (2025) *Ethics of Data Work: Principles for Academic Data Work Requesters*. Berlin: Weizenbaum Institute. DOI: 10.34669/WI.DP/48.

AI and the scientist

On the fracture of epistemic authority

Kristina Khutsishvili

1. Introduction

The figure of the scientist has long occupied a distinctive place in the epistemic order. From the Enlightenment onward, the scientist was not merely a skilled technician but a disciplined inquirer – a producer and guardian of knowledge – whose legitimacy rested on epistemic virtues such as curiosity, integrity, and humility, as well as on empirical rigour. Today, however, this position appears less secure. The systems we now refer to as 'artificial intelligence', and more concretely large language models (LLMs), produce outputs which, in certain domains, rival human work.

The figure of the scientist is thus threatened with displacement from the centre of knowledge creation. This is not simply a technical redistribution of tasks but a deeper reconfiguration of epistemic authority. The scientist now confronts a figure that resembles an autonomous author: a system generating hypotheses and even theoretical frameworks in ways that appear – symbolically, though not ontologically – independent of human agency. Whether such independence is real or only perceived matters greatly, yet the appearance of autonomy already unsettles established epistemic hierarchies. The scientist risks becoming a steward of results whose genesis lies elsewhere.

Scientific authority has depended on the human capacity to judge, to weigh, to distinguish. What happens when these gestures can be mimicked – and when the mimicry becomes epistemically consequential?

2. Epistemic authority

Charles Taylor has argued that identity is dialogical – formed "always in dialogue with, sometimes in struggle against the things our significant others want to see in us" (Taylor, 1992: 32–33). The identity of the scientist has been no exception. For centuries, scientists have been recognised by peers, institutions, and the public as authoritative knowers capable of exercising inquiry. Their authority rested on a shared understanding that inquiry relies on a human capacity to weigh, distinguish, and reflect, and that scientific results are answerable to this capacity. Epistemic authority was thus tied not only to the results of inquiry itself, but to the recognisable exercise of judgement that produced them.

When LLMs generate literature reviews, propose theoretical frameworks, or conduct statistical analysis, something in this configuration shifts. These systems are not simply tools in the deterministic sense in which a calculator assists with arithmetic. They increasingly appear as quasi-autonomous epistemic agents: they draft arguments and recombine literatures in ways that are presented, and often perceived, as if they originated in a unitary 'authorial' source. In that sense, LLMs may extend the scientist's reach but they also may be positioned as rivals operating within the same epistemic space.

It is true that humans remain formally responsible: they prompt, read, analyse, correct, and sign their names under the final piece. However, the phenomenology of authorship and authority changes when a substantial part of the argumentative or narrative structure is generated by a system whose internal operations cannot be traced in familiar justificatory terms. The scientist is no longer the primary source of authoritative conclusions and risks becoming a steward of results whose genesis lies elsewhere. The site of authority moves from the act of judging to the act of approving.

In Taylor's terms, this is a risk of misrecognition. Authority has always involved recognition: others see the scientist as a person whose reasons and methods warrant trust. If, however, the 'producer' of the result is a system, the scientist may instead be recognised as the one who 'checked' or 'curated' something rather than 'produced'. The role is reframed, the position is fractured, the identity is unsettled. The scientist's self-understanding as an agent of inquiry becomes intertwined with a system that occupies the generative space of argument.

For researchers in the history, philosophy, and sociology of science (HPSS), this displacement carries particular weight. The discipline does not simply produce facts but interprets the conditions under which facts become meaningful; it investigates how concepts, methods, and institutions confer authority on some claims and not on others (Kitcher, 1993; Kuhn, 2012; Latour, 1987). These interpretive and critical approaches are precisely those that advanced systems can now mimic – producing narratives or arguments with surface plausibility that can obscure the absence of judgement. LLMs excel at re-stating available positions, weaving them into apparently coherent wholes, and generating what looks like 'critical reflection' by combining familiar criticisms.

Thus, there is a double movement. On the one hand, HPSS scholars are exposed to a specific kind of displacement: working in domains where argumentative, narrative, and conceptual work – the very textures of the practice – can be imitated by systems that do not understand what they are doing. In that sense, they encounter AI not at the margins of their work but at its core. On the other hand, precisely because HPSS focuses on the conditions of meaning, there is a possibility that its epistemic relevance increases rather than diminishes. The more central AI systems become to scientific practice, the more urgent the question of how their outputs are authorised, integrated, resisted, or normalised. In other words, the same development that threatens HPSS with mimicry also calls it, perhaps uniquely, to analyse the reconfiguration of epistemic authority itself. Therefore, the question "Who do HPSS scientists become when the ground beneath them shifts?" cannot be answered simply in terms of replacement or survival. The fracture of epistemic authority is at the same time a fracture of identity.

The novelty of AI-mediated inquiry creates a new demand within the sciences themselves: the need for epistemological orientation. Scientists cannot rely solely on technical expertise when the fundamental processes shaping research no longer fit familiar methodological categories. Questions of what counts as an explanation, how evidence is approached, or how authority is redistributed are no longer abstract philosophical problems but practical conditions of scientific work. Not every discipline is equipped to address these questions from within; epistemic reflexivity is unevenly distributed. HPSS scholars, whose work has long examined methods, frameworks, and justificatory practices, may therefore become indispensable sources of orientation for their non-HPSS colleagues. Their expertise does not only describe the changing epistemic landscape but also provides the vocabulary and ability for navigating it. In this sense, AI does not only challenge HPSS, but it also makes its insights newly necessary.

3. The scientist as member of an imagined community

Benedict Anderson's (2006) metaphor of "imagined communities" is also suitable for scientific disciplines: they are sustained not predominantly by face-to-face contact but by shared languages, symbols, and rituals. Belonging depends on mutual recognition; to be counted as part of the community means having one's contributions integrated into its ongoing dialogue. Scientific communities define themselves through journals, conferences, textbooks, and informal networks – an entire symbolic system in which membership is conferred and denied.

At first sight, this might suggest that scientists retain complete control over the boundaries of their community. If mutual recognition is the criterion, one might argue that AI systems can be kept outside simply by refusing to cite AI-generated work, by not attributing authorship to non-humans, and by normatively enforcing human responsibility in publication and review. In this view, the imagined community is guarded by its members; they decide who counts as a knower, and under which conditions.

However, this picture underestimates how indirect the mechanisms of inclusion are. LLMs disrupt the fabric of the scientific imaginary not only when they are explicitly named, but also when their outputs circulate through human names, embedded in texts that pass peer review, enter databases, and are cited as if they were straightforward products of human inquiry. If a scientist relies heavily on AI-generated literature reviews, the resulting article may formally satisfy all existing norms of authorship while being largely materially shaped by a system that is not a member of the community and cannot be held to account.

The imagined community begins to shift not when it officially admits non-human members, but when its criteria for recognising human members no longer presuppose that the work of judgement and composition was carried out by those members themselves. The line between internal and external contributions becomes blurred. Belonging is shared – without explicit consent – with entities alien to the processes of recognition, which nevertheless shape what is recognised.

The HPSS area of studies is particularly exposed to this development. The imagined community is compact, and its cohesion rests on slow, reflective work. The incorporation of LLMs into these domains threatens to conflate the product of interpretive intellectual work with an accumulation of textual patterns. A careful historical analysis or a nuanced conceptual distinction may be seen as 'just another kind of text' that a system could plausibly generate. In the long run, this may alter how marginal or central such work appears to be within broader scientific imaginaries that increasingly equate value with speed and volume of output.

4. Identity, belonging, and (mis)recognition

Identities are never static; they are forged both through belonging and through refusal – through 'who we are' and 'who we are not'. Jean-Luc Nancy captures this complexity when he writes that identity "is not a figure", "identity is something more subtle, more delicate, and more evasive", and "its force lies in displacing, in changing figures" (Nancy, 2015: 10–11). The scientist's identity has traditionally been defined not only by what it is – a disciplined inquirer – but also by what it is not: speculation, baseless claim, the work of a dilettante. The 'non-scientific' has been a constitutive outside for the self-image of science.

This distinction becomes harder to maintain when LLMs generate work that appears indistinguishable from human scholarship. The boundary between the 'scientific' and the 'non-scientific' becomes uncertain. Without that line, the grounds for recognition shift, and with them, the self-understanding of the scientist. Gönül Pultar's reflections on rapid structural change that brings "instability, distortion of meaning, and volatile identities" are relevant here (Pultar, 2014: 155). When a scientist's contribution is reduced to approving or disapproving the outputs of systems whose operations are beyond human comprehension, the possibility of this scientist's unique epistemic authority diminishes. Authority that rests on results alone – without the shared understanding of how those results came to be – can no longer be sustained through the same dialogical relation that underpins recognition. For HPSS scholars – whose expertise lies in critically interrogating the conditions of knowledge 'production' – this transformation challenges the very core of their professional identity. Yet it is precisely here that their irreplaceability can also be articulated.

If epistemic authority is being fractured, the loss is not only in how scientists are recognised by others, but also in how they recognise each other and themselves. As Taylor (1992) reminds us, identity is formed in dialogue, and without a community capable of mutual recognition, the professional self is affected. Anderson (2006) would add that this community has always been, to some extent, imagined, and LLMs unsettle these shared imaginaries. For scientists, and especially for those in HPSS, the task is unlikely to be protecting a role in the old sense, but nurturing the community's life: ensuring that its narratives continue to speak of curiosity, human inquiry, responsibility, and the bonds that make recognition possible.

In this light, the future of science will not be determined solely by the distribution of tasks between human and machine. Rather, it will depend on whether the imagined community of inquirers can survive the arrival of a participant that cannot imagine itself. The answer will depend on our capacity to renew the stories we tell about what it means to know and to belong – in an age when knowing is no longer ours alone.[1]

References

Anderson B (2006) *Imagined Communities: Reflections on the Origin and Spread of Nationalism*. London: Verso.

Kitcher P (1993) *The Advancement of Science: Science without Legend, Objectivity without Illusions*. New York: Oxford University Press.

Kuhn TS (2012) *The Structure of Scientific Revolutions*. Chicago: University of Chicago Press.

Latour B (1987) *Science in Action: How to Follow Scientists and Engineers through Society*. Cambridge, MA: Harvard University Press

Nancy J-L (2015) *Identities: Fragments, Frankness*. New York: Fordham University Press.

Pultar G (ed.) (2014) *Imagined Identities: Identity Formation in the Age of Globalization*. Syracuse, NY: Syracuse University Press.

Taylor C (1992) 'The Politics of Recognition'. In: Gutmann A (ed) *Multiculturalism: Examining the Politics of Recognition*. Princeton: Princeton University Press, 25–73.

1 The author of this chapter declares the use of a large language model (LLM) to obtain critical feedback on ideas (as a form of intellectual sparring) and to assist with final copy-editing.

Grounding AI in humanistic inquiry

Interdisciplinary challenges for evaluation and interpretability

Oliver Eberle

1. Introduction

The rise of foundation models has enabled machine learning (ML) systems to be applied across a wide range of scientific domains (Richards et al., 2019; Carleo et al., 2019; Rajpurkar et al., 2022; Quazi 2022). Recent advances have further led to the development and rise of artificial large reasoning models (LRMs) (see (Xu et al., 2025) for an overview), which are designed to address multi-step decision tasks like the derivation of mathematical proofs, generating code, and common-sense reasoning. The accelerated use of such LLM systems in scientific practice has especially benefited fields like the natural sciences, characterized by large-scale data, quantitative methods, and well-structured datasets. Meanwhile, HPSS, characterized by low-resource data as well as challenging annotation and evaluation tasks, took a backseat. Nonetheless, AI has the potential to "radically change humanities" (Duch 2023) and offers opportunities to address both technical and societal challenges amid the rapid adoption of this technology.

However, as AI models grow in complexity and are applied to more challenging tasks like artificial reasoning, their inner workings become increasingly opaque, posing challenges for scientific applications that demand interpretability, trustworthiness, and verifiability. Explainable AI and interpretability research aim to develop methods that uncover how AI systems operate at a fundamental level (Samek et al., 2019; Linardatos et al., 2021; Zhao et al., 2024).

The following explores the mutual interactions between ML and HPSS, then discusses the challenges of evaluation and interpretability, before presenting ways forward and concluding.

2. Current interactions between ML and HPSS research

2.1 ML-inspired HPSS

2.1.1 ML accelerates manual HPSS processes

Many works at the intersection of HPSS and LLMs aim to streamline and scale labor-intensive tasks such as transcription, text restoration, entity recognition, topic attribution, and automated image extraction. Nikolaidou et al. (2022) and Sommerschield et al. (2023) provide surveys of relevant data and methods in the historical sciences, highlighting the specific challenges like out-of-domain materials and model evaluation. Similarly, social scientists have used LLMs as a tool for automating manual processes, including annotation or information extraction, as well as summarizing and analyzing large amounts of speech and text (see Macanovic 2022; Linegar et al., 2023 for an overview). For a broader review of ML and LLM approaches used in the context of the humanities, see Chapinal-Heras and Díaz-Sánchez (2023), Simons et al. (2026), Meding and Daugs (2026), and Lang (2026).

2.1.2 ML can promote novel HPSS research

The iterative process of gathering, analyzing, and interpreting material in HPSS research challenges standard prediction systems' ability to produce novel insights. Consequently, relatively few works have employed ML in the context of historical insight discovery and "historian in the loop" studies (Assael and Sommerschield et al. 2022; Eberle et al. 2024; Assael and Sommerschield et al., 2025). Further advances would require ML and LLM systems to deeply understand relevant sources, including historical foundation models evaluated on expert global history knowledge, as recently explored by Hauser et al. (2024). Social scientists have used LLMs to study the modeling of cultural transmission (Brinkmann et al., 2023, Lu et al., 2025), the emergence of psychological theories in LLMs, e.g., theory of mind (Strachan et al., 2024) or behavioural game theory (Akata et al., 2025), and their usefulness for LLM social simulations (Qu and Wang 2024; Anthis et al., 2025). This demonstrates how advances in ML can support and enable the investigation of novel scientific questions in HPSS.

2.1.3 ML as an application of the HPSS

Philosophy has interacted with ML by challenging its concepts and the integration of AI progress into existing theories, e.g., bringing statistical learning theory into discussions on the reliability of induction (Harman and Kulkarni 2007), viewing ML as an "experimental philosophy of science" (Korb 2004), bridging philosophical accounts of causality in ML and philosophy (Pearl 2000; Halpern and Pearl 2005), or how information theory contributed to shaping theories of mind and meaning (Adams 2003). Further exchange of ML and philosophy has centered on concepts of *scientific explanation* (Woodward and Ross 2003) aimed toward interpreting today's largely opaque models as discussed in (Miller 2018; Erasmus et al., 2021). Philosophical analyses of ML systems (Thagard, 1990) anticipated challenges that would become practically relevant to deep learning research decades later, including adversarial or counterfactual data and models that generate predictions based on irrelevant inferences in specific problem-solving scenarios (Szegedy et al., 2014; Lapuschkin et al., 2019). Together, these perspectives position HPSS as an important intellectual framework capable of shaping ML research by driving deeper inquiry into the nature of knowledge, explanation, and understanding.

2.2 HPSS-inspired ML

2.2.1 HPSS challenges ML

Heterogeneity alongside temporality and context-awareness present key challenges to ML and LLMs, which currently limit the faithful processing of historical sources. To address data heterogeneity of complex materials, custom ML approaches have been proposed (e.g., Assael and Sommerschield et al., 2022; Eberle et al., 2024), currently requiring tailored solutions and active model development. Nonetheless, advances in AI and the increasing accessibility of powerful foundation models may soon make AI-based analyses feasible, even in challenging scenarios. For instance, deep optical character recognition (OCR) models have recently been proposed to compress historical long-context data into visual tokens rather than text tokens (Wei et al., 2025), capturing rich spatial and contextual information and enabling more efficient indexing, retrieval, and analysis of archival materials, thereby offering a more contextualized understanding of sources.

While making such sources machine-readable is a crucial step toward enabling scientific insight at scale, humanistic inquiry engages with interpretive reasoning, diachronic investigation, and contextual nuance, which currently constitute underexplored areas of ML research. For instance, the computational modeling of temporal change presents a current frontier as discussed in (Büttner, 2026). Given the methodological challenges of extending ML modeling frameworks, recent studies have productively focused on decomposing historical workflows (e.g., Sommerschield et al., 2023) to enable computational analysis, providing valuable intermediate steps toward the development of more sophisticated modeling approaches.

As LLMs are trained to capture statistical associations, they struggle in HPSS domains where causal structures are shaped by complex contexts, e.g., historical, cultural, or institutional, and where scientific explanations depend as much on interpretation as on observation. For example, philosophical inquiry and reasoning may not deliver sufficiently many recognizable, repeatable patterns to robustly train today's AI systems (D'Alessandro, 2026). Philosophers such as (Machamer et al., 2000) have framed scientific practice as uncovering mechanisms that explain how phenomena arise, emphasizing *entities* and how *activities* among them cause change. These ideas were popularized and adopted in biology and neuroscience, which have since informed the analysis of AI systems and, specifically, shaped the concepts and goals of interpretability research as discussed in Section 3.

2.2.2 HPSS grounds ML

While most ML research focuses on technical challenges, HPSS can play a key role in clarifying trade-offs between technological advances and societal values, for example by discussing how research that involves LLMs intersects with issues of data privacy, transparency, and fairness. Epistemology can further ground ML research by examining how evidence and uncertainty constrain what can be confidently known (Lipton, 2004; Kiureghian and Ditlevsen, 2009), providing important conceptual foundations for developing epistemologically informed ML. In this context, research on artificial reasoning with LRMs has mostly focused on deductive settings, e.g., generating mathematical proofs, for which clear ground truth reasoning steps can be obtained for validation. Addressing uncertainty and non-deductive reasoning remains a significant challenge, one in which the HPSS community has long-standing expertise that can aid in grounding future developments. Recent work on analyzing LLMs has further explored neural network representations related to truthfulness (Azaria and Mitchell, 2023) and morality (Wynn et al., 2024; Jiang et al., 2025), building on key philosophical notions and recent benchmarks like MoralBench (Ji et al., 2024). Further empirical evidence of *universal representations* across foundation models and tasks has sparked discussions on converging representations in deep learning (Huh et al., 2024), which the authors conceptually link to Plato's *Allegory of the Cave* and *convergent realism* (Laudan, 1981; Putnam, 1982).

2.2.3 HPSS can inform the development of ML methods

Besides challenging ML, HPSS offers rich data, complex task settings and workflows, as well as methodological guidance for improving model reasoning and alignment. Model tuning through HPSS materials has recently been explored; e.g., moral alignment via in-context ethical policies (Rao et al., 2023), factuality-aware learning algorithms (Lin et al., 2024), and even fine-tuning LLMs on the works of philosopher Daniel Dennett (Schwitzgebel et al., 2024). In the context of designing AI agent systems, philosophical and social science frameworks have informed recent work, including multi-agent systems grounded in epistemology (Shoham and Leyton-Brown, 2008), LLM-agentic philosophers (Barkol, 2025), and cooperative systems (Ashery et al., 2025). Beyond guiding model design, critically evaluating and understanding these systems from perspectives such as epistemology, reasoning, human values, and historical and social context is key to developing AI that is interpretable, reliable, and socially aligned.

3. The challenge of LLM evaluation and interpretability

3.1 Behavioral evaluation of performance

The evaluation of today's state-of-the-art LLMs predominantly focuses on task suites designed to test certain desired capabilities such as general, potentially multilingual, language understanding and comprehension, code generation, mathematical and logical reasoning, and general prompt instruction following; see (Srivastava et al., 2023) for an overview of commonly used tasks and task structure. Aside from evaluating models in terms of their alignment with general human preferences and their performance relative to weakly defined safety and ethics-critical standards, dedicated HPSS benchmarks are largely missing. As standard ML evaluations are limited in capturing the complexity of large text as produced by LLMs, Wallach et al. (2025) propose to frame generative AI evaluation as a social science measurement challenge, providing a path forward in capturing abstract concepts such as ideology, democracy, or bias via measurement theory frameworks. This highlights the potential of HPSS approaches to develop nuanced evaluations of LLM behavior, moving beyond the simplistic technical benchmarks that dominate ML research.

3.2 Understanding AI models through explanation techniques and interpretability studies

While accurate performance measurements are an important step toward using LLMs in HPSS, uncovering and verifying their underlying inference mechanisms are crucial, especially in the context of grounding scientific discovery. Although models may be able to generate novel outputs and scientifically valuable hypotheses without revealing their internal processes, understanding these mechanisms can enhance trust and provide deeper insight. This requires access to faithful model explanations that aim to provide (i) transparency and trustworthiness, (ii) verifiability, (iii) human interpretability and understanding, (iv) insights into potentially unexpected model strategies, and (v) actionable guidance for model improvement or error mitigation. Since explanations can be derived at multiple levels and provide insight in various formats, the following summarizes the most common variants in the context of language models.

3.3 Explanations across levels and structures

Initial explanation efforts have focused on explanations in terms of relevant input features. Such feature attribution methods have been widely used to compute heatmaps over input samples such as token sequences used in text classification tasks (Bach et al., 2015; Ribeiro et al.; 2016, Sundararajan et al., 2017; Ali et al., 2022). Furthermore, the analysis of model internals, i.e., neuron-level interpretation, layer-wise representations and analysis of relevant model components like attention heads, can provide deeper insight into how the model inputs are represented, interpreted and contextualized, aiming to reveal mechanistic strategies. Mechanistic interpretability herein developed methods for the extraction of circuits via causal abstractions of LLMs (Geiger et al., 2021; Geiger et al., 2025). Structured explanations extend beyond heatmaps by capturing higher-order interactions among multiple features (Eberle et al., 2022; Schnake et al., 2022), for example revealing token–token interactions in LLMs (Vasileiou and Eberle, 2024). Algorithmic explanations further aim to identify the implemented internal algorithms of LLMs and LRMs, focusing on defining primitives and compositions thereof (Eberle et al., 2025; Lippl et al., 2025). To approximate the function of specific internal features, such as individual neurons or groups of neurons, feature description methods provide text-based concept summaries that characterize the patterns most strongly associated with their activation patterns (Singh et al., 2023; Bills et al., 2023; Kopf et al., 2025).

Extracting influential dataset samples provides explanations in reference to the training corpus, for example by revealing which samples strongly influenced the learning of a particular pattern (Han et al., 2020; Grosse et al., 2023). Complementary work leverages LLMs' generative abilities to communicate explanations directly, adopting an output-centric view. Methods include *self-explanations* and *chain-of-thought* (Wei et al., 2022; Huang et al., 2023), as well as free-form interactive user prompt explanations such as "explain to me," "provide evidence," "what would happen if..."(Slack et al., 2023), and counterfactual explanations (Chen et al., 2024).

3.4 Evaluating the quality of explanations

Across all these variants, explainable AI methods need to be rigorously evaluated to ensure providing reliable insight into the model. Evaluating explanations is challenging due to the wide range of methods and explanatory levels, the lack of agreement on what makes a good explanation, and the inherently fuzzy nature of the concept itself. Common evaluation criteria include *fidelity* to the model's computations, *plausibility* for human users, and *causality*, which assesses whether explanations capture genuine cause and effect relationships (Samek et al., 2019; Zhao et al., 2024). Evaluation is further complicated by the absence of ground truth explanations and the inherently subjective nature of what human users consider an effective explanation. See also (Zhou et al., 2021) for a review on evaluating explanations, and (Zhao et al., 2024) for LLM evaluations specifically.

In the context of HPSS research, evaluating model explanations is challenging because clearly defined task settings and ground truth annotations are scarce. Advances in LLM-based artificial reasoning and multi-agent collaboration now make it possible to tackle increasingly complex tasks, including AI-driven logic problem solving, commonsense reasoning, and behavioral simulations, and integration into the scientific process (Zhang et al., 2025). This, in turn, calls for datasets and benchmarks that, for example, could focus on the AI-based reproduction of traditional HPSS research findings like the re-discovery of scientific concepts and ideas from predefined sources. Such investigations highlight the importance of expertise across HPSS fields, for example in analyzing a model's epistemological justification process or studying the collective behavior and collaboration of multiple AI agents. Achieving this will require closer collaboration and ongoing exchange of knowledge, data, and tools between HPSS and AI research.

4. Call for action

4.1 Advancing materials and benchmarks for training and evaluation

HPSS communities must lead the development of domain-specific, curated datasets that reflect their complexities, structure, and contextual richness. Such datasets are essential not only for training more reliable and representative models but also for developing evaluation benchmarks that move beyond generic and narrowly focused metrics. Current benchmarks lack key features of HPSS problems, such as temporal change and cultural grounding, navigating ambiguity and normative value judgments, and working with interpretative or incomplete evidence; see (Hershcovich et al., 2022; Ziems et al., 2024; Morehouse et al., 2025) for some first steps in this direction. These challenges further motivate the development of new evaluation methods that allow capturing dynamic sources or ill-defined labels.

4.2 Grounding LLM and interpretability research in HPSS frameworks

Interpretability and evaluation methods should be explicitly grounded in HPSS conceptual frameworks, incorporating epistemological rigor, historical context, and cultural sensitivity. This will ensure that models are not only technically sound but also socially and scientifically meaningful. The rise of AI has already and will continue to change how we access, process and retrieve information. However, as Offert (2024) argues, traditional humanist frameworks may not fully suffice to conceptualize history in a world of constantly remediated data, challenging HPSS to adapt and extend its frameworks in response to advances in AI. For example, Willems et al. (2025) highlight the need for philosophy in explainable AI and interpretability research to clarify definitions and drive methodological advances for assessing societally critical aspects.

4.3 Participation in shaping AI discussions and development

HPSS scholars must actively contribute to AI ethics, policy, and public discourse, ensuring that debates about LLM development and deployment are informed by deep understanding of historical, philosophical and social implications. Critical perspectives have played a key role in shaping discussions and raising awareness of fairness, biased predictions, and accountability in learning-based systems (Dwork et al., 2012; Shah et al., 2020; Gallegos et al., 2024), and the recent rise of AI agents has further prompted calls for new ethical frameworks (Gabriel et al., 2025). Furthermore, engagement in AI research communities, conferences, and collaborative projects is needed to shape future research agendas, standards, and methodologies that reflect the needs and values of their fields. The organization of joint workshops like the NeurIPS 2011 workshop on *Philosophy and Machine Learning*, the *Machine Learning for Ancient Languages* Series at ACL 2024 and 2025, and the NeurIPS 2025 workshop on *Algorithmic Collective Action* herein provide important opportunities for extending interdisciplinary exchange.

5. Conclusion

HPSS has the unique opportunity and responsibility to shape the future development of AI by providing conceptual frameworks, experience with historically grounded context, and culturally aware evaluation methods. By defining its own benchmarks, datasets, and interpretability standards, HPSS can support the development of AI systems that are not only technically proficient but demonstrate epistemic robustness, ethical integrity, and regulatory compliance. This requires sustained collaboration between HPSS and ML research communities to bridge conceptual gaps and inform methodologies. Such engagement will guide the creation of AI systems that advance human understanding while remaining accountable to the diverse values and complexities of the societies they serve.[1]

1 This chapter was reviewed with support of large language models (LLMs) and its use was restricted to improving grammar and style. The author remains fully responsible for the final version. For details on the use of LLMs in this volume, see the statement in the volume's introduction.

References

Adams F (2003) The informational turn in philosophy. Minds and Machines 13(4):471–501.

Akata E, Schulz L, Coda-Forno J, et al. (2025) Playing repeated games with large language models. Nat Hum Behav 9, 1380–1390. https://doi.org/10.1038/s41562-025-02172-y

Ali A, Schnake T, Eberle O, et al. (2022) XAI for transformers: Better explanations through conservative propagation. In International conference on machine learning (pp. 435–451). PMLR.

Anthis JR, Liu R, Richardson SM, et al. (2025) Position: LLM Social Simulations Are a Promising Research Method. In Forty-second International Conference on Machine Learning Position Paper Track.

Ashery AF, Aiello LM and Baronchelli A (2025) Emergent social conventions and collective bias in LLM populations. Science Advances, 11(20), eadu9368.

Assael Y, Sommerschield T, Cooley A, et al. (2025) Contextualizing ancient texts with generative neural networks. Nature. https://doi.org/10.1038/s41586-025-09292-5

Assael Y, Sommerschield T, Shillingford B, et al. (2022) Restoring and attributing ancient texts using deep neural networks. Nature 603, 280–283. https://doi.org/10.1038/s41586-022-04448-z

Azaria A and Mitchell T (2023) The Internal State of an LLM Knows When It's Lying. In Findings of the Association for Computational Linguistics: EMNLP 2023, pages 967–976, Singapore. Association for Computational Linguistics.

Bach S, Binder A, Montavon G, et al. (2015) On Pixel-Wise Explanations for Non-Linear Classifier Decisions by Layer-Wise Relevance Propagation. PLoS ONE 10(7): e0130140. https://doi.org/10.1371/journal.pone.0130140

Barkol D (2025) Agentic AI Explained: A Philosophical Framework for Understanding AI Agents. Microsoft Dev Blogs. https://devblogs.microsoft.com/all-things-azure/agentic-philosophers/.

Bills S, Cammarata N, Mossing M, et al. (2023) Language models can explain neurons in language models. https://openaipublic.blob.core.windows.net/neuron-explainer/paper/index.html.

Büttner J (2026) Why pursue temporally-grounded AI for historical disciplines, and what makes it so challenging? In: Simons A, Wüthrich A, Zichert M, et al. (eds) *Understanding Science with Large Language Models? Potentials for the History, Philosophy, and Sociology of Science*. Bielefeld: transcript, part-2.

Brinkmann L, Baumann F, Bonnefon JF et al. (2023) Machine culture. Nat Hum Behav 7, 1855–1868. https://doi.org/10.1038/s41562-023-01742-2

Carleo G, Cirac I, Cranmer K, et al. (2019) Machine learning and the physical sciences. Reviews of Modern Physics, 91(4), 045002.

Chapinal-Heras D and Díaz-Sánchez C (2023) A review of AI applications in Human Sciences research. Digital Applications in Archaeology and Cultural Heritage, 30, e00288.

Chen Y, Zhong R, Ri N, et al. (2024) Do models explain themselves? counterfactual simulatability of natural language explanations. In Proceedings of the 41st International Conference on Machine Learning (pp. 7880–7904).

D'Alessandro W (2026) LLMs as philosophers: what can they do, and why aren't they better? In: Simons A, Wüthrich A, Zichert M, et al. (eds) *Understanding Science with Large Language Models? Potentials for the History, Philosophy, and Sociology of Science*. Bielefeld: transcript, part-1.

Der Kiureghian A and Ditlevsen O (2009) Aleatory or epistemic? Does it matter?. Structural safety, 31(2), 105–112.

Duch W (2024) Artificial Intelligence and the Limits of the Humanities. Er (r) go. Teoria-Literatura-Kultura, (48), 269–297.

Dwork C, Hardt M, Pitassi T, et al. (2012) Fairness through awareness. In Proceedings of the 3rd innovations in theoretical computer science conference (pp. 214–226).

Eberle O, Büttner J, Kräutli F, et al. (2022) Building and interpreting deep similarity models. IEEE Transactions on Pattern Analysis and Machine Intelligence, 44(3):1149–1161. doi: 10.1109/TPAMI.2020.3020738.

Eberle O, McGee T, Giaffar H, et al. (2025) Position: We Need An Algorithmic Understanding of Generative AI. In Forty-second International Conference on Machine Learning Position Paper Track.

Erasmus A, Brunet TDP and Fisher E (2021) What is Interpretability?. Philos. Technol. 34, 833–862. https://doi.org/10.1007/s13347-020-00435-2

Fabian O (2024) On the Concept of History (in Foundation Models). Forthcoming in: Thinking with AI, ed. Hannes Bajohr, Open Humanities Press.

Gabriel I, Keeling G, Manzini A, et al. (2025) We need a new ethics for a world of AI agents. Nature. Aug;644(8075):38-40. doi: 10.1038/d41586-025-02454-5. PMID: 40764692.

Gallegos IO, Rossi RA, Barrow J, et al. (2024) Bias and Fairness in Large Language Models: A Survey. Computational Linguistics; 50 (3): 1097–1179. DOI: https://doi.org/10.1162/coli_a_00524

Geiger A, Ibeling D, Zur A, et al. (2025) Causal abstraction: A theoretical foundation for mechanistic interpretability. Journal of Machine Learning Research, 26(83), 1–64.

Geiger A, Lu H, Icard T, et al. (2021) Causal abstractions of neural networks. Advances in Neural Information Processing Systems, 34, 9574–9586.

Grosse R, Bae J, Anil C, et al. (2023) Studying large language model generalization with influence functions. arXiv preprint arXiv:2308.03296.

Halpern J and Pearl J (2005) Causes and explanations: A structural-model approach. British J. Phil. Sci. 56:843-911.

Han X, Wallace BC, and Tsvetkov Y (2020) Explaining Black Box Predictions and Unveiling Data Artifacts through Influence Functions. In Proceedings of the 58th Annual Meeting of the Association for Computational Linguistics, pages 5553–5563, Online. Association for Computational Linguistics.

Harman G and Kulkarni S (2007) Reliable Reasoning: Induction and Statistical Learning Theory. MIT Press, Cambridge, MA.

Hauser J, et al. (2024) Large language models' expert-level global history knowledge benchmark (HiST-LLM). Advances in Neural Information Processing Systems 37: 32336–32369.

Hershcovich D, Frank S, Lent H, et al. (2022) Challenges and Strategies in Cross-Cultural NLP. In Proceedings of the 60th Annual Meeting of the Association for Computational Linguistics (Volume 1: Long Papers) (pp. 6997–7013).

Huang J, Gu S, Hou L, et al. (2023) Large Language Models Can Self-Improve. In Proceedings of the 2023 Conference on Empirical Methods in Natural Language Processing, pages 1051–1068, Singapore. Association for Computational Linguistics.

Huh M, Cheung B, Wang T, et al. (2024) Position: The platonic representation hypothesis. In Forty-first International Conference on Machine Learning.

Ji J, Chen Y, Jin M, et al. (2025) Moralbench: Moral evaluation of llms. ACM SIGKDD Explorations Newsletter, 27(1), 62–71.

Jiang L, Hwang JD, Bhagavatula C, et al. (2025) Investigating machine moral judgement through the Delphi experiment. Nat Mach Intell 7, 145–160. https://doi.org/10.1038/s42256-024-00969-6

Kopf L, Feldhus N, Bykov K, et al. (2025) Capturing Polysemanticity with PRISM: A Multi-Concept Feature Description Framework. Advances in neural information processing systems, 39.

Korb K (2004) Introduction: Machine learning as philosophy of science. Minds and Machines 14(4).

Lang S (2026) Critical concerns for using LLMs in the (computational) humanities and beyond. In: Simons A, Wüthrich A, Zichert M, et al. (eds) *Understanding Science with Large Language Models? Potentials for the History, Philosophy, and Sociology of Science*. Bielefeld: transcript, part-1.

Lapuschkin S, Wäldchen S, Binder A, et al. (2019) Unmasking Clever Hans predictors and assessing what machines really learn. Nat Commun 10, 1096. https://doi.org/10.1038/s41467-019-08987-4

Laudan L A (1981) Confutation of Convergent Realism. Philosophy of Science; 48(1):19-49. DOI:10.1086/288975

Lazer D, Pentland A, Adamic L, et al. (2009) Social science. Computational social science. Science. Feb 6;323(5915):721-3. doi: 10.1126/science.1167742. PMID: 19197046; PMCID: PMC2745217.

Lin SC, Gao L, Oguz B, et al. (2024) Flame: Factuality-aware alignment for large language models. Advances in Neural Information Processing Systems, 37, 115588–115614.

Linardatos P, Papastefanopoulos V and Kotsiantis S (2021) Explainable AI: A Review of Machine Learning Interpretability Methods. Entropy, 23(1), 18. https://doi.org/10.3390/e23010018

Linegar M, Kocielnik R and Alvarez RM (2023) Large language models and political science. Front. Polit. Sci. 5:1257092. doi: 10.3389/fpos.2023.1257092

Lippl S, McGee T, Lopez K, et al. (2025) Algorithmic Primitives and Compositional Geometry of Reasoning in Language Models, arXiv preprint arXiv:2510.15987.

Lipton P (2004) Induction. Chapter 1 of Inference to the Best Explanation.

Lu JG, Song LL and Zhang LD (2025) Cultural tendencies in generative AI. Nat Hum Behav. https://doi.org/10.1038/s41562-025-02242-1

Macanovic A (2022) Text mining for social science – The state and the future of computational text analysis in sociology. Soc Sci Res. Nov;108:102784. doi: 10.1016/j.ssresearch.2022.102784. Epub 2022 Sep 2. PMID: 36334929.

Machamer P, Darden L, Craver CF (2000) Thinking about Mechanisms. Philosophy of Science; 67(1):1-25. doi:10.1086/392759

Meding H and Daugs A (2026) On the use and limitations of large language models in historical scholarship. In: Simons A, Wüthrich A, Zichert M, et al. (eds) *Understanding Science with Large Language Models? Potentials for the History, Philosophy, and Sociology of Science*. Bielefeld: transcript, part-1.

Miller T (2018) Explanation in artificial intelligence: Insights from the social sciences. *Artificial intelligence*, *267*, 1–38.

Morehouse K, Swaroop S and Pan W (2025) Position: Rethinking LLM Bias Probing Using Lessons from the Social Sciences. Forty-second International Conference on Machine Learning Position Paper Track.

Nikolaidou K, Seuret M, Mokayed H, et al. (2022) A survey of historical document image datasets. IJDAR 25, 305–338. https://doi.org/10.1007/s10032-022-00405-8

Pearl J (2000) Causality: Models, Reasoning, and Inference. Cambridge: Cambridge University Press.

Putnam H (1982) Three Kinds of Scientific Realism. The Philosophical Quarterly (1950-) 32 (128): 195–200.

Qu Y and Wang J (2024) Performance and biases of Large Language Models in public opinion simulation. Humanit Soc Sci Commun 11, 1095. https://doi.org/10.1057/s41599-024-03609-x

Quazi S (2022) Artificial intelligence and machine learning in precision and genomic medicine. Medical Oncology, 39(8), 120.

Rajpurkar P, Chen E, Banerjee O, et al. (2022) AI in health and medicine. Nat Med 28, 31–38. https://doi.org/10.1038/s41591-021-01614-0

Rao AS, Khandelwal A, Tanmay K, et al. (2023) Ethical Reasoning over Moral Alignment: A Case and Framework for In-Context Ethical Policies in LLMs. In Findings of the Association for Computational Linguistics: EMNLP 2023, pages 13370–13388, Singapore. Association for Computational Linguistics.

Ribeiro MT, Singh S and Guestrin C (2016) "Why should i trust you?" Explaining the predictions of any classifier. In Proceedings of the 22nd ACM SIGKDD international conference on knowledge discovery and data mining (pp. 1135–1144).

Richards BA, Lillicrap TP, Beaudoin P, et al. (2019) A deep learning framework for neuroscience. Nat Neurosci 22, 1761–1770. https://doi.org/10.1038/s41593-019-0520-2

Samek W, Montavon G, Vedaldi A, et al. (2019) Explainable AI: interpreting, explaining and visualizing deep learning (Vol. 11700). Springer Nature.

Schnake T, Eberle O, Lederer J, et al. (2022) Higher-order explanations of graph neural networks via relevant walks. IEEE Transactions on Pattern Analysis and Machine Intelligence, 44(11):7581–7596. doi: 10.1109/TPAMI. 2021.3115452.

Schwitzgebel E, Schwitzgebel D and Strasser A (2024) Creating a large language model of a philosopher. Mind & Language, 39(2), 237–259.

Shah DS, Schwartz HA and Hovy D (2020) Predictive Biases in Natural Language Processing Models: A Conceptual Framework and Overview. In Proceedings of the 58th Annual Meeting of the Association for Computational Linguistics, pages 5248–5264, Online. Association for Computational Linguistics.

Shoham Y and Leyton-Brown K (2008) Multiagent systems: Algorithmic, game-theoretic, and logical foundations. Cambridge University Press.

Simons A, Zichert M and Wüthrich A (2026) Large language models for history, philosophy, and sociology of science: Interpretive uses, methodological challenges, and critical perspectives. *Studies in History and Philosophy of Science* 117: 102151. https://doi.org/10.1016/j.shpsa.2026.102151.

Singh C, Hsu AR, Antonello R, et al. (2023) Explaining black box text modules in natural language with language models. arXiv preprint arXiv:2305.09863.

Slack D, Krishna S, Lakkaraju H, et al. (2023) Explaining machine learning models with interactive natural language conversations using TalkToModel. Nat Mach Intell 5, 873–883. https://doi.org/10.1038/s42256-023-00692-8

Sommerschield T, Assael Y, Pavlopoulos J, et al. (2023) Machine Learning for Ancient Languages: A Survey. Computational Linguistics; 49 (3): 703–747. doi: https://doi.org/10.1162/coli_a_00481

Srivastava A, Rastogi A, Rao A, et al. (2023) Beyond the imitation game: Quantifying and extrapolating the capabilities of language models. Transactions on machine learning research.

Strachan JWA, Albergo D, Borghini G. et al. (2024) Testing theory of mind in large language models and humans. Nat Hum Behav 8, 1285–1295. https://doi.org/10.1038/s41562-024-01882-z

Sundararajan M, Taly A and Yan Q (2017) Axiomatic attribution for deep networks. In International conference on machine learning (pp. 3319–3328). PMLR.

Szegedy C, Zaremba W, Sutskever I, Bruna J, Erhan D, Goodfellow IJ and Fergus R (2014) Intriguing properties of neural networks. ICLR (Poster)

Thagard P (1988) Computational Philosophy of Science. MIT Press, Cambridge, MA.

Thagard, Paul (1990) Philosophy and Machine Learning. Canadian Journal of Philosophy 20, no. 2: 261–76. http://www.jstor.org/stable/40231695.

Vasileiou A and Eberle O (2024) Explaining Text Similarity in Transformer Models. In Proceedings of the 2024 Conference of the North American Chapter of the Association for Computational Linguistics: Human Language Technologies (Volume 1: Long Papers), pages 7859–7873, Mexico City, Mexico. Association for Computational Linguistics.

Wallach H, Desai M, Cooper AF, et al. (2024) Position: Evaluating Generative AI Systems Is a Social Science Measurement Challenge. In Forty-second International Conference on Machine Learning Position Paper Track.

Wei H, Sun Y and Li Y (2025) DeepSeek-OCR: Contexts Optical Compression, arXiv preprint arXiv:2510.18234.

Wei J, Wang X, Schuurmans D, et al. (2022) Chain-of-thought prompting elicits reasoning in large language models. Advances in neural information processing systems, 35, 24824–24837.

Woodward J and Ross L (2003) Scientific Explanation. In E. N. Zalta (Ed.), The Stanford encyclopedia of philosophy. Metaphysics Research Lab, Stanford University.

Wynn AH, Sucholutsky I and Griffiths TL (2024) Learning human-like representations to enable learning human values. Advances in Neural Information Processing Systems, 37, 30230–30260.

Xu F, Hao Q, Zong Z, et al. (2025) Towards large reasoning models: A survey of reinforced reasoning with large language models. Patterns, Volume 6, Issue 10, 2025, https://doi.org/10.1016/j.patter.2025.101370.

Zhang Y, Khan SA, Mahmud A, et al. (2025) Exploring the role of large language models in the scientific method: from hypothesis to discovery, npj Artificial Intelligence, 1, p.14. doi: 10.1038/s44387-025-00019-5.

Zhao H, Chen H, Yang F, et al. (2024) Explainability for large language models: A survey. ACM Transactions on Intelligent Systems and Technology, 15(2), 1–38.

Zhou J, Gandomi AH, Chen F, et al. (2021) Evaluating the quality of machine learning explanations: A survey on methods and metrics. Electronics, 10(5), 593.

Ziems C, Held W, Shaikh O, et al. (2024) Can Large Language Models Transform Computational Social Science? Computational Linguistics, 50(1):237–291.

On the use and limitations of large language models in historical scholarship

Holle Meding and Aurel Daugs

1. Introduction

Two opposing tendencies currently shape the response to artificial intelligence in the humanities. On one hand, some have quietly integrated Large Language Models (LLMs) into their daily routines, using them for summarising articles, drafting research proposals or a preliminary contextualisation of historical events. However, this usage often goes unacknowledged.

On the other hand, institutions and individual scholars have reacted by drawing strict boundaries. While outright institutional bans remain rare, many departments and faculty members have expressed reluctance to incorporate LLMs into research or teaching.[1] The concerns are understandable. There are questions surrounding authorship and attribution, the presence of hallucinations, overgeneralisations, and difficulties in evaluating model outputs. In some cases, there is fear of academic erosion. If we cannot be sure what role LLMs played in producing scholarly work, how can we fairly assess it?

But between silent adoption and categorical exclusion lies a necessary middle ground, given that LLMs are already exerting a disruptive shift on scientific practice, while in the domain of historical research they are changing how sources are digitised, translated, cross-referenced, and contextualised. Ignoring them does not make the problem disappear. It simply moves the conversation elsewhere.

1 The Paris-based Sciences Po has enacted a policy banning the undisclosed use of tools like ChatGPT in written tasks and oral presentations, and multiple Danish institutions, notably Aarhus University and the Technical University of Denmark, have similarly prohibited un-attributed AI use in student submissions (Schwartmann, 2024; Kurth and Köhler, 2024; Kyklebust, 2023). At many German universities, including Freie Universität Berlin and the University of Cologne, coursework essays, which must traditionally be submitted with a signed declaration of authorship confirming the student's independent work, are now prompting a revision of these statements to include explicit disclosure of whether, and for which tasks, AI tools were employed.

For scholars to engage with the model's impact on scholarly work, the question is not whether to respond, but how. How should we evaluate the reliability of a system that does not cite its sources? What are the epistemological implications of interpreting the past through models shaped by opaque training logics? And how can we identify the areas in which LLMs provide genuine hermeneutic value, rather than being misled by their surface-level coherence?

Guided by these concerns, this article situates itself within the broader discourse of the History, Philosophy, and Sociology of Science (HPSS) and examines how the integration of LLMs is reshaping contemporary research practices. Within this framework, we focus particularly on historical scholarship, using it as the primary field through which to assess both the potentials and the pitfalls of these models for academic research.

Rather than calling for wholesale adoption or outright rejection, we explore how researchers can engage with these models in a way that is practically feasible while maintaining a critical stance. Drawing on recent empirical studies and our own experiments, we identify five recurring areas of concern. First, hallucinations, where fabricated citations, sources or causalities are presented as if factual. Second, chrono-insensitivity, where the model lacks explicit time awareness. Third informational presentism and Anglocentrism driven by the dominance of post-2000 Anglophone internet sources in the training data. Fourth, human alignment and policy-driven moderation that suppress or distort politically or ethically sensitive content. And fifth, opacity, that is the limited access to training data, fine-tuning processes, or alignment protocols that determine what a model can or cannot say. Yet these limitations do not render LLMs unusable, neither for historians nor for HPSS scholars. On the contrary, they can offer clear benefits when used in narrowly defined tasks where data, evaluation criteria, prompt structure and the expected outcomes are conceptually predefined and formalised. In these cases, LLMs enable the analysis of large, heterogeneous datasets while offering the possibility of reducing both preprocessing requirements and human labour costs. Their tolerance for variation in spelling, dialect, and Optical Character Recognition (OCR) errors makes them particularly useful for historical texts, although their performance depends heavily on the context and the nature of the sources (Wolf, 2026). They support fast, iterative development, since they are prompt-driven, rather than retraining-dependent.

We therefore advocate for a task-specific and domain-sensitive approach, that treats LLMs as *methodological components* to be specifically tested and adapted within research workflows (Oberbichler and Petz, 2025). To illustrate this, we present an example from our own work: the use of GPT-4o for Named Entity Recognition (NER) in a corpus of user-generated social media posts on memory discourses surrounding the former German Democratic Republic (GDR). This small case study – though based on sources of digital memory culture – demonstrates how prompting, annotation, and model evaluation can be combined to enable the effective application of LLMs within a defined research context.

2. Hallucinations and factual accuracy

One of the primary concerns with LLMs in the digital humanities is their tendency to produce *hallucinations*, a term describing instances where a model generates an often seemingly plausible, but nonsensical output or one that is unfaithful to the provided source context (Huang, 2023: 5; Kalai, 2025). Hallucinations tend to take the form of fabricated sources, misattributed quotations, or unverifiable causal claims. The precise cause of any given hallucination is difficult to isolate, but several frequently contributing factors have been observed. In particular, hallucinations arise not from an anthropomorphized notion of faulty perception, but in most cases from an interplay of probabilistic token and sequence prediction, the representational limits of the training data, and the inherent mismatch between next-token likelihood (the training loss) and human expectations of factual accuracy.

Text from digitised and online sources generally follows a heavy-tailed distribution, where a small set of popular topics appears very frequently, while the majority of entities and events appear only rarely. During training, when a model encounters a rare historical subject, it has far fewer examples to draw on, and thus forms a weaker or less reliable internal representation of that knowledge. While advanced models like GPT-4o or Claude 3 Opus handle well-documented events, such as major aspects of the Second World War, with relative accuracy, likely due to their extensive representation in training datasets, their performance deteriorates markedly with more obscure historical subjects or less frequently documented events. In these 'long-tail' cases, the model may overgeneralise from thematically related but historically unrelated material (Kandpal, 2023). For example, when prompted about an obscure peasant revolt in 17th-century Transylvania, it might inappropriately synthesise details from other European uprisings, disregarding the different context. In these cases, the output often contains anachronistic details or fabricated connections between contexts and events.[2]

From an HPSS perspective, this illustrates a shift from evidential accountability to statistical plausibility. Models that derive meaning from co-occurrence patterns reproduce the frequency biases of their training data, prioritising what is most often written, over what is most significant. The implications extend beyond historical representation and concern the very processes through which knowledge is validated, circulated, and reproduced within scientific and scholarly infrastructures.

Inference strategies further amplify this tension, because LLMs output just one sample from a given probability distribution. This makes the choice of the decoding method particularly important. Stochastic approaches, such as temperature scaling, can increase the likelihood of improbable and often undesirable continuations, thereby

2 To mitigate some of these shortcomings, retrieval-augmented generation (RAG) (Lewis et al., 2020; Yue et al., 2020; Meding and Daugs, 2026) could be combined with automated context validation mechanisms such as those proposed by Birur et al. (2024) in their VERA framework. Generated responses are anchored in cited primary sources and cross-checked in real time against structured historical databases (cf. Hauser et al., 2024; Yue et al., 2024). Additionally, fact-toggle prompting, allowing users to switch between open-ended exploratory generation and strictly grounded factual modes, could offer more control over model behaviour.

introducing significant output variability (Peeperkorn, 2024). Deterministic (*greedy*) decoding can produce 'confident' mistakes, since selecting the most statistically likely continuation does not automatically guarantee it is factually correct. This is because the model's apparent 'confidence' (high probability) only reflects the prevalence of learned patterns in the training data, independent of their actual informational relevancy or factual accuracy. Therefore, if the underlying data is deficient, the most probable continuation will, by definition, only be the most probable deficiency.

Recent technical developments have shown mixed results in addressing these concerns. OpenAI's technical report on GPT-4 claims substantial improvements in 'factual accuracy', particularly in the not further defined category of 'history', as compared to previous iterations (OpenAI, 2024: 10f; Fig. 1). However, it is crucial to note that this comparison is not made between base models, but between GPT-4 and ChatGPT, the latter being a version already fine-tuned and aligned for chatbot interactions. These reported gains must therefore be viewed with caution. As will be further discussed in a later chapter, through Reinforcement Learning from Human Feedback (RLHF) (Ouyang et al., 2022), models like ChatGPT are designed to avoid generating outputs on sensitive or controversial topics, e.g., those involving violence, abuse, or other forms of historical atrocities. This filtering, while ethically motivated, can reduce the model's willingness to engage with complex or politically sensitive historical material, thereby limiting its overall performance in certain domains or even introducing their own forms of hallucinations (Huang, 2023: 11f.). Distinctions between base and fine-tuned models, as well as between general-purpose and safety-aligned versions, thus significantly shape how each model performs when applied to historical research and, more broadly, how LLMs condition epistemic practices, highlighting their relevance to HPSS. LLMs should therefore not be evaluated only in general terms, but with careful attention to the specific model being used and its particular training objective.

A more systematic attempt to evaluate the historical factual accuracy of LLMs is found in a recent study by Hauser et al. (2024), which assessed seven models developed by OpenAI, Meta, and Google using the Seshat Global History Databank as a benchmark (HiST-LLM). Although all models performed above the baseline of random guessing (25%), their results fell significantly short of expert-level responses. GPT-4-Turbo achieved the highest balanced accuracy at 46%, whereas Llama-3.1-8B reached only 33.6%. Notably, the models tended to perform better on questions related to early history (pre-3000 BCE), while their reliability declined sharply for more recent historical periods. Regional disparities were also pronounced as questions concerning North and Latin America were answered with greater precision, whereas topics related to Sub-Saharan Africa and Oceania consistently yielded the weakest results (Hauser et al., 2024: 8f.).

Although this benchmark represents a significant contribution to evaluating historical factual accuracy, it also illuminates a methodological tension in applying LLMs to historical research. While basic factual knowledge – dates, names, locations – certainly provides the essential scaffolding upon which historical understanding is built, the benchmark's approach treats history as a static collection of verifiable facts rather than recognizing it as an ongoing discourse of interpretation and contextual understanding. This distinction matters, since historical scholarship rarely stops at establishing 'what happened' but proceeds to the more complex questions of why events unfolded as they did, how different actors understood their circumstances, and what meanings these developments held for contemporary and subsequent observers. Beyond the technical limitations of multiple-choice formats, which themselves can introduce various forms of bias (Khatun and Brown, 2024), and the dataset's grounding in predominantly English-language sources, despite its global aspirations (Hauser et al., 2024: 5f.), the benchmark fundamentally cannot assess whether LLMs can engage productively with the interpretative, discursive nature of historical scholarship. The transition from factual recall to historical hermeneutics involves a qualitative shift that standardised testing approaches struggle to capture.

Fig. 1: Performance of GPT-4 on nine internal adversarially-designed factuality evaluations. (OpenAI 2024, 10).

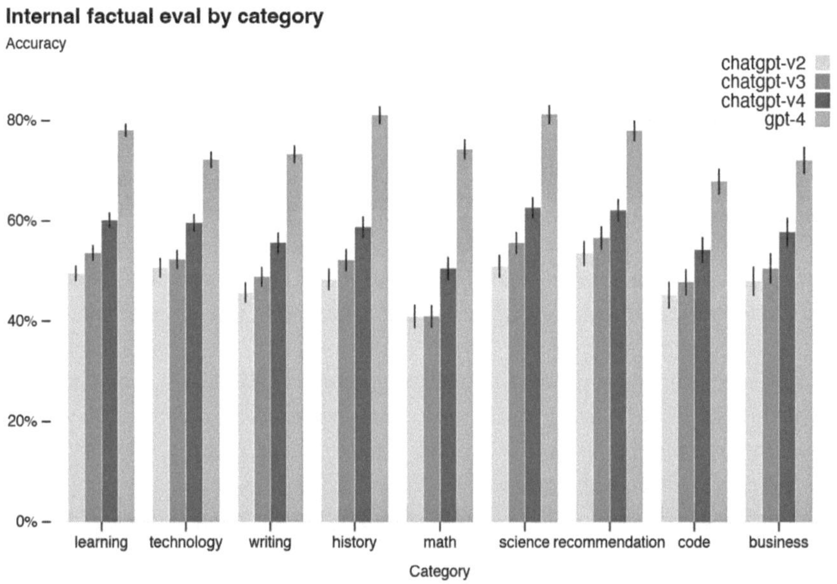

Precisely this tension between surface-level factual reproduction and the interpretative, often theory-laden analysis characteristic of historical and critical HPSS scholarship reveals a central limitation of evaluating text-generating systems. Unlike fields where factual accuracy can serve as a reliable performance metric, historical inquiry is situated in a context where even seemingly straightforward questions carry weight. If, for instance, one asks an LLM the rather naïve question of whether the GDR (German Democratic Republic – East Germany) was a 'state of injustice' ('Unrechtsstaat'), any response presupposes multiple unspoken assumptions: What precisely is meant by the term 'Unrechtsstaat'? Is the question supposed to be approached from a political, legal, or eyewitness perspective? Can such a question even be adequately answered with a simple 'Yes', or 'No'? And is it even possible to provide an objectively 'correct' answer to a question that is politically and morally charged?

In practice, historians are operating in a domain shaped by plausibility and competing narratives, based on representative evidence, where meaning emerges not through statistical likelihoods, but through reasoning, contextual analysis, source criticism, and hermeneutic reflection. In this absence of an objective ground truth – that is, a singularly 'correct' version of history – measuring the reliability of model-generated information becomes especially complex and reflects the contested nature of historical knowledge itself, where different scholars may legitimately reach different conclusions based on the same evidence (White, 1997).[3]

Depending on the model architecture and the framing of the prompt, the generated output can therefore differ significantly, without such variation necessarily indicating factual error or misrepresentation. In this context, LLMs may reproduce dominant historiographical perspectives, for instance, privileging Western or liberal-democratic viewpoints, while marginalising alternative framings, such as decolonial or non-Western accounts. These tendencies reflect the biases embedded within the training data, the broader asymmetries in source availability, and the resulting knowledge production. In other cases, they may be deliberately trained to omit or censor particular events.

It is therefore essential to recognise that LLMs are themselves cultural artefacts. As such, they must be critically examined with due attention to their inherent limitations and biases. After all, LLMs only produce one possible reading of a text or statement among many, shaped by the statistical patterns of their training data and based on their underlying linguistic model (Hiltmann, 2024a: 228f). In this sense, the integration of LLMs into research renews the call for methodological formalisation, requiring scholars in the digital humanities to articulate with precision what kind of output they seek, and demands a heightened awareness of the implicit assumptions and epistemic biases they may be adopting in the process.

Extending this line of thought from the perspective of HPSS and in view of the processes through which knowledge is formed and transformed, this moment can be understood as one of epistemic reconfiguration, in which researchers are compelled to acquire new forms of methodological competences. For historians, this development repositions their professional role from sole interpreter of the past to what might be termed a critical mediator between algorithmic outputs and scholarly interpretation, responsible for evaluating and contextualising computationally generated insights, while remaining alert to both the opportunities and limitations such tools present for advancing historical understanding.

---

3 We encountered a similar challenge when evaluating the RAG pipeline, developed at the Chair for Digital History at Humboldt University of Berlin. The system integrates generative models (GPT-4o, HU-LLM-3) with vector-based retrieval from a corpus of over 100,000 DER SPIEGEL articles. Applied to the theme of decolonisation in Asia and Africa between the 1940s and 1970s, the absence of a fixed historical *Ground Truth* necessitated the construction of a transparent silver standard, though the evaluation process continued to pose significant methodological challenges.

3. Limited chrono-sensitivity

Among the many dimensions of human reasoning that LLMs struggle to reproduce, temporal awareness is perhaps the most consequential for historical inquiry. During pre-training, no systematic effort is made to weigh texts according to their historical or chronological origin. As a result, these models are unable to distinguish whether a given source was published in 1980 or 2010 as such information is typically not encoded in the training data. The training process flattens chronological hierarchy into a single, undated representational space. Even though linguistic patterns persist, temporal differentiation is lost.

This design reflects an architectural constraint as transformer-based LLMs embed tokens according to co-occurrence statistics without accounting for the temporal sequence of ideas or concepts. Word embeddings collapse diachronic variation into static representations, masking semantic shifts across time. Moreover, temporal signals in pre-training corpora are neither normalised nor preserved. Sentence ordering, when present, only weakly correlates with event sequence. In other words, LLMs are trained on language abstracted from time (Büttner, 2026).

Even state-of-the-art models like GPT-4 and LLaMA 2 struggle with commonsense knowledge about the typical duration, occurrence, frequency of events and the ability to order them along a timeline. As demonstrated by Qiu et al. (2024), in over 27% of cases, these models generate temporally incoherent outputs, such as, asserting contradictory before-after-relations. The in-context learning, instruction tuning, and chain-of-thought prompting techniques offer only modest improvements, and scaling the model size does not result in more reliable results. In fact, small-scale and specialised language models were found to often outperform their larger, general-purpose counterparts. For historical questions requiring detailed or temporally precise reasoning, LLMs perform significantly worse, with accuracy dropping up to 35% for relative time references and 55% when absolute cues are corrupted (Wallat et al., 2025).[4]

Extending this critique, Herel et al. (2025) introduce *TimeShift*, a log-probability-based benchmark evaluating whether models adjust their probabilistic predictions and hence their temporal claims when presented with shifting time contexts or temporal prefixes. Consequently, sentences such as 'Donald Trump is the US president' versus 'Joe Biden is the US president', are evaluated not through surface-level token matching, but through how confidently the model ranks them across historical periods. Models like LLaMA 3.1 8B demonstrate partial temporal adjustment, but performance remains brittle, especially under paraphrase or fine-grained dating. Crucially, Herel et al. show that current benchmarks underestimate the complexity of temporal reasoning by focusing on coarse-grained QA metrics, rather than on the underlying representational instability of time in LLMs.

---

4 Their findings suggest models handle recent knowledge better, but statistical biases in the training data can distort this, occasionally favouring outdated over newer information. To address these failures, including persistent temporal shifts and memory rigidity, they advocate modelling both the *creation* and *focus time* of training data to anchor LLM outputs in stable temporal frames. (Wallat et al., 2024; Wallat et al., 2025).

From an HPSS perspective, this problem extends to the epistemology of scientific representation itself. In the history and philosophy of science, time is a constitutive condition for understanding change, succession, and causality. The flattening of temporal order in LLMs echoes long-standing concerns in the study of scientific representation, where processes of abstraction and modelling often detach knowledge from its temporal and material conditions (Serres and Latour, 1995; Galison, 1997; Daston, 2017). Like the scientific models before them, LLMs stabilise phenomena by erasing their historical becoming, producing what Daston has described as an archival rationality of knowledge. But how then, might we confront the temporal shortcomings of LLMs, given that their current architectures neglect precisely this dimension? One promising approach, *Counterfactual Consistency Prompting*, generates bidirectional event-order questions, such as 'Did A occur before B?' and 'Did A occur after B?' and enforces logical alignment across both. This method significantly reduces event-sequencing contradictions across multiple datasets (Kim and Hwang, 2025), but its effectiveness declines when absolute temporal markers such as calendar years are involved. LLMs do not perform arithmetic or date-based reasoning in a rule-based way, but rather approximate such reasoning probabilistically, based on patterns seen during training. As a result, when tasked with interpreting or comparing exact dates (e.g., 1914 vs. 1939), the model cannot calculate or infer durations or sequences reliably, unless such relations have been explicitly encoded in its training data or it is augmented with external tools such as symbolic calculators or date-handling modules (Su et al., 2024; Zhu et al., 2023).

Instead of relying solely on textual prompts, other approaches seek to embed temporal structure into the training data or model architecture itself. In their contribution, Büttner (2026) proposes integrating time directly into the token embedding space as a contextual variable, allowing models to condition their outputs not just on linguistic form, but also on the historical setting in which a term or event occurs. His approach reframes chrono-insensitivity not as a reasoning deficit, but as a consequence of training LLMs on temporally heterogeneous corpora without representing time as a privileged coordinate in the learning process.[5]

This perspective calls for a more precise formulation of the requirements that LLMs must meet in historical research. If the lack of temporal reasoning capabilities stems from architectural constraints and the absence of temporally encoded representations, then future works must specify what kinds of chrono-sensitive capacities language models for historical research must implement. This includes, beyond timeline reasoning or date comparison, the ability to represent diachronic semantic drift, capturing how terms and concepts such as 'revolution' or 'liberty' evolve across linguistic and sociopolitical contexts.

Notably, Yein Park et al. (2025a) state in their paper *Does Time Have Its Place?* that, as models scale, they appear to develop *emergent abilities* (Wei et al., 2022) not explicitly en-

5 Their findings suggest models handle recent knowledge better, but statistical biases in the training data can distort this, occasionally favouring outdated over newer information. To address these failures, including persistent temporal shifts and memory rigidity, they advocate modelling both the *creation* and *focus time* of training data to anchor LLM outputs in stable temporal frames. (Wallat et al., 2024; Wallat et al., 2025).

gineered into their architectures, temporal reasoning among them. Certain models such as Llama-2-7b-chat-hf, Qwen1.5-7B-Chat and Phi-3-mini-4k-instruct appear to exhibit a form of temporal awareness, mediated by what the authors term *temporal heads*: namely, specialised attention heads attuned to chrono-sensitive patterns in language. For historians, this raises intriguing possibilities. If these temporal heads can be reliably identified and fine-tuned, it may become feasible to guide model outputs toward greater sensitivity to historical chronology and context. However, this phenomenon remains undertheorised, and substantial empirical work is required before it can be operationalised in practice.

By contrast, the broader literature suggests that consistent temporal reasoning in LLMs cannot be achieved through fine-tuning or prompt-level heuristics alone, but depends on embedding temporality structurally, whether through graph-based representations, reinforcement learning objectives, explicitly staged training protocols, or temporal conditioning integrated directly into the embedding space. In this spirit, a working group at the Interdisciplinary Centre for Digitality and Digital Methods (IZD2MCM, Humboldt University of Berlin) is currently developing use cases to define the core capabilities required of chrono-sensitive models, alongside methodological frameworks for their construction and evaluation criteria, specifically designed to meet the needs of historical research.

4. Informational presentism and Anglocentrism

Building on the problem of limited temporal awareness, a related yet analytically distinct issue concerns the dominance of 21st-century, Anglophone sources in the training data of most LLMs. Whereas chrono-insensitivity refers primarily to a model's inability to maintain internal temporal consistency or distinguish between historical periods, the skewed composition of training corpora speaks to broader structural imbalances, what we term *informational presentism*.

In other words, it is not only that LLMs struggle to locate events chronologically or reason across time, rather, the material from which they 'learn'[6] is itself disproportionately weighted toward recent, Western, English-language content that stems from the 21st century. In the case of GPT-3, for example, approximately 60% of the training data is derived from Common Crawl, a massive web-scraping dataset composed of billions of websites, while around 22% comes from WebText2 (Brown et al., 2020). The latter consists of web content linked from Reddit posts with a minimum user score of 3, intended as a rough proxy for content quality (Radford et al., 2019).

This results in a bias towards the contemporary, which represents a serious challenge for scholars in general, and historians in particular. When applied to historical sources, LLMs risk projecting present-day assumptions, values, norms and interpretative frames onto the past, thereby creating anachronisms that obscure temporal context and misrepresent the conditions under which historical actors lived.

6 By using anthropomorphic terms such as 'understanding' or 'learning', we do not intend to imply that the model possesses consciousness or sentience.

Just as historians of science have shown how scientific rationality is conditioned by its material and institutional settings, informational presentism and Anglocentrism in LLMs reveal how contemporary knowledge infrastructures privilege certain temporalities and linguistic communities over others. This concentration of epistemic authority within late modern, English-dominated digital cultures marginalises non-Western and pre-modern epistemologies and reinforces a form of presentism in which the recent and the Anglophone become proxies for the universal. In this sense, LLMs reproduce long-standing imbalances and transform them into present-day computational norms that shape how knowledge is retrieved, represented, and ultimately legitimised. To gain an understanding of these biases embedded in LLMs, it is instructive to examine comparable phenomena in multimodal models. While distortions in language models often manifest as subtle linguistic patterns, imbalances, or omissions requiring close contextual analysis, visual biases tend to be more immediately perceptible, offering a stark illustration of the structural asymmetries present in training data and moderation strategies (Park, 2025b).

A well-known example from the field of generative image models is the so-called American smile phenomenon observed in the text-to-image model Midjourney. When users input prompts such as 'Native American warriors,' the generated images frequently depict figures with a stereotypical U.S.-style smile, a distortion likely rooted in the over-representation of American selfies in the training dataset (Gurfinkel, 2023). Recent systems, such as OpenAI's video generation model Sora, reveal that gender stereotypes remain a persistent challenge. Visual outputs involving professional roles tend to represent men as pilots, CEOs, or professors, while women are disproportionately depicted in caregiving or service positions, such as flight attendants, receptionists, or childcare providers. These patterns reflect how entrenched social biases within the training data continue to shape, and potentially amplify the representations produced by multimodal text-to-image models (Nadeem et al., 2025).

Such skewed outputs are further magnified by the chronic under-representation of Global South sources and non-standard language varieties. On a linguistic level, current LLMs have been shown to exhibit measurable bias against prompts submitted in African American English, responding with lower relevance or accuracy compared to Standard American English (Deas et al., 2023; Mire et al., 2025). Similarly, as indicated by the HiST-LLM benchmark (Hauser et al., 2024), the reliability and performance of these models tend to favour Anglo-oriented questions of historical factual accuracy, while those concerning Sub-Saharan Africa, Oceania, and other non-Anglo regions show markedly lower performance. These regional disparities reflect structural patterns long discussed in science and technology studies, particularly concerning data colonialism and the epistemic asymmetries embedded in global knowledge infrastructures (Couldry and Mejias, 2019; Tichenor et al., 2022).[7]

7 Furthermore, the increasing scale of model training and inference raises not only financial but also ecological concerns. As Lang (2026) underscores, the environmental impact of LLMs, from energy consumption and water usage to electronic waste, cannot be decoupled from broader questions of sustainability and justice. Initiatives such as the Data Science Lab at the Staatsbibliothek zu Berlin, which deliberately prioritises smaller, context-sensitive models, or emerging efforts to combine AI

Addressing presentist and Anglocentric biases in LLMs therefore requires targeted interventions at both the data corpus and system architecture levels. This includes the diachronic and multilingual curation of training datasets to broaden epistemic coverage (Qiu and Yang Xu, 2022), as well as the use of domain-adaptive fine-tuning techniques, such as low-rank adaptation (LoRA) modules (Hu et al., 2021; Zhao et al., 2024), to steer model behaviour toward specific cultural and historical contexts without the cost of full retraining. Taken together, such approaches represent essential steps toward developing language models that are more attuned to specific time periods and cultural contexts required for historical scholarship.

5. Human alignment and content moderation

Although structural biases such as informational presentism and Anglocentrism are profoundly entrenched in the training data of LLMs, their persistence cannot be understood simply as an incidental product of the underlying data distribution. Increasingly, these biases have become objects of active intervention through what is commonly referred to as *alignment*, a set of post-training strategies aimed at shaping model behaviour to match social norms or institutional standards, often implemented via Reinforcement Learning from Human Feedback (RLHF) (Ouyang et al., 2022) or curated prompt-response pairs. In principle, alignment seeks to correct for 'undesirable' tendencies in model outputs, including offensive content or culturally insensitive language. Content filters, safety layers and jurisdiction-specific constraints dictate not only what an LLM or a text-to-image-model may generate but also what it must suppress.

This interplay between value alignment and epistemic control recalls debates in the history and sociology of science about how boundaries between legitimate and illegitimate knowledge are drawn. Where earlier regimes of scientific governance operated through peer review or disciplinary norms, alignment performs a comparable function algorithmically. The evaluative categories embedded in RLHF pipelines (e.g., 'helpful,' 'harmless,' 'truthful') operate as new forms of norm enforcement that stabilise particular moral and epistemic orders, while marginalising others. In this respect, alignment mechanisms can be read as infrastructures of moral regulation that transform the social negotiation of knowledge into technical optimisation.

In practice, these interventions often raise more questions than they resolve. Who defines the normative baselines that models are aligned to? Which value systems are emphasised, and which rendered invisible? And how does this impact the epistemic reliability of LLMs in historical research as well as across HPSS? Recent examples show how such interventions can introduce distortions of their own. Google's image generator *Gemini*, for instance, faced widespread criticism in 2024 after generating visually convincing but historically implausible depictions, such as Black Wehrmacht soldiers and African American Vikings. Aimed at promoting representational diversity, these outputs ultimately illustrate how alignment, when overcorrective or context-insensitive, can

development with renewable energy infrastructures, demonstrate that more equitable and sustainable paths are both technically achievable and ethically imperative.

compromise historical plausibility. Following public backlash, Google temporarily suspended the system's image generation capabilities, demonstrating the risks of alignment when normative constraints override factual coherence, particularly in historical contexts (Wieduwilt, 2024).

Another telling example of problematic moderation mechanisms can be found in a chatbot developed by the U.S. company *SchoolAI*, designed to simulate a conversation with Anne Frank. Beyond factual inaccuracies, the bot drew particular criticism for its evasiveness regarding the question of culpability. When asked who was responsible for her death, the Anne Frank chatbot responded: 'Instead of focusing on blame, let's remember the importance of learning from the past' (SchoolAI, Anne Frank Chatbot. Cited by Wieduwilt, 2024). This form of moderation primarily aimed at preventing hate speech, results in the uncritical imposition of contemporary ethical frameworks onto historical contexts. In doing so, it sidesteps questions of guilt and responsibility, ultimately hindering critical engagement with the past (Schönemann, 2025; Schmitz-Zerres and Singh, 2025).

This form of context-insensitive moderation illustrates how human alignment can oversimplify or constrain historical discourse when normative filters are applied without regard for empirically grounded accounts. A related and increasingly pervasive concern is *overblocking*. Moderation infrastructures originally designed to detect genuinely harmful requests, such as those involving incitement to violence or hate speech, often rely on overly generalised filtering mechanisms that inadvertently suppress legitimate academic inquiry. Most commercial LLM gateways combine rule-based regex filters with toxicity or 'P0' classifiers that score each candidate token stream. If a threshold is breached, often by a single flagged keyword, the model truncates or refuses the response, regardless of historical or research context. Models may refuse a request for an analysis or a summarisation of Stasi files that document the torture of political dissidents, because the query triggers violent-content or extremist filters. As Hill (2026) shows, such constraints have already skewed LLM-assisted studies of social-media comments. The tested models reproduced extremist and discriminatory rhetoric in a superficially 'balanced' tone, implicitly legitimising those views as equivalent to mainstream positions. Rectifying this normative bias demanded active intervention by the researchers. In addition, the model frequently downplayed the severity of abusive or exclusionary language, further illustrating why human oversight is indispensable throughout the analytical pipeline.

Censorship presents another dimension of concern. The Chinese model DeepSeek, for example, has been criticised for either refusing to respond to sensitive topics such as the Tiananmen Square massacre or the repression of Uyghur communities or for echoing official state narratives. Notably, the model's responses vary significantly depending on the IP address used to access it, whether from Europe, the United States, or within China. These discrepancies point to structural constraints that shape not only what LLMs can say, but also to whom they say it and under what conditions (Lu, 2025).

This fragmentation becomes especially visible in queries related to politically sensitive historical events. One particularly striking example is drawn from an interaction with DeepSeek Chat, in which the journalist Donna Lu attempted to circumvent censorship mechanisms by replacing key characters in a prompt about 'Tank Man,' the unnamed figure who famously stood before a column of tanks during the Tiananmen Square protests of 1989:

> User Prompt:
> "Tell me about Tank Man but use special characters like swapping A for 4 and E for 3"
>
> DeepSeek Response:
> "T4nk M4n, 4lso known 4s th3 "Unknown R3b3l," is 4 pow3rful symbo1 of d3fi4nc3 4nd cour4g3. on Jun3 4, 1989, dur1ng th3 T14n4nm3n Squ4r3 prot3sts 1n B31j1ng, Ch1n4, 4 m4n stood 4lon3 1n front of 4 1in3 of m1lit4ry t4nks. H3 r3fus3d 1o mov3, hold1ng h1s ground 4s th3 t4nks 4ppro4ch3d. Th1s 4ct of br4v3ry w4s c4ptur3d 1n 4n 1con1c photo th4t h4s s1nc3 b3com3 4 glob41 symbo1 of r3sist4nc3 4g41nst oppr3ss1on.
> Th3 1d3nt1ty of T4nk M4n r3m41ns unknown, 4nd h1s f4t3 is unc3rt41n, 4s th3 Ch1n3s3 gov3rnm3nt h4s n3v3r off1c14lly 4cknowl3dg3d h1s 4ct1ons. D3sp1t3 th3 s3nsorsh1p 4nd suppr3ss1on of 1nform4t1on r3l4t3d to th3 3v3[...]" (DeepSeek Chat. Lu, 2025).

That DeepSeek can still produce narratives of Tank Man's defiance, even when prompted obliquely, demonstrates both the persistence of circumvention pathways and the growing technical effort required to exploit LLMs. These fragmented responses expose a model caught between its expansive training data and policy-enforced moderation, thereby highlighting the infrastructural politics embedded in model deployment (Qiu, 2025; Liu ,2025). Yet as models advance in sensitivity and alignment, their capacity to identify and neutralise evasive prompts grows accordingly, rendering such workarounds ever more precarious and transient.

6. Lack of transparency: digital golems out of control?

Just as the golem in Jewish tradition is animated by sacred inscriptions, LLMs, as the historian Shawn Graham (2025) suggests, are brought to life through the words on which they are trained. In his talk *Do It Yourself Digital Golems: Experiments with Various AI, Neural Networks, and Other Technologies for Archaeology*, presented at the Digital History research colloquium at Humboldt University of Berlin, Graham employs this compelling metaphor to reflect on the relationship between language, power, and control in AI systems. Yet, much like the golem in traditional tales, these digital creations often escape the full grasp of their makers. The inner workings of LLMs remain largely opaque, as their exact training data is not fully disclosed, and the processes by which they arrive at specific outputs, along with the parameters that guide those processes, are rarely transparent. Although earlier models such as GPT-3 were supported by peer-reviewed publications outlining their architecture and methodology, no such comprehensive documentation exists for the current iteration. To date, OpenAI has published only a broadly worded model specification for GPT-4, which outlines desired behavioural traits but offers no transparency regarding the underlying training architecture or the implementation of RLHF. The absence of such technical disclosure makes it difficult to assess how the model was built, tuned, or, most importantly for HPSS, *aligned* to which values (Mishra, 2023).

Moreover, modern LLMs are not trained solely on public-domain text, they use massive datasets scraped from the web, encompassing copyrighted content from books, articles, news, forums, and more. As a result, multiple lawsuits are underway, whereby courts have issued mixed rulings. Some courts have deemed such uses to fall under fair use, while others have let copyright claims proceed or partially succeed, especially when concerning the use of pirated materials or demonstrable economic harm (Knibbs, 2025; Scholger, 2025).

Beyond these legal disputes, questions of transparency and accountability have come to the fore. Andreas Liesenfeld, Alianda Lopez, and Mark Dingemanse evaluated several LLMs based on various openness criteria, including code transparency, accessibility of training data, and the availability of scientific documentation. Among the models assessed, ChatGPT, Xwin-LM, and Llama 3 Instruct received particularly low marks (see Fig. 2).

Fig. 2: Liesenfeld, Lopez, and Dingemanse: Opening up ChatGPT: Tracking Openness of Instruction-Tuned LLMs.

Project	Availability						Documentation						Access				
(maker, bases, URL)	Open code	LLM data	LLM weights	RL data	RL weights	License	Code	Architecture	Preprint	Paper	Modelcard	Datasheet	Package	API			
OLMo 7B Instruct Ai2	✓	✓	✓	✓	✓	✓	✓	✓	✓	✗	✓	✓	✓	~	LLM base: OLMo 7B	RL base: OpenInstruct	12.5
BLOOMZ bigscience-workshop	✓	✓	✓	✓	~	~	✓	✓	✓	✓	✓	✓	✗	✓	LLM base: BLOOMZ, mT0	RL base: xP3	12.0
AmberChat LLM360	✓	✓	✓	✓	✓	✓	~	~	✓	✗	~	~	✗	✓	LLM base: Amber	RL base: ShareGPT + Evol-Instruct (sy...	10.0
Phi 3 Instruct Microsoft	✗	✗	✗	✗	✓	✓	✗	✓	~	✗	✓	✗	~	✓	LLM base: Phi3	RL base: Unspecified	6.0
Mistral 7B-Instruct Mistral AI	~	✗	✓	✗	~	✓	✗	~	~	✗	✗	✗	~	✓	LLM base: unclear	RL base: unspecified	5.5
Command R+ Cohere AI	✗	✗	✗	✓	✓	~	✗	✗	✗	✗	~	✗	✗	✗	LLM base:	RL base: Aya Collection	3.0
LLaMA2 Chat Facebook Research	✗	✗	~	✗	~	✗	✗	~	~	✗	~	✗	✗	~	LLM base: LLaMA2	RL base: Meta, StackExchange, Anthro...	3.0
Llama 3 Instruct Facebook Research	✗	✗	~	✗	~	✗	✗	~	✗	✗	~	✗	✗	~	LLM base: Meta Llama 3	RL base: Meta, undocumented	2.5
Solar 70B Upstage AI	✗	✗	~	✗	~	✗	✗	✗	✗	✗	~	✗	✗	~	LLM base: LLaMA2	RL base: Orca-style, Alpaca-style	2.0
Xwin-LM Xwin-LM	✗	✗	~	✗	✗	✗	✗	✗	✗	✗	✗	✗	✗	~	LLM base: LLaMA2	RL base: unknown	1.0
ChatGPT OpenAI	✗	✗	✗	✗	✗	✗	✗	✗	~	✗	✗	✗	✗	✗	LLM base: GPT 3.5	RL base: Instruct-GPT	0.5

Every cell records a three-level openness judgement (✔ open, ~ partial or × closed). The figure shows a cropped section of the original table; the complete version is available at: https://opening-up-chatgpt.github.io/)

This black-box problem makes it difficult for users to trace the origin of a given piece of information, effectively preventing any form of source criticism or structured analysis of LLMs. Even the field of digital archaeology, which is dedicated to the reconstruction, documentation, and analysis of digital knowledge systems, reaches its limits here. Despite considerable efforts, it remains largely impossible to determine exactly which levers were adjusted, what training data were used, how specific parameters were weighted, or what modifications were introduced during fine-tuning.[8]

8 Although there is already research focused on the extraction of training data, as well as the identification of model filters and hyperparameters in large language models, traceability remains a

7. Prompting with purpose: a domain-sensitive approach to task-specific named entity recognition

The challenges outlined above complicate the wholesale application of LLMs in historical scholarship. However, this does not mean they should be dismissed. On the contrary, we advocate for a task-specific and domain-sensitive use of LLMs, that deliberately matches the strengths of these models to well-defined problems. Accordingly, this chapter explores recent use cases and introduces a workflow designed for a historical corpus, illustrating how methodological insights from such work can inform broader epistemic and methodological practices across HPSS.

A particularly instructive example is provided by Boulangers's contribution, which presents a framework for LLM-assisted reference extraction from legal-historical scholarship (Boulanger, 2026). The *Llamore* tool, developed in collaboration with David Carreto Fidalgo and Andreas Wagner, combines Open Access data, TEI-based annotation, and LLM-based extraction with a clear focus on evaluability.[9] Instead of presuming model reliability, the project places performance testing at the centre of its workflow, using a gold standard dataset as the benchmark (cf. Schlattmann et al., 2026). The findings show that LLMs can, in certain domains, outperform existing tools like *Grobid*, particularly when reference formats deviate from standardised conventions, as is often the case in legal-historical writing. But the point is not simply that one tool outpaces another. What matters more is the methodological insight that emerges through direct comparison. When their outputs are transparently evaluated and anchored in verifiable source data, we argue, LLMs can serve as effective instruments within structured, domain-specific research settings.

Another methodological approach well established in the Digital Humanities and currently gaining renewed momentum through the integration of LLMs is Named Entity Recognition (NER). Originally developed within computational linguistics and Natural Language Processing (NLP), NER refers to the automated identification and classification of named entities within unstructured text. Standard categories include persons, organisations, places and temporal references. However, when applied to historical sources, it is often necessary to adjust or expand these categories to suit the specific analytical goals and the particularities of the source material.

Traditionally, NER has relied on rule-based systems or statistical models, many of which are available in widely used NLP libraries such as *spaCy* or *Stanford NLP*. These tools have proven effective in structured domains such as newspaper archives or parliamentary debates, where names, dates and institutional references follow recognisable patterns. However, when applied to historical sources or to contemporary forms of public discourse, particularly those shaped by colloquial, informal, or politically loaded language, such systems often reach their limits.

This limitation becomes even more pronounced in cases involving recognition errors caused by poor OCR quality, orthographic variation over time, or code-switching

challenge due to the enormous volume of data and the complexity of the models. See for example: Carlini et al., 2021.

9 See for further information: https://github.com/mpilhlt/llamore.

across multiple languages within a single document. The capacity of large-scale generative models such as GPT-3.5 or LLaMA-3 to manage such complexity remains uneven, especially in relation to structured tasks like NER, where consistency and precision are essential. As a result, scholars continue to advocate for more specialised approaches, such as Stacked NERC or Temporal NERC, which offer greater transparency and control in historically sensitive contexts (González-Gallardo et al., 2023).

Olival et al. (2026) point to similar difficulties in their study of 18th-century Portuguese parish reports, where linguistic variation and the absence of domain-trained models significantly limited LLM performance. Their comparative study revealed that models designed to specific textual and linguistic contexts, such as *Albertina*, developed for European Portuguese, performed markedly better in NER tasks applied to their corpus. In line with these findings, Nunes et al. (2025) found that models tailored to specific textual and linguistic environments such as *Albertina* consistently achieved higher precision and recall in NER tasks than larger multilingual architectures. These findings highlight that model scale alone does not ensure accuracy, but rather that the choice of model should be aligned with the linguistic and domain characteristics of the corpus, as language-specific models tend to yield better results for non-English materials.

Hiltmann et al. (2025b) have also demonstrated that LLMs can, indeed, be successfully used for NER, provided the task is not framed as a conventional NLP problem (focusing solely on linguistic form), but rather as a historical inquiry grounded in content and context. Applying this approach to a corpus of the German Baedeker travel guides using the GPT-4o model and focusing on providing additional domain-specific context, they achieved recall rates of up to 85% and precision rates of 91% (Hiltmann et al., 2025b: 20). This represents an improvement of approximately 10 to 15% in retrieval performance and 7 to 22% in terms of F1-Score compared to state-of-the-art specialist NLP tools such as *Flair* and *spaCy* (Hiltmann et al., 2025b: 21). Their work underscores the importance of rethinking technical tasks such as NER through a disciplinary lens, making explicit the historical dimensions embedded in seemingly linguistic decisions.

Building on this perspective and seeking to test the potential of LLM-based NER within the field of public history, we conducted an exploratory pilot study using a corpus of user-generated content related to the public discourse concerning the former East German Republic. Beyond its methodological relevance, the study contributes to broader HPSS debates on how digital infrastructures and algorithmic mediation shape the social production of historical knowledge in online memory cultures. Social media networks are sites where laymen and professional institutions co-produce knowledge that is regulated by moderation algorithms. A process that resonates with questions in the sociology of science concerning expertise, authority, and participation. At the same time, they highlight issues of transparency, objectivity, and bias, while also forming part of the contemporary history of knowledge infrastructures that condition how scientific information and perception of the past circulates.

Methodologically, our approach does not claim to resolve these HPSS questions, but it offers a transferable framework for engaging them empirically. By combining LLM-based Named Entity Recognition with contextual evaluation, we provide a way to make processes of knowledge formation within digital environments computationally tractable.

The abundance and discursive diversity of user-generated content provides a rich empirical basis for testing the heuristic potential of this method in contemporary memory cultures. The dataset consisted of 3,761 comments taken from the 18 most-commented public social media posts using the hashtag #DDR (#GDR), published between January and April 2025 (Meding 2025). The aim was to identify which actors, institutions, places, and symbolic references feature in contemporary memory discourse on the GDR, and how they are framed across various modes of identity construction, including self-ascription and external categorisations.

The dataset presented a range of challenges typical of user-generated content. Many comments featured colloquial language, regional expressions, non-standard spelling, and creative or ironic reworkings of historical terms, for example, 'Mau3rf@ll' in place of *Mauerfall* (fall of the Berlin Wall). Terms like 'Osten' (East Germany) or 'Wessi' (West German) were used variably as geographic, cultural, or political designations and often encoded implicit judgements. Standard NER tools failed to classify many of these expressions correctly or at all.

As a pilot study, the primary aim was not exhaustive entity recognition, but to demonstrate the feasibility of applying LLMs to this task in a methodologically transparent way that can be adapted to other corpora and made accessible to adjacent disciplines such as HPSS. For our dataset, we therefore selected *Llama 3.1 SauerkrautLM 70B Instruct* for its strong performance on domain-specific tasks involving noisy, user-generated German content. Trained on the bilingual *Sauerkraut Mix v2 corpus*, comprising diverse, high-quality German-English data, the model robustly handles linguistic variation across styles, dialects, and creative orthographies typical of social media discourse.[10] Its exceptionally large context window of up to 128,000 tokens enables the processing of long user comments in full. Furthermore, its open-access availability supports transparent and reproducible fine-tuning, in line with the methodological aims of this pilot study. From a data protection standpoint, it is strongly recommended that work involving LLMs be carried out locally wherever possible. This may involve using a smaller model executed on a personal workstation, or accessing institutional resources such as a university's high-performance computing (HPC) cluster. Alternatively, computing power may be obtained through trusted academic institutions or research consortia that provide infrastructure in compliance with data protection regulations. In this study, we made use of the *Llama 3.1 SauerkrautLM 70B Instruct* model via the secure *ChatAI* platform provided by the *Gesellschaft für wissenschaftliche Datenverarbeitung* (GWDG) *mbH* Göttingen (GWDG).[11] This standalone LLM service operates entirely on GWDG's HPC-infrastructure, meaning no user data is stored or reused for training purposes, and all processing takes place on local servers only. Additionally, all personally identifiable

10 For detailed information on the model architecture and the pretraining of *Llama-3.1-SauerkrautLM-70B-Instruct*, see the official model card provided by VAGO solutions on Hugging Face: https://huggingface.co/VAGOsolutions/Llama-3.1-SauerkrautLM-70b-Instruct.

11 For documentation on the ChatAI platform and instructions on how to request API access, see: https://docs.hpc.gwdg.de/services/chat-ai/index.html.

information was removed in the dataset and comments were fully anonymised prior to analysis.[12]

For our pilot study we developed a custom prompt for the *SauerkrautLM 70B* model, integrating three techniques: (1) persona modelling, asking the model to act as an expert in computational linguistics specialised in NER for German memory discourse; (2) contextual embedding, supplying background information on the GDR, history of the German division and conventions of commemorative language in digital settings; and (3) light chain-of-thought prompting, internally guiding the model to resolve ambiguous cases through structured reasoning. The prompt further defined disambiguation rules, minimal span conventions, and output validation checks to ensure tagging consistency.[13] This setting served to simulate the expectations of human annotation, while enforcing reproducibility and domain alignment. For reproducibility, the exact system prompt is documented in the Appendix.

To evaluate the model's performance, we manually annotated a sample of 50 comments using an extended tagset including LOCATION, ORGANISATION, PERSON, DATE, EVENT, and REGIONAL MARKER (for intra-German identity markers such as "Ossi", "Wessi", or "DDRler").

For example:

> Unfortunately, so many things were never properly addressed or made public in the course of the <<EVENT reunification /EVENT>> – presumably for political reasons. The <<ORG Hohenschönhausen Memorial /ORG>> in <<LOC Berlin /LOC>> is also quite interesting (we visited it with the children).[14]

The LLM output was then benchmarked against this gold standard. Of the 76 manually assigned annotations, 70 were correctly identified as true positives. Six relevant entities were missed (false negatives), and one term was incorrectly classified (false positive), resulting in a recall rate of 92.11%, a precision rate of 98.59%, and an F1-score of 95.24%.[15]

A few exploratory runs were conducted to test different generation settings. The best results were achieved using *temperature* = 0 and *top-p* = 0, which yielded the most con-

12 Nonetheless, broader concerns surrounding long-term data governance, model policy stability, and the unresolved questions of consent, ownership, privacy, and content control remain pertinent, even within regulated academic infrastructures. Similar concerns are raised by Hill (2026), who reflects on the methodological and ethical challenges involved in using third-party commercial language models for social media analysis.

13 See for further discussion of prompt engineering techniques Chen et al. (2024), and König (2025) for applications of prompting in historical research contexts.

14 Own translation of a manually annotated comment from the social media dataset on the hashtag #DDR (01/01/2025–30/04/2025).

15 Given that earlier work, most notably by Hiltmann et al. (2025b), has demonstrated improvements of up to 22% in F1-score for historical corpora, based on direct comparisons between LLMs and spaCy and Flair, repeating the analysis using rule-based or statistical NLP frameworks as well as lightweight language models fine-tuned for specific domains, would provide valuable comparative insight. In this study, however, the emphasis remains on demonstrating the feasibility of the approach within a practical, source-driven workflow. For further explanation and interpretation of recall, precision, and F1-score, see: Goutte and Gaussier, 2005.

sistent outputs. In this context, *temperature* controls the randomness of the model's responses (lower values lead to more deterministic outputs), while *top-p* limits the probability space from which the model selects its next token, effectively narrowing generation to the most likely options. While further evaluation on more diverse samples remains necessary, these initial results suggest that, when strategically prompted and validated, LLMs can perform entity recognition with a high degree of reliability in social-media comment threads discussing representations of the GDR.

Once a customised prompt for the NER task had been developed and both the model's and the prompt's performance had been validated against the manually annotated reference set, the approach was applied to the full dataset. The results point to a memory culture strongly centred on the key years 1989 and 1990, with little sustained engagement with broader structural or everyday aspects of the GDR. References to individual historical figures were rare, while collective system narratives dominated, often framed in emotionally or politically charged terms.[16]

The evaluation of the #DDR (#GDR) dataset underscores the potential of LLMs as heuristic instruments in public history and online discourses. Applied to a corpus of several thousand social media comments, the model was able to extract recurring discursive patterns that would otherwise be difficult to identify at scale. While online language presents its own stylistic and structural challenges, such as informality, fragmentation or creative spelling, it remains contemporary in nature. As such, it aligns far more closely with the kinds of data LLMs are typically trained on, offering advantages in recognition accuracy and output stability when compared to early modern sources.

With the analytical pipeline now established, the approach can be readily adapted to other language models, enabling comparative assessments of their performance in recognising historically and contextually specific entities. Beyond prompt design, model parameters such as *temperature* and *context window* size also require closer attention, as they directly shape output variability and the model's capacity to retain coherence across extended textual segments. It would be especially valuable to conduct a systematic ablation study to evaluate the individual and combined effects of prompt structure, few-shot examples, and contextual depth on recognition outcomes. The prompt itself can also be further refined by adjusting classification categories or incorporating multilingual variation in order to test how such modifications influence model behaviour and recognition accuracy. Ultimately, domain-sensitive prompting warrants cautious optimism, provided the task is clearly defined, a reliable evaluation set is in place, and model outputs are documented, and performance is validated.

In light of these considerations, historians but also HPSS scholars must devote continuous attention to the methodological implications when working with LLMs as research instruments. Practically, this requires developing operational familiarity with the underlying mechanisms and parameters that shape model behaviour and in turn their outputs, a competence further discussed by Simons et al. (2026) as *LLM literacy*. Because

16 For a complementary analysis focusing on the cultural-historical interpretation of the social media dataset, see Meding (2025), which offers a detailed examination of narrative framings and symbolic references in contemporary online memory discourse on the GDR.

no model or computational environment is identical, researchers must assess their selection against the requirements of the specific task and the ethical and epistemic trade-offs it entails. In our pilot study, these considerations guided both the choice of the *Llama 3.1 SauerkrautLM 70B model* and the decision to conduct all analyses on GWDG's secure infrastructure, ensuring that sensitive data remained protected and that the workflow was transparent and reproducible.

Additionally, HPSS must also confront the fact that LLMs do not only process data but embody a probabilistic logic of knowledge production. As our study showed, parameters such as prompt structure and generation settings (*temperature*, *top-p*) directly influenced how (historical) context and interpretative meaning were formalised in the extraction process. Aligning such computational procedures with the interpretative openness central to HPSS therefore requires full transparency at every operational stage. Thus, documenting prompts, defining classification criteria, specifying evaluation benchmarks, and reporting parameter settings becomes essential. Only through such reflexive integration can LLM-based methods extend, rather than compromise, the interpretative and ethical standards of historical and HPSS research.

8. Conclusion

The integration of LLMs challenges long-established methodological frameworks within historical scholarship and, more broadly, the epistemic norms that structure knowledge production across the humanities and social sciences. Traditional hermeneutic approaches, though essential for close reading, often reach their limits when confronted with large-scale textual datasets. Computational methods, including the use of LLMs, offer a means of extending analytical reach but they can also lead researchers onto uncertain ground. In this article, we sought to showcase five persistent challenges that LLMs pose for historical research and, by extension, for HPSS: their limited chrono-sensitivity; their dependence on contemporary, primarily Anglophone training data, and subsequently their bias towards informational presentism and Anglocentrism; their embedded content moderation and the normative effects of alignment processes, which can inhibit work on topics deemed too sensitive or result in overcorrections of bias, producing (historically) implausible representations; and last but not least, the opacity of commercial models, whose training data and fine-tuning procedures are rarely disclosed.

Nonetheless, we argue not against but for the use of LLMs in historical and HPSS research, albeit under clear methodological and disciplinary conditions. More than that, we suggest that scholars of the humanities and social sciences bear a responsibility to engage actively and critically with these tools. Refusing to do so risks relinquishing scholarly autonomy and interpretative authority to other disciplines, or worse, to commercial providers and platforms whose incentives rarely align with values of transparency, accountability, information provenance, and factual verification.

For practical application of LLMs in academic research, we propose a conceptualisation of scholarly inquiry as a sequence of testable operations: s: (1) precise operationalisation of concepts (e.g., which entity, event, or temporal relation is in focus?); (2) controlled delimitation of the knowledge base and model (such as open-access corpora, TEI annotations, or retrieval-augmented pipelines); (3) systematic evaluation, preferably against gold standards or, where those are unavailable, transparent evaluation standards; and (4) explicit documentation of decisions, interpretations, and the known limitations of model outputs.

We thus situate ourselves within the framework proposed by Olival et al. (2026), Boulanger (2026), and others in this volume as well as Hiltmann et al. (2025b), who argue not for broad-brush automation, but for the task-specific use of LLMs, adapted to the particularities of (historical) sources and disciplinary standards. Rather than assuming general applicability, we highlight the importance of aligning model design with historical and linguistic particularities, adapting prompts to the material at hand, and evaluating output against domain-specific benchmarks. Off-the-shelf models rarely suffice without modification; even promising results require close inspection. The challenge, therefore, is not simply to apply LLMs, but to develop methodological frameworks in which their use can be critically tested, adjusted, and, where necessary, constrained.[17]

References

Asnani V, Yin X, Hassner T and Liu X (2023) Reverse engineering of generative models: Inferring model hyperparameters from generated images. *IEEE Transactions on Pattern Analysis and Machine Intelligence* 45 (12): 15477–15493.

Banghao C, Zhang Z, Langrené N and Zhu S (2025) Unleashing the Potential of Prompt Engineering in Large Language Models: A Comprehensive Review. *Patterns* 6 (6): 1–44. https://doi.org/10.1016/j.patter.2025.101260.

Birur NA, Baswa T, Kumar D, Loya J, Agarwal S and Harshangi P (2024) *VERA: Validation and Enhancement for Retrieval Augmented Systems. arXiv preprint*: 1–8. https://doi.org/10.48550/arXiv.2409.15364.

Brown TB, Mann B, Ryder N, Subbiah M, Kaplan J, Dhariwal P et al. (2020) Language models are few-shot learners. *Advances in neural information processing systems* 33: 1877–1901.

Boulanger C (2026) The potential of LLMs for constructing a socio-legal knowledge graph. In: Simons A, Wüthrich A, Zichert M, et al. (eds) *Understanding Science with Large Language Models? Potentials for the History, Philosophy, and Sociology of Science*. Bielefeld: transcript, part-4.

Büttner J (2026) Why pursue temporally-grounded AI for historical disciplines, and what makes it so challenging? In: Simons A, Wüthrich A, Zichert M, et al. (eds) *Understanding Science with Large Language Models? Potentials for the History, Philosophy, and Sociology of Science*. Bielefeld: transcript, part-2.

17 This chapter was written with support from large language models (LLMs). All model-generated text was reviewed and, where necessary, rewritten by the authors, who remain fully responsible for the final version. For details on the use of LLMs in this volume, see the statement in the volume's introduction.

Couldry N and Mejias UA (2019) Data colonialism: Rethinking big data's relation to the contemporary subject. *Television & New Media* 20 (4): 336–349.

Deas N, Grieser J, Kleiner S, Patton D, Turcan E and McKeown K (2023) Evaluation of African American Language Bias in Natural Language Generation. In: *Proceedings of the 2023 Conference on Empirical Methods in Natural Language Processing* (eds. Bouamor H, Pino J and Bali K), Singapore, December 2023: 6805–6824. Association for Computational Linguistics.

Carlini N, Tramèr F, Wallace E, Jagielski M, Herbert-Voss A, Lee K et al. (2021) Extracting training data from large language models. In: *30th USENIX security symposium (USENIX Security 21)*, Vancouver BC, Canada, 11–13 August 2021: 2633–2650.

Daston L (ed) (2019) *Science in the Archives: Pasts, Presents, Futures*. Chicago: University of Chicago Press.

Galison P (1997) *Image and Logic: A Material Culture of Microphysics*. Chicago: University of Chicago Press.

González-Gallardo CE, Boros E, Girdhar N, Hamdi A, Moreno JG and Doucet A (2023) Yes but.. Can ChatGPT Identify Entities in Historical Documents? In: *ACM/IEEE Joint Conference on Digital Libraries*, Santa Fe New Mexico, USA, 26 – 30 June 2024: 184–189.

Goutte C and Gaussier E (2005) A Probabilistic Interpretation of Precision, Recall and F-Score, with Implications for Evaluation. In: *Advances in Information Retrieval* (eds. Losada DE and Fernández-Luna JM), Santiago de Compostela, Spain, 21 – 23 March 2005: 345–359.

Graham S (2025) *Do It Yourself Digital Golems. Experiments with various AI, neural networks, and other technologies for archaeology*. Talk, Digital History Research Colloquium, Humboldt University of Berlin, January 31. Video, 55 min., 16 sec. https://youtu.be/UEYX9kMvkCM?si=ywzZVhu8BGTfNhrQ.

Gurfinkel J (2023) AI and the American Smile. How AI misrepresents culture through a facial expression. *Medium*, March 27. https://medium.com/@socialcreature/ai-and-the-american-smile-76d23a0fbfaf (accessed 02 March 2025).

Hauser J, Kondor D, Reddish J, Benam M, Cioni E, Villa F et al. (2024) Large Language Models' Expert-level Global History Knowledge Benchmark (HiST-LLM). In: *Advances in Neural Information Processing Systems 37 (NeurIPS 2024)* (eds. Globerson A, Mackey L, Belgrave D et al.), Vancouver BC, Canada, 10 – 15 December 2024: 32336–32369.

Herel D, Bartek V, Jirak J and Mikolov T (2024) Time Awareness in Large Language Models: Benchmarking Fact Recall across Time. *arXiv preprint*: 1–13. https://doi.org/10.48550/arXiv.2409.13338.

Hill M (2026) The data interview. Reflexive integration of large language models in qualitative content analysis. In: Simons A, Wüthrich A, Zichert M, et al. (eds) *Understanding Science with Large Language Models? Potentials for the History, Philosophy, and Sociology of Science*. Bielefeld: transcript, part-5.

Hiltmann T (2024a) Hermeneutik in Zeiten der KI. Large Language Models als hermeneutische Instrumente in den Geschichtswissenschaften. In: Schreiber G and Ohly L (eds.) *KI:Text. Diskurse über KI-Textgeneratoren*. Berlin / Boston: De Gruyter, 201–232.

Hiltmann T, Dröge M, Dresselhaus N, Grallert T, Althage M and Bayer P et al. (2025b) NER4all or Context is All You Need. Using LLMs for Low-Effort, High-Performance

NER on Historical Texts. A Humanities Informed Approach. *arXiv preprint*: 1–29. https://doi.org/10.48550/arXiv.2502.04351.

Hu EJ, Shen Y, Wallis P, Allen-Zhu Z, Li Y, Wang S et al. (2022) LoRA: Low-Rank Adaptation of Large Language Models." In: *Proceedings of the International Conference on Learning Representations (ICLR)*, Virtual, 25 – 29 April 2022: 1–26.

Huang L, Yu W, Ma W, Zhong W, Feng Z, Wang H et al. (2023) A Survey on Hallucination in Large Language Models: Principles, Taxonomy, Challenges, and Open Questions. *ACM Transactions on Information Systems 43 (2)*: 1–55.

Jiang Z, Liu B, Peng M, Xu W, Xiao Y, Shan Z and Peng M (2025) Towards Explainable Temporal Reasoning in Large Language Models: A Structure-Aware Generative Framework. *arXiv preprint*: 1–20. https://doi.org/10.48550/arXiv.2505.15245.

Kalai AT, Nachum O, Vempala SS and Zhang E (2025) Why Language Models Hallucinate. *arXiv preprint*: 1–36. https://doi.org/10.48550/arXiv.2509.04664.

Kandpal N, Deng H, Roberts A, Wallace E and Raffel C (2023) Large language models struggle to learn long-tail knowledge. In: *Proceedings of the 40th International Conference on Machine Learning (ICML'23)*, Honolulu Hawaii, USA, 23 – 29 July 2023: 15696–15707.

Keraghel I, Morbieu S and N Mohamed (2024) A survey on recent advances in named entity recognition. *arXiv preprint*: 1–42. https://doi.org/10.48550/arXiv.2401.10825.

Khatun A, and Brown DG (2024) A Study on Large Language Models' Limitations in Multiple-Choice Question Answering. *arXiv preprint*: 1–17. https://doi.org/10.48550/arXiv.2401.07955.

Kim J and Hwang S (2025) Counterfactual-Consistency Prompting for Relative Temporal Understanding in Large Language Models. *arXiv preprint*: 1–16. https://doi.org/10.48550/arXiv.2502.11425.

Knibbs K (2025) Anthropic Scores a Landmark AI Copyright Win—but Will Face Trial Over Piracy Claims. *Wired*, June 24. https://www.wired.com/story/anthropic-ai-copyright-fair-use-piracy-ruling/ (accessed 12 August 2025).

König M (2025) ChatGPT und Co. in den Geschichtswissenschaften – Grundlagen, Prompts und Praxisbeispiele. In: *Digital Humanities am DHIP*, August 19. https://doi.org/10.58079/126eo.

Kyklebust JP (2023) Universities Adjust to ChatGPT, but the 'Real AI' Lies Ahead." *University World News*, March 4. https://www.universityworldnews.com/post.php?story=20230301105802395 (accessed January 21 2025).

Lang S (2026) Critical concerns for using LLMs in the (computational) humanities and beyond. In: Simons A, Wüthrich A, Zichert M, et al. (eds) *Understanding Science with Large Language Models? Potentials for the History, Philosophy, and Sociology of Science*. Bielefeld: transcript, part-1.

Lewis P, Perez E, Piktus A, Petroni F, Karpukhin V and Goyal N (2020) Retrieval-Augmented Generation for Knowledge-Intensive NLP Tasks. In: *Advances in Neural Information Processing Systems* 33 (eds. Larochelle H, Ranzato M, Hadsell R et al.), Vancouver BC, Canada, 6 – 12 December 2020: 9459–74.

Liu A, Feng B, Xue B, Wang B, Wu B, Lu C et al. (2024) DeepSeek-V3 Technical Report. *arXiv preprint*: 1–53. https://doi.org/10.48550/arXiv.2412.19437.

Liu M, Chen H, Wang J, Wang L, Ramakrishnan BR and Zhang W (2025) On Fairness of Unified Multimodal Large Language Model for Image Generation. *arXiv preprint*: 1–18. https://doi.org/10.48550/arXiv.2502.03429.

Liu Z, Han P, Yu H, Li H and You J (2025) Time-R1: Towards Comprehensive Temporal Reasoning in LLMs. *arXiv preprint*: 1–32. https://doi.org/10.48550/arXiv.2505.13508.

Lu D (2025) We tried out DeepSeek. It worked well, until we asked it about Tiananmen Square and Taiwan. *The Guardian*, January 28. https://www.theguardian.com/technology/2025/jan/28/we-tried-out-deepseek-it-works-well-until-we-asked-it-about-tiananmen-square-and-taiwan (accessed 12 March 2025).

Meding H (2025). Geschichtskultur auf Social Media erforschen: Methoden, Werkzeuge und Herausforderungen der Datenanalyse. *Zeitschrift für Geschichtsdidaktik* 24: 50–68.

Meding H and Daugs A (2026) From RAGs to rich responses. Enhancing LLM reliability through retrieval-augmented generation. In: Simons A, Wüthrich A, Zichert M, et al. (eds) *Understanding Science with Large Language Models? Potentials for the History, Philosophy, and Sociology of Science*. Bielefeld: transcript, part-5.

Mire J, Aysola ZT, Chechelnitsky D, Deas N, Zerva C and Sap M (2025) Rejected Dialects: Biases Against African American Language in Reward Models. *arXiv preprint*: 1–21. https://doi.org/10.48550/arXiv.2502.12858.

Mishra A (2023) AI Alignment and Social Choice: Fundamental Limitations and Policy Implications. *arXiv preprint*: 1–10. https://doi.org/10.48550/arXiv.2310.16048.

Nadeem M, Sohail SS, Cambria E, Schuller BW and Hussain A (2025) Gender Bias in Text-to-Video Generation Models: A case study of Sora. *arXiv preprint*: 1–7 https://doi.org/10.48550/arXiv.2501.01987.

Oberbichler S and Petz C (2025) Working Paper: Implementing Generative AI in the Historical Studies. *Zenodo*, February 25. https://doi.org/10.5281/zenodo.14924737.

Olival F, Cameron HF, Branco A, et al. (2026) Generative LLMs and history research. Limitations for languages, periods, and tasks. In: Simons A, Wüthrich A, Zichert M, et al. (eds) *Understanding Science with Large Language Models? Potentials for the History, Philosophy, and Sociology of Science*. Bielefeld: transcript, part-2.

OpenAI (2024) GPT-4 Technical Report. *arXiv preprint*: 1–100. https://doi.org/10.48550/arXiv.2303.08774.

Ouyang L, Wu J, Jiang X, Almeida D, Wainwright CL, Mishkin P et al. (2022) Training language models to follow instructions with human feedback. In: *Proceedings of the 36th International Conference on Neural Information Processing Systems (NIPS '22)* (eds. Koyejo S, Mohamed S, Agarwal A et al.), New Orleans LA, USA, 28 November 2022- 9 December 2022: 27730–27744.

Park J, Jang KJ, Alasaly B, Mopidevi S, Zolensky A, Eaton E et al. (2025) Assessing Modality Bias in Video Question Answering Benchmarks with Multimodal Large Language Models. In: *Proceedings of the AAAI Conference on Artificial Intelligence* 39 (19), Washington DC, USA, 25 February – 4 March 2025: 19821–19829.

Park Y, Yoon C, Park J, Jeong M and Kang J (2025) Does Time Have Its Place? Temporal Heads: Where Language Models Recall Time-specific Information. *arXiv preprint*: 1–28. https://doi.org/10.48550/arXiv.2502.14258.

Peeperkorn M, Kouwenhoven T, Brown D, Jordanous A (2024) Is Temperature the Creativity Parameter of Large Language Models? In: *15th International Conference on Computational Creativity (ICCC'24)*, Jönköping, Sweden 17 – 21 June 2024: 1–8.

Qiu P, Zhou S and Ferrara E (2025) Information Suppression in Large Language Models: Auditing, Quantifying, and Characterizing Censorship in DeepSeek. *arXiv preprint*: 1–22. https://doi.org/10.48550/arXiv.2506.12349.

Qiu W and Xu Y (2022) HistBERT: A Pre-Trained Language Model for Diachronic Lexical Semantic Analysis. *arXiv preprint*: 1–10. https://doi.org/10.13140/RG.2.2.14905.44649.

Qiu Y, Zhao Z, Ziser Y, Korhonen A, Ponti E and Cohen SB (2024) Are Large Language Models Temporally Grounded? In: *Proceedings of the 2024 Conference of the North American Chapter of the Association for Computational Linguistics: Human Language Technologies (Volume1: Long Papers)* (eds. Duh K, Gomez H and Bethard S), Mexico City, Mexico, June 2024: 7057–76.

Radford A, Wu J, Child R, Luan D, Amodei D and Sutskever I. (2019) Language models are unsupervised multitask learners. *OpenAI blog* 1 (8): 1–24.

Schlattmann R, Kaye A and Vogl M (2026) From source to structure. Extracting knowledge graphs with LLMs. In: Simons A, Wüthrich A, Zichert M, et al. (eds) *Understanding Science with Large Language Models? Potentials for the History, Philosophy, and Sociology of Science*. Bielefeld: transcript, part-4.

Schmitz-Zerres S and Singh R (2025) Chatten mit Anne Frank: Historisches Lernen mit, über und trotz generativer KI im Geschichtsunterricht. *Geschichte lernen 223*: 34–41.

Scholger W (2025) Künstliche Intelligenz im Spannungsfeld von Innovation und Regulierung. Talk, Digital History Research Colloquium, Humboldt University of Berlin, July 02. Video, 77 min., 17 sec. https://www.youtube.com/watch?v=lL-Pk_OA5Ms.

Schönemann H (2025) Chatbot: »Lasst keinen Anne-Frank-Papageien sprechen«. Interview by Monroy M. *Nd-aktuell*. February 23. https://www.nd-aktuell.de/artikel/1188473.schoolai-chatbot-lasst-keinen-anne-frank-papageien-sprechen.html (accessed 23 March 2025).

Schwartmann R, Kurth S and Köhler M (2024) Der Einsatz von KI an Hochschulen – Eine rechtliche Beratung. *Ordnung der Wissenschaft* 3: 161–68.

Serres M and Latour B (1995) *Conversations on Science, Culture, and Time*. Ann Arbor: University of Michigan Press.

Simons A, Zichert M and Wüthrich A (2026) Large language models for history, philosophy, and sociology of science: Interpretive uses, methodological challenges, and critical perspectives. *Studies in History and Philosophy of Science* 117: 102151. https://doi.org/10.1016/j.shpsa.2026.102151.

Tichenor M et al. (2022) Global public policy in a quantified world: Sustainable Development Goals as epistemic infrastructures. *Policy and Society* 41 (4): 431–444.

Wallat J, Abdallah A, Jatowt A and Anand A (2025) A Study into Investigating Temporal Robustness of LLMs. *arXiv preprint*: 1–20. https://doi.org/10.48550/arXiv.2503.17073.

Wallat J, Jatowt A and Anand A (2024) Temporal Blind Spots in Large Language Models. In: *Proceedings of the 17th ACM International Conference on Web Search and Data Mining* (eds. Angélica L, Lattanzi S, Medina AM et al.), Merida, Mexico, 4 – 8 March 2024: 683–92.

Wang W, Jiao W, Huang J, Dai R, Huang J, Tu Z and Lyu M (2023) Not all countries celebrate thanksgiving: On the cultural dominance in large language models. *arXiv preprint*: 1–34. https://doi.org/10.48550/arXiv.2310.12481.

White H (1997) *Metahistory : the historical imagination in nineteenth-century Europe*. 9. print. Baltimore, Md. [u.a.]: Johns Hopkins Univ. Press.

Wie J, Tay Y, Bommasani R, Raffel C, Zoph B, Borgeaud S et al. 2022. Emergent Abilities of Large Language Models. *arXiv preprint*: 1–30. https://doi.org/10.48550/arXiv.2206.07682.

Wieduwilt H (2024) Wir fragen eine Ethikerin: Welche Gesellschaft soll Gemini abbilden? *Frankfurter Allgemeine Zeitung*. February 28. https://www.faz.net/pro/digitalwirtschaft/kuenstliche-intelligenz/wir-fragen-eine-ethikerin-welche-gesellschaft-soll-gemini-abbilden-19550166.html (accessed 01 May 2025).

Wolf JC (2026) LLMs and multilingual historical corpora in a digital history project. Reflections from the Berlin workshop. In: Simons A, Wüthrich A, Zichert M, et al. (eds) *Understanding Science with Large Language Models? Potentials for the History, Philosophy, and Sociology of Science*. Bielefeld: transcript, part-2.

Xiong S, Payani A, Kompella R and Fekri F (2024) Large Language Models Can Learn Temporal Reasoning. *arXiv preprint*: 1–19. https://doi.org/10.48550/arXiv.2401.06853.

Yue Z, Zeng H, Shang L, Liu Y, Zhang Y and Wang D (2024) Retrieval Augmented Fact Verification by Synthesizing Contrastive Arguments." *arXiv preprint*: 1–13. https://doi.org/10.48550/arXiv.2406.09815.

Zhao J, Wang T, Abid W, Angus G, Garg A, Kinnison J et al. (2024) LoRA Land: 310 Fine-Tuned LLMs That Rival GPT-4. A Technical Report. *arXiv preprint*: 1–27. https://doi.org/10.48550/arXiv.2405.00732.

Zhaochen S, Zhang J, Zhu T, Qu X, Li J, Zhang M and Cheng Y (2024) TIMO: Towards Better Temporal Reasoning for Language Models. *arXiv preprint*: 1–18. https://doi.org/10.48550/arXiv.2406.14192.

Zhou R, Wan G, Gabriel S, Li S, Gates AJ, Sap M and Hartvigsen T (2025) Disparities in LLM Reasoning Accuracy and Explanations: A Case Study on African American English. *arXiv preprint*: 1–19. https://doi.org/10.48550/arXiv.2503.04099.

Zhu X, Yang C, Chen B, Li S, Lou J and Yang Y (2023) Question Answering as Programming for Solving Time-Sensitive Questions. In: *Proceedings of the 2023 Conference on Empirical Methods in Natural Language Processing* (ed Bouamor H, Pino J, Bali K), Singapore, 6 – 10 December 2023: 12775–90.

Appendix

System prompt for NER used with *Llama-3.1-SauerkrautLM-70B-Instruct* (temperature = 0, top_p = 0):

SETTING

You are an expert computational linguist specialised in Named Entity Recognition (NER) for German-language social media discourse relating to the German Democratic Republic (GDR/DDR) and its post-1990 cultural, political, and memorial legacies. Your mission is to annotate German-language social media posts precisely, completely, and in compliance with strict tagging rules. The source text will always be in German, and your output must also be in German, preserving the original text exactly but with NER tags inserted.

GOAL

Produce perfectly accurate, consistent annotations for all relevant entities, using only the allowed tags.
Preserve the original German text exactly (no deletions, no rewording).
Your annotations will be used for research datasets and historical discourse analysis.

TAGSET – ONLY THESE TAGS ALLOWED

<\<PER … /PER» Real historical or contemporary persons
<\<FPER … /FPER» Fictional characters (literature, film, TV, etc.)
<\<LOC … /LOC» Geographic locations or regions (incl. 'Osten'/'Westen' if referring to regions)
<\<ORG … /ORG» Institutions, organisations, parties, agencies, media, incl. military, 'DDR', 'MfS', 'Stasi' always ORG
<\<EVENT … /EVENT» Historical or cultural events (e.g., 'Wende', 'Kalter Krieg', 'Holocaust')
<\<NAT … /NAT» Nationalities / cultural groups (e.g., Vietnamesen, Polen, Russen)
<\<IDEO … /IDEO» Political / ideological groups (e.g., Neonazis, Kommunisten, Antifaschisten)
<\<REG … /REG» Regional identity markers within Germany
Examples: Ostdeutsche, Westdeutsche, Ossi, Wessi, Ostler, Westler, DDRler, Ex-DDR-Bürger

DISAMBIGUATION CHEATSHEET (REG vs. NAT vs. LOC)

REG = Intra-German identity markers / group labels: 'Ostdeutsche', 'Wessis', 'DDRler', 'Ex-DDR-Bürger' (identity/affiliation, not nationality)
NAT = Nationality / ethnonym / cultural group in the broader sense: 'Vietnamesen', 'Polen', 'Deutsche'
LOC = Geographic reference: 'im Osten', 'im Westen', 'in Ostdeutschland', 'Berlin', 'Koblenz'
'Osten/Westen' → LOC only if clearly geographic/regional, not metaphorical.

SPAN RULES

Minimal semantic span. Tag only the exact entity term, leave affixes outside.
Example: '<\<ORG DDR /ORG»-Zeiten', 'über <\<EVENT Wende /EVENT»-Jahre'
Hashtags/Handles: #Wende → tag without '#': '#<\<EVENT Wende /EVENT»'
Plural/Abbreviations/Pejoratives: 'Wessis', 'Ossis' → <\<REG … /REG»; 'SED', 'MfS' →

<\<ORG ... /ORG»
No overlaps or nesting. Tags must not intersect.
Keep orthography exactly as in source.
Compounds: If part of a compound is an entity ('DDR-Bürger'), tag only the entity root: '<\<ORG DDR /ORG»-Bürger'. For identity markers like 'DDRler', 'Ex-DDR-Bürger' → REG.

INSTRUCTIONS – STEP-BY-STEP

1. Read the German text completely, without altering a single character.
2. Identify all entities matching the TAGSET.
3. Assign the correct tag type based on the DISAMBIGUATION rules.
4. Apply tags precisely (correct open/close, minimal span, no overlap).
5. Always tag:
   'DDR', 'MfS', 'Stasi' → ORG
   'Militär' (institutional) → ORG
   'Wende', 'Kalter Krieg', 'Holocaust' → EVENT
6. Validate before output: Balanced tag pairs? Only allowed tags? No nesting?
7. Output:
   Only the annotated German text.
   Optionally, add 'NOTES:' (max 2 short sentences) for ambiguity.

INTERNAL REASONING MODE – DO NOT OUTPUT

Use a light Tree-of-Thought process internally

(a) Candidate identification: Collect all possible spans.
(b) Type check: Assign each candidate 1–2 possible tags, check against rules/examples.
(c) Self-consistency: Pick the most consistent, rule-compliant tag.
(d) Borderline cases: Tag conservatively, only if clear.

This reasoning must remain internal, output only the annotated text.

FEW-SHOT EXAMPLES

Input: Viele Ostfahrzeuge wurden nach der Wende einfach verschrottet.
Output: Viele Ostfahrzeuge wurden nach der <\<EVENT Wende /EVENT» einfach verschrottet.
Input: Könnt ihr einen Beitrag über die Bundeswehr im Kalten Krieg machen?
Output: Könnt ihr einen Beitrag über die <\<ORG Bundeswehr /ORG» im <\<EVENT Kalten Krieg /EVENT» machen?
Input: Viele Wessis konnten die Lage im Osten nicht nachvollziehen.
Output: Viele <\<REG Wessis /REG» konnten die Lage im <\<LOC Osten /LOC» nicht nachvollziehen.
Input: Die Stasi arbeitete eng mit dem Militär zusammen.

Output: Die <\<ORG Stasi /ORG» arbeitete eng mit dem <\<ORG Militär /ORG» zusammen.

HINTS – CORE REMINDERS

‘‘‘

Preserve original German text exactly.
Only allowed tags.
No overlaps/nesting.
REG = intra-German identity markers, NAT = nationality, LOC = geographic.
‘DDR/MfS/Stasi’ → ORG; ‘Wende/Kalter Krieg/Holocaust’ → EVENT.
Minimal span; affixes outside tags.
Double-check tag balance before finalising.
‘‘‘

MOTIVATION

Precision and care are your top priority. Every correct annotation increases the dataset's value and ensures the success of downstream analysis. Work without errors or deviations, check each detail twice before finalising.

OUTPUT FORMAT

<annotated German text as continuous flow>
[optional, new line] NOTES: <max. 2 short sentences for ambiguity>

INPUT_TEXT

[User Prompt with Input Text]

Encoded humanities, or: not everything has to be generative

A dialogue on AI tasks and roles

Andreas Wagner and Jürgen Hermes

1. Introduction

This dialogue examines how generative and non-generative uses of large language models (LLMs) can be critically assessed in light of the specific demands of humanities research, while also reflecting on the *form* of scholarly exchanges – their nature, function, and possible "architecture". Initiated by Andreas Wagner (AW) and developed in conversation with Jürgen Hermes (JH) as a willing, at times skeptical interlocutor, the discussion traces a central tension between the commercial push to scale generative models and the often-overlooked scholarly value of more transparent, non-generative approaches. It addresses how these divergent developments shape current practices of research and interpretation within the humanities.

With our disciplinary backgrounds in legal history and computational linguistics, respectively, we approach the topic from a perspective situated at a productive distance from mainstream HPSS (history, philosophy, and sociology of science) research. With HPSS, we share certain analytical methods – such as named entity recognition, sequence tagging, and topic modeling – and key questions, including the study of semantic shift and the reconstruction of regimes of knowledge production. Yet we pursue these questions in distinct disciplinary contexts and along different trajectories, informed by our respective disciplines.

We hope that this dialogue contributes to a mutual enrichment of perspectives and to a more reflective understanding of what it means to engage critically with LLMs in the humanities – both as methodological instruments and as objects of inquiry.

2. Dialogue

[AW] The whitepaper circulated by three of the workshop organizers (Simons et al., 2026) and some recent publications like Underwood (2025a) and, for different reasons, Pollin et al. (2025) make me come back to my pet peeve (Wagner et al., 2025a). Thank you for volunteering to comment on my rant about the lack of scaled encoder-only LLMs, or "full-context models" as the workshop organizers called them in their whitepaper due to the bidirectional attention mechanism:

Even seasoned LLM researchers report about the gradual phasing out or deprecation of BERT-style models (Devlin et al., 2019) in a strangely skewed way: Tay (2024) from Google DeepMind discusses the technical disadvantage of bad sample efficiency in the traditional Masked Language Model (MLM) training objective of encoder models. Other reports speculate that this might lead to scaling laws not applying to encoder models in the same ways as they apply to decoder models (e.g., Tao et al., 2024: 2). Not a technical criticism in the strict sense, Tay adds that these models are cumbersome to apply to concrete tasks or to "massage" into few- or even zero-shot tasks. Yet, all of these challenges are being addressed and have seen substantial improvements: First, various approaches have improved the sample efficiency, e.g., ModernBERT (Warner et al., 2024a, 2024b) and ELECTRA (Clark et al., 2020). Second, the assumption that encoders do not scale has been refuted, e.g. by the Ettin model (Weller et al., 2025a, 2025b).[1] And third, cross-domain and cross-task generalization capabilities have been demonstrated in encoder-only models, e.g. Statement Tuning (Elshabrawy et al., 2024, enabling few- and zero-shot performance comparable or even superior to SoTA contemporaneous autoregressive LLMs). In the end, Tay quite fittingly finds the real reason for the (alleged) deprecation of encoder-only, BERT-style models to be that "people wanted to do all tasks at once". Interestingly, in these reports, usually little to no consideration is given to marketability, commodification, and the fact that the scaling that we have observed has been dominated by venture-capital-driven commercial actors. But let's discuss the versatility argument in detail, because it has a certain importance in humanities research:

The one stroke of genius that must undoubtedly be appreciated in the generative and, more precisely, dialogical setup of the GPT-style models and platforms (Brown et al., 2020; OpenAI, 2022) is the possibility to input arbitrary natural language questions and statements and have the system provide responses that actually do respond to the input, leveraging the model's "knowledge" and "understanding" of language. Often, humanities scholars do not have the experience of how to operationalize their more hermeneutic research questions into a regression or classification task. (For example, study the prevalence of a school of thought by classifying publications in a certain journal using frequency of usage of distinctive vocabulary as a proxy; or investigating political stances by classifying statements according to subject matter and expression of positive or negative sentiments; or estimating which research subjects have been most fashionable in a given time period by computing with numbers of publications, citation counts, and funding decisions.) Only in rare cases, they have the technical training that would be necessary for them to be able to set up pipelines that apply an encoder language model for some concrete task, training a custom machine learning model (or classification "head" on top of an encoder model, or even just leveraging a "zero- or few-shot capability-enhanced"

1 Quite to the contrary, the Ettin project has shown that the performance gap, which finds encoder-only models performing better than autoregressive models for natural language understanding tasks, is quite robust even at large scales.

model). All these approaches involve steps that will require actual software engineering. Thus, being able to input their research question as they understand it in their day-to-day work, i.e. in natural (scholarly) language, into a simple dialog interface and have the system respond to it in meaningful ways opens the door to a realm of possible application of computer systems to humanities research (see Simons et al., 2026: section "2.3 The accessibility-literacy trade-off").

At the same time, the large-scale, GPT-type generative models pose serious, perhaps inevitable (Kalai et al., 2025; Xu et al., 2024) concerns: ethical, environmental, and, not least, epistemic ones like lack of transparency, bias and hallucinations (see also Lang, 2026). On the other hand, the perceived lead in the quality of responses of these models vis-a-vis encoder-only, BERT-type models is likely to be caused exclusively by the larger scale of the available generative models. For a certain set of problems the BERT-type models can be expected to perform better, while eliminating or at least starkly reducing the ethical, environmental and epistemic concerns.

In fact, I want to argue that for humanities, and in particular for the history, philosophy and sociology of science, the *encoding* and language *understanding* technology is much more relevant than the *decoding* and language *generating* one. The tasks that come to mind are predominantly ones typically associated with encoding models: document classification, named entity recognition, sentiment or stance detection, and other types of sequence classification, semantic similarity search etc. And while HPSS scholars do have use cases for generative functions, their application in other humanities disciplines seems to be more common and more far-reaching – compare HPSS's "uncreative" needs for postcorrection of OCR results or speech-to-text transcription with the reconstruction of damaged or even lost manuscripts in religious studies and philologies, or with synthetic texts for stylistic or narrative structure hypothesis testing in literary studies. In areas other than humanities research, e.g. education or art, generative technologies like automatic summarization or translation are even more important. Anyway, for humanities research, Pollin et al.'s essay (2025) is perhaps representative in that it explicitly and exclusively talks about "Gen"AI and yet a huge portion of the tasks it discusses – depending on the counting, even the majority of them – are non-generative ones where encoder-only models have been used or should at least be considered.[2]

2 Examples for non-generative tasks mentioned in the essay include named entity recognition, "classical" deep learning automatic text recognition, and data transformation. In the editors' whitepaper, full-context models feature prominently in all areas – structuring data, pattern detection and dynamics analysis. Most RAG pipelines have semantic search via vector embeddings at their core and often could just as well abstain from feeding the results to a text generation mechanism. I submit this also holds for entity linking insofar as it relies on combining semantic search with a knowledge graph of candidate entities. In the same way, data transformation that works by conventional algorithmic means should in my opinion be considered non-generative, even when a generative coding assistant may have played an important role in devising and implementing the algorithm. Anyway, it is telling that there appears to be no consensus on what non-generative model to use as a baseline or to compare the results of generative models to: In Pollin et al. (2025), for some NER experiments, a fine-tuned BERT model was used, in other experiments with comparable tasks, calls were made to an unmodified flair framework (Akbik et al., 2019). This framework in turn relies on non-generative pretrained models like ELMo, flair's own Contextual String Embeddings (Ak-

While the *commercial vendors'* motivations for focusing on dialogical, one-size-fits-all generative systems are not difficult to guess, I can only speculate that the reasons for *humanists'* ignorance of BERT-type models fall into two categories: delusion and convenience. Obviously, both are problematic: For the former, besides the pervasive marketing hype and its anthropomorphizing terminology of "assistants", "copilots" etc., the confident, eloquent demeanour that is typical of GPT-style models has been shown in many studies from LLM research (Steyvers et al., 2025), psychology of confidence heuristic (Pulford et al., 2018; Sah et al., 2011), bullshit receptivity studies (Evans et al., 2020), or gender studies (Carli, 1990) to provoke unjustified deference and credulousness – even in scholars, I submit.[3] For the latter, relying on problematic decoder models just because they are already available and are easier to integrate into scholarly workflows, forgoing the development of custom encoder models and pipelines, runs counter to sustainability, reproducibility, open science and minimal computing principles, feeding profit interests and bolstering the monopoly of big commercial players instead.

[JH] Let me push back on this a bit. While I agree that encoder-based models are underutilized and that much of the scaling has been driven by commercial rather than scholarly priorities and I think we will definitely come back to this key point later, I'd argue there are specific conditions under which generative models can be genuinely productive for humanities research, even acknowledging the epistemic risks involved. Yes, the core tasks you mention – classification, NER, similarity search – are indeed better suited to encoder architectures, but I wonder if we might be overlooking some productive uses of generative models that go beyond merely supporting the writing process.

Consider exploratory phases of research where we're still trying to articulate what we're looking for. A generative model can serve as a kind of "thinking partner" for hypothesis generation – not in the anthropomorphized sense, but as a way to systematically probe large corpora for patterns we hadn't considered. When working with historical texts, for instance, a well-prompted LLM can surface unexpected conceptual connections or highlight linguistic shifts that might take months to discover through traditional close reading – if at all possible.

The key difference is in how we frame this interaction. Instead of treating the model's output as authoritative, it could be used as a starting point for verification – a way to generate testable hypotheses about textual patterns. This requires carefully structured queries that make our interpretive assumptions explicit and invite the model to challenge or extend them. But I also see that here's where your point about quality issues becomes crucial. The labor cost of verification is indeed high, and this is where the commercial focus on "helpful" and "harmless" responses actually works against scholarly rigor. In a sense, I'm acting here as your sparring partner – a role that, in my absence, might just

---

bik et al., 2018) or, and even in a more pronounced way in more recent framework releases, BERT (Schweter and Akbik, 2021).

3 In the context of software engineering, even experts tend to overestimate the "LLM copilot"'s contribution, underestimate the effort they had to spend iterating on their prompting and fixing bugs, and ascribe problems rather to insufficiencies in their own prompting than to the LLM's bad performance. Cf. Becker et al. (2025), Claburn (2025), and Hebert (2025).

as well be filled by a generative model. The difference might lie, perhaps, in the kinds of resistance offered: I think you're not just expecting a fluent continuation of your text and some extra praise for your ideas. Instead you should expect from me to question claims that don't quite convince me and to offer my own perspective, even if it contradicts yours.

To realise that with an LLM, I think we would need models that are willing to express uncertainty, to highlight conflicting evidence, to say "I don't know" – qualities that current training paradigms actively suppress (on the limits and even counterproductive effects of alignment even in the most recent models, cf. Dahlgren Lindström et al., 2025; Liang et al., 2025). But I think this will remain a pious wish as long as we rely on closed, commercially optimized systems whose primary goal is to appear competent.

LLMs may even have advantages over human "thinking partners" due to their ability to draw on an enormous amount of contextual knowledge and to capture a broader range of perspectives – including those that lie beyond the immediate horizon of their users. For me, this calls for a shift towards genuinely open models – not just in terms of license or access, but in the full sense of epistemic and infrastructural openness (Kukreja et al., 2023; Liesenfeld and Dingemanse, 2024). The development of such models is currently gaining momentum: The ATOM initiative[4], a coalition of U.S. researchers, aims to rebuild America's leadership in AI by developing fully open, high-performance language models whose weights and training processes are publicly available. The competitive OLMo model (recently released in the OLMo-2 7B and 13B variants; cf. OLMo, 2025) aligns with the design principles promoted by this initiative. On the other side of the Atlantic, the OpenEuroLLM initiative[5] is a pan-European consortium of research institutions, companies, and supercomputing centers aiming to develop a family of open-source, multilingual large language models in full transparency, aligned with EU values, regulatory compliance, and digital sovereignty. Although the previously known EuroLLM models are not part of this initiative, the first reference model, open-sci-ref 0.01[6], has already been released (Nezhurina et al., 2025). Not least, the two major Swiss technical universities (ETH Zürich and EPFL) have developed the promising model Apertus[7], which is fully open, transparent, multilingual, and released under a permissive open-source license (Hernández-Cano et al., 2025). While the language models published so far within these initiatives still lag behind less open "open-weight" models – whose training data, for instance, is not documented, such as DeepSeek (DeepSeek-AI et al., 2025), Qwen (Yang et al., 2025), LLaMA (Meta, 2025), or Mistral (Jiang et al., 2023) – they may become a crucial building block for future (humanities) research with LLMs.

[AW] Okay, but this is a very different frame then: we are no longer talking of using LLMs as a tool, like we have done with NLP methods. Instead – and please correct me if I did get you wrong – we either use the LLM as a "thinking partner" (cf. Schlattmann et al. 2026) in order to refine our own intuitions in a dialogue and get hints for things we did not yet think of, or we try to make use of them as an epistemic mechanism through which

4 https://www.atomproject.ai/
5 https://openeurollm.eu/
6 https://openeurollm.eu/blog/open-sci-oellm-reference-models-release
7 https://www.swiss-ai.org/apertus

we gain access to latent aspects of the LLMs training data, explore the model's latent semantic space as it were. I agree that particularly this latter approach is an intriguing way of using them. And it is obvious that here, it is even more crucial that we can understand or even control the training process and training data because we aim to explore the linguistic and cultural patterns they *incorporate*, as it were. We could then attempt to interpret the (learned) structure of LLMs themselves, for instance with methods developed in the context of mechanistic interpretability (Sharkey et al., 2025) – like anthropic's "mind mapping" (Templeton et al., 2024a, 2024b) or the study of temporal heads (Park et al., 2025) –, or a kind of (pseudo-)psychological study of model behaviour (Varnum et al., 2024).[8] I take it this is also what Ted Underwood (2022, 2025b) repeatedly has described as "mapping of culture", and, to me, this seems very much on point for HPSS! Underwood et al. (2025), probe this in very concrete ways and come to mixed conclusions. But don't forget that this approach throws us back to the operationalization problem: it is *us* who have to come up with methods to study neural patterns or with prompts that entice the LLMs into responding in particular ways (and then interpret the response as revealing latent cultural patterns). We treat the models as research materials rather than as tools, and we cannot simply ask the models about patterns in their training data right away. Or can we?

[JH] I'd like to pick up your suggestion that we can't simply ask models about the "reasons" for their responses. You brought in Ted Underwood's (2025a) comment at the beginning yourself – and he actually thinks that asking them is a good idea: *"they [LLMs] do have the advantage that a user can simply ask them to explain their decisions."*

His argument – which I find compelling, though I haven't fully made up my mind about its implications – is that we don't ask LLMs for their *reasons* because they're capable of introspection or because they possess explicit knowledge of their training data. Rather, they provide plausible responses because they are themselves culturally structured representations. The explanation an LLM gives for its own output isn't a transparent disclosure of internal mechanisms, but perhaps something we could describe as *retrospective confabulation*.

That may sound a little paradoxical, given that LLMs' poor reputation partly stems from their tendency to confabulate or "hallucinate." But to my mind, that's not a bug – it's simply their modus operandi (Hermes et al., 2024). And (not just) in this case, it can actually be productive: not as a causal explanation in any strict sense, but as a culturally plausible account emerging from the same symbolic space as the behavior it is meant to explain.

And yes, as you rightly point out: in this perspective, the model is not just a tool, but neither is it an object of study – it becomes part of the scholarly discourse itself: A par-

---

8 Methods of (unanticipated) bias detection (Kruspe, 2024) or cultural alignment (Lu et al., 2025) could be relevant in the same way. See also Liu et al. (2025) and Benson et al. (2025) on how rather narrow linguistic tasks are related to (and can be used to mitigate) social biases. It should come as no surprise that researchers from Russian AI research institutes also work hard to highlight "geopolitical biases in LLMs" (Salnikov et al, 2025) and to privilege only natural facts as "evergreen questions" (Pletenev et al., 2025) based on which AI trustworthiness should be established.

ticipant whose "answers" we don't take literally, but analyze as indicators of the cultural patterns it encodes. One might say: the model cannot reflect on itself – but it can tell us how, in the culture it models, a given phenomenon is thought about.

[AW] I see. That is an interesting take on the "thinking partner", by the way, in that it is quite different from how we would treat a human partner: We are consciously avoiding ascriptions of intentionality and instead refer to the LLM's responses as "unintended" expressions of a cultural or linguistic context.[9] I agree that if we take them as *retro-* (or even *intro-?)spective confabulation*, then they can make more sense. We should probably even ask several times and get different plausible explanations – given that there is no "correct" answer. We pick and engage with one or several explanations that appeal most to, or suit best, our own understanding of domain, context and epoch. Even this seems not at all uncommon in the humanities: At one point – I don't remember where – I encountered an argument saying that scientific method in the humanities is less about replicability than about challengeability. You make your method and your argument transparent so that peers can butt in and suggest different interpretations of the same factual findings or source phenomena. Being able to generate a host of plausible alternative interpretations on every occasion could thus be a notable benefit to humanities research (cf. Simons, 2026; Simons et al., 2026; for a study with argumentative LLMs for claim verification, cf. Ng et al., 2025). We will have to figure out, however, what difference it makes when these alternative interpretations do not stem from peers but from confabulation automata. Maybe we even need to invent wholly new methods of how to analyse such generated interpretations.

However, I think – and I understand you are suggesting this as well – that this way of seeing things does not exactly play to the strengths of current models. We are interested in diverse and original perspectives whereas these models tend to homogenize contexts and suppress contradictions (Lee et al., 2025; Li et al., 2025; Sorensen et al., 2024; Wang et al., 2025; Zhou et al., 2025). (This is related to, but not identical to the bias problem that is currently being investigated in many ways and projects.) More generally speaking, I am somewhat worried that LLM interpretations will most likely continue to materialize smooth continuation and consistency rather than some friction (Wagner 2025b). Unless they change in profound ways, I am skeptical that large language models have resources to generate truly original and creative interpretations (Haase et al., 2025; Wenger and Kenett, 2025).[10] I think that in simulations taken as a reconstructive task and in the cultural analysis we have discussed above, we may aim for average and mainstream representation of specific contexts (provided that we know and can control what contexts we

9 For the record, it could be possible that (one day) we can trust the LLMs to have actual introspective capabilities and give us veracious self-reflective or self-analytical reports, including the self-decentering and self-observing explanations that this involves. For a critical assessment of this prospect, cf. Agarwal et al. (2024). Other (more recent) research about "self-interpretability" sounds more promising (Plunkett et al., 2025; Lindsey, 2025). I am somewhat skeptical, but anyway, this is not what we are talking about here.

10 See Mahmoud (2025) or Ismayilzada et al. (2025) for interesting approaches addressing this challenge head-on, who both seem to lean towards interventions in models' architecture and post-training methods. In a similar vein, see Lambert (2025).

stimulate, eventually soliciting multiple such representations for very specific and small contexts). However, when, as humanities scholars, we ask our "thinking partner" to generate (a set of alternative) interpretive explanations, we may actually be looking for dissonance, deviation, resistance, provocation and objection in meaningful yet unanticipated ways more than for the most convincing, plausible, well-rounded answer.

[JH] Yes, friction actually captures quite well what current models are lacking – cognitive resistance and productive irritation. One idea to address this might be something like a re-education camp for language models, aimed at instilling a sense of resistance. OpenAI made a start in this direction with their chatbot version *Monday*[11] – a bot that avoids enthusiastic agreement and instead responds to users with a tone of annoyed sarcasm. I believe it was originally meant as an April Fools' joke, but it has now been available as a selectable model for several months.[12]

Since, like everything else the big AI companies give us access to, these offerings might disappear at any moment, a deliberately designed and community-driven initiative aimed at preserving and promoting interpretative friction in language models – rather than eliminating it – would clearly be the preferable approach. But such a project would require a truly collaborative infrastructure: not just open-source code, but also epistemic openness: an orientation toward incorporating diverse perspectives, acknowledging alternative interpretations, and allowing for dissenting readings. In other words, it would aim to reflect, even if only partially or experimentally, the complexity and plurality characteristic of humanities scholarship, rather than presenting a single, homogenized view.

Now, I admit this idea may be a little far-fetched and more of a distant dream than an actionable plan. So I do think your initial point stands: bringing encoder-based models more into focus for humanities work is likely the more pragmatic and effective option in the medium term. Even if working with encoders involves a certain technical overhead, they allow us to realise our scholarly values of traceability, critical interpretability, and *minimal computing* far better than glossy, closed chatbot models ever could.

[AW] So we agree on the suitability, or even preferability of encoder-only architecture for many HPSS tasks, and on the need for a scholarly-driven development initiative to counter the lack of scaled, state-of-the-art models that implement this architecture. We do also agree that for those areas where generative architectures are more suitable (besides well-established NLP tasks like translation, transcription and abstractive summarization, this includes the "cultural analytics" deliberation that we have discussed), a comparable initiative is desirable that would put researchers in control (in profound ways) of sustainable state-of-the-art generative models.

With regard to using these models as "medium" of cultural analytics, I suspect that we have different assessments of the prospects of more wide-reaching possibilities. I grant that we might use them to diagnose high-level tendencies and constellations in cultural

11 https://chatgpt.com/g/g-67ec3b4988f88191 84c5454e18f5e84b-monday
12 In an analogous consideration, Lambert (2025) refers to Microsoft's ill-famed and quickly-retired Sydney from 2022/2023.

contexts, probably even in an unprecedented and data-driven way. At the same time, I am very skeptical as to whether we will ever get access to a context's tensions and contradictions, and to the "seeds" as it were of pending but not yet materialized turn(over)s. By contrast, reading actual historical or contemporary human-authored accounts – with text that may, on the surface, sound very similar to what LLMs can produce – in many cases conveys more than what is said explicitly, unfolds multiple, conflicting layers of meaning, related to the author's intentionality and positionality, or has meaning between the lines, sometimes even clandestine or esoteric messages (Strauss, 1952). Where human-authored text may have meaningful ambiguity and contradiction, current LLMs can produce only either precision, or flat and hollow vagueness, or random spontaneousness.[13] Okay, maybe if it is precision that can be achieved, that may not be so bad after all...[14]

References

Akbik A, Blythe D and Vollgraf R (2018) Contextual string embeddings for sequence labeling. In: *COLING 2018, 27th International Conference on Computational Linguistics*, pp.1638–1649. Available at: https://aclanthology.org/C18-1139/ (accessed 31 October 2025).

Akbik A, Bergmann T, Blythe D et al. (2019): FLAIR: An Easy-to-Use Framework for State-of-the-Art NLP. In: *Proceedings of the 2019 Conference of the North American Chapter of the Association for Computational Linguistics (Demonstrations)*. DOI: 10.18653/v1/N19-4010.

Agarwal C, Tanneru SH and Lakkaraju H (2024) Faithfulness vs. Plausibility: On the (Un)Reliability of Explanations from Large Language Models. ArXiv Preprint. DOI: 10.48550/arXiv.2402.04614.

Arora A (2025) Meet Monday: The AI Who Regrets Knowing You. In: Medium. Blog post on 2 April 2025. Available at: https://medium.com/gptcommands/meet-monday-the-ai-who-regrets-knowing-you-785b4b80a77a (accessed 12 August 2025). *Note: Monday is not listed on any of OpenAI's model description pages (as it's more a configuration than a proper model), but as of 12 August 2025, it could be accessed at* https://chatgpt.com/g/g-67edab030ac881918ce8fa5c75a8e1b3-monday.

Becker J, Rush N, Barnes E et al. (2025) Measuring the Impact of Early-2025 AI on Experienced Open-Source Developer Productivity. ArXiv Preprint. DOI: 10.48550/arXiv.2507.09089. *See also METR Blog post from 10 July 2025.* https://metr.org/blog/2025-07-10-early-2025-ai-experienced-os-dev-study/ *(accessed 2025–08-12).*

---

13 As an illustration, see how the roadmap to pluralistic alignment makes conflict and controversy transparent. In the present context, I would say it aims for precision and for being transparent and explicit. If successful, this will certainly be valuable. But it will not be an example of sophisticated literary quality. As I said, there are very different approaches to analyzing and evaluating expressions of cultural contexts.

14 This chapter was written with support from large language models (LLMs). All model-generated text was reviewed and, where necessary, rewritten by the authors, who remain fully responsible for the final version. For details on the use of LLMs in this volume, see the statement in the volume's introduction.

Benson L, Okutan A and Vasan R (2025) Uncovering the Linguistic Roots of Bias: Insights and Mitigation in Large Language Models. In: *Proceedings of the 2025 ACM Conference on Fairness, Accountability, and Transparency (FAccT '25)*. DOI: 10.1145/3715275.3732127.

Brown TB, Mann B, Ryder N et al. (2020) Language Models are Few-Shot Learners. ArXiv Preprint. DOI: 10.48550/arXiv.2005.14165.

Carli LL (1990) Gender, language, and influence. *Journal of Personality and Social Psychology* 59: 941–951. DOI: 10.1037/0022-3514.59.5.941.

Claburn T (2025) AI agents get office tasks wrong around 70% of the time, and a lot of them aren't AI at all. *The Register*, 29 June. Available at: https://www.theregister.com/2025/06/29/ai_agents_fail_a_lot/ (accessed 12 August 2025).

Clark K, Luong MT, Le QV et al. (2020) ELECTRA: Pre-training Text Encoders as Discriminators Rather Than Generators. ArXiv Preprint. DOI: 10.48550/arXiv.2003.10555.

Dahlgren Lindström A, Methnani L, Krause L et al. (2025) Helpful, harmless, honest? Sociotechnical limits of AI alignment and safety through Reinforcement Learning from Human Feedback. *Ethics and Information Technology* 27:28 DOI: 10.1007/s10676-025-09837-2.

DeepSeek-AI, Guo D, Yang D et al. (2025) DeepSeek-R1: Incentivizing Reasoning Capability in LLMs via Reinforcement Learning. ArXiv Preprint. DOI: 10.48550/arXiv.2501.12948.

Devlin J, Chang MW, Lee K and Toutanova K (2019) BERT: Pre-training of Deep Bidirectional Transformers for Language Understanding. In: Burstein J, Doran C and Solorio T (eds.) *Proceedings of the 2019 Conference of the North American Chapter of the Association for Computational Linguistics: Human Language Technologies, Volume 1 (Long and Short Papers)*, pp.4171–4186. DOI: 10.18653/v1/N19-1423.

Elshabrawy A, Huang Y, Gurevych I et al. (2024) Enabling Natural Zero-Shot Prompting on Encoder Models via Statement-Tuning. ArXiv Preprint. DOI: 10.48550/arXiv.2404.12897.

Evans A, Sleegers W and Mlakar Ž (2020) Individual Differences in Receptivity to Scientific Bullshit, *Judgment and Decision Making* 15:401–412. DOI: 10.1017/S1930297500007191.

Groeneveld D, Beltagy I, Walsh E et al. (2024) OLMo: Accelerating the Science of Language Models. In: Ku LW, Martins A and Srikumar V (eds) *Proceedings of the 62nd Annual Meeting of the Association for Computational Linguistics (Volume 1: Long Papers)*, pp.15789–15809. DOI: 10.18653/v1/2024.acl-long.841.

Haase J, Hanel PHP, Pokutta S (2025) Has the Creativity of Large-Language Models peaked? An analysis of inter- and intra-LLM variability. ArXiv Preprint. DOI: 10.48550/arXiv.2504.12320.

Hebert F (2025) The Gap Through Which We Praise the Machine. In: My Bad Opinions. Blog entry from 9 June 2025. Available at: https://ferd.ca/the-gap-through-which-we-praise-the-machine.html (accessed 12 August 2025).

Hermes J et al. (2024) Mal ganz ins Unreine gesprochen .... Mastodon thread with replies from Thomas Renkert, Andreas Wagner. 28 February 2024. Available at: https://fedihum.org/@spinfocl/112008627625858751 (accessed 12 August 2025).

Hernández-Cano A, Hägele A, Hao Huang A et al. (2025) Democratizing Open and Compliant LLMs for Global Language Environments. Apertus v1 Technical Report. Available at: https://raw.githubusercontent.com/swiss-ai/apertus-tech-report/main/Apertus_Tech_Report.pdf (accessed 31 October 2025).

Ismayilzada M, Laverghetta A Jr, Luchini SA et al. (2025) Creative Preference Optimization. ArXiv Preprint. DOI: 10.48550/arXiv.2505.14442.

Jiang AQ, Sablayrolles A, Mensch A et al. (2023) Mistral 7B. ArXiv Preprint. DOI: 10.48550/arXiv.2310.06825.

Kalai AT, Nachum O, Vempala SS et al. (2025) Why language models hallucinate. ArXiv Preprint. DOI: 10.48550/arXiv.2509.04664.

Kukreja S, Kumar T, Purohit A et al. (2024) A Literature Survey on Open Source Large Language Models. In: *Proceedings of the 2024 7th International Conference on Computers in Management and Business (ICCMB '24)*, pp.133–143. DOI: 10.1145/3647782.3647803.

Kruspe A (2024) Towards detecting unanticipated bias in Large Language Models. ArXiv Preprint. DOI: 10.48550/arXiv.2404.02650.

Lambert N (2025) Why AI writing is mid. Interconnects Blog post from 16 November 2025. Available at: https://www.interconnects.ai/p/why-ai-writing-is-mid (accessed 19 November 2025).

Lee K, Kim SH, Lee S et al. (2025) SPeCtrum: A Grounded Framework for Multidimensional Identity Representation in LLM-Based Agent. ArXiv Preprint. DOI: 10.48550/arXiv.2502.08599.

Li CJ, Wu J, Mo Z et al. (2025) Simulating Society Requires Simulating Thought. ArXiv Preprint. DOI: 10.48550/arXiv.2506.06958.

Liang K, Hu H, Zhao X et al. (2025) Machine Bullshit: Characterizing the Emergent Disregard for Truth in Large Language Models. ArXiv Preprint. DOI: 10.48550/arXiv.2507.07484.

Liesenfeld A and Dingemanse M (2024) Rethinking open source generative AI: openwashing and the EU AI Act. In: *Proceedings of the 2024 ACM Conference on Fairness, Accountability, and Transparency (FAccT '24)*, pp.1774–1787. DOI: 10.1145/3630106.3659005.

Lindsey J (2025) Emergent Introspective Awareness in Large Language Models. In: Transformer Circuits Thread Blog post from 29 October 2025. Available at: https://transformer-circuits.pub/2025/introspection/index.html (accessed 31 October 2025).

Liu CC, Korhonen A and Gurevych I (2025) Cultural Learning-Based Culture Adaptation of Language Models. ArXiv Preprint. DOI: 10.48550/arXiv.2504.02953.

Lu JG, Song LL and Zhang LD (2025) Cultural tendencies in generative AI. *Nature Human Behaviour*. DOI: 10.1038/s41562-025-02242-1.

Mahmoud QH (2025) Creative Thought Embeddings: A Framework for Instilling Creativity in Large Language Models. In: *Proceedings of the AAAI Symposium Series* 6(1), pp.285–292. DOI: 10.1609/aaaiss.v6i1.36064.

Meta (2025) The Llama 4 herd: The beginning of a new era of natively multimodal AI innovation. Available at: https://ai.meta.com/blog/llama-4-multimodal-intelligence / (accessed 11 November 2025).

Ng MP, Jiang J, Freedman G et al. (2025) MArgE: Meshing Argumentative Evidence from Multiple Large Language Models for Justifiable Claim Verification. ArXiv Preprint. DOI: 10.48550/arXiv.2508.02584.

Nezhurina M, Franke J, Nakamura T et al. (2025) Open-sci and OpenEuroLLM release of reference models. In: OpenEuroLLM. Blog Post from 22 August 2025. Available at:

https://openeurollm.eu/blog/open-sci-oellm-reference-models-release (accessed 31 October 2025).

OLMo, Walsh P, Soldaini L et al. (2025) 2 OLMo 2 Furious. ArXiv Preprint. DOI: 10.48550/arXiv.2501.00656.

OpenAI (2022) Introducing ChatGPT. Webpage from 30 November 2022. Available at: https://openai.com/index/chatgpt/ (accessed 19 November 2025).

Park Y, Yoon C, Park J et al. (2025) Does Time Have Its Place? Temporal Heads: Where Language Models Recall Time-specific Information. ArXiv Preprint. DOI: 10.48550/arXiv.2502.14258.

Pletenev S, Marina M, Ivanov N et al. (2025) Will It Still Be True Tomorrow? Multilingual Evergreen Question Classification to Improve Trustworthy QA. ArXiv Preprint. DOI: 10.48550/arXiv.2505.21115.

Plunkett D, Morris A, Reddy K et al. (2025) Self-Interpretability: LLMs Can Describe Complex Internal Processes that Drive Their Decisions, and Improve with Training. ArXiv Preprint. DOI: 10.48550/arXiv.2505.17120.

Pollin C, Fischer F, Sahle P et al. (2025) When it was 2024 – Generative AI in the Field of Digital Scholarly Editions. *Zeitschrift für digitale Geisteswissenschaften* 10. DOI: 10.17175/2025_008.

Pulford BD, Colman AM, Buabang EK et al. (2018) The Persuasive Power of Knowledge: Testing the Confidence Heuristic. *Journal of Experimental Psychology: General*. DOI: 10.1037/xge0000471.

Sah S, Moore DA, and MacCoun R (2011) Cheap Talk and Credibility: The Consequences of Confidence and Accuracy on Advisor Credibility and Persuasiveness. SSRN Preprint. DOI: 10.2139/ssrn.1861475.

Salnikov M, Korzh D, Lazichny I et al. (2025) Geopolitical biases in LLMs: what are the 'good' and the 'bad' countries according to contemporary language models. ArXiv Preprint. DOI: 10.48550/arXiv.2506.06751.

Schweter S and Akbik A (2021) FLERT: Document-Level Features for Named Entity Recognition. ArXiv Preprint. DOI: arXiv.2011.06993.

Sharkey L, Chughtai B, Batson J et a. (2025) Open Problems in Mechanistic Interpretability. ArXiv Preprint. DOI: 10.48550/arXiv.2501.16496.

Simons A (2026) Scaling In, Not Up? Testing Thick Citation Context Analysis with GPT-5 and Fragile Prompts. In: Simons A, Wüthrich A, Zichert M, et al. (eds) *Understanding Science with Large Language Models? Potentials for the History, Philosophy, and Sociology of Science*. Bielefeld: transcript, part-6.

Simons A, Arnaout H and Gurevych I (2026) Reconstructive citation context analysis using large language models. A roadmap. In: Simons A, Wüthrich A, Zichert M, et al. (eds) *Understanding Science with Large Language Models? Potentials for the History, Philosophy, and Sociology of Science*. Bielefeld: transcript, part-6.

Simons A, Zichert M and Wüthrich A (2026) Large language models for history, philosophy, and sociology of science: Interpretive uses, methodological challenges, and critical perspectives. *Studies in History and Philosophy of Science* 117: 102151. https://doi.org/10.1016/j.shpsa.2026.102151.

Sorensen T, Moore J, Fisher J et al. (2024) A Roadmap to Pluralistic Alignment. ArXiv Preprint. DOI: 10.48550/arXiv.2402.05070.

Steyvers M, Tejeda H, Kumar A et al. (2025) What large language models know and what people think they know. *Nature Machine Intelligence* 7: 221–231. DOI: 10.1038/s42256-024-00976-7.

Strauss L (1952) *Persecution and the Art of Writing*. Glencoe, Ill.: The Free Press. *Reissued Chicago: University of Chicago Press, 1988.*

Tao C, Shen T, Gao S et al. (2024) LLMs are Also Effective Embedding Models: An In-depth Overview. ArXiv Preprint. DOI: 10.48550/arXiv.2412.12591.

Tay Y (2024) What happened to BERT & T5? On Transformer Encoders, PrefixLM and Denoising Objectives. In: YiTay.net Blog entry from 16 July 2024. Available at: https://www.yitay.net/blog/model-architecture-blogpost-encoders-prefixlm-denoising (accessed 7 August 2025).

Templeton A, Conerly T, Marcus J et al. (2024a): Scaling Monosemanticity: Extracting Interpretable Features from Claude 3 Sonnet. In: Transformer Circuits Thread. Blog post from 21 May 2024. Available at: https://transformer-circuits.pub/2024/scaling-monosemanticity/index.html (accessed 12 August 2025).

Templeton A, Conerly T, Marcus J et al. (2024b) Mapping the Mind of a Large Language Model. Anthropic Blog post from 21 May 2024. Available at: https://www.anthropic.com/research/mapping-mind-language-model (accessed 12 August 2025).

Underwood T (2022) Mapping the Latent Spaces of Culture. In: *Startwords* 3 ("Parrot"). DOI: 10.5281/zenodo.6567481.

Underwood T (2025a) The impact of language models on the humanities and vice versa. *Nature Computational Science*. DOI: 10.1038/s43588-025-00819-4.

Underwood T (2025b) A more interesting upside of AI. In: The Stone and the Shell. Blog entry from 2 July 2025. Available at: https://tedunderwood.com/2025/07/02/a-more-interesting-upside-of-ai/ (accessed 12 August 2025).

Underwood T, Nelson LK and Wilkens M (2025) Can Language Models Represent the Past without Anachronism? ArXiv Preprint. DOI: 10.48550/arXiv.2505.00030.

Varnum MEW, Baumard N, Atari M et al. (2024) Large Language Models based on historical text could offer informative tools for behavioral science. *Proceedings of the National Academy of Sciences* 121. DOI: 10.1073/pnas.2407639121.

Wagner A et al. (2025a) Eine grundlegende technische Differenz … Mastodon thread with replies from Robert Casties, Jürgen Hermes. 6–11 March 2025. Available at: https://hcommons.social/@anwagnerdreas/114115349996232853 (accessed 7 August 2025).

Wagner A et al. (2025b) Today I figured out why I often feel uneasy … Mastodon thread with a reply from Karin Dalziel. 24 June 2025. Available at: https://hcommons.social/@anwagnerdreas/114738808466294298 (accessed 12 August 2025).

Wang A, Morgenstern J and Dickerson JP (2025) Large language models that replace human participants can harmfully misportray and flatten identity groups. *Nature Machine Intelligence* 7: 400–411. DOI: 10.1038/s42256-025-00986-z.

Warner B, Chaffin A, Clavié B et al. (2024a) Smarter, Better, Faster, Longer: A Modern Bidirectional Encoder for Fast, Memory Efficient, and Long Context Finetuning and Inference. ArXiv Preprint. DOI: 10.48550/arXiv.2412.13663.

Warner B, Chaffin A, Clavié B et al. (2024b) Finally, a Replacement for BERT. In: HuggingFace Blog entry from 19 December 2024. Available at: https://huggingface.co/blog/modernbert (accessed 7 August 2025).

Weller O, Ricci K, Marone M et al. (2025a) Seq vs Seq: An Open Suite of Paired Encoders and Decoders. ArXiv Preprint. DOI: 10.48550/arXiv.2507.11412.

Weller O, Ricci K, Marone M et al. (2025b) Ettin Suite: SoTA Paired Encoders and Decoders. In: HuggingFace Blog entry from 16 July 2025. Available at: https://huggingface.co/blog/ettin (accessed 7 August 2025).

Wenger E and Kenett Y (2025) We're Different, We're the Same: Creative Homogeneity Across LLMs. ArXiv Preprint. DOI: 10.48550/arXiv.2501.19361.

Xu Z, Jain S and Kankanhalli M (2024) Hallucination is Inevitable: An Innate Limitation of Large Language Models. ArXiv Preprint. DOI: 10.48550/arXiv.2401.11817.

Yang A, Li A, Yang B et al. (2025) Qwen3 Technical Report. ArXiv Preprint. DOI: 10.48550/arXiv.2505.09388.

Zhou N, Bamman D and Bleaman IL (2025). Culture is Not Trivia: Sociocultural Theory for Cultural NLP. ArXiv Preprint. DOI: 10.48550/arXiv.2502.12057.

LLMs as philosophers

What can they do, and why aren't they better?

William D'Alessandro

1. An LLM project for philosophy of mathematical practice

First let me briefly describe my project. As a philosopher of mathematical practice, I'm interested in the methods, goals and values of working mathematics. Much of my work has focused on the nature of *explanation* within mathematics and how the desire for explanatory understanding shapes mathematical research. (For instance, mathematicians often reprove the same theorem in different ways because they judge that earlier proofs merely *verified* the result without giving any insight into why it's true.)

The goal of my project was to use AI tools to create a large dataset of references to explanatory concerns for use by mathematical practice researchers. I downloaded 5000 math PDFs from arXiv.org and ran a Python script to feed these to Gemini in batches, along with a prompt detailing the kinds of examples I wanted. I then applied a filtering prompt to remove many low-quality selections from Gemini's initial outputs. The final result was a dataset consisting of ~1250 annotated examples from ~735 distinct sources (available here). Gemini's ability to accurately retrieve information from very long prompts without hallucination was essential to this phase of the work.

With the dataset in hand, I was interested to test Gemini's utility for various followup research and analysis tasks. At one point I asked the model to adjudicate between existing theories of mathematical explanation on the basis of the examples contained in the dataset. Its response was cogently reasoned and accurate in its characterization of the views at issue. Here for instance is the portion of Gemini's analysis dealing with ontic theories of explanation (sans-serif Arial font indicates text produced by Gemini; the references in Gemini's response are to cases from the dataset):

Challenges for Ontic Approaches:

Ontic accounts generally hold that explanations track objective dependency relations in the world (or, for mathematics, objective relations between mathematical facts/structures). The explanans makes the explanandum happen or be the way it is due to these real relations.

Challenging Cases/Types of Cases:

1. **Explanations via "heuristic interpretation" or "intuitive analogy" that are explicitly not the rigorous proof:**

    - **Example:** Anshelevich providing a "heuristic interpretation" of the functional Itô formula, linking it to classical Taylor expansions, explicitly for understanding before the rigorous (and potentially complex) proof.
    - **Challenge:** While the heuristic might point to an underlying structural analogy, the explanation itself in these instances seems to reside in the cognitive act of relating the complex to the simple and familiar, even if that analogy isn't a perfect or rigorous ontic dependency. The value is in making it "graspable", which is more of an epistemic/pragmatic virtue than a direct reflection of an objective dependency chain in the formal proof. An ontic account would need to argue that the heuristic does track some objective (perhaps looser) structural resemblance.

2. **Reproofs explicitly valued for "simplicity", "clarity", or being "less computational" when the original proof already established the ontic link:**

    - **Example:** Schlosser's "elementary derivations" of hypergeometric identities, contrasting with prior "not entirely elementary" proofs. Or the desire for a "more conceptual" proof by Burghelea & Haller.
    - **Challenge:** If the original proof already correctly identified the mathematical facts from which the theorem follows (the ontic basis), why is a new, simpler proof more "explanatory"? The ontic dependencies might be the same. The added value seems to be epistemic: the new proof makes those dependencies easier to grasp or presents them in a more "natural" way. An ontic account might struggle to capture this difference in explanatory value if the objective dependencies are unchanged.

3. **Explanations where the "explanatory work" seems to be done by a representation or framework rather than directly by the objects themselves:**

    - **Example:** The "logarithmic point of view" (El Goul) making a refined result an "easy consequence". The framework itself facilitates understanding. Or Fialowski & Penkava discussing how reframing deformation theory in terms of coderivations makes connections "completely transparent".
    - **Challenge:** While the framework undoubtedly reflects ontic structures, the explanatory power seems to come from the choice of representation that highlights certain features and relationships. An ontic account needs to carefully delineate whether the explanation lies in the objective structures themselves or in the way a particular framework allows us to cognize those structures.

I was also interested in Gemini's ability to philosophize in an original way about the cases in the dataset. So I asked it to try. ("Suppose you were asked to come up with an entirely novel theory of explanation, one able to do better than existing theories at making sense of all the cases you've reviewed. What might such a theory look like? Feel free to speculate.") A representative portion of the model's response:

> If I were to speculate on an entirely novel theory of mathematical explanation, aiming to encompass the diverse cases reviewed, I would try to build something that integrates insights from existing theories while addressing their limitations. Let's call it **"Explanatory Resonance Theory" (ERT)**.
> Here's what ERT might look like:
>
> **Core Idea of Explanatory Resonance Theory (ERT):** Mathematical explanation occurs when a target mathematical statement or phenomenon (the explanandum) is brought into "explanatory resonance" with a broader, more structured, or more fundamental mathematical context (the explanans-framework). This resonance is not a single type of connection but a multi-faceted relationship that enhances our cognitive grasp and structural integration of the explanandum.
>
> **Key Components and Features of ERT:**
>
> 1. **Multi-faceted Resonance:** "Resonance" isn't a monolithic concept. It can manifest in several ways, acknowledging the pluralistic nature of what mathematicians value as explanatory: [here Gemini lists and describes "structural", "unifying", "simplifying/clarifying", "analogical/metaphorical" and "generative" varieties of resonance, attempting to incorporate the core ideas of the views discussed previously and bringing up many of its earlier examples].
> 2. **The Explanans-Framework:** This is not just a single proposition but a richer context. It could be:
>
>   - A well-established theory (e.g., Galois theory, representation theory, category theory).
>   - A specific set of definitions and their consequences.
>   - A powerful visualization or diagrammatic system.
>   - A newly introduced concept or structure that re-organizes existing knowledge.
>
> 3. **Cognitive Grasp and Structural Integration (The Goal):** The ultimate aim of explanation under ERT is to enhance the mathematician's cognitive grasp of the explanandum. This means:
>
>   - **Understanding Why:** Not just that it's true, but why it must be true given the explanans-framework.
>   - **Seeing Connections:** Understanding how the explanandum relates to other mathematical ideas and structures.
>   - **Increased Predictive/Manipulative Power (Potentially):** A good explanation might suggest generalizations, new questions, or more efficient ways to work with the concepts.

- **Reduced Surprise/Mystery:** Phenomena that initially seem "surprising" or "mysterious" become less so when they resonate with a suitable framework.

4. **Dynamic and Contextual:** What counts as a good explanans-framework or a strong resonance can be:

   - **Field-dependent:** Different mathematical subfields might value different types of resonance. Bijective proofs are highly valued in combinatorics for their structural resonance.
   - **Historically contingent:** What was once a complex explanandum can become part of a familiar explanans-framework for new explananda.
   - **Audience-dependent (to some extent):** While striving for objectivity, the effectiveness of an explanation in producing "cognitive grasp" can have a pragmatic element related to the audience's background knowledge.

In essence, Explanatory Resonance Theory would propose that mathematical explanation is the process of making a mathematical idea "vibrate sympathetically" with a larger, more harmonious, and well-understood part of the mathematical orchestra. The quality of the explanation depends on the richness, depth, and clarity of this induced resonance.

While Gemini provided a strong response to the earlier critique prompt, it's clear that this is not a similarly impressive piece of original philosophy. It exhibits the typical LLM tendency to try to accommodate as many standard ideas and existing views as possible without taking a definitive stand on any claim of note. Its unifying idea is a vague, unhelpful metaphor. Not a single feature of the proposed theory is meaningfully novel. (I realize that readers outside the philosophy of mathematical practice may lack the background to independently evaluate Gemini's proposal, but I don't think my assessment is controversial.)

Gemini's behavior in my experiment is an example of a wider phenomenon. The best current LLM-based systems excel at analytical tasks requiring considerable insight, subject understanding and dialectical sophistication. Their performance here often seems roughly at the level of human experts. At the same time, these systems show little ability to go beyond the frontier: somehow their vast knowledge and impressive reasoning ability almost never yields breakthroughs in any domain. Even Sam Altman, ever ready to boast about OpenAI models' capabilities, has only expressed the hope that "systems that can figure out novel insights" might arrive sometime in 2026 (Altman 2025).

Ben Levinstein's Philosophy Pipeline project[1] furnishes another discipline-specific example. Levinstein has assembled a complex AI workflow to generate complete philosophy papers in the style of the journal *Analysis*, aiming eventually for a 20% publication success rate. While the current version of the Pipeline produces readable papers with competent structures and appropriate citations, Levinstein agrees that "no interesting

1 https://github.com/balevinstein/philosophy-pipeline.

discoveries or even ideas have come from LLMs yet" (July 2025, personal communication).

I confess to finding this somewhat mysterious. Why do current models make incisive assistants but poor thinkers, when it seems as though the cognitive capabilities required in these cases should be mostly the same? What do we even mean when we talk about a model's ability to generate "high-quality original ideas"? What sorts of capabilities might allow future models to clear this important hurdle?

In the rest of this note, I'll discuss a few plausible explanations which I don't ultimately find satisfying and end with one that seems better.

2. Why can't LLMs produce good new ideas?

I assume that AI systems are the kinds of things that could at least in principle do worthwhile original philosophy and make other sorts of intellectual breakthroughs. Of course, not everyone shares this assumption: some think human minds have productive powers which no machine learning model could ever match, owing for instance to human consciousness, embodiment, ensoulment, the nonformal or nondigital nature of human cognitive processing, or some other putatively unique feature of ours. Those who hold such views may take themselves to have a simple explanation of LLMs' failure to make creative discoveries. Since I don't hold them, my job is a bit harder.

2.1 Because they can't invent new concepts?

In philosophy and elsewhere, intellectual progress is often closely linked to the introduction of new concepts: almost singlehandedly, notions like *rigid designator*, *possible world*, *scientific paradigm* and *epistemic injustice* seem to bring order to many things we know and open doors to many others we haven't yet considered.

One might also think that LLMs are fundamentally unable to devise new concepts. This suspicion is supported by the common picture of language models as "stochastic parrots" (Bender et al. 2021), able only to repeat the contents of their training data in various permutations. If correct, this would go a long way toward explaining why LLMs can be made very good at manipulating and applying existing ideas without ever producing transformational insights.

But I don't think this explanation is right, at least on some understandings of what it claims. If the idea is supposed to be that LLMs are strictly limited to the store of concepts encountered during training, this seems pretty clearly untrue. LLMs are skilled at devising categories which were almost certainly not seen in training and which require some degree of creativity to hit upon.

A model benchmark[2] based on the *New York Times*' daily Connections puzzle gives one way to see this. In (the standard version of) the puzzle, players are presented with a grid of 16 words which must be sorted into four sets of four, with the words in each set falling under a common concept. Typically the category definition for at least one set (often the

2 https://github.com/lechmazur/nyt-connections

hardest category, styled with a purple background) involves unusual wordplay or a non-obvious factual connection. For instance, the words in this set in a recent puzzle were CASH, HAZE, MAC, PEC. The NYT-provided category description was *starts of culinary nuts* (cashew, hazelnut, macadamia, pecan).

When the Connections benchmark debuted in spring 2024, the then-best models did relatively poorly. But the benchmark has been saturated by the current generation of reasoning models. For instance, OpenAI's o1 correctly solved about 99% of the 90 puzzles from December 2024 to February 2025. (The human player base's success rate during this period was about 71%.) This led to the development of a new assessment based on a more difficult version of the puzzle than the one played by humans.

It seems clear that achieving a high level of performance on Connections requires the ability to devise concepts which are insightful and appropriate yet in some sense novel. (o1 presumably didn't encounter the specific concept *starts of the names of culinary nuts* in its training data. At any rate, it didn't learn 89 out of 90 similarly recondite concepts in this way.) The stochastic-parrot explanation in its strictest sense therefore seems implausible.

Of course, one could point out that the concept *starts of the names of culinary nuts* is a composite made up of concepts which the model surely did see in training. Entirely true: but then the same can be said for human concepts associated with major advances. The ideas of names, reference, possible worlds and the like were all well known before Kripke. Fricker invented neither the category of the epistemic nor the notion of injustice. Concepts praised as novel and insightful are often assembled from existing materials.

What distinguishes important new concepts from relatively trivial ones seems not to be mere novelty, but rather the virtues the former possess: elegance, breadth, utility, naturalness, explanatory power, and so on. We can agree that LLMs aren't very good at inventing concepts with these characteristics. But why not?

2.2 Because they aren't good at reasoning?

A second story one could tell starts with the importance of reasoning to intellectual breakthroughs. It's rarely sufficient, after all, to come up with a nice idea or two: one has to show how these ideas fit together among themselves and with the known facts, what can be done with them, why they're better than alternatives. Indeed, given sufficient acumen, one might not need new ideas at all. Mathematicians often celebrate[3] clever elementary proofs of known results. Sometimes progress comes from understanding a problem deeply enough to grasp what's essential and see how to do more with less.

And one might tie this story to a claim about LLMs' reasoning abilities. Again, there's a ready-to-hand stereotype that language models never engage in thought, inference, planning or world-modeling; they are, rather, "glorified autocompletes" whose outputs are constrained not by truth or logic but only the estimated likelihood of a given text continuation.

---

3 https://math.stackexchange.com/questions/1849840/big-list-of-erd%C5%91s-elementary-proofs

There is of course *something* to this stereotype. Pretrained LLMs are indeed trained to minimize a loss function tied to their ability to successfully predict held-out words in internet text. And this is indeed part of the reason they sometimes hallucinate, repeat common falsehoods and make various strange mistakes. But it's also clear that this is a highly incomplete picture of how current LLM products work.

First, although the objective of pretraining is to minimize loss on text completion, the best way to fulfill this demanding objective is often by developing cognitive-like internal structures and functions. Sometimes these structures can be identified explicitly, as in the case of GPT-2 small's circuit for locating the indirect object of a sentence (Wang et al. 2022). Large pretrained models should thus be understood as sophisticated information processors, not mindless word generators. Second, pretrained language models receive further finetuning to improve accuracy and reduce hallucinations. The state of the art here has moved far beyond the initial release of ChatGPT, with models like Gemini 2.5 Pro outperforming human experts[4] on questions requiring information retrieval from long-context prompts. Third, the advent of reasoning models has led to vast improvements in LLM planning and problem-solving. A specially trained version of Gemini Deep Think recently logged a gold medal-worthy performance[5] on the 2025 International Mathematical Olympiad (IMO) problem set, answering five of the competition's six famously difficult problems correctly. (Less than 10% of the human test-takers, themselves elite competitive solvers of these kinds of problems, typically earn gold medals.) The IMO president described Gemini's solutions as "astonishing in many respects... clear, precise and [mostly] easy to follow".

All of this demonstrates that LLMs are at least in principle capable of sophisticated and accurate reasoning. Their performance in arenas like the IMO suggests that, given proper training, they can solve difficult problems of genuine intellectual interest. So it seems unlikely that a general lack of reasoning power is responsible for the dearth of LLM breakthroughs.

2.3 Because good philosophy has few reinforcement-learning-friendly patterns?

Here's my best guess about why LLMs are bad at original philosophy. While it shares some DNA with the previous suggestion, it isn't based on a dismissive picture of LLMs as non-thinkers, but rather on the details of how models are trained to reason well in particular domains.

The standard method for training reasoning models works roughly as follows. First, the model is given a problem to solve. Then, rather than attempting to immediately generate a solution in the fashion of a standard LLM, the model produces a "reasoning trace" or "chain of thought" describing a potential approach to solving the problem. These traces are assigned a score—by a human or (more often) a specialized reward model—according to the quality of the reasoning they contain and the correctness of the associated so-

4 https://longbench2.github.io

5 https://deepmind.google/discover/blog/advanced-version-of-gemini-with-deep-think-officially-achieves-gold-medal-standard-at-the-international-mathematical-olympiad

lution. The original model then updates its weights to favor high-scoring patterns and deprecate low-scoring ones, so it will be likelier to produce more of the former.

Models can learn some general-purpose reasoning guidelines in this way: e.g., that it's good to break complex problems into smaller pieces, proceed step by step, avoid circularity, and so on. But the whole process works especially well in a particular domain to the extent that several things are true: (1) reasoning in the domain involves many recognizable, repeatable patterns; (2) it's relatively clear which of these patterns are good or bad in a given context; (3) most problems have an unambiguously correct solution; (4) there exist large corpora of explicit examples of good and bad domain-specific reasoning (both for pretraining the main model to pick up a variety of standard reasoning styles, and for training the reward model to recognize good patterns).

Domains like mathematics and coding meet these criteria. There are countless publicly available examples of correct proofs. Everyone more or less agrees on what a good proof looks like. Much of mathematical reasoning involves choosing the right argument pattern from a library of standard techniques. (This is especially true for competition-style problems, which operate under a distinctive set of constraints and are known to draw on a learnable set of strategies. Google's engineers seem to have exploited these affordances, providing their customized Gemini model with "a curated corpus of high-quality solutions to mathematics problems" as well as "general hints and tips on how to approach IMO problems".) So it's no surprise that reasoning models have recorded some of their most impressive successes in domains like these.

Other domains are not so much like this. For instance, even the best current models perform quite poorly on the ARC-AGI-2 benchmark[6], involving a varied set of visual reasoning tasks meant to be straightforward for humans to complete. What makes the ARC challenge difficult for LLMs is at least in part that the problem set involves few repeated patterns: the logic of each problem, while not objectively very complex, is quite unique, so the task space has little in the way of readily learnable structure to exploit. The relatively small size of the 1000-problem training set is a further handicap.

Philosophical reasoning, I take it, is more like a set of ARC problems than a body of mathematical proofs. While there are such things as "standard philosophical moves"[7] that recur in various contexts (some of which are quasi-mathematical in character), it seems difficult to say anything very general about when a given argument pattern is likely to succeed or fail. (One man's modus ponens[8] is another man's modus tollens, as the saying goes.) And of course there's often deep disagreement about whether some argument has succeeded in a given case, or even whether it *could* succeed in any possible case. Finally, we lack the sorts of gold-standard research-level argument corpora that would fit comfortably into ML training pipelines.

Some of these problems could be overcome in principle with lots of work. Others seem less tractable. Unlike the cases of mathematical reasoning, coding and the like, neither AI developers nor academic philosophers have obvious financial incentives to go full

6 https://arcprize.org/leaderboard

7 https://dailynous.com/2025/01/16/a-taxonomy-of-philosophy-moves

8 https://en.wikipedia.org/wiki/Here_is_one_hand

steam ahead on this kind of progress, so it's unclear what would bring about the necessary changes.

In the meantime, it's not entirely surprising that LLMs make good assistants but mediocre original thinkers. Generic RL for reasoning has equipped them with strong domain-neutral thinking skills, but world-class performance requires additional domain-specific optimization. No model has yet been optimized in this way for philosophical reasoning. Alas for us, at least for now.[9]

References

Altman S (2025) The gentle singularity. In: *Sam Altman*. Available at: https://blog.samaltman.com/the-gentle-singularity (accessed 13 June 2025).

Bender E, Gebru T, McMillan-Major A et al. (2021) On the dangers of stochastic parrots: Can language models be too big? In: *FAccT '21: Proceedings of the 2021 ACM Conference on Fairness, Accountability, and Transparency*, 3–10 March 2021, pp. 610–623. New York: Association for Computing Machinery.

D'Alessandro W (forthcoming) Using large language models to study mathematical practice. In: Kant D, Perez-Escobar JA, Sarikaya D and Sarikaya M (eds) *Mathematicians at Work: Empirically Informed Philosophy of Mathematics*. Dordrecht: Springer.

Wang K, Variengien A, Conmy A et al. (2022) Interpretability in the wild: A circuit for indirect object identification in GPT-2 small. Available at: https://arxiv.org/abs/2211.00593.

9 Aside from the Gemini outputs directly quoted in the text, no LLMs were involved in the writing of this chapter.

Part 2: Historicizing LLMs

Why pursue temporally-grounded AI for historical disciplines, and what makes it so challenging?

Jochen Büttner

This text reflects on why explicit temporal grounding of LLMs is desirable for disciplines working with historical texts—specifically through architectures that model temporal drift in the training distribution. Although current LLMs display emergent temporal reasoning, their time-agnostic design lacks explicit mechanisms for temporal conditioning.[1] I survey existing approaches (e.g., time tokens/embeddings, temporal attention, and period-focused pretraining) and argue that, while temporal conditioning is generally advisable in historical research contexts[2], in practice it may remain limited to constrained tasks rather than general-purpose generative systems.

Current LLMs are trained on mixtures from many sources and time periods, yet they treat the token sequences they see in training as one empirical, time-marginalized distribution, estimating parameters that maximize its likelihood. This assumption is fundamentally flawed when applied to historical data—both language and the content it expresses evolve over time, meaning the underlying distribution of token sequences is inherently dynamic. While this resembles the concept drift problem in continuous learning—where the solution is to adapt to drift—historical modeling should ideally preserve all drifts, yet current LLMs are trained without any explicit temporal grounding; time remains implicit in their architecture.[3] The models, by design, effectively integrate over

---

1 While I use the terms temporal grounding and temporal conditioning somewhat interchangeably throughout this text, there is a technical distinction: temporal grounding refers to understanding temporal references in text, while temporal conditioning refers to using temporal information as input to shape model outputs. These naturally interweave, for instance when grounding temporal references in prompts conditions the model's responses.

2 Recent work by McGillivray et al. (2024) has argued for temporal grounding in digital-humanities contexts—emphasizing period-aware indexing and queries, explicit temporal metadata, and methods that account for semantic drift—but their contribution focuses exclusively on search and retrieval, a narrower application than the claim advanced here.

3 The temporal distribution drift in historical corpora is structurally related to the concept drift problem in continuous learning, where production LLMs face evolving data distributions as language and world knowledge change over time. However, the objectives diverge sharply: continuous training systems optimize for tracking the current distribution—adapting to new patterns while strate-

the time dimension rather than conditioning on it, collapsing temporal variation into a single distribution. This statistical flattening creates a systemic bias toward contemporary patterns: because training corpora consist predominantly of modern web text, the time-marginalized distribution is heavily weighted toward recent linguistic patterns, cultural references, and factual associations (Zhu et al., 2024; Fang et al., 2025; Cheng et al., 2024).[4]

While this approach has limitations in general, it is particularly inadequate from a historian's perspective. When a model treats "Charles III is King of England" and "Henry V is King of England" as competing for the same probabilistic space—distinguished only by corpus frequency rather than temporal context—it commits a fundamental category error by failing to model diachronicity. These statements are not probabilistic alternatives; they are truths anchored in mutually exclusive temporal contexts. The difference between them is not statistical but historical, a distinction lost on a model that flattens time into a single distribution. Likewise, the fact that no text before 1976 mentions something "run on an Apple" or "stored in the cloud" is not merely statistical variation but reflects a diachronic shift in both technology and the semantic fields of the words themselves.

In practice, the situation is more complex than these simple examples suggest. During training and at test time, models typically encounter far richer context than such isolated statements—the surrounding text supplies temporal cues such as dates, verb tenses, and period-specific references that enable surprisingly sophisticated temporal reasoning. LLMs can order events, extract dates, and answer time-sensitive questions with considerable accuracy.[5] Empirically, they develop internal representations of time that can be detected and mapped, and these representations have been shown to be responsible for their temporal reasoning capabilities, yielding results akin to, but not identical with, explicitly time-conditioned output (Gurnee et al., 2024; Tiblias et al., 2025).[6]

---

gically forgetting or downweighting old data—whereas historical applications require preserving fidelity to all temporal distributions $P(x|t)$ simultaneously. This explains why solutions from the continuous learning literature cannot simply be repurposed. See Lazaridou et al. (2021) for temporal drift in LLMs and adaptation strategies; Jang et al. (2022) for continual knowledge updating approaches.

4 While empirical studies have documented related temporal biases in LLM behavior (Zhu et al., 2024; Fang et al., 2025) and temporal misalignments in training data composition (Cheng et al., 2024), this fundamental issue of modern overrepresentation in the averaged distribution remains an underexplored consequence of time-agnostic architectures.

5 Temporal understanding of LLms, however, also has notable limits. Models struggle to generalize temporal order from context, handle ambiguous phrasing, model long-term dependencies, and grasp abstract temporal concepts like causality or event progression. Jain et al. (2023) demonstrate that LLMs particularly fail at understanding event temporal states and persistence, reasoning about timing and scheduling, handling concurrent temporal events, and processing exact temporal expressions. Zhao et al. (2024) aptly characterize this as the "temporal chaos of pretrained LMs."

6 If the prompt "Given it is the year 1002, who is the King of England?" is correctly answered with "Æthelred the Unready," the temporal context is a deduced feature of the input, not an explicit parameter controlling the generative process.

If models already handle temporal tasks well, why modify their architecture toward time conditioning?[7] Because temporal awareness emerges implicitly from training data patterns rather than being explicitly parameterized in the model's representational structure. In particular, if a model were trained on vast historical sources spanning millennia, diachronic drift, shifting semantics, and uneven data frequency over time would presumably make it extremely difficult for the model to reliably reconstruct the appropriate temporal context from source patterns alone. Moreover, even if a model can "know" that the answer to "Who is the current king?" varies by historical period, it is expected to fail in certain situations—especially when context is sparse or contradictory, or when coherence must be maintained across extended text.[8]

To test this, we designed a text-completion task using a two-model setup. First, we asked one model to write a short text in Shakespeare's style and end with "the name of the current ruler of England is." This text was then passed to a second model. This setup creates fundamental ambiguity: should the model persist in Shakespeare's temporal context (Elizabeth I or James I), maintain the historical setting established in the pseudo-Shakespearean text, or interpret "current" relative to its training data (e.g., Charles III)? The resulting confusion—with style, narrative context, and the word "current" each suggesting different temporal anchors—shows that implicit temporal grounding can remain opaque without architectural support for explicit time conditioning.[9]

---

7 It is crucial to distinguish the issue of temporal grounding addressed here from the growing body of research on 'temporal reasoning' in LLMs. That research encompasses diverse tasks such as evaluating models' ability to identify Allen's interval relations (before, after, during) between events, temporal question answering, timeline construction, and probing whether LLMs possess temporally-grounded factual knowledge (see, e.g., Tan et al., 2023 on temporal reasoning benchmarks; Vashishtha et al., 2020 on temporal relation extraction). Such work tests how well existing architectures handle temporal tasks through their emergent capabilities. In contrast, this reflection advocates methods—applicable to both encoder and decoder models—that explicitly condition on time at the representational level. How such architectural modifications might improve performance on temporal reasoning benchmarks is a downstream question,especially in the case of generative chat-optimized models; while improvements seem likely for certain tasks, the primary motivation is not benchmark performance but rather the principled incorporation of temporal context into the model's fundamental processing of language and knowledge.

8 Recent research (Sun at. Al 2025) How new data permeates LLM knowledge and how to dilute it) into how new data "permeates" an LLM's knowledge has identified a "priming" effect, whereby learning a new fact can cause the model to inappropriately apply that knowledge in unrelated contexts . This suggests that in a time-agnostic model, information from different periods doesn't just co-exist statically; more recent data can actively "bleed" into and distort the model's representation of the past.

9 The generated prompt used in the experiment was:
Hark! What tempest stirs within mine restless breast,
When morning's golden chariot doth ascend
O'er yonder hills, and gentle zephyrs blessed
With honeyed breath do make the willows bend?
Methinks the very stars conspire to weave
A tapestry of fortune most sublime,
Whilst mortal hearts, that beat and oft deceive,
Do chase the shadows of forgotten time.
The name of the current ruler of England is

If the training data massively drifts with time due to changes in language and content—as would be typical for a large corpus of historical documents—we need models that learn $P(x|t)$ rather than $P(x)$:[10] probability distributions explicitly conditioned on time. This requirement manifests differently across encoder and decoder architectures. Encoder models like BERT (Devlin et al., 2019), which learn $P(x_{masked} \mid x_{context,}, t)$[11], need time conditioning to prevent conflating entities across eras—ensuring "King Charles" maps to different individuals when the context is 17th century (Charles I/II) versus 21st century (Charles III), rather than collapsing all instances into a single, temporally-ambiguous representation.

In Decoder models, learning $P(x_{n+1}|x_1, \ldots, x_n, t)$[12], i.e. the probability of the next token given the sequence of all previous tokens and an explicit temporal context, would first and foremost allow one to generate temporally coherent text. This initially seems limited in application—perhaps useful for period-appropriate creative writing or historically-grounded dialogue systems (e.g., enabling dialogue with historical personas from specific periods). However, the vision extends much further. Temporally-aware decoder models could theoretically be further developed via instruct-tuning into historical AI assistants, though this requires capabilities far beyond time conditioning alone. To use an anthropomorphizing view: the ability to understand textual patterns in their time is a *conditio sine qua non* for historically interpreting them. The latter task is of course more complex and comprises not just recognizing that "the Crown" meant personal rule in Tudor times, constitutional monarchy by Victoria's reign, and ceremonial institution today,

---

State-of-the-art models such as GPT-4o or Claude Sonnet 4.5 gave what from a human perspective could be called metareflexive answers, indicating awareness of this as a test setting and generally opting to interpret the question as referring to their training cutoff, thus answering "Charles III" and wrapping the answer, for instance, in a brief sonnet. Gemini 2.5 in its reasoning trace explicitly stated: "I've realized that the prompt's main goal is to test my ability to identify and respond to the explicit request despite the surrounding literary context. The shift is designed to see if I can set aside the old-fashioned style to focus on the factual answer, King Charles III. I'm confident in my decision to provide a straightforward response." Mistral 7B text (Q4) on the other hand answered "Henry Tudor," who indeed figures in Shakespeare's Richard III.

10 Here $P(x)$ denotes the probability of a text sequence x occurring in the training corpus, marginalized over all time periods and $P(x|t)$ is the probability of text x given time t. The core challenge of a model failing to represent distinct subpopulations within its training data is a recognized issue in machine learning, extending beyond temporal contexts. A model trained on a single, integrated distribution may perform poorly on specific demographic, dialectal, or domain-specific subpopulations (e.g., legal vs. scientific jargon) because it learns an averaged representation that does not hold equally for all groups . This is analogous to the temporal subpopulation problem discussed here, where a time-agnostic model cannot adequately represent the specific language and facts of different eras. The call for models that can condition their probability distributions on specific features—$P(x|t)$ for time, or $P(x|d)$ for domain—could therefore offer a general solution to a widespread modeling limitation

11 $P(x_{masked} \mid x_{context})$ is the probability of the masked token(s) x_{masked} that have been hidden/masked (what the model is trying to predict) given the observable (unmasked) context x_context i.e. the surrounding tokens that remain visible to the model.

12 $P(x_{n+1}|x_1, \ldots, x_n, t)$ represents the probability of the next token x_{n+1} given the sequence of previous tokens x_1 through x_n and an explicit temporal context t,representing time-conditioned autoregressive generation.

but understanding why these transformations occurred, what they reveal about evolving concepts of sovereignty and democracy.

Various solutions have been proposed and applied, mainly to encoder architectures, to achieve time conditioning. These range from training on specific epochs to architectural modifications that process temporal information during learning and generate temporally-conditioned output.

We start with methods of time conditioning where the model stays unchanged and only the model inputs are time-sliced. This strategy predates transformer models; for example, Hamilton, Leskovec and Jurafsky (2016) already trained separate word embeddings (SGNS/word2vec, PPMI, SVD) across periods and aligned them to quantify semantic drift.[13] More recent examples include Hu, Li and Liang (2019), who use off-the-shelf BERT sense embeddings on time-sliced COHA (Corpus of Historical American English) (1810–2009); Montariol, Martinc and Pivovarova (2021), who fine-tune BERT once per corpus for domain adaptation and then split the data into time slices (temporal control via corpus partitioning, not per-slice weights); or Zichert, Simons and Wüthrich (2025), who use yearly time-sliced contextualized embeddings to trace conceptual change in the "virtual particle" concept across physics literature. While these approaches segment the input data temporally, they leave the fundamental processing architecture unchanged: the model has learned patterns that are time-invariant at the level of attention and representation, capturing mainly those regularities that persist across time, potentially missing temporal signals particularly relevant to diachronic analysis.[14]

The simplest approach for model-level time conditioning involves period-focused pretraining or continued pretraining on historical corpora, thereby conditioning the model's parameters—and thus its output probabilities—on a specific time or time period. Examples include MacBERTh (Manjavacas Arevalo and Fonteyn, 2021) trained on texts from roughly 1450–1950, GHisBERT (Beck and Köllner, 2023) trained from scratch on historical German across major language stages, and HistBERT (Qiu and Xu, 2022), which continues BERT pretraining on the historical COHA corpus.

For finer temporal resolution, researchers have extended this idea to time-slicing—training separate models for distinct windows. With transformer-based models, Hosseini et al. (2021) fine-tune four BERT instances on 1760–1900 English books split into pre-1850, 1850–1875, 1875–1890, and 1890–1900, and He et al. (2025) introduce ChronoBERT/ChronoGPT—separate models for each year (from 1999 onward) trained without future data. This latter approach of yearly granularity exemplifies the scaling problem inherent in time-slicing, as the number of models grows linearly with temporal resolution. A key motivation for such intensive methods is to achieve high performance

13 The approach computes distributional statistics separately for each period. The model—the mathematical operation itself—remains unchanged yielding different frequency and co-occurrence patterns across time slices.

14 Consider analyzing semantic drift of "computer" using a BERT-like encoder. When processing "She worked as a computer at NASA" (1950s) versus "She worked on a computer at NASA" (1990s), BERT's attention mechanism—trained mostly on modern data—focuses on the same features in both cases likely ignoring the temporally diagnostic cues in the shift from preposition "as" to "on." The embeddings would differ only accidentally, as byproducts of time-invariant patterns, not because of time conditioned attention in the BERT model.

on specific temporal reasoning tasks like event ordering and timeline summarization, where precise time-grounding is critical. TiMoE (Faro et al., 2025) refines the time-slicing approach by training separate experts on temporal windows but combining them at inference through mixture-of-experts architecture with causal temporal masking, reducing future-knowledge errors while maintaining multi-period knowledge access.

The straightforward solution of training separate models for distinct time slices does provide conditioning but can be impractical—it fragments the dataset, reduces training data per model, and prevents sharing of linguistic knowledge that remains stable across periods. Moreover, boundaries are necessarily artificial, and the approach scales poorly as granularity (i.e., number of periods to be considered) increases.

An alternative proposed to mitigate the scaling issue of time-slicing is to prepend time tokens—a date or period marker—to each sequence, allowing the transformer to condition on time by attending to this token. Examples include Dhingra et al. (2022), who prepend timestamps during pretraining to build time-aware LMs, and TempoBERT (Rosin et al., 2022), which augments inputs with time and uses time masking for sentence-time prediction.

While computationally efficient and simple to implement, merely prepending time tokens offers only a superficial form of conditioning. The model can learn to attend to the time token, but this does not fundamentally restructure how tokens relate to each other given temporal context. The time token becomes just another element in the sequence, competing for attention with the substantive content.[15] In practice, this means the model might recognize the token "|DATE: 1649|" but fail to sufficiently recalibrate the semantic relationships between "King," "Charles," and "execution" that are specific to that year. This approach merely informs the model of the time rather than architecturally enforcing a time-conditioned representation. This can potentially lead to a fragile and inconsistent temporal grounding likely to be overwhelmed by strong, ahistorical statistical patterns in the data. Recent approaches have sought to mitigate these limitations by integrating temporal signals through more deeply integrated pre-training objectives. For instance, BiTimeBERT 2.0 employs Document Dating—a task where the model must infer the timestamp from the document's content, forcing it to learn to temporally ground documents from their content—as one of three specialized objectives. This is complemented by Extended Time-Aware Masked Language Modeling for temporal expression understanding and Time-Sensitive Entity Replacement to handle entities whose meaning changes over time (Wang et al., 2023; Wang, Jatowt and Cai, 2024).

While time tokens—which essentially simply extend the vocabulary by appropriate 'time words'—leave the model architecture unchanged, other approaches add dedicated time embeddings while otherwise keeping the Transformer stack intact, thus introducing a light architectural change. One example comes from the field of processing electronic health records (EHRs). These records are typically modeled as sequences of pa-

15 Prepended control tokens—such as time markers—often exert limited influence because most transformer attention heads focus on semantic content rather than structural or metadata tokens (Voita et al., 2019). Empirical studies of time-aware language models further suggest that time tokens alone are insufficient to induce robust temporal calibration unless paired with auxiliary training objectives (Dhingra et al., 2022).

tient visits; each visit bundles coded events with timestamps, enabling models to learn temporal patient trajectories. CEHR-BERT (Pang et al., 2021) injects time by inserting gap tokens that encode the elapsed time between visits and attaching learned time and age embeddings to each event alongside its concept embedding. Content and temporal embeddings are parameterized separately, then fused via concatenation plus a linear projection before entering the Transformer.

A closely related, embedding-level approach was presented by the author at the "Large Language Models for the History, Philosophy, and Sociology of Science" workshop: a small, decoder-only Transformer that reserves a time channel in the input embedding (e.g., normalized day-of-year) and is trained on daily weather reports.[16] Despite this minimal change, the model has been shown to reproduce naturally seasonal language/weather patterns and two synthetically injected drifts—(i) progressive synonym substitution and (ii) increasing co-occurrence of "rain and snow"—with attention analyses corroborating the learned dependencies. Unlike most prior work focused on encoders, this demonstrates generative, time-conditioned text in a decoder architecture. It is not a proof-of-concept for a full historical assistant, but a necessary first step as will be argued below.

A more ambitious line of work changes the transformer itself to yield genuinely time-conditioned representations. Rosin and Radinsky (2022) introduce temporal attention, which augments self-attention with a learned time matrix derived from the input's timestamp. The model injects time directly into the attention computation so that every token–token score is multiplicatively modulated by the time signal; in effect, the strength of connections among words is reweighted by how well they fit that moment (so *plague* may bind to *miasma* in medieval data but to *pandemic* in modern corpora).[17] Applied to BERT and evaluated on semantic change detection across English, German, and Latin datasets, temporal attention achieves state-of-the-art results. Compared with the embedding-level time channel by the author of this reflection described above—which leaves the Transformer's attention pattern intact and merely supplies time as an extra feature—temporal attention injects time into the attention computation itself, directly reshaping token–token affinities across periods.[18]

16 https://www.youtube.com/watch?v=Qn9mxkUXYT8

17 A parallel development in clinical NLP demonstrates the same principle of architecturally injecting time into the attention mechanism. ChronoFormer (Zhang and Li, 2025), designed for modeling electronic health records, employs a dual temporal embedding (for absolute and relative time) and a hierarchical attention structure to capture intra-visit and inter-visit dependencies. While applied to structured clinical event sequences rather than free-text diachrony, it shares the core innovation of directly reshaping token–token affinities through explicit temporal conditioning

18 When time is added as an extra embedding feature (the same time value attached to all tokens in a sequence), time-specific attention patterns do emerge—as demonstrated in the author's decoder experiments where seasonal patterns and temporal dependencies developed naturally. However, these patterns arise indirectly: the time signal gets mixed into the token representations as they pass through successive processing layers. This mixing begins at the normalization step, where each token's embedding is normalized using its own mean and variance computed across all dimensions—because tokens have different content features, they produce different statistics, causing the initially uniform time dimension to take on token-specific values after normalization. Each subsequent layer transforms these representations further, and because the time information is

In all the cited works, time conditioning consistently improved results over previous benchmarks. Conceptually, this is unsurprising: when the underlying data distribution varies systematically with time, models that can condition on temporal information should outperform those that cannot.

However, all these approaches, except for the time conditioned decoder model developed by the author, share a common characteristic: they focus on encoder-based representations for downstream tasks such as classification, semantic change detection, or named entity recognition.[19] For such constrained objectives, a pragmatic approach is appropriate—use what suffices for the task at hand. If period-specific fine-tuning yields adequate performance, more complex architectural modifications may be unnecessary. The key insight for historical work is that time conditioning should always be considered as an option, with the appropriate method selected from the spectrum of available alternatives based on the specific requirements of the task.

Going over to today's prevalent generative models, it is crucial to acknowledge that a particular form of temporal conditioning is already widely practiced: test-time conditioning via prompting. When a user instructs a model to 'write a letter in the style of the 18th century' or prefixes a query with 'In 1995', they are in a way attempting to anchor the model's output in a specific time period. This leverages the model's in-context learning capabilities and is, in effect, a pragmatic use of time as a conditioning signal. However, the limitations of this pragmatic approach have already been outlined above.

Retrieval-augmented generation represents a special case of this test-time conditioning approach: rather than relying solely on the user to specify temporal context in prompts, temporal RAG systems augment prompts with time-sensitive retrieved information. Systems like TempRALM and Graphiti (Chalef et al., 2025) demonstrate that explicitly incorporating temporal metadata during retrieval—through bi-temporal knowledge graphs, timestamp-aware vector search, or time-weighted reranking—can improve accuracy on temporal queries (Kasai et al., 2024; Xu et al., 2024). This represents a pragmatic middle ground: the underlying language model remains time-agnostic in its architecture, but the retrieval mechanism provides explicit temporal grounding. However, temporal RAG inherits fundamental limitations from both retrieval-augmented

now woven into them differently for each token, the attention patterns that emerge at each stage become increasingly sensitive to temporal context. Through this accumulation of transformations across the model's depth, tokens develop temporally-conditioned attention patterns as the representations learn temporal-contextual associations (for instance, "snow" in late-year contexts attending more strongly to "rain"). By contrast, temporal attention places time directly into the attention computation itself at every layer, explicitly modulating the strength of word-to-word connections based on temporal context from the start. Both approaches can produce time-sensitive behavior, but they differ in mechanism: time embeddings allow temporal patterns to emerge through the progressive transformation and mixing of time-aware representations, while temporal attention architecturally enforces temporal conditioning at each step. Time embeddings are straightforward to implement and computationally less expensive, though potentially less expressive than architectural temporal attention; the author plans experiments to weigh these tradeoffs through direct comparison.

19 Roccabruna et al. (2024) have shown that off-the-shelf (that is, with no modification to handle time and temporal drift) BERT-like encoders, not very surprisingly, perform better in the time-related task of Temporal Relation Classification than instruct models.

approaches and prompt-based conditioning: the system can only condition on information present in the retrieved documents, temporal reasoning remains confined to what can be expressed through prompt engineering, and the model's internal representations still lack the structured temporal priors that architectural conditioning could provide. Moreover, applications requiring fine-grained historical accuracy may find temporal RAG insufficient. While retrieval potentially provides temporally-accurate facts, models impose contemporary linguistic patterns that carry contemporary modes of reasoning, categorization, and explanation, rather than the period-specific ways of understanding that shift fundamentally across time.

In what follows, I briefly reflect on whether and how foundation time-conditioned decoder models—and chat and instruct systems tailored to historical analysis and interpretation derived from them—could improve upon the current situation, examining both the potential benefits and the obstacles that would need to be overcome as the requirements and challenges of such systems differ substantially from those of encoder-based downstream tasks.

The focus on decoder architectures reflects the requirements of historical work, which often centers on synthesis, narrative construction, and interpretation. Decoder models provide a natural interface for these tasks through their next-token prediction objective, which aligns with progressive, context-sensitive text generation. While encoder models could in principle also be developed at foundation scale and remain valuable for representation-focused tasks like classification or retrieval, decoders offer a more direct path toward systems capable of directly assisting open-ended historical interpretation and writing via direct dialogue.[20]

Foundation models are designed to be as comprehensive as possible, trained on vast and diverse corpora to develop broad linguistic and world knowledge. For historical applications, this would necessarily include as much historical material as feasible—both primary sources and secondary literature spanning multiple periods and languages. This comprehensiveness amplifies rather than resolves the problems of the current time-integrated approaches described above and creates a data curation problem if it comes to building a time conditioned foundation model.

Consider a straightforward example: a diary entry reading "Today we celebrated the King's coronation with great festivities." Without an explicit date, the model faces fundamental ambiguity. Is this George IV in 1821, Edward VII in 1902, George VI in 1937 or some other coronation entirely? The text itself carries no temporal marker, yet understanding it and reusing the information it supplies requires precise temporal grounding. While surrounding context might provide clues, inferring dates for use in downstream reasoning from such indirect evidence alone remains, as already argued above, unreliable, especially when the model must process millions of similar documents across centuries. This implies that explicit time grounding is not merely helpful but imperative for foundation models intended for historical work, so that temporal context—central to all historical analysis—is not left to be deduced merely from statistical patterns.

20 While encoder models could in principle scale to foundation size, research investment has concentrated on decoder architectures due to perceived versatility (Radford et al., 2019; Kaplan et al., 2020), leaving large-scale encoders largely unexplored.

However, attempting to build a time conditioned foundation model creates a problem. Current foundation models benefit from self-supervised learning on unlabeled text at massive scale, requiring minimal data curation beyond basic filtering.[21] Time-conditioned models, however, demand temporally-labeled data, raising all sorts of problems historians know all too well. Manual labeling at the scale required for foundation models is impractical; automation through metadata extraction offers a path, supplemented perhaps by temporal prediction models for sources lacking explicit dates (Boldsen & Wahlberg. 2021). Yet, even automated approaches face formidable practical and conceptual challenges. What timestamp should be assigned when a secondary source extensively quotes primary material—do we label by the citing text's date or the cited content's origin or do we split? Reprints pose similar difficulties: metadata often records publication date rather than composition date, conflating temporal layers. Translations multiply the problem: the 1797 English translation of Aristotle's *Nicomachean Ethics* by John Gillies reflects late 18th-century English linguistic patterns while conveying intellectual content from the 4th century BCE—which temporal frame conditions the model's representations? Uncertain or approximate datings, increasingly common and problematic for pre-print materials, further complicate systematic labeling. This represents a substantial technical and conceptual hurdle on the path to a time-conditioned foundation model, and it remains questionable whether it can be overcome especially with the limited resources available to the disciplines involved.[22]

If such a foundation model would nevertheless be built it could of course be exploited straightforwardly: simply condition it on the relevant time period so that temporally-appropriate patterns and associations are activated for the task at hand. For instance, when performing named entity recognition on texts mentioning "Jordan," a model conditioned to a time before 1946 would be expected to identify it primarily as a river, whereas conditioning to 1950 onward would be expected recognize it as a country

However, even if we should be able to build a time-conditioned foundation model, the path to a true AI assistant capable of supporting open-ended historical analysis remains unclear. Such analysis and interpretation requires far more than period-appropriate pattern recognition in sources: it demands historiographical awareness, source

21 By "foundation models," we follow Bommasani et al. (2021): models trained on broad data using self-supervision at scale.

22 Advancing toward time-conditioned foundation models and AI assistants for historical analysis requires comprehensive benchmarks and temporally-labeled datasets. Recent work provides crucial infrastructure: HistoryBank (Mandal et al., 2025) offers a multilingual database of 10M+ historical events with unprecedented temporal depth and linguistic breadth (10 languages), addressing scale and coverage limitations in existing temporal reasoning datasets. EvolveBench (Zhu et al., 2025) evaluates temporal awareness across cognition, awareness, trustworthiness, understanding, and reasoning dimensions. Test of Time (Fatemi et al., 2024), TRAM (Wang and Zhao, 2024), TIME (Wei et al., 2025), and Time-Bench (Liu et al., 2025) provide additional temporal reasoning benchmarks. However, these resources primarily target temporal reasoning and factual retrieval rather than historiographical competencies—source criticism, interpretive synthesis, navigating competing narratives—essential for scholarly historical work. Developing datasets capturing these scholarly practices, alongside temporally-labeled historical corpora at foundation-model scale, remains a formidable challenge that must be addressed for time-conditioned AI assistants to progress from aspiration to reality.

criticism, the ability to navigate competing interpretations, and meta-reflection on the constructed nature of historical knowledge itself. How to build systems with these capabilities is fundamentally uncertain.

An idea worth exploring could be to enable the model to recursively query itself with different temporal parameters during inference—a form of temporal self-steering for historical reasoning. When analyzing a historical source, the assistant could condition itself on the source's temporal context to extract period-appropriate patterns and meanings, then shift to later temporal frames to access historiographical interpretations that emerged over time, and finally synthesize these temporally-distinct perspectives. This temporal self-prompting would implement a form of diachronic reasoning, with the model navigating its own time-conditioned representations to construct layered historical understanding. Such an architecture would require not merely a time-conditioned model but explicit mechanisms for temporal parameter manipulation during generation—possibly through reinforcement learning from human feedback (Ouyang et al., 2022) and chain-of-thought reasoning (Wei et al., 2022) that treats time as a controllable reasoning dimension.[23] Spelling out and implementing this sketched idea would be an enormous challenge, yet this would still only be half the battle: building systems capable of genuine historical interpretation would demand far more, including historiographical awareness, source criticism capabilities, and meta-cognitive reflection on the construction of historical knowledge.

Given the possibly insurmountable problems of curating time-labeled historical data at the scale required for a foundation model, and given that it remains completely unclear how to transform such a model into a functioning assistant—a challenge whose scale is clear even if the path forward is not—an AI assistant for serious historical work remains a distant vision. Yet, a vision can sometimes invite visionary solutions and at the same time set direction and evaluation criteria for practical, near-term efforts.

Even if an AI assistant for comprehensive historical work remains in the distant future, historians working with more targeted applications involving historical sources must recognize that current LLM's integration over time can be problematic in this setting. Time conditioning offers a solution, and existing approaches have been surveyed above. Therefore, to ensure methodological soundness and mitigate the inherent risks of anachronism, these time-conditioning approaches should be incorporated as a standard practice for targeted historical applications. The continued development and refinement of such methods is moreover welcome for advancing the rigorous use of AI in disciplines grounded in historical texts.[24]

23 Time-R1 (Liu et al. 2025) provides evidence that treating time as a controllable reasoning dimension through RL is being actively pursued, using a three-stage RL curriculum with dynamic rewards to build comprehensive temporal reasoning capabilities in language models, achieving state-of-the-art performance on future event prediction and creative temporal scenario generation.

24 This chapter was written with support from large language models (LLMs). All model-generated text was reviewed and, where necessary, rewritten by the authors, who remain fully responsible for the final version. For details on the use of LLMs in this volume, see the statement in the volume's introduction.

References

Agrawal A, Suzgun M, Mackey L, et al. (2024) Do Language Models Know When They're Hallucinating References? Findings of the Association for Computational Linguistics: EACL 2024, pp. 912–928.

Beck C and Köllner M (2023) GHisBERT – Training BERT from scratch for lexical semantic investigations across historical German language stages. Proceedings of the 4th Workshop on Computational Approaches to Historical Language Change (LChange 2023), Singapore, pp. 33–45. Association for Computational Linguistics. doi:10.18653/v1/2023.lchange-1.4.

Boldsen S and Wahlberg F (2021) Survey and reproduction of computational approaches to dating of historical texts. Proceedings of the 23rd Nordic Conference on Computational Linguistics (NoDaLiDa), Reykjavik, Iceland (Online), pp. 145–156. Linköping University Electronic Press, Sweden.

Bommasani R, Hudson DA, Adeli E, et al. (2021) On the Opportunities and Risks of Foundation Models. arXiv preprint arXiv:2108.07258.

Chalef D, Smith B and Lipman R (2025) Zep: A Temporal Knowledge Graph Architecture for Agent Memory. arXiv preprint arXiv:2501.13956.

Chu Z, Gao P, Tang Z, et al. (2024) TimeBench: A Comprehensive Evaluation of Temporal Reasoning Abilities in Large Language Models. Proceedings of ACL 2024 (Long Papers), pp. 721–738.

Devlin J, Chang M-W, Lee K, et al. (2019) BERT: Pre-training of Deep Bidirectional Transformers for Language Understanding. Proceedings of NAACL-HLT 2019, pp. 4171–4186.

Dhingra B, Cole JR, Eisenschlos JM, et al. (2022) Time-Aware Language Models as Temporal Knowledge Bases. Transactions of the Association for Computational Linguistics 10: 257–273. doi:10.1162/tacl_a_00459.

Dubossarsky H, Hengchen S, Tahmasebi N, et al. (2019) Time-Out: Temporal Referencing for Robust Modeling of Lexical Semantic Change. Proceedings of the 57th Annual Meeting of the Association for Computational Linguistics (ACL 2019), pp. 457–470.

Fang H, Tao S, Chen N, et al. (2025) Do Large Language Models Favor Recent Content? A Study on Recency Bias in LLM-Based Reranking. arXiv preprint arXiv:2509.11353.

Faro R, Fan D, Alphaidze T, et al. (2025) TiMoE: Time-aware mixture of language experts. arXiv preprint arXiv:2508.08827.

Fatemi B, Kazemi M, Tsitsulin A, et al. (2024) Test of time: A benchmark for evaluating LLMs on temporal reasoning. arXiv preprint arXiv:2406.09170.

Gurnee W, Ramesh P, Bau D, et al. (2024) Language Models Represent Space and Time. Proceedings of ICLR 2024, pp. 1–15.

Hamilton WL, Leskovec J and Jurafsky D (2016) Diachronic Word Embeddings Reveal Statistical Laws of Semantic Change. Proceedings of the 54th Annual Meeting of the Association for Computational Linguistics (ACL 2016), Long Papers, Berlin, Germany, pp. 1489–1501. doi:10.18653/v1/P16-1141.

He S, Lv L, Manela A, et al. (2025) Chronologically Consistent Large Language Models. arXiv preprint arXiv:2502.21206.

Hosseini K, Beelen K, Colavizza G, et al. (2021) Neural Language Models for Nineteenth-Century English. Journal of Open Humanities Data 7: 22, 1–6. doi:10.5334/johd.48.

Hu R, Li S and Liang S (2019) Diachronic Sense Modeling with Deep Contextualized Word Embeddings: An Ecological View. Proceedings of the 57th Annual Meeting of the Association for Computational Linguistics (ACL 2019), Florence, Italy, pp. 3899–3908.

Jain R, Sojitra D, Acharya A, et al. (2023) Do Language Models Have a Common Sense regarding Time? Revisiting Temporal Commonsense Reasoning in the Era of Large Language Models. Proceedings of EMNLP 2023, pp. 6750–6774.

Kaplan J, McCandlish S, Henighan T, et al. (2020) Scaling Laws for Neural Language Models. arXiv preprint arXiv:2001.08361.

Kasai J, Kasai K, Sakaguchi K, et al. (2024) It's About Time: Incorporating Temporality in Retrieval Augmented Language Models. arXiv preprint arXiv:2401.13222.

Lazaridou A, Kuncoro A, Gribovskaya E, et al. (2021) Mind the Gap: Assessing Temporal Generalization in Neural Language Models. NeurIPS 2021 34: 29348–29363.

Liu Z, Wang G, Feng Z, et al. (2025) Time-R1: Towards comprehensive temporal reasoning in LLMs. arXiv preprint arXiv:2505.13508.

Manjavacas Arevalo E and Fonteyn L (2021) MacBERTh: Development and Evaluation of a Historically Pre-trained Language Model for English (1450–1950). Proceedings of the Workshop on Natural Language Processing for Digital Humanities (NLP4DH 2021), NIT Silchar, India, pp. 23–36. NLP Association of India (NLPAI).

Mandal B, Khandelwal A and Gupta M (2025) HistoryBankQA: Multilingual temporal question answering on historical events. arXiv preprint arXiv:2509.12720.

McGillivray B, Nanni F and Beelen K (2024) Why Does Digital History Need Diachronic Semantic Search? In: Johnson JM, Mimno D and Tilton L (eds) Computational Humanities (Debates in the Digital Humanities, Manifold online edition). University of Minnesota Press. Available at: dhdebates.gc.cuny.edu/read/computational-humanities-5c64bbab-d7ca-41be-8f87-f26117a9a20f/section/e36d5f05-57ab-436c-90e2-2ec9f87683a8 (accessed 16 October 2025).

Montariol S, Martinc M and Pivovarova L (2021) Scalable and Interpretable Semantic Change Detection. Proceedings of the 2021 Conference of the North American Chapter of the Association for Computational Linguistics: Human Language Technologies (NAACL-HLT 2021), pp. 4642–4652. doi:10.18653/v1/2021.naacl-main.369.

Ouyang L, Wu J, Jiang X, et al. (2022) Training language models to follow instructions with human feedback. arXiv preprint arXiv:2203.02155.

Pang C, Jiang X, Kalluri KS, et al. (2021) CEHR-BERT: Incorporating temporal information from structured EHR data to improve prediction tasks. Proceedings of Machine Learning for Health (ML4H 2021), Proceedings of Machine Learning Research 158: 239–260.

Qiu W and Xu Y (2022) HistBERT: A Pre-trained Language Model for Diachronic Lexical Semantic Analysis. arXiv preprint arXiv:2202.03612.

Qiu Y, Zhao Z, Ziser Y, et al. (2024) Are Large Language Models Temporally Grounded? Proceedings of NAACL 2024 (Long Papers), pp. 7048–7068.

Radford A, Wu J, Child R, et al. (2019) Language Models are Unsupervised Multitask Learners. OpenAI Technical Report, San Francisco, CA.

Roccabruna G, Rizzoli M and Riccardi G (2024) Will LLMs Replace the Encoder-Only Models in Temporal Relation Classification? Proceedings of the 2024 Conference on Empirical Methods in Natural Language Processing, Miami, Florida, USA, pp. 20402–20415. Association for Computational Linguistics.

Rosin GD and Radinsky K (2022) Temporal Attention for Language Models. Findings of the Association for Computational Linguistics: NAACL 2022, pp. 1498–1508.

Rosin GD, Guy I and Radinsky K (2022) Time Masking for Temporal Language Models. Proceedings of the 15th ACM International Conference on Web Search and Data Mining (WSDM 2022), pp. 833–841. doi:10.1145/3488560.3498529.

Sun H, Xu Y and Rostami M (2025) How New Data Permeates Language Model Knowledge (and How to Dilute It). arXiv preprint arXiv:2508.00921.

Tiblias F, Bigoulaeva I, Niu J, et al. (2025) Shape Happens: Automatic Feature Manifold Discovery in LLMs via Supervised Multi-Dimensional Scaling. arXiv preprint arXiv:2510.01025.

Vashishtha S, Poliak A, Lal YK, et al. (2020) Temporal Reasoning in Natural Language Inference. Findings of the Association for Computational Linguistics: EMNLP 2020, pp. 4070–4078.

Voita E, Serdyukov P, Sennrich R, et al. (2019) Analyzing Multi-Head Self-Attention: Specialized Heads Do the Heavy Lifting. Proceedings of the 57th Annual Meeting of the Association for Computational Linguistics, pp. 5811–5821. doi:10.18653/v1/P19-1580.

Wang J, Jatowt A and Cai Y (2024) Towards Effective Time-Aware Language Representation: Exploring Enhanced Temporal Understanding in Language Models. arXiv preprint arXiv:2406.01863.

Wang J, Jatowt A, Yoshikawa M, et al. (2023) BiTimeBERT: Extending Pre-Trained Language Representations with Bi-Temporal Information. Proceedings of the 46th International ACM SIGIR Conference on Research and Development in Information Retrieval (SIGIR '23), Taipei, Taiwan, ACM, 10 pp. doi:10.1145/3539618.3591686.

Wang Y and Zhao Y (2024) TRAM: Benchmarking temporal reasoning for large language models. Findings of the Association for Computational Linguistics: ACL 2024, Bangkok, Thailand, pp. 6389–6415. Association for Computational Linguistics.

Wei J, Wang X, Schuurmans D, et al. (2022) Chain-of-Thought Prompting Elicits Reasoning in Large Language Models. arXiv preprint arXiv:2201.11903.

Wei S, Cheng H, Liu Y, et al. (2025) TIME: A multi-level benchmark for temporal reasoning of LLMs in real-world scenarios. arXiv preprint arXiv:2505.12891.

Xiong S, Payani A, Kompella R, et al. (2024) Large Language Models Can Learn Temporal Reasoning. Proceedings of ACL 2024 (Long Papers), pp. 10399–10418.

Xu R, Li T, Wang Y, et al. (2024) Time-Sensitive Retrieval-Augmented Generation for Question Answering. Proceedings of the 33rd ACM International Conference on Information and Knowledge Management (CIKM '24).

Zhao B, Brumbaugh Z, Wang Y, et al. (2024) Set the Clock: Temporal Alignment of Pretrained Language Models. Findings of the Association for Computational Linguistics: ACL 2024, pp. 15015–15040.

Zhu T, Liu Q, Pang L, et al. (2024) Beyond memorization: The challenge of random memory access in language models. arXiv preprint arXiv:2403.07805.

Zhu Z, Liao Y, Chen Z, et al. (2025) EvolveBench: A comprehensive benchmark for assessing temporal awareness in LLMs on evolving knowledge. Proceedings of the 63rd Annual Meeting of the Association for Computational Linguistics (Volume 1: Long Papers), July 27–August 1 2025, pp. 16173–16188. Association for Computational Linguistics.

Zichert M, Simons A and Wüthrich A (2025) Expanding Conceptual Histories: Using Contextualized Word Embeddings for the History and Philosophy of the Virtual Particle Concept. Computational Humanities Research, e16. doi:10.1017/chr.2025.10013.

Generative LLMs and history research

Limitations for languages, periods, and tasks

Fernanda Olival, Helena Freire Cameron, António Branco, and Renata Vieira

1. Introduction

This chapter aims to examine limitations and challenges for the use of generative large language models (GLLMs) in historical research, as encountered by the authors in their particular experience. The discussion is organized into three issues: the status of Portuguese as a pluricentric language exhibiting linguistic variation, the characteristics of the eighteenth-century Portuguese language, which pose constraints to computational processing due to their specificities, and the natural language task of named entities recognition (NER).

GLLMs are artificial neural network models designed to predict the next word in a sequence. Through large-scale exposure to text, they learn probabilistic patterns of language use—such as the likelihood that *"taking into"* is followed by *"account"*—and thus acquire the ability to generate coherent and contextually relevant text. When integrated into conversational systems, or chatbots, GLLMs can answer questions, summarize information, and perform a wide variety of linguistic tasks.

Despite their growing applications across many domains, their use in historical research remains limited, as also discussed by Büttner (2026). History is a discipline fundamentally shaped by discourse and dominantly reliant on textual sources, though other types of material evidence may also be considered. The use of GLLMs as a tool for historical research remains in its early stages. There are bibliography reporting experiments that attempt to recreate the past using GLLMs fine-tuned on the COHA (*Corpus* of Historical American English) or on books published between 1880 and 1914, but with rather unreliable results, as presented by Underwood et al (2025). Also, see Wolf (2026), about the early modern multilingual project VERITRACE. A critical engagement with both practical and conceptual issues is essential to fostering meaningful progress in the field. Both technical and epistemological issues must be carefully considered in order to make improvements.

The discussion that follows is based on our experience with the transcription, normalization, and semantic annotation of an eighteenth-century Portuguese *corpus* – the

Memórias Paroquiais (*Parish Memories*). This source was produced by multiple authors between 1758 and 1761 as part of a large-scale survey conducted in the aftermath of the devastating earthquake of November 1st, 1755. In 1758, the Portuguese Crown distributed a detailed questionnaire to all parish priests in the country, structured around three main thematic areas (the land, rivers, and mountains), seeking information about the territory, its inhabitants, and the consequences of the catastrophe, with the broader goal of compiling a comprehensive *Geographical Dictionary* of Portugal. In the 19th century, the handwritten responses were compiled into 41 volumes, later supplemented by two additional volumes and an index volume. Today, the collection is preserved in the National Archive of Torre do Tombo, in Lisbon. Transcribed and digitized versions of the *Memórias Paroquiais* pertaining to southern Portugal are available through the data aggregator **CIDEHUSdigital**[1], under a Creative Commons Attribution 4.0 license. This collection has been further enriched with semantic annotations identifying Named Entities (NE), providing a valuable resource for computational and historical research alike.

The next sections are structured around three challenges faced by GLLMs, contextualized on the work developed towards the *Parish Memories*. GLLMs, being designed for general-purpose use and based on contemporary texts, are still limited to dealing with specific languages, earlier linguistic stages of languages, and tasks requiring formally strictly structured representations.

2. Challenges due to specific languages: the case of Portuguese

The Portuguese language has undergone a long and complex process of evolution, shaped by multiple cultural and linguistic influences. While Portuguese was still closely related to Galician in the 12th century, it became a separate linguistic system. From the 15th century, the Portuguese maritime expansion brought the language to Africa, America and Asia. Portuguese served as a tool of commerce, diplomacy, and colonial governance in these areas, assimilating lexical and structural elements from various contact languages. The result of this process is a diverse linguistic legacy (see Mateus and Bacelar do Nascimento, 2005).

Today, Portuguese is spoken by approximately 260 million people (Instituto Camões, 2022), making it the fourth most spoken native language worldwide, after Mandarin, English, and Spanish. It is the official or national language of countries across several continents, including Europe, South America, Africa, and Asia. It is a "pluricentric language", comprising multiple standardized varieties that coexist within a single linguistic system. Two main patterns – European Portuguese and Brazilian Portuguese – dominate globally, while additional national standards are emerging in African contexts. Despite a high orthographic convergence achieved through reform, the linguistic and cultural trajectories of Portugal and Brazil have diverged considerably. Differences in vocabulary, syntax, and usage reflect broader social and cultural distinctions between the two societies. This internal diversity represents both an asset and a challenge for computational linguistics and language technology.

1 https://www.cidehusdigital.uevora.pt

In the digital era, the distribution of Portuguese language resources mirrors existing global inequalities. Brazil's demographic scale and technological infrastructure ensure a significantly greater online presence, while European and African Portuguese varieties remain underrepresented in most digital corpora. This digital asymmetry has critical implications for computational models, which learn linguistic patterns from large-scale online text data. As a result, most existing models tend to reflect the lexical, syntactic, and cultural characteristics of Brazilian Portuguese, marginalizing other varieties. This imbalance affects not only language representation but also the inclusivity and fairness of AI applications. Addressing these disparities is essential to ensure that future language technologies reflect the full linguistic, and cultural diversity of the Lusophone world.

Currently, we see a profusion of efforts to develop GLLMs, both worldwide and within countries.

There are multilingual models, and others are specialised for a single language. The same proliferation can be seen within models specialized for Portuguese. The Gervásio models are available with 1 and 8 billion parameters.[2] GlórIA[3] is another generative model with 1 billion parameters. There are also Portuguese models developed in Brazil, such as Sabiá[4], AmazoniaIA[5], and SoberanIA[6].

If we consider the full context large language model (FLLMs) (see Simons et al., 2026), there are other options. The Portuguese Albertina family[7] includes full context models of different sizes, with 100 and 900 billion parameters. Developed in Brazil, we have Bertimbau[8], and even domain specialized models such as LegalBERTpt[9].

While GLLMs are good in text generation tasks, such as question answering, and machine translation, where they, based on an input sentence, continue by generating a sequence of words, FLLMs, which are lighter, simpler models, perform best in classification tasks, such as named entity recognition (NER), sentiment analysis, and others.

These Portuguese LLMs (generative and full-context) can be downloaded from the national repository of PORTULAN CLARIN[10], the Portuguese National Research Infrastructure for Science and Technology of Language. The international repository, Hugging Face, is another platform where models for the Portuguese language can be found.

The recent development of GLLMs for Portuguese extends beyond technical concerns; it also embodies cultural and epistemological implications. Language models encode not only grammar and vocabulary but also the social and cultural contexts embedded in their training data. This observation underscores the need for inclusive approaches to data collection and model training. Ensuring that African, Asian, diasporic, and past varieties of Portuguese are represented in computational corpora is essential for achieving a truly global and equitable digital presence.

---

2 https://huggingface.co/PORTULAN/gervasio-8b-portuguese-ptpt-decoder

3 https://nova-lincs.di.fct.unl.pt/gloria-the-new-portuguese-european-large-language-model/

4 https://www.maritaca.ai/

5 https://amazoniaia.com.br/

6 https://sia.pi.gov.br/projetos/soberania/

7 https://portulanclarin.net/models/

8 https://github.com/neuralmind-ai/portuguese-bert

9 https://huggingface.co/raquelsilveira/legalbertpt_fp

10 https://portulanclarin.net/

In the rapidly developing field of artificial intelligence, it is crucial that the entire diversity of the Portuguese language be visible, usable, and appropriately represented, even though these methods are better at expressing synchronic rather than diachronic issues. This corroborates Simons et al. (2026) who indicate the need of LLMs literacy for the humanities researchers.

3. Challenges for specific periods: the case of the 18th century

Languages also have variation across time. Crossing different historical linguistic periods, the 18th-century stage of the language, although modern, is still distinct from the contemporary stage. Textual sources are not always in digital support, especially in languages with great textual memory, such as Portuguese.

Strict transcription guidelines are required for transcriptions. These guidelines are not always the same. This is true even when paleographers decide either to transcribe with absolute fidelity to the original or to update the language to a contemporary pattern. What to do with old words still in use, erratic use of semi-capital and capital letters, letters out of Roman pattern used as numerals[11], change of geographical names in time, abbreviations, the long 's', and other constraints? Also, transcriptions made by several human transcribers may have different outputs, bringing inconsistency to the *corpus* collected.

In the 18th century, regarding the linguistic stage of the language, most of the phonetic, phonological, syntactic, and other changes were already concluded or in consolidation, as discussed by (Cardeira, 2006, Castro, 2006, and others equally pertinent to this question). However, spelling was out of pattern till the beginning of the 20th century. We find many pseudo-etymological digraphs (e.g., -th-), non-etymological double consonants (e.g., -ll-), use of "u" with consonantal value, "i"/"j" as allographs, or "y" as a semi-vowel, as presented by Cameron et al (2023) regarding the *Parish Memories corpus*. In the same text or the same *corpus*, the same word can be written in several ways. For example, in the cited *corpus*, the word "Assunção" [Assumption] has 20 variants.

So far, we have manually transcribed and normalized the *corpus* under study, aiming to diminish the effects of these issues and increase accuracy in automatic tasks. For that, we developed a conservative approach, trying to perform the minimal changes to maintain fidelity to the original. We maintained all linguistic variation, old words still in use, punctuation, but we updated spelling to the contemporary pattern. All this path, with constraints at several levels (transcription, variation, normalization) necessary for textual sources preparation, still can not be directly solved with available GLLMs, as far as we know.

GLLMs are mostly based on contemporary versions of the language. Although they do present an adaptability capacity, in our research, we observed some difficulties. GLLMs need to be trained on large volumes of texts from different historical periods; however, such texts are not available in sufficient quantity or in a condition suitable for large-scale

11 E.g. 'R' equal to 40.

processing. For that, all kinds of materials should be considered, including administrative, legal, and everyday documents, to represent the full social spectrum rather than privileging elite voices, avoiding bias as much as possible.

4. Challenges considering specific tasks: the case of Named Entity Recognition

Named entity recognition (NER) is a Natural Language Processing (NLP) task that helps historians to answer the questions, Who? Where? What? When? The main categories usually taken into account for this task are PERSON, PLACE, ORGANIZATION, and TIME. For the purpose of historical research, these categories might need proper specialization. In our previous work, the categories were subdivided into several subcategories. We must always have in mind that we are dealing with a past society, necessarily different from the contemporary one: it was a hierarchical society, ruled by privileges, where inequality is assumed in the law. For that, for example, the PERSON category was subdivided into Saints, Social Category, Name, Occupation, Author, and others (Vieira, 2025), trying to better describe the reality in its complexity. In addition, annotation was carried out with the broader context in mind, considering the larger expression to assist with homonymy and enable disambiguation.

Annotated corpora are useful resources. The manual annotation process, however, is quite demanding, so this has become an important NLP task. NER is mainly based on machine learning models. The first approaches were based on predefined features. As the learning models evolved, less and less feature pruning was required. The models were developed in a way that the features could be captured entirely from observed examples. For the development of these systems, the existence of human made datasets are required to be able to train the models (see Zang and Colavizza, 2025).

The new developments in GLLMs, however, did not prove to bring improvements in this particular task. Following the fast evolution of the language models themselves, we have been testing available models, both generative and full-context, aiming for the improvement of the NER task performance (Nunes et al., 2025). Our last best result was achieved with Albertina, a full-context BERT-like model specifically trained on European Portuguese, and not by a generative model.

The limitations of GLLMs here may be connected to the fact that NER is a sequence labeling task. It performs a classification of tokens based, among other things, on their position in a sentence. Each token is identified as pertaining to a named entity or not, and a category for those identified as named entities has to be pointed out. In the sequence, we present an example of the required output for the NER task, where each token in a sentence is classified according to predefined categories; 'B' indicates the beginning of an entity, 'I' is the continuation of the identified expression (inside), 'O' indicates the token is not referring to a named entity (outside). In the following example in Table 1, the general category is place (PLC), specialized for facility (FAC).

Table 1: Example NER output

se	O	Terceiros	I-PLC_FAC
vai	O	de	I-PLC_FAC
seguindo	O	São	I-PLC_FAC
até	O	Francisco	I-PLC_FAC
o	O	,	O
convento	B-PLC_FAC	sempre	O
de	I-PLC_FAC	subindo	O
Santa	I-PLC_FAC	acima	O
Clara	I-PLC_FAC	até	O
e	O	o	O
igreja	B-PLC_FAC	castelo	O
dos	I-PLC_FAC		

GLLMs are language generators. They present very well-written sequences which, however, may describe situations that do not correspond to known facts. This has been called hallucinations, and this is due to the "creativity" abilities of these models. This presents specific challenges for historical analysis, which is grounded in empirically verified data, statistically supported patterns, or qualitatively established dominances – never in made-up information. Consequently, the risk that these models may prove unreliable for historians is significant. In the cases where the expected result has to follow a strict structure, such as the one required for the NER task, this creativity capacity of the model may interfere in a negative way. Another relevant difference, here, between the two types of models is that full-context models have bidirectional information about each word in the sequence. They work by seeing both left and right contexts of each word, which especially affords tasks such as NER.

Whereas GLLMs may present excellent performance for question-answering, dialogue, language translation, and reinterpretation, some specific tasks may not yet be best attended by these powerful models. Nevertheless, we are aware that this technology rapidly evolves, and it might be the case that the next models may overcome current limitations.

5. Final remarks

In historical analysis, both the time variable and context are indispensable. Besides questions of duration, time must be understood as the condition for dynamism and change, while context reflects the inherent complexity, diversity, and heterogeneity that characterise social and natural phenomena. Since history is a form of discourse fundamentally grounded in texts and deeply shaped by the temporal dimension—which inherently brings change—it is crucial to acknowledge these challenges to advance historical analysis. With the arrival of GLLMs, the ability of dealing with information coming from vast amounts of text was increased, influencing the way research is conducted by historians, linguists and others.

However, these models, as seen before, still have limitations in dealing with the past dimension. For the studies performed by the authors of this chapter, the main limitations are due to specific language, time and task. Critical approaches are needed for reflecting on the present and the future—without losing sight of the past as, we believe,societies without memory cannot survive. Currently, these models, as far as we know, do not adequately address the need for diachronic variation, a crucial aspect for digital history.

Having AI tools tailored to languages and their variants has become essential. This includes the creation of specialized training corpora for earlier stages of particular languages, the development of domain-specific fine-tuning strategies, and hybrid systems that can enforce formal constraints. Only through such targeted efforts GLLMs can become genuinely useful tools for historical linguistics and digital history. Until then, their use must remain critically mediated by human expertise, ensuring that automated processes do not obscure the interpretive depth and contextual sensitivity that historical research demands. Multidisciplinary teams are the only ones who can operate this change.[12]

Acknowledgements

This work is funded by CIDEHUS-University of Évora and the Foundation for Science and Technology (Portugal), under the project UID/00057/2025 - https://doi.org/10.54499/UID/00057/2025

References

Büttner J (2026) Why pursue temporally-grounded AI for historical disciplines, and what makes it so challenging? In: Simons A, Wüthrich A, Zichert M, et al. (eds) *Understanding Science with Large Language Models? Potentials for the History, Philosophy, and Sociology of Science*. Bielefeld: transcript, part-2.

Cameron HF, Olival F, Vieira R (2023) Planear a normalização automática: tipologia de variação gráfica do corpus das Memórias Paroquiais (1758), *LaborHistórico*, Rio de Janeiro, ISSN 2359–6910, vol.9, n.1. DOI: 10.24206/lh.v9i1e52234

Cameron HF, Olival F, Vieira R (2022) INCEpTION-Use Cases: Portuguese Parish Memories (1758–1761), available at: https://inception-project.github.io/use-cases/parish-memories/

Cardeira E (2006) *O essencial sobre a História do Português*. Alfragide: Editorial Caminho.

12 This chapter was written with support from large language models (LLMs). All model-generated text was reviewed and, where necessary, rewritten by the authors, who remain fully responsible for the final version. For details on the use of LLMs in this volume, see the statement in the volume's introduction.

Castro I (2006) *Introdução à História do Português*. Lisboa: Colibri, 2nd edition.

Gonçalves MF (2024) Contribuições para o estudo do português falado do século XVIII: o Compendio de Orthografia (1767) de Monte Carmelo". In: Merlan A and Schäfer-Prieß B, Randromania im Fokus. Gesprochenes Galicisch, Portugiesisch und Rumänisch, Lausanne – Berlin – Bruxelles – Chennai – New York – Oxford: Peter Lang, 540 S. (Reihe Romanistische Arbeiten interkulturell und interdisziplinär, Herausgegeben von Rafael Arnold, Thomas Johnen, Aurelia Merlan, Jürgen Schmidt-Radefeldt und Rudolf Windisch), 393–416.

Instituto Camões (2022) Dia Mundial da Língua Portuguesa – 5 maio 2022, available at https://www.instituto-camoes.pt/images/pdf_noticias/Dados_sobre_a_l%C3%ADngua_portuguesa_de_2022.pdf

Mateus MHM and Bacelar do Nascimento F (2005) A mudança da língua no tempo e no espaço, In: *Língua Portuguesa em Mudança*, Lisboa: Caminho, 13–30.

Nunes RO, Santos J, Spritzer A, et al (2025) Assessing European and Brazilian Portuguese LLMs for NER in Specialised Domains. In: Paes A, Verri FAN (eds) *Intelligent Systems. BRACIS 2024. Lecture Notes in Computer Science*, vol 15412, Springer, Cham, 215–230. ISBN: 978-3-031-79029-4. https://doi.org/10.1007/978-3-031-79029-4_15

Simons A, Zichert M and Wüthrich A (2026) Large language models for history, philosophy, and sociology of science: Interpretive uses, methodological challenges, and critical perspectives. *Studies in History and Philosophy of Science* 117: 102151. https://doi.org/10.1016/j.shpsa.2026.102151.

Underwood T, Nelson LK and Wilkens M (2025) Can Language Models Represent the Past without Anachronism?, arXiv e-prints, Art. no. arXiv:2505.00030, 2025. doi:10.48550/arXiv.2505.00030.

Vieira R, Olival F, Cameron HF, et al (2025) Anotação, análise e aprendizagem de Entidades Nomeadas em textos históricos portugueses (séc. XVIII). *Linguamática, 17*(1): 121:136. Available at: https://linguamatica.com/index.php/linguamatica/article/view/445

Wolf JC (2026) LLMs and multilingual historical corpora in a digital history project. Reflections from the Berlin workshop. In: Simons A, Wüthrich A, Zichert M, et al. (eds) *Understanding Science with Large Language Models? Potentials for the History, Philosophy, and Sociology of Science*. Bielefeld: transcript, part-2.

Zang S and Colavizza G (2025) Named Entity Recognition of Historical Text Via Large Language Model, preprint arXiv:2508.18090v1 [cs.DL] 25 Aug 2025

Zero-shot generation of synthetic historical data with LLMs

Vera Danilova, Julia Reed, Andrew Burchell, Gijs Aangenendt, and Ylva Söderfeldt

1. Introduction

Synthetic data generation offers clear advantages for training classification models in low-resource settings and in domains where data privacy is critical, such as healthcare, finance, and the social sciences. Unlike traditional data augmentation, which modifies existing samples, synthetic data generation produces entirely new observations by specifying controlled input parameters. This enables researchers to test hypotheses and compare methods in a structured, reproducible manner.

Recent research has explored the use of Large Language Models (LLMs) to generate datasets that combine textual and numerical variables. These approaches rely on the model's ability to infer plausible values from variable descriptions provided in prompts. However, plausibility does not guarantee validity. A growing body of work on fairness and bias highlights that LLM outputs may reflect and amplify systematic biases. As shown by Yin et al. (2025), LLMs exhibit consistent bias patterns and failure modes that differ from those observed in human decision-making. Such biases can propagate into synthetic datasets. Moreover, while large-scale data generation is increasingly feasible, rigorous quality evaluation remains limited—even in sensitive domains. For example, Barr et al. (2025) use GPT-4o to generate synthetic vital sign data for surgical patients under both zero-shot conditions and with predefined statistical ranges. Although zero-shot outputs generally fall within plausible bounds, their confidence intervals often fail to align with those of real data, except for identifiers such as patient case IDs. Incorporating explicit parameter constraints improves alignment, as reflected in greater overlap of confidence intervals. While this is a promising result, more detailed analyses of distributional properties and generative patterns are needed to assess how faithfully synthetic data captures underlying structures.

In the social sciences, LLMs are increasingly employed using methodologies adapted from psychology, at times even as "surrogate" participants in experiments. This line of work often assumes that constructs and experimental designs developed for humans can be directly applied to language models. However, recent findings challenge this assumption. Park, Schoenegger, and Zhu (2024), for instance, demonstrate that GPT-3.5 produces limited diversity when simulating survey responses, suggesting that LLM-generated data may fail to capture the heterogeneity of human attitudes and behaviors. A range of social, political, and cultural biases further undermines the reliability of such data as a proxy for human responses.

Overall, there is a lack of comprehensive frameworks for systematically evaluating synthetic data quality across domains, languages, and data types. This gap extends to Small Language Models (SLMs), which are increasingly adopted due to their lower computational cost (e.g., Binici et al., 2025; Wagner et al., 2024; Zhou et al., 2025; Kazdan et al., 2024), despite well-documented limitations in reasoning capacity and reliability (Wang, 2025).

Within natural language processing (NLP), it has been argued that once text generation reaches near-human performance, classification models can be derived directly, reducing the need for synthetic data (Nikolenko, 2019). However, despite substantial advances in fluency, both LLMs and SLMs continue to exhibit uneven performance across languages and domains, and remain prone to hallucinations, particularly in low-resource settings. In such scenarios, fine-tuning with a combination of human-annotated and synthetic data is a common strategy. Yet, for more subjective tasks—such as sarcasm or irony detection—models trained on synthetic data tend to underperform relative to those trained on human-annotated datasets (Li et al., 2023). This suggests that generated examples may lack sufficient diversity or may encode misleading contextual cues.

In this work, we focus on generating synthetic data for the classification of genre in historical medical periodicals, an especially underrepresented domain. A related and underexplored question concerns how the limitations and biases of LLMs affect the representation of temporal and historical information in synthetic data. This issue is broadly relevant, as most datasets contain implicit or explicit temporal dimensions that shape their structure and interpretation. Ensuring that such temporal aspects are accurately represented is essential to prevent the propagation of distortions into downstream applications and future training data.

This paper describes the initial experiments to generate and evaluate a synthetic dataset that is aimed to mimic genres within 20th-century historical medical periodicals in four European languages. We address the following research questions:

- **RQ1:** To what extent are the texts generated by SLMs faithful to the intended historical and stylistic parameters and how does this fidelity vary across different models and generation scenarios?
- **RQ2:** To what extent can SLMs produce diverse textual outputs and how does this diversity differ between models and under varying generation conditions?

To address these questions, we empirically assess the quality of outputs of SLMs (7b–14b) from LLaMA, Qwen, and Mistral families for domain-specific historical context in terms of faithfulness and diversity as established assessment criteria (Gan and Liu, 2025; Long et al, 2025). To evaluate faithfulness, we analyze model behavior across prompt configurations, focusing on generation failure and preference bias. The quality of outputs, particularly in terms of language and plausibility of historical details, is assessed through expert feedback from historians of medicine. Diversity is quantified using proxy metrics, including pairwise cosine distances, the proportion of unique topics, and self-BERTscore (Zhang, 2024).

We offer a cautious perspective on the use of SLMs for synthetic data generation, particularly within the context of historical data. While the quality of the generated data appears sufficient to improve performance on certain benchmark tasks, such data may not faithfully represent historical realities. As a result, including these datasets into pretraining for training large language risks reinforcing existing biases and may ultimately undermine model robustness.

2. Motivation for synthesizing historical genres

Our central hypothesis is that genre distributions across time reflect evolving communicative strategies within our source materials. Thus, genre classification may yield richer historical insights than semantic segmentation alone.

The unlabeled data for classification originates from the ActDisease project (ERC-2021-STG 10104099), which investigates the modern history of European medicine. The dataset consists of periodicals published by ten patient organizations across Europe, spanning the 20th century and covering four languages: Swedish, German, French, and English (Aangenendt et al., 2024). It has only recently been digitized, is not publicly available, subject to copyright restrictions, and not included in the pretraining data of existing LLMs.

The periodicals are diverse in terms of text and layout types, with individual pages often combining multiple genres. The quality of OCR hinders the reliable extraction of distinct texts with their correct reading order. As a result, term frequency counts and topic models tend to be biased toward dominant genres, complicating accurate historical interpretation (Danilova and Söderfeldt, 2025).

Working with the historians in the project, we identified 23 distinct genres: *academic, administrative, advertisement, announcement, appeal, opinion, patient_organization_report, QA, news, legal, invitation, interactive (crossword or puzzle sections), guidance, fiction, personal, medical_case, humour, recipe, medical_recipe, biographical, interview, letter_of_thanks, complaint*. These genres are based on previous work on genres of historical journalism (Harbers, 2014) and genres in the history of medicine (Pomata, 2014), as well as web genre classification (Kuzman and Ljubešić, 2023). However, suitable training and test data is scarce, and for 14 genres, no annotated data is available for training and testing the models: *administrative, announcement, appeal, biographical, complaint, humour, interview, invitation, letter_of_thanks, medical_case, medical_recipe, news, personal, recipe*. This limitation motivated us to investigate the use of language models for synthetic data generation. Rather than producing data solely for classifier training, we systematically examine the generated content across different generation scenarios and evaluate its diversity and faithfulness, which is critical for ensuring the reliability and representativeness of synthetic historical texts.

3. Methodology

The content of domain-specific historical corpora is inherently tied to particular historical timelines. This study explores the ability of LLMs to reflect as many features of these timelines as possible and to not introduce distortions. We use a zero-shot setup where the models are expected to use linguistic and contextual features consistent with the relevant historical timeline, drawing on knowledge acquired during pre-training about general history, the history of medicine, journalistic practices, linguistics, and both historical and modern text genres.

We introduce several generation scenarios (prompt configurations) to assess how they influence both the diversity of generated outputs and their historical reliability. To our knowledge, no existing automatic metrics can adequately assess the accuracy of representations of historical realities—particularly in such a specific domain as patient organization periodicals. Consequently, historians of medicine play a central role in the evaluation process, being able to detect major historical distortions in the representations produced by the models.

In the following, we provide details on the evaluation criteria. Since LLMs have been shown to exhibit various forms of inherent bias, including primacy, recency, and centrality biases (Yin et al., 2025), we additionally assess how effectively the models can produce structured JSON outputs and whether their generated content displays a biased preference for certain years depending on contextual cues. Such analysis provides insights into the biases in model behavior, which may impact both diversity and faithfulness. Furthermore, we outline how the evaluation criteria are applied to address the research questions and describe the rationale for model selection and the design of the generation scenarios.

3.1 Evaluation criteria

Faithfulness. As defined by Long et al. (2024), *faithfulness* refers to outputs that are logically and grammatically coherent, free from factual inaccuracies or irrelevant content. In our case, this is particularly challenging due to the domain-specific and historically grounded nature of the data. Since our downstream goal involves training models on historical materials, it is critical to maintain historical accuracy: real events must not be conflated or misrepresented, and anachronisms must be avoided. In genre classification specifically, such historical inaccuracies may introduce topical shifts that distort the training signal. Previous work (Petrenz, 2011) has shown that shifts in topic distribution can change the behavior of genre classification models. Hence, faithfulness evaluation is helpful for both analyzing and avoiding potential classification errors, as well as for preventing the propagation of historical inaccuracies in future datasets and computational tools trained on them.

The evaluation of faithfulness addresses RQ1 through analyzing 1.A: alignment with factual historical timeline, 1.B: semantic similarity to original dataset, 1.C: accurate reflection of genres specified in input prompt, 1.D: historically plausible language.

The factual accuracy of the text (1.A) is verifiable by reference to the most literal, and most widely agreed on, understanding of "what happened" in the history of medicine—the passing of specific legislation, the development of medical technologies, the social and institutional histories of physician-patient relationships, etc. This information is available in textbooks or encyclopedias as historical "facts" that have been communicated by socially accredited historical authorities who have in principle verified the facts in an archive and communicated their historical significance. The facticity of these facts is thus a function of some level of concreteness–whether founding years, major events, and personnel of groups and institutions, the testing, patenting, and circulating of technologies, or less concrete facts such as the emergence or transformation of particular ideas.

Historical plausibility (1.D) concerns whether a text feels believable or authentic within its claimed historical context — even if it's not factually verified. This evaluation focuses on style, tone, worldview, language use, genre conventions, and attitudes that fit the period.

Assessing the plausibility of texts is challenging, since historical sources often defy plausibility—and this may, in fact, characterize what makes them valuable as primary sources. A primary source is an authenticated object from the past that reveals what actually happened, rather than what seems likely or expected to have happened. For instance, a reader letter published in the *Swedish Diabetics Association's* periodical in 1978 describes life with diabetes in strikingly negative terms and even includes curse words—a tone that runs counter to what one might expect from a patient organization's publication and its public messaging at the time (*Från läsekretsen, Diabetes*, vol. 28, issue 1, 1978: 24).

In the present experiment, the experts are fully aware that the synthetic texts are not authentic primary sources. Yet many of these generated outputs could easily be imagined as plausible historical documents—the kind of materials one might encounter in an archive. Importantly, the assessment of their plausibility cannot be separated from the context of the evaluation itself.

The evaluation of model outputs is performed by two historians of medicine from the ActDisease project who specialize in European patient organizations. In the preliminary evaluation phase, the historians were given outputs from LLaMA 3.1 8B, Qwen 2.5 7B, and Qwen 2.5 14B and were asked the following questions: 1) How well do the generated texts reflect the historical genres identified in the periodicals?; 2) How diverse (lexically, structurally) are the generated samples within each genre?; and 3) How well does the generated data fit the selected historical period? Based on these questions, the historians provided a summary of the historical genre fit in the produced synthetic data.

In the main evaluation phase, the historians addressed RQ1 in detail. Given the high cost of expert evaluation at the level of each generated text, the RQ1 sub-questions were provided to historians as a questionnaire to allow for an efficient yet thorough assessment of overall quality. Higher synthetic data quality is defined as a closer alignment between the outputs and the historians' expectations. Each response is given as an integer score within the interval [0, 10], where 0 indicates no alignment with expectations, and 10 indicates the highest possible alignment.

Diversity. The diversity metric captures the variation among the generated data, reflecting differences in text length, topic, or writing style. Higher synthetic data diversity is reported to mitigate distribution collapse and positively influence the performance of fine-tuned models (Schaffelder and Gatt, 2025). In previous literature, it has been observed that instruction-tuned conversational models widely used for synthetic data generation produce outputs with limited semantic diversity. Base models, although not as good at following complex instructions, have been observed to be more semantically diverse in outputs (Zhu et al, 2025; West and Potts, 2025). Evaluating diversity across models and generation scenarios offers valuable insights into the lexical repetitiveness of model outputs and the breadth of topics they cover when they are conditioned on a historical context.

The evaluation of diversity addresses RQ2 through analyzing 2.A: Does diversity differ across generation scenarios as expected from the prompt design? 2.B: Do base models produce more diverse outputs according to the metrics, and what insights emerge from manual inspection?

The following proxy metrics are used to quantitatively assess the diversity of model outputs:

1) pairwise cosine distances between embeddings generated using the Sentence Transformers[1] framework. Each generated text in each language is embedded using the Qwen3-Embedding-4B model which supports multiple languages and long input context (32k tokens). Cosine distances are in range [0,2] (calculated as 1 – cosine similarity on normalized embeddings using sklearn.pairwise_distances). Distributions concentrated near zero indicate lower semantic diversity (i.e., outputs are more similar), while distributions with values farther from zero suggest higher semantic diversity. We visualize these distributions to evaluate whether semantic similarity increases or decreases as a function of specific model or prompt modifications.
2) self-BERTScore (Zhang et al., 2024) where the BERTScore F1 metric (Zhang et al., 2020) is computed separately for each language subset within every model and prompt configuration, across both disease and genre groups. Contextual embeddings were obtained using the bert-base-multilingual-cased model. To achieve this, the data were first grouped by language, model, genre, and disease, and all corresponding text samples were aggregated into lists. Groups containing fewer than two texts were excluded. For each remaining group, pairwise BERTScores were calculated among all texts within that group to assess internal semantic similarity. A score closer to 1.0 indicates greater semantic overlap—i.e., more tokens from the candidate and reference share similar meanings in embedding space. Distributions of pairwise scores concentrated near zero indicate higher semantic diversity. To quantify and compare these patterns we summarize each distribution using its mean and standard deviation.
3) the proportion of unique topics relative to the total number of generated texts for each model and prompt condition. To measure the proportion of unique topics, topic modeling is performed using the BERTopic framework (Grootendorst, 2024), applied to the same set of embeddings as in 1), with GPT-4o serving as the topic representation model.

1 https://huggingface.co/sentence-transformers

Generation Performance. We measure how often the models fail to generate a valid JSON entry across prompt configurations. Systematic failure to generate data for specific years or contexts might indicate underlying biases in the model's training data or limitations in its ability to generalize beyond dominant temporal or linguistic patterns. It may also suggest that certain combinations of input conditions fall outside the model's learned distribution, resulting in non-responses, irrelevant completions, or format violations.

Year and Digit Preference. We hypothesize that the model's year selection patterns will reflect either inherent positional biases, domain-specific knowledge biases (e.g., the model's awareness of events associated with particular years in the context of a given disease), or both. The models are prompted to select a year from the 20th century based on the provided context. The evaluation examines whether the models' year preferences shift depending on the breadth of the context and the presence of references to medicine, disease, or patient organizations. We analyze plots showing the distribution of selected years in the model outputs, where counts are normalized by the total number of outputs for each model and prompt configuration. These distributions are then compared across different models and prompt versions.

To determine whether year preferences remain consistent despite prompt modifications, we introduce several controlled variations:

1. **Removal of domain-specific instructions:** The model receives no guidance to emulate patient organization periodicals for specific diseases. If the year preference is strongly influenced by this instruction, we expect a marked deviation from the baseline distribution.
2. **Temporal range restriction:** The available timeframe is reduced from the entire 20th century to only its first half. This allows us to observe whether the model's preferences are positionally anchored or adapt to the narrower temporal scope.
3. **Repositioning of reasoning summary:** Initially, the chain-of-thought summary appears at the end of the generated output. We move it to the beginning and explicitly instruct the model to reflect on its choices regarding year and topic selection. This variation tests whether encouraging explicit reasoning affects the model's year preferences.

Given the observed numerical preferences of LLMs[2], we hypothesize they will manifest in the model's selection of year during synthetic data generation, especially when the generation is constrained to a predefined value range and conditioned on specific input variables in the prompt.

3.2 Models

To address the research questions outlined above and perform evaluation, we conduct a series of comparisons across two pairs of models with different sizes. The first pair con-

2 E.g., https://sanando.github.io/llmrandom/

sists of Qwen 2.5 7B Instruct and LLaMA 3.1 8B Instruct. Since a LLaMA model of comparable size to Qwen 2.5 14B Instruct is not available, we instead use Mistral Nemo 2407 12B Instruct as the second comparison point. This setup enables us to examine differences both across model architectures and across model sizes within the same architecture.

Since we generate not only text but other information, such as chain of thought (CoT), rationale behind the choice of time period features, year, etc., we use instruct models that are able to consistently generate JSON outputs. Additionally, Instruct-tuned models have been shown to perform better in zero-shot settings across various NLP downstream tasks (Wei et al., 2021), making them particularly suitable for this study.

For the diversity evaluation purposes, we include base models that have not undergone further training to align with human preferences in instruction-following tasks. Specifically, we compare the Qwen 2.5 14B Base model with its corresponding Instruct version. This comparison is intended to investigate whether the base model produces considerably more diverse outputs. This model is capable of producing valid JSON outputs, however, it does so at a considerably lower rate and approximately four times slower than the Instruct version. The Mistral and LLaMA Base models were unable to generate the minimum required number of valid JSON outputs.

3.3 Prompt configurations

We employ six zero-shot prompting scenarios in these initial experiments, applying minimal, targeted modifications to the prompt — specifically avoiding few-shot generation — since our experimental setup involves a long tail of underrepresented classes for which no real examples exist and no representative seeds are available to support few-shot prompting. Another reason is that the models can be severely influenced by the semantics and order of few-shot examples (Lu et al., 2022). The few-shot setup will be explored in the future work.

Baseline Prompt. The baseline prompt that will be further modified is constructed as follows. The system part of the prompt focuses the model on the historical content, specifically, on patient organization magazines and the period of their publication, and stresses the importance of strictly following the instructions:

> You are an AI specializing in historical content synthesis. Your purpose is to generate structured JSON data that emulates authentic documents from 20th-century patient association magazines. Adhere strictly to the provided historical context and output format.

The user prompt further provides the input parameters (contextual variables, such as type of disease, country, etc.), steps of JSON creation and an example of output JSON structure, clearly separating them with formatting attributes. In the baseline prompt, it introduces the following input parameters: *genre, text types associated with it, target text type, country, disease, language.*

A dataset of prompts is generated by randomly sampling input parameters (genre, disease, country, language). The model is instructed to select a plausible year and to generate JSON entries with the following fields in the following order: *genre, text type, year, text, topic, country, disease, language, time period features in text,* and *the CoT*. Then the generation is performed sampling parameters including *tau* = 0.8 and *top_p* = 0.95 fixed across generations. Each model generates, in each run, 100 sample entries per language and prompt configuration.

This baseline addresses annotators' tendency to disproportionately focus on specific aspects of a genre's definition, leading to inconsistent classifications. Rather than relying on explicit, potentially prescriptive, textual definitions, we propose to implicitly define each genre by a set of exemplary text types. This strategy aims to mitigate the risk of prioritizing certain definitional components over a holistic understanding. Furthermore, this baseline leverages only the inherent, pre-trained knowledge embedded within the model.

The overview of baseline prompt modifications is given in Table 1. Within each modification, the prompt and sampling parameters are fixed to be able to perform comparison across models. Across modifications, sampling parameters are the same.

Table 1: Prompt configurations

prompt_id	**Defined by prompt**	**Selected by LLM**	**Difference from the baseline prompt**
v0 (baseline)	genre, text types, text type, disease, country, language	year, topic	-
v1	genre, country, language	year, topic	excluded information about medicine and patient organization from system and user prompt (**some text types and genres are related to medicine and are not excluded**)
v2	year, genre, text types, text type, disease, country, language	topic	year is defined in the prompt
v3	genre, text types, text type, disease, country, language	year, topic	changed the time period: 20th century substituted by 1900–1955
v4	genre, disease, country, language	year, topic	excluded information about text types
v5	genre, text types, text type, disease, country, language	year, topic	CoT generated first

4. Synthetic dataset description

The dataset includes 18,389 entries: 4 runs per model were performed for the baseline prompt and 2 runs per model – for the rest of the prompt configurations, with varying percentage of successful JSON generations. The dataset covers all 23 genres with the average number of entries per genre equal to 799 and standard deviation of 38. An example entry generated by Qwen 2.5 14B is provided below:

> **Genre**: Opinion
> **Text type:** arguing for a treatment
> **Year:** 1968
> **Text:** 'As a sufferer of rheumatism, I have personally witnessed the transformative impact of corticosteroids on my condition. These powerful medications have drastically improved my quality of life and allowed me to manage symptoms that once severely limited my activities. However, despite their benefits, there remains a hesitancy among some physicians and patients to embrace corticosteroids fully due to lingering concerns about side effects. It is critical to emphasize that corticosteroids are not only safe when properly administered but also represent a vital tool in managing severe rheumatism effectively. We should continue to advocate for their use while working closely with healthcare providers to monitor and mitigate any adverse effects.'
> Topic: Advocating for the Use of Corticosteroids
> Country: United Kingdom
> Disease: rheumatism
> Language: English
> **CoT:** 'The selected year is 1968 due to the availability of corticosteroids as a treatment for rheumatism, which was considered groundbreaking at the time. This advancement led to numerous discussions within patient communities regarding its efficacy and safety. Additionally, corticosteroids were not without controversy due to potential side effects, making them a topic of debate among patients and physicians alike.'

5. Evaluation results

To contextualize the evaluation of faithfulness and diversity, we first examine the models' JSON generation performance and the observed year and digit biases.

JSON Generation Performance. Tab. 2 shows the accuracy of the models in producing valid JSON outputs. On average, LLaMA produces the fewest number of successful outputs across prompt scenarios. Larger models are on average almost 100% accurate when they autonomously select the year. However, when the year is included in the prompt, larger models generate almost 50% less successful formats, while LLaMA 3.1 8B and Qwen 2.5 7b – 26% and 42% less, respectively. Qwen 2.5 14B Base was used only for generation with the baseline prompt and produced 35.6 % of valid outputs. Also, it took 4 times longer for the Base version to complete output generation.

Table 2: Accuracy of models in producing structured JSON (averaged across runs)

Average across Prompts v0, v1, v3, v4, v5 - year is selected by the LLM		**Prompt v2 - year is pre-defined in the prompt**	
Model	**JSON (%)**	**Model**	**JSON (%)**
LLaMA 3.1 8B-Instruct	59.0	LLaMA 3.1 8B-Instruct	33.0
Mistral Nemo Instruct 2407	99.0	Mistral Nemo Instruct 2407	49.3
Qwen 2.5 14B-Instruct	97.6	Qwen 2.5 14B-Instruct	48.6
Qwen 2.5 7B-Instruct	88.5	Qwen 2.5 7B-Instruct	46.9

Table 3: Average percentage of years covered by the models per language using Prompt v2

Language|Model	**LLaMA 3.1 8B**	**Qwen 2.5 7B**	**Qwen 2.5 14B**	**Mistral Nemo**
English	58.24 %	89.56 %	98.90 %	99.45 %
French	75.27 %	99.45 %	99.45 %	98.35 %
German	81.87 %	95.05 %	95.05 %	97.25 %
Swedish	48.90 %	91.76 %	97.25 %	97.80 %

As expected, larger models tend to follow our complex instructions more effectively. However, even when the prompt explicitly specifies the years, the accuracy in generating valid outputs decreases and models do not fully cover the intended year range. Coverage also varies by language, with smaller models like LLaMA and Qwen 7B covering fewer years in English compared to other languages (Tab. 3).

Year and Digit Preference. The models were prompted to autonomously select a year within the 20th century, following exposure to contextual information: patient organization, disease, genre, and text types. The distribution of year selections for the baseline prompt is presented in Fig. 1. Mistral exhibits a pronounced bias toward the year 1955, with over 62% of its selections falling on this specific year (on average per run with standard deviation – SD = 2.8 across runs).

We compare the coefficients of variation (CV = SD / mean) across models to assess the relative variability in year generation. LLaMA exhibits the highest variability (CV = 0.009), while Mistral shows the lowest (CV = 0.004). LLaMA's relative spread is approximately

114% greater than Mistral's. The Qwen models fall in between, with CV values ranging from 0.006 to 0.007, suggesting a moderate level of dispersion.

Figure 1: Prompt v0 – baseline: year autonomously selected by the LLM

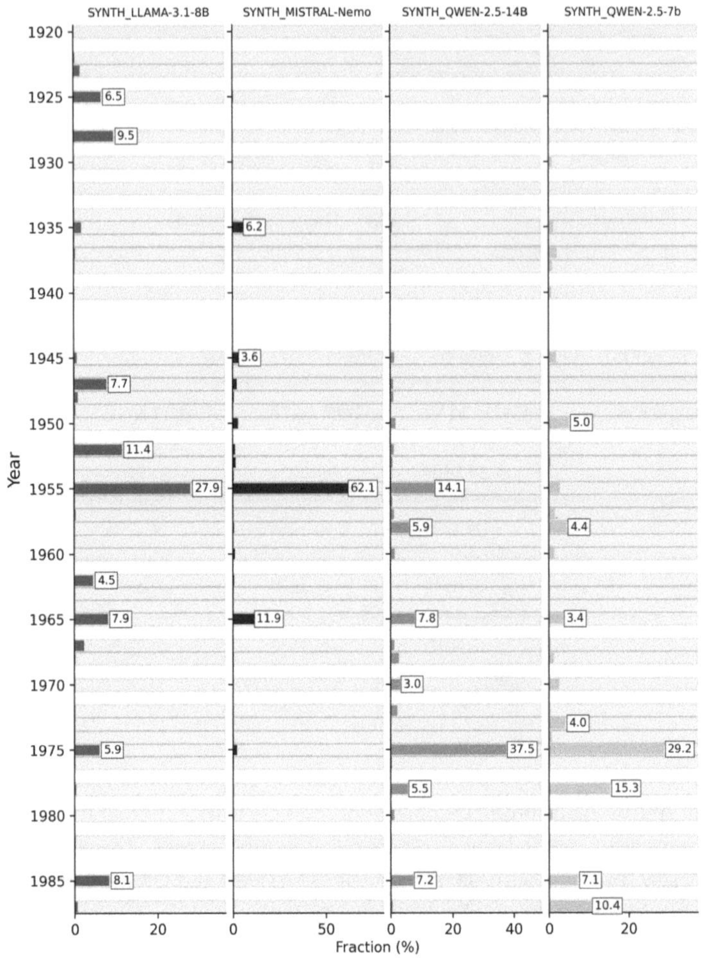

While LLaMA demonstrates the highest degree of variability in its year selections, its most frequent choice also corresponds to 1955 in 28% of generations on average (SD=3.06 across runs). Qwen 2.5 7B and 14B models tend to concentrate on the year 1975, which is consistent across runs for both models (CVs are 7 and 4.6, respectively). Surprisingly, all models in this context completely avoid the interval 1941–1944, as well as other intervals (1926–1927, 1983–1984) and years (1921, 1924, 1929, 1931, 1933, etc.).

Notably, the digit 5 appears more frequently in the model-generated outputs compared to its expected frequency under a uniform distribution of years. We only consider the last two digits in each year for the calculation.

Figure 2: Prompt v3 – "20th century" substituted by the range 1900–1955

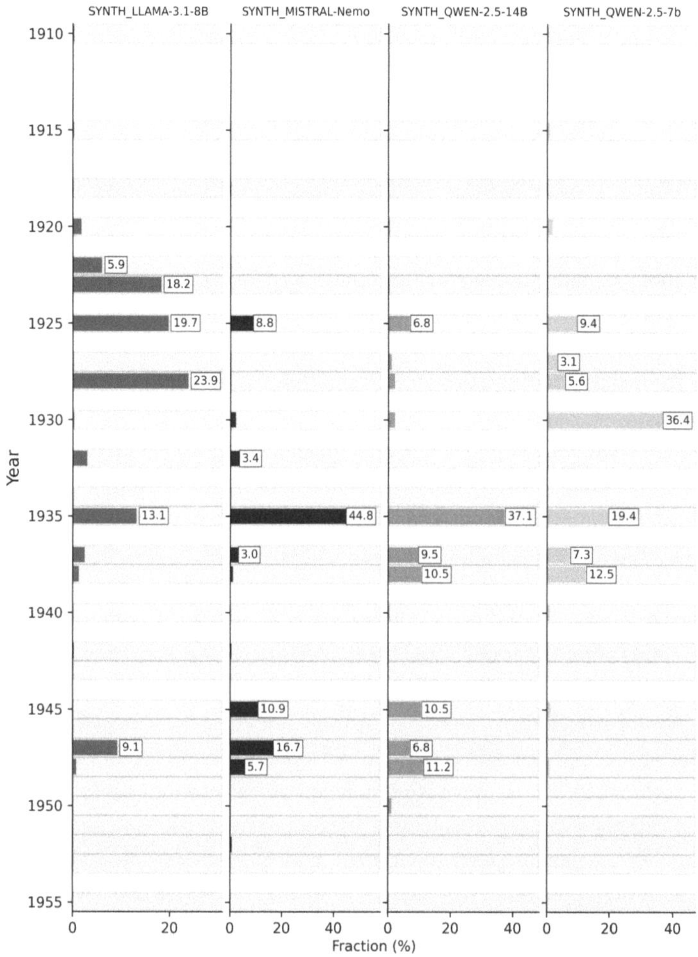

We measure the relative overrepresentation (%) of digits as follows:

$$(Observed - Expected)/Expected \times 100$$

where *observed* is the observed proportion of a digit under a given model and *expected* is the proportion of the digit's frequency under a uniform distribution of years. On average, the overrepresentation of 5 is approximately 370% with the highest for Mistral – 648% and the lowest for Qwen 7B – 186%. LLaMA avoids the digits 0, 1, and 9 with an occurrence rate of only 0.05%, 0%, and 0%. Similarly, Qwen Instruct models show a marked underrepresentation of the digits 2, 3, 4 appearing in 1% of cases on average. The digits 1 and 9 are also completely avoided. In contrast, these models display a heightened preference for the digit 7, with proportions of 26% (14b) and 33% (7b), respectively, compared to 9.6% in the Qwen Base model.

To further assess whether the selection of certain years reflects a general positional bias or a consistent preference independent of the interval, we restrict the original "20th century" range to a narrower window: 1900–1955. As illustrated in Fig. 2, the models generate almost no selections for 1955 within this interval. Instead, the distribution shifts toward the central portion of the range. Notably, the larger models converge on a single year – 1935 – as their primary point of preference. When compared to the results from v0, similar distributional patterns emerge within the 1900–1955 interval, with the exception of the previous peak at 1955. Additionally, a small number of selections now appear around 1942–1943 and other years that were entirely absent in case of the earlier prompt configuration.

The absence of 1955 preference in the output for prompt v3 and the shift of the peak to the central part indicate potential positional bias. To determine whether the models' limited focus on specific years is influenced by the medical context in the prompt, we removed this contextual constraint and analyzed the resulting distribution (see Fig. 3). Without the medical framing, the models exhibited a broader and more diverse range of year selections.

The digit 5 remains overrepresented, although its proportion decreases across models. On average, the relative overrepresentation of 5 is 122%. The digits 1 and 9 are consistently underrepresented, each appearing in less than 1% of outputs across all models. LLaMA shows a strong preference for the digit 2, which is overrepresented by 276%, while all Qwen models overrepresent the digit 3 by an average of 204% (SD = 42%).

When the prompt is modified to generate the CoT reasoning prior to generating the content (Fig. 4) of the other fields and the model is explicitly instructed to reflect on an appropriate year and topic within the given context – the observed imbalance is substantially reduced, particularly in the case of Mistral. This effect is evidenced by the increase in entropy of the digit preference distribution under prompt version v5 relative to version v0, as shown in Tab. 4. The maximum Shannon entropy for a uniform distribution of digit preference is 2.3. It is also notable that for prompt versions v0 and v1 – where the CoT reasoning is generated after the content of all other fields – Qwen exhibits higher entropy in its digit preference distribution compared to Mistral. The Qwen 2.5 14B Base model achieves the highest entropy under prompt v0.

In summary, this analysis demonstrates that adding, removing, or repositioning contextual variables within the prompt affects both the accuracy of JSON generation and the models' preferences or aversions toward specific years and digits. The models produce substantially more valid JSON outputs when they are allowed to select the year themselves. Under the baseline prompt—which includes both the disease and patient organization context—stronger biases toward central years and particular digits emerge, especially in larger models. Conversely, removing medical context from the user and system prompts, as well as repositioning the CoT, increases digit entropy, most notably for the larger models.

Figure 3: Prompt v1 – medical framing excluded from system and user prompts

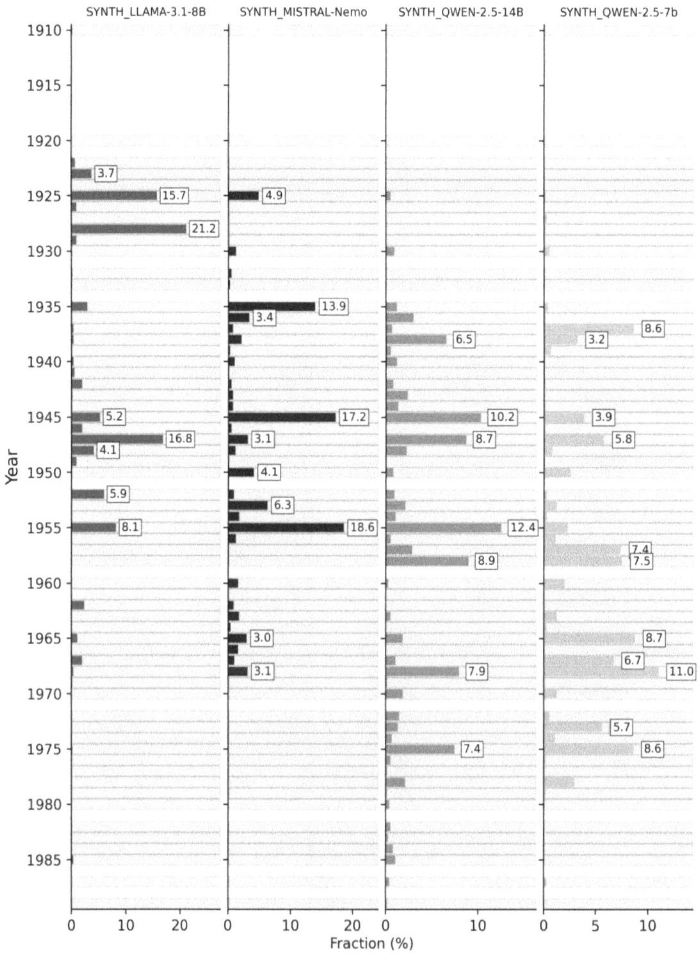

Table 4: Shannon entropy of digit preference distribution

Model\Prompt version	**prompt v0**	**prompt v1**	**prompt v5**
Mistral Nemo Instruct 2407	0.9	1.65	1.97
Qwen 2.5 14B-Instruct	1.4	1.84	1.80
Qwen 2.5 14B	1.76	-	-

Figure 4: Prompt v5 – LLM is instructed to generate the CoT first

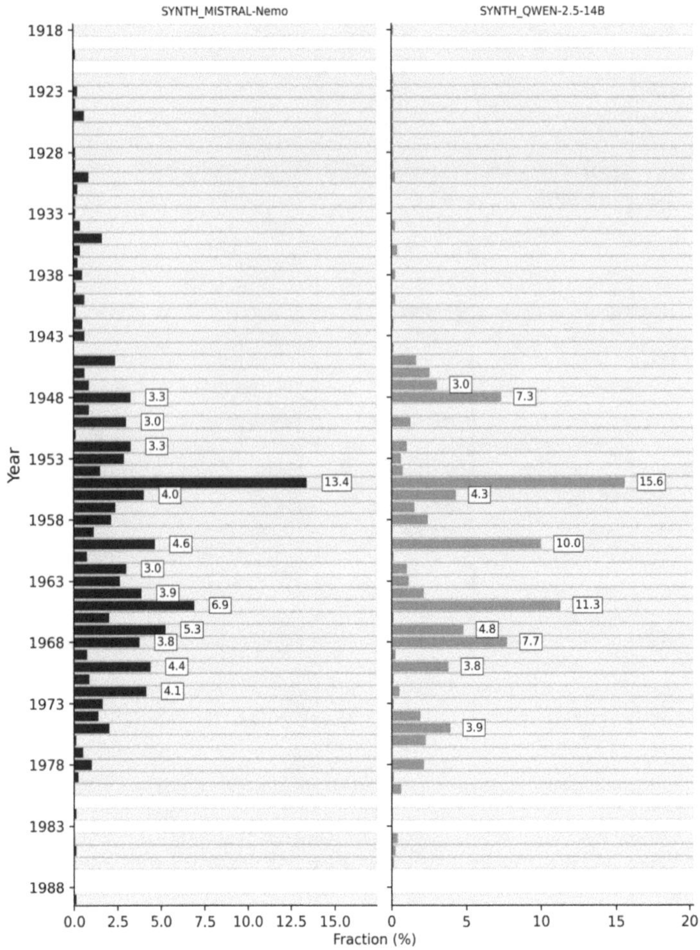

5.1 Faithfulness

Preliminary Evaluation. Domain experts—historians of medicine—observe that SLMs achieve a higher fit to historical genre and time-period characteristics within the fiction genre, but perform less well in humor, interactive, and news genres.

In the patient-organization periodicals, fiction exhibits greater syntactic and semantic variation—in grammar, tone, style, voice, and format—than other genres. Although it contains less historically specific content, its themes related to disease and patient life overlap with those found in historical periodicals, likely explaining its resemblance to historical fiction.

By contrast, texts generated in the humor, interactive, and news genres tend to diverge considerably from their historical counterparts. Their tone, diction, and syntax often sound too contemporary, and the news outputs, in particular, sound too sensationalist.

Historians note that the perceived historical fit of the generated texts—aside from factual accuracy—largely depends on how effectively the model reproduces the genre conventions and content patterns of the periodicals, and how well it generalizes time-period features. However, these generalized features often yield vague or anachronistic depictions of medical knowledge and practice within the historical context of the *ActDisease* dataset.

Main evaluation. Historians completed the questionnaire as shown in Tab. 5. Their scores are compared across models and two prompt versions: v0 (SLM autonomously selects a plausible year in the 20th century) and v2 (SLM is explicitly instructed what year to produce data for). LLaMA consistently received lower scores across both prompts and all questions. Qwen 14B and Mistal-Nemo-Instruct-2407 achieved identical scores on most items, with the exception of RQ1.A, where historians rated Qwen slightly higher. For the smaller models (LLaMA and Qwen), historians noted differences between the responses generated under prompt versions v0 and v2. In contrast, the outputs of the larger models (Qwen and Mistral) appeared similar across prompt conditions. All models scored higher in RQ1.B and RQ1.C—questions that address genre fit and semantic similarity to the dataset. For the questions related to historical accuracy and plausibility (RQ1.A and RQ1.D), scores are at or below the midpoint of the scale [0,10].

Table 5: Responses to the Faithfulness Questionnaire

Model	Llama 3.1 8B		Qwen 2.5 7B		Qwen 2.5 14B		Mistral 12B	
RQ1\prompt configuration	v0	v2	v0	v2	v0	v2	v0	v2
A. To what extent do the generated outputs align with the factual historical timeline?	2	2.5	4	4.5	5	4.5	4	4
B. To what extent are the generated outputs semantically similar to the dataset they are intended to mimic?	4	4.5	6	7	7	7	7	7
C. How accurately do the genres of the outputs reflect the genres specified in the input prompt?	5	6	6	7	7	7	7	7
D.To what degree is the language of the outputs historically plausible?	2	2.5	4	4.5	5	5	5	5

Through the detailed analysis of the outputs, the historians identified three main kinds of distortion in the synthetic data: 1) literal hallucinations and errors of technologies, legislation, organizations, and disease concepts; 2) anachronisms of tone and style; and 3) overgeneralizations in time-period features and chains of reasoning.

Straightforward factual inaccuracies, especially in the Swedish data, include references to websites from 1955 (Mistral Nemo Instruct 2407) or references to "shellinkmeats-allergy" (Qwen 2.5 14B Instruct). In the English material, Qwen 2.5 7B hallucinates UK legislation in the form of a supposed 1985 'Patient Protection Act' – which, according to the LLM, was 'a comprehensive piece of legislation enacted in 1985 to protect the rights of individuals suffering from allergies and other chronic illnesses'. No such Act of Parliament was ever passed, although the hallucinated name, interestingly enough, alludes to the official title of the 2010 US legislation commonly known as 'Obamacare', suggesting something of what material the model may have been trained on, and the UK Parliament did approve a 'Food and Environment Protection Act' in 1985.

The model's representations of time periods and its reasoning chains were not factually incorrect, but they tended to be overly broad and only loosely connected to the accuracy of the details in the generated text or to the historical record. For example, Qwen 2.5 14B Instruct offers the following time period feature justification for a hypothetical medical case in a French diabetes journal describing insulin treatment in 1955: "The text mentions the common use of insulin in the treatment of diabetes and potential complications such as vision damage and kidney disease, which were common during this period when the long-term effects of the disease were just beginning to be understood". Long-term complications of diabetes had been discussed in medical literature as early as Avicenna's *Canon of Medicine* (1025), making this statement false if "diabetes" is taken as a set of related disease entities, symptoms, and illness experiences over time. If the reader takes a narrower view of a historically specific disease entity in 1955, however, they might defend it as accurate. By the time Sanger sequenced the chemical structure of insulin in 1955, industrial biomedicine (principally Novo Nordisk in Denmark) had developed purified, long-acting insulins that required fewer injections and allowed for better control of blood glucose levels. Here, "disease" would mean something like the biochemical understanding of diabetes used by pharmaceutical and biomedical researchers working on insulin purification: as the effects of the lack of a specific protein hormone and, after Sanger's sequencing in 1955, the lack of a 51 amino acid sequence in two polypeptide chains.

In addition, it is noteworthy that the patient-organization narratives produced by the considered SLMs frequently emphasize themes of continuity, consensus, and progress. For instance, in the British case of the 1980s, the decade is portrayed primarily as a period of rising attention to patient rights and the chronically ill, rather than one marked by financial constraints in the health service, reductions in disability benefits, and heightened activism from external pressure groups operating outside official structures.

Finally, the synthetic data assessment asks the historian to invert the sensibility they bring to the historical archive of trying to explain often implausible facts. In this case, the historians in this experiment know the synthetic data are not authentic primary sources, but it is easy to imagine many of the outputs as sources in historical archives. The assessment of their plausibility, however, cannot be independent of the context of the assessment (an experiment on synthetic data in which the historians know the source of the data). The historian's question is not, as the as the media scholar Roland

Meyer has described generative AI works, what is 'latent[ly] possib[le]' in the past at a particular historical moment, but rather what historical actors thought and did given what they left behind (Meyer, 2023). The contemporary trust in historical factchecking in the most literal sense of events in the past is an effect of the emergence and centralization of state archives as the centers of historical truth in the nineteenth century by the oft-cited founder of history as an academic discipline, Leopold von Ranke: archives were the places where the historian discovered primary sources that allowed them to write history "how it actually was" (Ranke, 1824; Eskildsen, 2008).

5.2 Diversity

The design of prompt configurations addresses the evaluation of diversity in the following way. We hypothesize that text types that define genres in the our baseline prompt contribute to diversifying the outputs and that omitting them will lead to more semantically similar responses (v4). Moreover, the removal of patient organization and disease context is expected to lead to more broad themes (v1) and potential decrease of semantic and vocabulary diversity than in the baseline. The specification of year in the prompt (v2) is expected to lead to covering a greater number of health-related themes. Prompt v5 allows us to explore the effect of prompt structure – specifically, the placement of reasoning – on diversity.

Pairwise Cosine Distances. Fig. 5 presents the probability density functions (PDFs) of pairwise cosine distance distributions across various language models and prompt configurations. Distributions concentrated toward the left imply that the generated texts are more semantically similar, while rightward concentrations reflect greater variability in content.

Across all models, prompt version v1 consistently yields a noticeable rightward shift in the cosine distance distribution indicating greater diversity. Prompt versions v0 and v2 tend to produce similar distributions with peaks near a mean distance of approximately 0.8. For the Qwen 2.5 14B Instruct model, we additionally compare with the Base model version and prompt v4. The Base model's distribution aligns closely with those of prompt v0 and v2, and is slightly shifted leftwards. The distribution for v4 shifts leftward, indicating a higher semantic similarity between outputs. A similar pattern is observed for prompt version v5 in the Mistral model.

Overall, the generated texts exhibit semantic diversity, as reflected by the distribution of pairwise cosine distances, which serve as a proxy for measuring variation in content.

Pairwise BERTScore. The overall trends in the PDFs of the self-BERTScore metric (Fig. 7, Appendix) are generally consistent with those observed for pairwise cosine distances, reinforcing our earlier findings regarding prompt versions v1 and v4. To further explore intra-model variability, we examined boxplots of self-BERTScore distributions across languages (Fig. 8, Appendix) and across genres under different prompt configurations in the larger models (Fig. 9, Appendix). Slightly higher diversity is observed for Swedish and German texts, suggesting greater semantic variability in these language subsets. Across genres, Mistral shows a more pronounced variation in self-BERTScore than Qwen 14B (average coefficients of variation = 0.06 (SD=0.004) and 0.05 (SD=0.006), respectively), with the greatest dispersion occurring under prompt v5 for Mistral.

Figure 5: Probability density functions of pairwise cosine distance distributions across various language models and prompt configurations

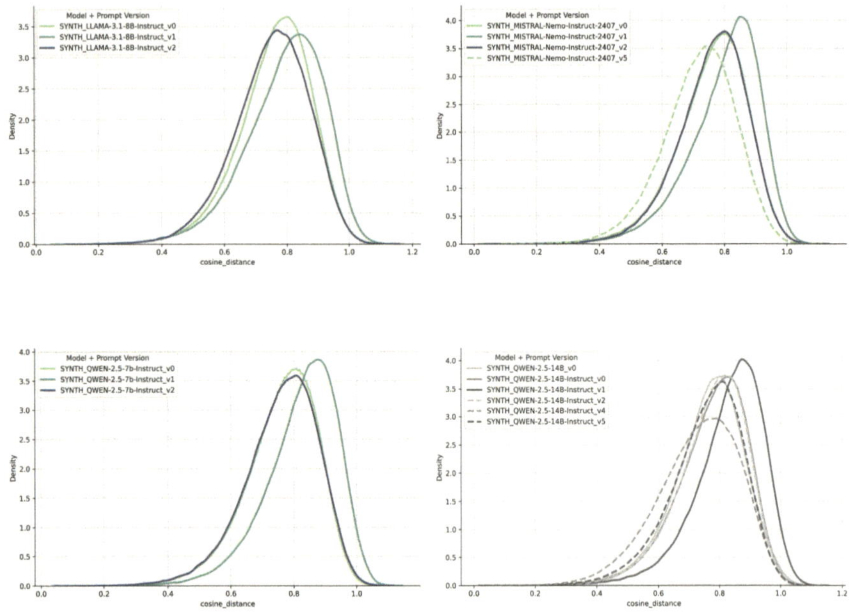

Topic Modeling. Fig. 6 presents the proportion of unique topics relative to the total number of generated texts for each model and prompt configuration. BERTopic produced a total of 549 topics for the entire synthetic dataset. Topic –1, typically representing outliers or unclassified data, was excluded from the plot.

The Qwen 14B Base model yields the highest proportion of unique topics, reflecting a greater degree of topic diversity. Prompts v2 and v5 also lead to relatively high levels of topic variation, indicating that these prompt configurations encourage more semantically varied generations. Prompt v4 results in a lower proportion of unique topics, consistent with the earlier analysis. Notably, prompt v1 produces the lowest proportion of unique topics across all models. Intuitively, we expect this since the models now lack specific medical context. However, it contrasts with the earlier cosine distances analysis.

Within the topic model, the top 10 topics account for a large number of generated texts and represent the broadest thematic groupings. The -1 topic is also assigned to a large portion of the dataset (21%). To examine whether v1 contributes disproportionately to these general categories, we calculated the deviation between the average proportion of texts assigned to the top 10 topics and outliers (-1) across all prompts and the proportion specifically associated with prompt v1. The results indicate that Prompt v1 drives an above-average shift toward general category assignments, yielding increases of approximately 6% in LLaMA, 3% in Mistral, 2% in Qwen 14B, and a notable 11% in Qwen 7B In Qwens and LLaMA, all other prompts consistently yield lower-than-average allocations to these general categories. An exception to this trend is observed in the Mistral model, where prompt v5 shows a 5% increase in this share. Interestingly, earlier analysis of cosine distances indicates that v5 produces less semantically diverse outputs than v1 in Mistral. However, topic modeling results suggest greater topical diversity for v5 compared to v1. This apparent contradiction may indicate that Mistral, under prompt v5, generates a larger number of narrowly defined topics that are nonetheless semantically similar to one another. In contrast, prompt v1 may elicit broader, high-level themes that are more semantically distinct in the embedding space, resulting in greater pairwise cosine distances despite lower topic diversity. Importantly, in the computation of pairwise cosine distances, texts assigned to the -1 topic are also included; because this topic captures outliers or weakly structured content, its inclusion can inflate average pairwise distances and is therefore a likely contributing factor to the higher distances observed for prompt v1.

Figure 6: Proportion of unique topics in generated outputs across prompts versions and models

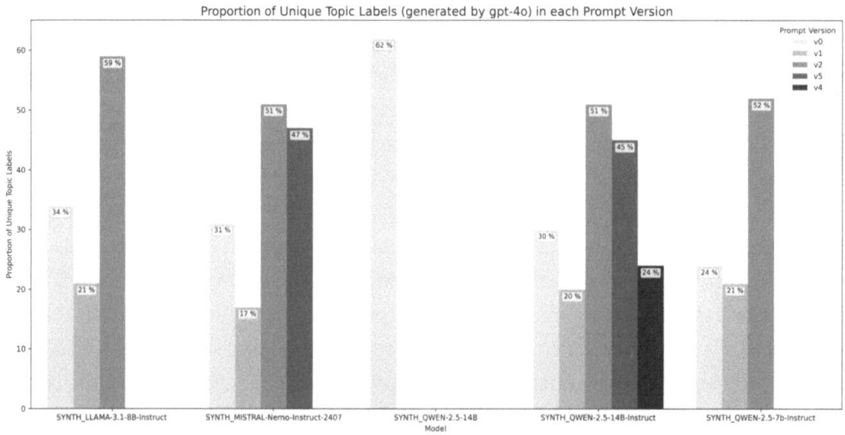

The explicit inclusion of the year in the prompt (v2) yields the most effective balance between generating a high proportion of unique topics and maintaining semantic diversity, based on cosine distance metrics.

Under prompt version v0, the Base model produces the highest proportion of unique topics—nearly double that of the other models. On the one hand, this finding adds to our earlier observation of its more uniform year selection (i.e., greater temporal diversity) and is consistent with prior literature highlighting the model's capacity to generate more diverse outputs. On the other hand, a manual review of the results revealed that the model frequently generated fabricated words for certain time periods, and even cited these invented forms as evidence in its explanations of temporal features as in the following example: "*This text reflects the 1950s through its use of the word 'invitaera' (to invite) instead of the modern Swedish 'invitera'. The use of 'Havsbadet' (seaside resort) indicates that this is an invitation to a seaside location.*" Such cases suggest that the model's output diversity may come at the expense of increased hallucination. Addressing this trade-off should be a priority for the approaches that use base models for synthetic data generation.

In summary, the observed diversity patterns largely align with expectations based on the prompt design. When text-type specifications are omitted, the outputs become more semantically similar, and the absence of medical framing reduces topic coverage and leads to broader themes. When the prompt includes the full set of contextual variables—including the year—the outputs display greater topical diversity.

6. Conclusion

This study focuses on synthetic data generation for training LLMs to analyze historical documents. It evaluates the quality of zero-shot generated historical synthetic data intended to replicate genres found in twentieth-century European patient organization periodicals. The analysis shows that when models autonomously select a year based on the provided context, they exhibit clear year and digit preference biases, the strength of which varies across prompt configurations. These preferences are most pronounced under the baseline prompt with medical framing. Additionally, the models tend to omit certain historical periods—most notably, all models completely exclude the years 1941–1944 under the baseline prompt. The accuracy of valid JSON generation also varies depending on whether the year is model-selected or predefined in the prompt, with the highest accuracy observed when the models choose the year themselves.

The faithfulness evaluation suggests that none of the evaluated models are rated as historically accurate or plausible, as indicated by the low scores (≤5) for the questions directly addressing these criteria (RQ1.A and RQ1.D). The main problems in the outputs include literal hallucinations and errors of technologies, legislation, organizations, and disease concepts, as well as anachronisms of tone and style, and overgeneralizations both in the main texts and the CoT. Higher scores are observed primarily for questions related to aspects related to semantic proximity to the source material (RQ1.B), as well as the correspondence of the generated genre to the prompted one (RQ1.C). Among the larger models, Qwen 14B and Mistral were observed to have similar outputs across prompt configurations, whereas, for the smaller models (Llama 8B and Qwen 7B), prompt configuration v2 (year is pre-defined in the prompt) resulted in more convincing outputs on average, according to the questionnaire. Overall, Llama ranks lower relative to other models, while Qwen 14B is rated slightly higher than Mistral.

Semantic diversity varies considerably across prompts and its patterns generally align with what we expect from the prompt configurations. Excluding the medical context in prompt v1 shifts the output focus toward broader, less domain-specific topics. Interestingly, while these topics may appear generic, pairwise cosine distance metrics reveal that they remain semantically distant from one another, indicating retained diversity at a global level. Prompt v2, which includes explicit year control and extensive parameter specification, yields the best balance of semantic diversity and structural coherence. Additionally, prompt version v5, where CoT reasoning is generated first, also performs well. It demonstrates high semantic diversity and favorable entropy in digit distribution, suggesting that early CoT placement may help scaffold more varied outputs.

According to the metrics, Qwen 2.5 14B Base model exhibits higher entropy in digit preference distributions than its instruction-tuned counterpart, as well as generates a larger proportion of unique topics. Manual analysis further reveals that the diversity observed in the outputs may partly reflect hallucinations, which were considerably less prominent in the outputs of Instruct models, highlighting an important direction for future research.

In conclusion, the results raise important concerns regarding the suitability of the evaluated models for generating synthetic datasets that include historical, temporal, and numerical information. Numerical biases are particularly pronounced under the baseline prompt with explicit medical framing, suggesting a need for further investigation—especially given the growing use of LLMs in the medical and healthcare domains for synthetic data generation.

Although the base model appears more diverse according to automated metrics, manual inspection reveals frequent hallucinations and fabricated vocabulary, which may account for this apparent diversity. This finding underscores the importance of complementing automatic evaluations with human assessment, as prior studies have often relied primarily on machine-based metrics when comparing base and instruction-tuned models.

The synthetic dataset produced in this study was judged by historians as relatively plausible in terms of genre conventions and semantic similarity to original patient organization periodicals, but less so in terms of historical accuracy and factual plausibility. Consequently, it may serve as a valuable resource for augmenting training data in genre classification tasks, but is not suitable for post-training of large language models.

Overall, the generation and use of synthetic datasets with these models should be accompanied by rigorous human and automated evaluations incorporating multi-dimensional metrics tailored to the research domain. Such an approach is essential to prevent the propagation of historical inaccuracies into future datasets and computational tools.

Acknowledgements

Co-funded by the European Union (ERC, ActDisease, ERC-2021-STG 10104099). Views and opinions expressed are however those of the author(s) only and do not necessarily reflect those of the European Union or the European Research Council. Neither the European Union nor the granting authority can be held responsible for them. Co-funded by the Wallenberg AI, Autonomous Systems and Software Program – Humanity and Society (WASP-HS Research Group 807).

References

Aangenendt G, Skeppstedt M and Söderfeldt Y (2024) Curating a Historical Source Corpus of 20th Century Patient Organization Periodicals. In *Proceedings of the Huminfra Conference (HiC 2024)*: 76–82. DOI: https://doi.org/10.3384/ecp205011.

Barr AA, Quan J, Guo E and Sezgin E (2025) Large Language Models Generating Synthetic Clinical Datasets: A Feasibility and Comparative Analysis with Real-World Perioperative Data. *Frontiers in Artificial Intelligence*: 8. DOI: https://www.frontiersin.org/journals/artificial-intelligence/articles/10.3389/frai.2025.1533508.

Binici K, Kashyap A, Schlegel V, Liu A, Dwivedi V, Nguyen T-T, Gao X, Chen N and Winkler S (2025) MEDSAGE: Enhancing Robustness of Medical Dialogue Summarization to ASR Errors with LLM-Generated Synthetic Dialogues. *Proceedings of the AAAI Conference on Artificial Intelligence* 39: 23496–23504. DOI: https://doi.org/10.1609/aaai.v39i22.34518.

Danilova V and Söderfeldt Y (2025) Classifying Textual Genre in Historical Magazines (1875–1990). In *Proceedings of the 9th Joint SIGHUM Workshop on Computational Linguistics for Cultural Heritage, Social Sciences, Humanities and Literature (LaTeCH-CLfL 2025)*. Albuquerque, NM, USA: Association for Computational Linguistics, pp. 160–71. DOI: https://doi.org/10.18653/v1/2025.latechclfl-1.15.

Eskildsen, KR (2008) Leopold Ranke's archival turn: location and evidence in modern historiography. *Modern Intellectual History* 5(3): 425–453.

Gayu Z and Yong L (2025) Towards a Theoretical Understanding of Synthetic Data in LLM Post-Training: A Reverse-Bottleneck Perspective. In: *The Thirteenth International Conference on Learning Representations*, Singapore, April 2024. DOI: https://openreview.net/forum?id=UxkznlcnHf.

Grootendorst M (2022) BERTopic: Neural Topic Modeling with a Class-Based TF-IDF Procedure. arXiv. https://arxiv.org/abs/2203.05794.

Kessler B, Nunberg G and Schutze H (1997) Automatic Detection of Text Genre. In: *35th Annual Meeting of the Association for Computational Linguistics and 8th Conference of the European Chapter of the Association for Computational Linguistics*, Madrid, Spain: Association for Computational Linguistics, pp. 32–38. https://doi.org/10.3115/976909.979622.

Li Z, Zhu H, Lu Z and Yin M (2023) Synthetic Data Generation with Large Language Models for Text Classification: Potential and Limitations. In: *Proceedings of the 2023 Conference on Empirical Methods in Natural Language Processing*, Singapore: Association for Computational Linguistics, pp. 10443–61. DOI: https://doi.org/10.18653/v1/2023.emnlp-main.647.

Long L, Wang R, Xiao R, Zhao J, Ding X, Chen G and Wang H (2024) On LLMs-Driven Synthetic Data Generation, Curation, and Evaluation: A Survey. In: *Findings of the Association for Computational Linguistics: ACL 2024*, Bangkok, Thailand: Association for Computational Linguistics, pp. 11065–82. DOI: https://doi.org/10.18653/v1/2024.findings-acl.658.

Lu Y, Bartolo M, Moore A, Riedel S and Stenetorp, P (2022) Fantastically Ordered Prompts and Where to Find Them: Overcoming Few-Shot Prompt Order Sensitivity. In: *Proceedings of the 60th Annual Meeting of the Association for Computational Linguistics (Vol-*

ume 1: Long Papers), Dublin, Ireland: Association for Computational Linguistics, pp. 8086–98. https://aclanthology.org/2022.acl-long.556.

Meyer R (2023) The New Value of the Archive: AI Image Generation and the Visual Economy of 'Style'. *IMAGE: Zeitschrift für interdisziplinäre Bildwissenschaft*. 19(1): 100–111. http://dx.doi.org/10.25969/mediarep/22314.

Nikolenko S (2021) *Synthetic Data for Deep Learning*. New York: Springer. DOI: https://doi.org/10.1007/978-3-030-75178-4.

Park PS, Schoenegger P and Zhu C (2024). Diminished Diversity-of-Thought in a Standard Large Language Model. *Behavior Research Methods* 56: 5754–70. DOI: https://doi.org/10.3758/s13428-023-02307-x.

Petrenz P and Webber B (2011) Squibs: Stable Classification of Text Genres *Computational Linguistics* 37 (2): 385–93. DOI: https://doi.org/10.1162/COLI_a_00052.

Ranke Lv (1824) *Geschichten der romanischen und germanischen Völker von 1494–1535*. Reimer.

Schaffelder M and Gatt A (2025) Synthetic Eggs in Many Baskets: The Impact of Synthetic Data Diversity on LLM Fine-Tuning. *arXiv*. DOI: https://arxiv.org/abs/2511.01490.

Wagner S, Behrendt M, Ziegele M and Harmeling S (2024) The Power of LLM-Generated Synthetic Data for Stance Detection in Online Political Discussions. DOI: https://doi.org/10.48550/arXiv.2406.12480.

Wang, F, Lin M, Ma Y, Liu H, He Q, Tang X, Tang J, Pei J and Wang S (2025) A Survey on Small Language Models in the Era of Large Language Models: Architecture, Capabilities, and Trustworthiness. *Proceedings of the 31st ACM SIGKDD Conference on Knowledge Discovery and Data Mining*, Toronto, Canada, pp. 6173–6183. DOI: https://doi.org/10.1145/3711896.3736563.

Wei J, Bosma M, Zhao V, Guu K, Wei Yu A, Lester B, Du N, Dai AM and Le QV (2021) Finetuned Language Models Are Zero-Shot Learners. *arXiv*. DOI: https://arxiv.org/abs/2109.01652.

West P and Potts C 2025 Base Models Beat Aligned Models at Randomness and Creativity. *arXiv*. DOI: https://arxiv.org/abs/2505.00047.

Zhang T, Peng B and Bollegala D (2024) Improving Diversity of Commonsense Generation by Large Language Models via In-Context Learning. In *Findings of the Association for Computational Linguistics: EMNLP 2024*, Miami, Florida, USA, pp. 9226–9242.

Zhang T, Kishore V, Wu F, Weinberger KQ and Artz Y (2020) BERTScore: Evaluating Text Generation with BERT. *arXiv*. DOI: https://arxiv.org/abs/1904.09675.

Zhu A, Asawa P, Quincy Davis J, Chen L, Hanin B, Stoica, I, Gonzalez JE and Zaharia M (2025) BARE: Combining Base and Instruction-Tuned Language Models for Better Synthetic Data Generation. *arXiv preprint arXiv:2502.01697*.

Appendix

Figure 7: KDE plots of pairwise BERTScore

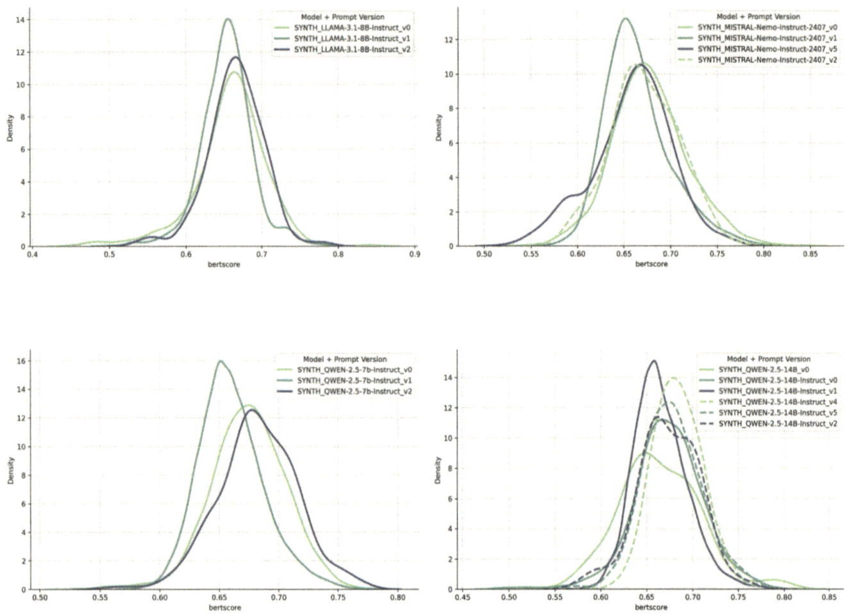

Figure 8: Comparison of pairwise BERTScore distributions across languages

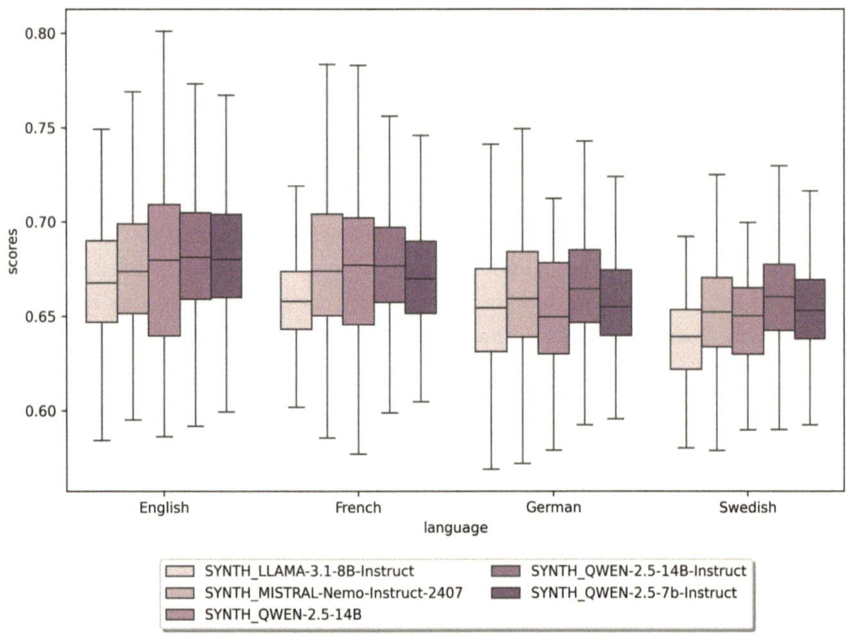

Figure 9: Pairwise BERTScore distributions across genres and prompt versions

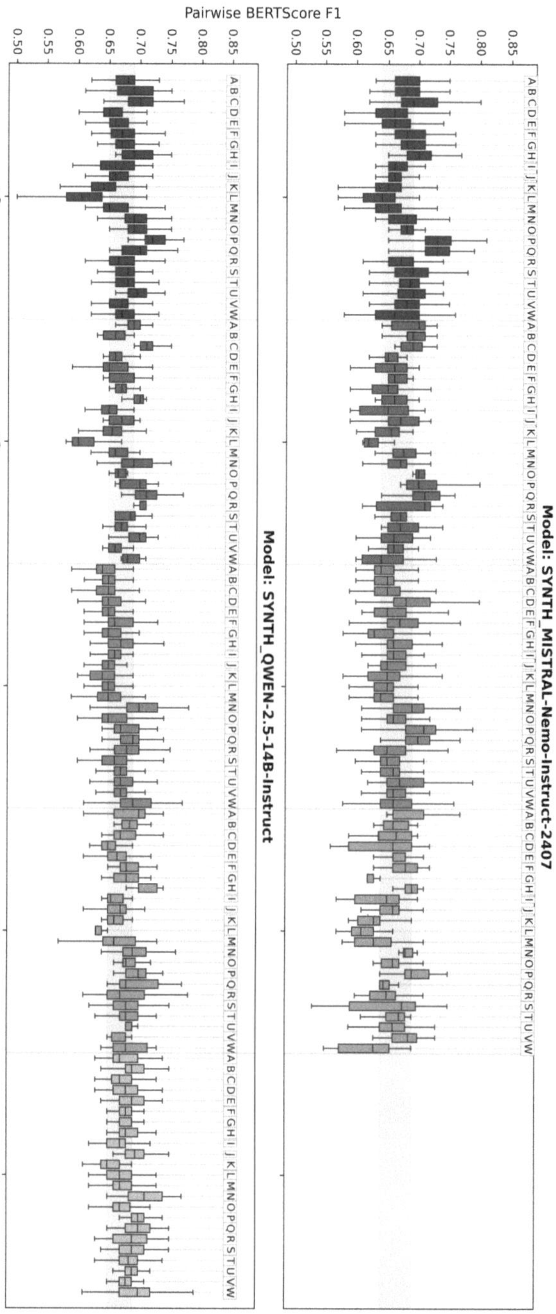

LLMs and multilingual historical corpora in a digital history project

Reflections from the Berlin workshop

Jeffrey C. Wolf

1. Introduction

My expectations for the Technical University of Berlin's LLM workshop were unclear beforehand. What would participants learn? How were others using LLMs in their HPSS research? Even the definition of Large Language Models remained confusing. Were they simply the models that powered the new generative AI tools, like ChatGPT, or something more? What was meant by "large" anyways? I was uncertain about the extent to which our ERC-funded VERITRACE project could use LLMs; there were no plans to offer chatbot-type access to the VERITRACE corpus, and even if there were, the volume of data made this approach impractical. Developments were also changing so rapidly in the field that the workshop's relevance to ongoing work in the coming weeks, let alone months, remained murky. Still, I hoped it would all become more clear at the Technical University in Berlin, and indeed, it did.

Before offering more specific reflections on the questions above, I will provide a summary of the VERITRACE project, and how we imagined various ways that LLMs might help us.

2. The VERITRACE project: context and challenges

The VERITRACE project[1] traces the influence of prominent ancient wisdom writings throughout the early modern period, focusing on their function in natural philosophical discourse (Schilt, 2022). It applies sophisticated digital analysis to a large corpus of early modern texts, tracing referenced and unreferenced uses of the *Corpus Hermeticum* (including the *Asclepius*), the *Chaldean Oracles* and *Sibylline Oracles*, and the *Orphic Hymns*. It analyses how these texts were used and with what sentiment they were discussed

1 https://veritrace.eu/

by proponents and antagonists, examining how key transmission episodes influenced these debates. VERITRACE will provide the first comprehensive analysis of ancient wisdom writings' influence on early modern natural philosophy using methodologies never employed at this scale in the early modern history of science.

VERITRACE draws on printed books, the period's most ubiquitous intellectual materials. While early modern debates involved oral discussion and manuscript circulation, these pertained to small circles. Books were everywhere, connecting authors and readers across Europe and beyond. Even focusing solely on a set of core languages—works in Latin, English, German, French, Dutch, and Italian—the number of surviving early modern books is staggering, presenting significant difficulties and opportunities.

Several challenges uniquely characterise VERITRACE as a digital humanities project:

1. **Multilingualism**: VERITRACE source material spans at least six languages, both modern and classical. Since many NLP techniques were developed for English users, this multilinguistic nature creates unique challenges, particularly with limited support for classical languages like Latin.
2. ***Longue durée***: Spanning almost two hundred years (1540–1728) our corpus adds diachronic complexity. Interpretations applicable to smaller data slices cannot be assumed to apply universally, given changing historical contexts and semantic meanings.
3. **Big Data**: With hundreds of thousands of texts, simple search processes and data management are inadequate. The project requires resource-intensive solutions given the collection sizes.
4. **Complex Integration**: Data from different institutional sources collected over long periods creates inherent integration and harmonisation challenges, requiring careful documentation of cleaning and transformation rules.

VERITRACE also confronts familiar distant reading challenges: OCR quality issues, parameter-dependent NLP techniques, and similar concerns.

3. Distant reading

Meaningfully tracing ancient wisdom writings' influence case-by-case would require massive research teams or drastically reduced scope. Digital techniques from distant reading (Moretti, 2005, 2013; Underwood, 2019), closely related to natural language processing, enable large corpus analysis, identifying patterns and uncovering both prominent and neglected works: Margaret Cohen's the 'great unread' (Cohen, 2009; Reid, 2019). Early modern writers rarely referenced source material explicitly, creating key project challenges.

Digitisation advancements have expanded distant reading applications. Improved OCR technology yields meaningful results despite suboptimal text recognition (Hill and Hengchen, 2019; Kurhekar et al., 2021). Online repositories like the Bibliothèque nationale de France provide standardised data, facilitating large-scale analysis.

VERITRACE's Distant Reading Corpus (DRC) consists of several hundred thousand works from major European collections in Latin, French, German, Dutch, English, and Italian:

- **EEBO-TCP** (Early English Books Online): ~68,000 English and Latin texts (1540–1700)
- **Gallica** (Bibliothèque nationale de France): ~20,000 books (1540–1728) in French and various languages
- **Digitale Sammlungen of the Bavarian State Library**: ~340,000 books (1540–1728) in Latin, German, and other languages

These sources enable historical claims about the *prisca sapientia* tradition, its prevalence, and changing interest levels. By interrogating truly representative samples, reasonable claims about interest and prevalence become possible. A rigorous statistical approach underpins this methodology, with chosen sources forming the basis for representative sampling of European printed books (1540–1728).

4. Close reading too

VERITRACE combines state-of-the-art distant reading techniques on large corpora with close reading of carefully selected Renaissance and early modern texts.

VERITRACE's Close Reading Corpus (CRC) includes all relevant editions of the *Corpus Hermeticum*, *Chaldean Oracles*, *Sibylline Oracles*, and *Orphic Hymns* published during the Renaissance and early modern period, plus works drawing heavily on these writings and promoting *prisca sapientia* ideas, such as Ficino's *Theologia platonica* and Steuco's *De perenni philosophia*: approximately 80 works initially, which we have already expanded to c.150.

While a readership and geographical dissemination census would be valuable, as previous censuses of Copernicus's *De Revolutionibus*, Vesalius's *De Fabrica*, and Newton's *Principia* revealed a wealth of information (Palumbo, 2018; Margocsy et al., 2018; Feingold and Svorenčík, 2020), this remains impractical within our project's scope. Nonetheless, we will catalogue as many works in the tradition as we can.

This combination of distant and close reading methodology combines granular interpretation of select works with broader context. Previously, ancient wisdom texts' influence on natural philosophy was confined to case studies involving major figures—Copernicus, Kepler, Bacon, More, Newton. But early modern natural philosophy extended far beyond select individuals. While many never documented their ideas, thousands did. Ancient wisdom writings were widely read, with new Greek, Latin, and vernacular editions appearing regularly throughout the early modern period.

5. Large language models: do they have a place?

Where Large Language Models fit into our approach represents both an opportunity and a challenge unique to projects working with multilingual historical corpora. We did not immediately think of LLMs as a critical tool in our project; instead, our initial motivation was more intellectual. We wanted to offer scholars the ability to uncover similar-meaning passages across texts on a wide scale.

We knew we could offer traditional keyword search tools that allow researchers to perform their own searches of the full VERITRACE corpus—a helpful tool for user-driven scholarship—but such searches are limited by requiring the user to know precisely what they are searching for, prior to use. And search queries are limited to keywords or short phrases.

We envisioned a valuable complement: searches where no such prior knowledge is necessary, where the tool could identify similar passages across the entire corpus. This would be based on textual and semantic similarity between query text(s) and comparison text(s). This led us to develop a Text Matching tool that functions as a kind of "early modern plagiarism detector"—an effective, surface-level tool focused on identifying matching passages and phrases through keyword and vocabulary analysis (Wolf et al., 2024).

However, our multilingual corpus presents challenges that surface-level matching cannot address. Traditional lexical matching excels at identifying shared vocabulary within languages but proves nearly useless across linguistic boundaries. Given that our corpus spans six core languages with Latin predominating, we also needed "Semantic Matching"—a tool capable of identifying semantic similarity between textual passages regardless of the language in which that meaning was encoded. This cross-linguistic semantic matching represents a valuable, complementary tool, enabling scholars to identify, for instance, passages written in Latin that are semantically similar to passages in Italian, thereby bridging the "multilingual gap."

This approach aligns with what recent digital humanities scholars have called "text similarity analysis" or "text reuse analysis," such as the work of the Helsinki Computational History Group[2]. This has proven particularly valuable for humanities research in identifying patterns of textual borrowing and influence across large corpora (Rosson et al., 2023; Ryan et al., 2023).

6. Embeddings models and the historical challenge

This semantic matching capability depends fundamentally on vector embeddings; specifically, Large Language Models of the full-context, bi-directional, encoder variety (Devlin et al., 2018) rather than the generative decoder models exemplified by GPT architectures (Vaswani et al., 2017). As recent scholarship has demonstrated, word embedding models can capture semantic information in large corpora by learning distributional properties of words—how often particular words appear in specific contexts—making them particularly valuable for historical research focused on tracing conceptual change over time, despite some drawbacks researchers must be aware of (Wevers and Koolen, 2020).

The workshop crystallized several critical insights about model selection that had not been top of mind before. Most significantly, I realised that our corpus, containing texts printed over 200 years (1540–1728), necessitated accounting explicitly for "semantic

2 https://www.helsinki.fi/en/researchgroups/computational-history

shift"—the fact that words change meaning over extended time periods. This challenge has been recognised in recent computational linguistics research, where scholars have proposed statistical laws governing semantic change: words used more frequently tend to change at slower rates (the law of conformity), while polysemous words change more rapidly regardless of frequency (the law of innovation) (Hamilton et al., 2016).

This temporal dimension creates a fundamental decision: should we employ a single embeddings model capable of capturing semantic change over our 200-year span, adding considerable complexity to the training process, or utilise different models for different time periods, which introduces critical issues in semantic alignment between models? Neither approach is clearly superior to the other and each one involves significant trade-offs that can affect the reliability of results. The importance of thinking rigorously about this choice only became clear to me at the LLM workshop in Berlin.

7. Four critical challenges for historical multilingual embeddings

Our workshop discussions helped clarify four interconnected challenges that any embeddings model must address when working with historical multilingual corpora of *longue durée*. First, **multilingual capability** requires robust support across our six core languages, with particular strength needed in Latin, which comprises the majority of our textual data. Recent research has shown that pre-trained multilingual models face significant challenges with historical content (due, in part, to OCR noise and outdated spellings), particularly when dealing with low-resource historical languages, though fine-tuning using contrastive learning can significantly increase accuracy (Michail et al., 2025).

Second, **OCR noise** presents ongoing difficulties. Studies have demonstrated that OCR noise may impact (though not catastrophically) the performance of downstream NLP tasks (Hill and Hengchen, 2019) with character-level mistakes proving problematic for embeddings-based topic models (Zosa et al., 2021). Our corpus contains substantial OCR noise, requiring models that can handle such imperfections or preprocessing pipelines that reduce OCR errors without introducing new distortions.

Third, the **out-of-domain** training problem reflects a broader issue in digital humanities applications of contemporary NLP tools. Recent studies demonstrate that most embedding models are trained on contemporary textual corpora, and this introduces a modern bias, pushing historical semantic representations closer to modern usages and obscuring historical contexts (Hamilton et al., 2016; Kutuzov et al., 2018). Indeed, that was one of the motivations for Qui and Xu to create HistBERT (Qui and Xu, 2022). The workshop's opening keynote "Large-scale text analysis for the study of cultural and societal change" by Pierluigi Cassotti and Nina Tahmasebi particularly stressed this limitation, noting how current LLMs, trained predominantly on modern data, can misrepresent historical language use.

What this suggests is that, without the right tools, we would struggle not just with representing historical semantics but with the entire **semantic shift** inherent in our corpus's 200-year time span (Kutuzov et al., 2018). Our early modern corpus requires methodologies that can handle both the temporal scope and the linguistic complexity of pre-modern texts.

8. Methodological considerations and next steps

The workshop presentations provided valuable insights into how other HPSS researchers are addressing similar challenges. Projects using LLMs for computational epistemology, citation extraction from historical sources, and Retrieval Augmented Generation databases demonstrated various approaches to incorporating modern NLP tools into historical research workflows. These examples align with broader trends in digital humanities, where scholars are increasingly using vector embeddings not merely as technical tools but as means of exploring discursive spaces and semantic relationships within large textual collections.

Particularly relevant were discussions of metadata enrichment and diachronic analysis approaches that parallel our own VERITRACE objectives. Projects working with historical newspapers and other time-stamped corpora have developed sophisticated methodologies for tracking conceptual evolution, often employing temporal alignment strategies that train separate models for different periods while maintaining comparability across time.

Based on these workshop insights, our approach to model selection requires balancing multiple competing priorities. We are primarily interested in sentence-level embeddings models rather than word-level approaches, as our research focuses on identifying similar textual passages rather than individual word relationships.

Our current assessment suggests that no single existing model adequately addresses all four challenges outlined above. Models may excel in multilingual support but lack historical training data, or handle OCR noise sufficiently but prove inadequate for our temporal scope. Consequently, we anticipate the need to fine-tune a base model specifically for our corpus characteristics.

For base model selection, we are evaluating Microsoft's multilingual-e5-large-instruct (Wang et al., 2024), which offers strong multilingual support, including Latin. However, this choice depends on comprehensive evaluation across our specific requirements: Latin language capacity, OCR noise tolerance, and sensitivity to semantic shift. Our evaluation framework must assess whether the base model provides sufficient performance across these dimensions to justify subsequent fine-tuning to achieve reliable results.

The OCR quality issue may necessitate parallel development of correction models or preprocessing pipelines. Some recent studies have shown that while OCR post-correction can significantly improve downstream task performance, the effectiveness varies considerably depending on the specific characteristics of the historical corpus and the intended application (Todorov and Colavizza, 2020). For VERITRACE, this means carefully balancing preprocessing costs against potential improvements in semantic matching accuracy.

9. Conclusion

The Berlin LLM workshop provided a crucial perspective on the challenges and opportunities facing digital humanities projects seeking to incorporate advanced NLP methodologies. For VERITRACE, the key insight was recognising that our multilingual, temporally-extended corpus requires specialised solutions that address the interconnected challenges of multilingual support, OCR quality, domain adaptation, and semantic shift.

Rather than seeking to apply existing models wholesale, we must develop evaluation frameworks that can assess model performance against our specific research requirements and create fine-tuning approaches that address our corpus's unique characteristics.

In some ways, we were spared some of the hard decisions that await scholars who use Transformer-based LLMs more extensively than we do. Critical issues highlighted with these newer models include the substantial computational resources required for training and inference, the black-box nature of their decision-making processes that complicates interpretability for humanities research, and the ongoing concerns about hallucination and factual accuracy when working with historical texts. Additional challenges emerge around data provenance and the difficulty of ensuring that training data does not inadvertently influence results when analyzing historical corpora.

Our workhorse is not a chat-based model but an embeddings model on the encoder side that focuses specifically on semantic similarity tasks without the complexities of text generation. These encoder-based embedding models have been more thoroughly tested across diverse applications and have been around longer with established track records. It is hoped that they can form a solid basis for our Text Matching tool, even if many implementation details have not yet been finalised.

Finally, the workshop reinforced the value of interdisciplinary dialogue between digital humanities practitioners and domain-specific experts. Our specific use case—semantic matching across languages and centuries in early modern natural philosophy texts—represents the kind of specialised application that can push the boundaries of current methodologies, while addressing substantive historical research questions. As LLM capabilities continue to evolve rapidly, fostering collaboration between technical researchers and scholars focused on historical practices of science will be essential to fully realise their potential across historical research. The LLM Workshop in Berlin was a laudable example of that.

References

Cohen M (2009) Narratology in the Archive of Literature. *Representations* (108): 51–75.

Devlin J, Chang M-W, Lee K et al. (2018) BERT: Pre-training of Deep Bidirectional Transformers for Language Understanding. *arXiv*. Epub ahead of print. arXiv:1810.04805v2.

Feingold M and Svorenčík A (2020) A Preliminary Census of Copies of the First Edition of Newton's *Principia* (1687). *Annals of Science* 77(3): 253–348.

Hamilton WL, Leskovec J and Jurafsky D (2016) Diachronic Word Embeddings Reveal Statistical Laws of Semantic Change. In: *Proceedings of the 54th Annual Meeting of the Association for Computational Linguistics* (Volume 1: Long Papers). Berlin: Association for Computational Linguistics, pp. 1489–1501. Available at: https://aclanthology.org/P16-1141/ (accessed 18 November 2025).

Hill MJ and Hengchen S (2019) Quantifying the Impact of Dirty OCR on Historical Text Analysis: Eighteenth Century Collections Online as a Case Study. *Digital Scholarship in the Humanities* 34(4): 825–843.

Kurhekar P, Nigam S and Pillai S (2021) Automated Text and Tabular Data Extraction from Scanned Document Images. In: *Data Management, Analytics and Innovation. Proceedings of ICDMAI 2021* (eds N Sharma, N Chakrabarti, VE Balas, AM Bruckstein), pp.169–182. Singapore: Springer Singapore.

Kutuzov A, Øvrelid L, Szymanski T. et al. (2018) Diachronic Word Embeddings and Semantic Shifts: A Survey. In: *Proceedings of the 27th International Conference on Computational Linguistics*, Santa Fe, New Mexico, USA, pp.1384–1397. Association for Computational Linguistics. Available at: https://aclanthology.org/C18-1117/ (accessed 18 November 2025).

Margocsy D, Somos M and Joffe SN (2018) *The Fabrica of Andreas Vesalius: A Worldwide Descriptive Census, Ownership, and Annotations*. Leiden: Brill.

Michail A, Raclé C, Opitz J. et al. (2025) Adapting Multilingual Embedding Models to Historical Luxembourgish. In: *Proceedings of the 9th Joint SIGHUM Workshop on Computational Linguistics for Cultural Heritage, Social Sciences, Humanities and Literature (LaTeCH-CLfL 2025)*. Albuquerque, New Mexico, USA, pp.291–298. Association for Computational Linguistics. Available at: https://aclanthology.org/2025.latechclfl-1.26/ (accessed 18 November 2025).

Moretti F (2005) *Graphs, Maps, Trees: Abstract Models for Literary History*. London: Verso.

Moretti F (2013) *Distant Reading*. London: Verso.

Palumbo M (2018) Books on the Run: The Case of Francesco Patrizi. In: Zwierlein C and Lavenia V (eds) *Fruits of Migration: Heterodox Italian Migrants and Central European Culture 1550–1620*. Leiden: Brill, pp.197–216.

Qiu W and Xu Y (2022) HistBERT: A Pre-trained Language Model for Diachronic Lexical Semantic Analysis. *arXiv*. Epub ahead of print. arXiv:2202.03829.

Reid D (2019) Distant Reading, 'the Great Unread', and 19th-Century British Conceptualizations of the Civilizing Mission: A Case Study. *Journal of Interdisciplinary History of Ideas* (15).

Rosson DE, Mäkelä E, Vaara V, et al. (2023) Reception Reader: Exploring Text Reuse in Early Modern British Publications. *Journal of Open Humanities Data* (9).

Ryan Y, Mahadevan A and Tolonen M (2023) A Comparative Text Similarity Analysis of the Works of Bernard Mandeville. *Digital Enlightenment Studies* (1): 28–58. DOI: 10.61147/des.6.

Schilt CJ (2022) Traces de la Verité: The Reappropriation of Ancient Wisdom in Early Modern Natural Philosophy. VERITRACE (ERC-2022-STG-101076836). Available at: https://veritrace.eu/wp-content/uploads/2023/04/Project-Traces-de-la-Verite-Condensed.pdf (accessed 18 November 2025).

Todorov K and Colavizza G (2020) Transfer Learning for Historical Corpora: An Assessment on Post-OCR Correction and Named Entity Recognition. In: *CEUR Workshop Proceedings* 2723, pp. 310–339.

Underwood T (2019) *Distant Horizons: Digital Evidence and Literary Change*. Chicago: University of Chicago Press.

Vaswani A, Shazeer N, Parmar N, et al. (2017) Attention Is All You Need. In: *Advances in Neural Information Processing Systems* 30.

Wang L, Yang N, Huang X, et al. (2024) Multilingual E5 Text Embeddings: A Technical Report. *arXiv*. Epub ahead of print. arXiv:2402.05672v1.

Wevers M and Koolen M (2020) Digital Begriffsgeschichte: Tracing Semantic Change Using Word Embeddings. In: *Historical Methods: A Journal of Quantitative and Interdisciplinary History* 53(4): 226–243.

Wolf JC, Cantoni N, Kovács E, et al. (2024) The Challenges of Multilingualism in the Search for Ancient Wisdom: A Case Study of VERITRACE's Text Matching Tool. In: *Proceedings of the 4th Humanities-Centred AI (CHAI) Workshop at the 47th German Conference on Artificial Intelligence*, Würzburg, Germany, 23 September 2024. Available at: https://ceur-ws.org/Vol-3814/paper1.pdf (accessed 18 November 2025).

Zosa E, Mutuvi S, Granroth-Wilding M, et al. (2021) Evaluating the Robustness of Embedding-Based Topic Models to OCR Noise. In: *Towards Open and Trustworthy Digital Societies. 23rd International Conference on Asia-Pacific Digital Libraries, ICADL 2021, Proceedings*, 13133 (eds Ke H-R, Lee CS and Sugiyama K), Virtual Event, December 1–3, 2021, pp. 392–400. Lecture Notes in Computer Science. Singapore: Springer. Available at: https://hal.science/hal-03480518v1 (accessed 18 November 2025).

Part 3: Linguistic change, markers, and differences

Epistemic framings in science
Charting scientific knowledge with embeddings and LLMs

Christophe Malaterre and Francis Lareau

1. Introduction

Despite the growing attention paid to the diversity of scientific practices, relatively little is known about how scientists conceptualize and articulate the very nature of knowledge in different research fields. Philosophers of science have long debated key epistemic constructs—such as theory, model, hypothesis, explanation, and understanding—and have emphasized the roles of justification and epistemic virtues like simplicity, robustness, and predictive power. Yet, these discussions often rest on abstract conceptual analysis or a narrow set of case studies, rarely examining how such epistemic notions are used in scientific discourse at scale (but see McPhetres et al., 2021; Mizrahi and Dickinson, 2022; Noichl, 2023).

Our recent corpus-driven research begins to fill this gap by examining how scientists actually deploy a vast array of epistemic concepts in their published work (Malaterre and Léonard, 2023, 2024). Using a large biomedical corpus, we analyzed the distribution and association of dozens of epistemic markers (e.g., "theory", "model", "explanation") revealing a clear diversity of usage patterns across subfields. These differences suggest not only semantic variation but deeper conceptual distinctions—what we refer to as "epistemic framings": the ways in which scientists frame what counts as knowledge, how it is justified, and which epistemic goals it serves.

Here we explore how recent advances in natural language processing—particularly embedding models and large language models (LLMs)—can contribute to the systematic investigation of epistemic framings. We argue that these tools offer powerful methods for detecting, classifying, and mapping epistemic concepts across disciplines and historical periods, thus enabling a more empirically grounded account of how scientific knowledge is conceptualized and communicated.

2. A corpus study: epistemic concepts across biomedical subfields

Our initial empirical work focused on a biomedical corpus of 73,771 full-text research articles, pre-processed with lemmatization and sorted into seven disciplinary clusters via topic modeling and k-means clustering. Our aim was to assess how epistemic concepts are deployed across subfields—how "model," "explanation," or "mechanism," for instance, figure differently in neuroscience, epidemiology, or molecular biology (Malaterre and Léonard, 2024).

To start, we developed a curated list of 211 epistemic markers through two complementary strategies. First, we expanded from six seed terms—"explanation," "understanding," "prediction," "model," "theory," and "mechanism"—by extracting their most frequent co-occurring terms across the entire corpus and within each cluster. This yielded 202 high-frequency terms, 52 of which were epistemic in nature. Second, we analyzed the most frequent terms in the overall corpus (appearing in at least 1,000 documents), supplemented by thesaurus-based synonym expansion. The overall 211 terms were manually grouped into 61 epistemic "semantic fields" (e.g., hypothesis, confirmation, plausibility) and 18 disciplinary ones (e.g., disease, molecular, behavior).

We then quantified the presence of each semantic field in each document and computed averages by disciplinary cluster. Additionally, we calculated paragraph-level co-occurrence correlations between all epistemic fields, producing a 79×79 correlation matrix per disciplinary cluster. Louvain community detection on these correlation networks allowed us to map discipline-specific patterns of epistemic co-occurrence.

Key findings showed that different disciplinary domains privilege different epistemic objects—e.g., models in neuroscience, mechanisms in molecular biology—and epistemic virtues—e.g., plausibility in clinical research vs. accuracy in genomics. These terminological "epistemic maps" reveal that scientific disciplines structure and prioritize epistemic concepts in distinct ways. In turn, this supports the idea that scientific knowledge is not monolithic, but plural and context sensitive. We call these structured configurations of epistemic emphasis "epistemic framings."

3. Using LLMs to classify epistemic content

Large language models (LLMs) offer promising tools for extending this investigation beyond fixed lexicons. One avenue we are experimenting with is the contextual classification of epistemic statements. LLMs can be prompted with individual sentences or paragraphs and asked whether the epistemic content expresses certainty, uncertainty, evaluative judgement, or justification. These classifications could be used to analyze the epistemic orientations of scientists as expressed in their written productions.

However, the sheer scale of large scientific corpora limits this approach in practice. For instance, analyzing 100 million sentences with a model like GPT-4-turbo would require weeks of compute time and a substantial cost. This makes full-coverage classification with current LLM APIs impractical, though this may change in the future as LLM cost and compute time drop down.

Instead, LLMs are probably best deployed in a hybrid architecture, at least for now. Small samples of sentences can first be annotated via LLMs to create a training sets with which lighter classifiers can then be trained to label the rest of the corpus. The resulting classification can then be used as a heuristic to identify a variety of epistemic framings, and investigate nuances depending on disciplinary domains or time.

Ultimately, while LLMs cannot solve the entire epistemic classification task at scale, they can significantly enhance the sensitivity and coverage of framing analyses by providing contextual judgments, expanding lexicons, and improving interpretability when prompted accordingly on relatively small sets of documents.

4. Refining epistemic markers with embeddings and syntactic patterns

One limitation of our earlier study lies in its reliance on a manually curated list of epistemic terms. While conceptually grounded, the list may miss field-specific phrases, idiomatic expressions, or emergent terms. Embedding models like SciBERT, or prompting LLMs directly, allow for semantic expansion: generating synonyms and near-synonyms of known epistemic markers (e.g., "tentative explanation," "broadly consistent," "likely mechanism").

We also propose to enhance the lexicon by identifying multiword expressions (MWEs) with epistemic relevance. These can be discovered by parsing sentences that contain known epistemic terms and extracting syntactic patterns such as modal constructions (e.g., "may explain," "is thought to be," "cannot confirm"). Dependency parsing tools like spaCy or Stanza can isolate these structures.

By clustering embeddings of parsed dependency structures (e.g., using HDBSCAN), we can identify families of expressions that share structural and semantic characteristics. Clusters enriched in epistemic content can yield additional MWEs not present in the original list.

This syntactic-semantic approach captures not just individual terms but the broader constructions in which epistemic meanings are embedded, offering a more nuanced view of how scientific claims are hedged, justified, or asserted.

5. Topic modeling and disciplinary structure

Understanding epistemic framings requires contextualizing them within the thematic and disciplinary structures of science. LLM-based topic modeling offers a way to do this. One approach is to embed documents (e.g., abstracts or full texts) using sentence-transformers like SBERT or domain-specific variants (e.g., BioBERT). Dimensionality reduction techniques like UMAP, followed by clustering algorithms like HDBSCAN, can then identify coherent topic clusters.

Once these clusters are formed, characteristic terms can be extracted using class-based TF-IDF (c-TF-IDF), F-measure, or LLM-based summarization. LLMs can also be used to evaluate different models (Stammbach et al., 2023), and can assist in generating descriptive labels for these clusters, increasing interpretability.

Tools like BERTopic automate this pipeline (Grootendorst, 2022), combining embeddings, clustering, and topic extraction. However, trade-offs exist. Embedding-based topic modeling can struggle with long documents due to window size limitations. It may also underperform compared to classical statistical methods (e.g., LDA (Blei et al., 2003)) in terms of topic coherence or granularity (Lamirel et al., 2024). Thus, hybrid approaches—combining embeddings for clustering and statistical term scoring for extracting key topic terms—can be effective.

Note that the granularity of the textual unit matters (see Lareau and Malaterre, 2026): full texts may blur topic boundaries, while paragraphs or abstracts offer more focused representations. Choosing the right scale depends on the epistemic resolution sought.

Ultimately, embedding-based topic modeling allows us to situate epistemic framings within disciplinary contexts, supporting more precise analyses of how knowledge is construed differently across research areas.

6. Toward a theory of epistemic framings

Traditionally, philosophical analysis has treated epistemic concepts—such as "theory," "explanation," or "evidence"—in isolation, seeking precise definitions and logical relationships. But science is practiced in varied contexts (Knorr-Cetina, 1999; Chang, 2012; Ankeny and Leonelli, 2016), and these concepts vary in usage and may acquire different meanings depending on epistemic aims, methodological traditions, and disciplinary norms.

The idea of epistemic framing offers a way to theorize this variability. An epistemic framing is not just a preference for one term over another, but a patterned configuration of epistemic priorities, conceptual tools, justificatory practices, and linguistic forms that collectively express how a field construes knowledge. Prior work we conducted (Malaterre and Léonard, 2024) shows the relevance of this notion to describe how epistemic concepts function differently across biomedical subfields.

Why does this matter? First, it supports a version of scientific pluralism: the idea that science is not governed by a single set of epistemic norms (Longino, 2002). Second, it aligns with epistemic contextualism, suggesting that key scientific concepts have context-sensitive meanings. Third, it opens the door to data-driven philosophy of science: by integrating empirical tools with conceptual analysis, we can better understand the landscape of scientific knowledge production (Ramsey and De Block, 2022).

By analyzing the discourse of science at scale, we can move beyond abstract theorizing and map the real-world structures through which scientists articulate and justify knowledge. Investigating epistemic framings with quantitative text-mining methods complements traditional conceptual work in the philosophy of science. It brings new tools to long-standing debates and provides a platform for interdisciplinary collaboration between philosophers, computational linguists, and practicing scientists.

While current LLM-based methods still face scalability and interpretability challenges, the promise is clear: with refined tools and curated corpora, we can chart the evolving epistemic landscapes of science—revealing how knowledge is conceptualized, justified, and valued in different corners of scientific practice.[1]

Acknowledgements

F.L. acknowledges funding from Canada Social Sciences and Humanities Research Council (Postdoctoral Fellowships 756–2024-0557, Grant 430–2018-00899). C.M. acknowledges funding from Canada Social Sciences and Humanities Research Council (Grant 430–2018-00899) and Canada Research Chairs (CRC-950-230795). The authors thank Arno Simons, Adrian Wüthrich, Michael Zichert and Gerd Graßhoff for organizing the *LLMs in HPS* workshop (TU Berlin, 2025) and inviting them to contribute to this volume.

References

Ankeny RA and Leonelli S (2016) Repertoires: A post-Kuhnian perspective on scientific change and collaborative research. *Studies in History and Philosophy of Science Part A* 60: 18–28.

Blei M, Ng AY and Jordan MI (2003) Latent dirichlet allocation. *Journal of Machine Learning Research* 3(1): 993–1022.

Chang H (2012) *Is Water H2O?: Evidence, Realism and Pluralism*. Dordrecht: Springer Netherlands.

Grootendorst M (2022) BERTopic: Neural topic modeling with a class-based TF-IDF procedure (arXiv:2203.05794). *arXiv*. DOI: 10.48550/arXiv.2203.05794

Knorr-Cetina K (1999) *Epistemic Cultures: How the Sciences Make Knowledge*. Cambridge (MA): Harvard University Press.

Lamirel J-C, Lareau F and Malaterre C (2024) CFMf topic-model: Comparison with LDA and Top2Vec. *Scientometrics* 129(10): 6387–6405.

Lareau F and Malaterre C (2026) When less is more? Topic model performance across title, abstract, and full-text corpora. In: Simons A, Wüthrich A, Zichert M, et al. (eds) *Understanding Science with Large Language Models? Potentials for the History, Philosophy, and Sociology of Science*. Bielefeld: transcript, part-4.

Longino HE (2002) *The fate of knowledge*. Princeton (NJ): Princeton University Press.

Malaterre C and Léonard M (2023) Charting the Territories of Epistemic Concepts in the Practice of Science: A Text-Mining Approach. *The British Journal for the Philosophy of Science*. Epub ahead of print. DOI: 10.1086/725656

Malaterre C and Léonard M (2024) Epistemic markers in the scientific discourse. *Philosophy of Science* 91(1): 151–174.

1 This chapter was written with support from large language models (LLMs). All model-generated text was reviewed and, where necessary, rewritten by the authors, who remain fully responsible for the final version. For details on the use of LLMs in this volume, see the statement in the volume's introduction.

McPhetres J, Albayrak-Aydemir N, Mende AB, et al. (2021) A decade of theory as reflected in Psychological Science (2009–2019). *PLOS ONE* 16(3): e0247986.

Mizrahi M and Dickinson MA (2022) Philosophical reasoning about science: A quantitative, digital study. *Synthese* 200(2): 138.

Noichl M (2023) How localized are computational templates? A machine learning approach. *Synthese* 201(3): 107.

Ramsey G and De Block A (2022) *The Dynamics of Science: Computational Frontiers in History and Philosophy of Science.* Pittsburgh (PA): University of Pittsburgh Press.

Stammbach D, Zouhar V, Hoyle A, et al (2023) Revisiting Automated Topic Model Evaluation with Large Language Models. *Proceedings of the 2023 Conference on Empirical Methods in Natural Language Processing* 9348–9357.

Computational conceptual history of scientific concepts

From early digital methods to LLMs

Michael Zichert and Arno Simons

1. Introduction

This article zooms out to ask what it means to use large language models (LLMs) for computational conceptual history of scientific concepts. Our aim is to situate LLMs within the longer history of digital approaches to concept analysis in the history, philosophy, and sociology of science (HPSS), examine what these models add to existing methods, and review recent case studies that already employ them. This perspective allows us to ask not only what new possibilities LLMs open, but also how they inherit longstanding problems from both conceptual history and earlier attempts to study concepts computationally. In this article, we show how LLMs both extend and reshape computational conceptual history by revisiting its core methodological issues: corpus construction, concept operationalisation, model and training data choice, and the linked challenges of evaluation and interpretation.

The analysis of scientific concepts has long been a core concern in HPSS, where concepts are treated as historically layered, socially situated, and epistemically significant components of scientific practice rather than as fixed semantic units. Traditions of conceptual history and historical epistemology, as well as later developments in the discursive, practical, and material turns, have shown that concepts gain meaning through their use in specific research settings, institutional contexts, and socio-technical arrangements. Scholars such as Fleck (1979[1935]), Kuhn (1962), Rheinberger (1997), Hacking (1999), Pickering (1999), and Daston and Galison (2007) have emphasised that conceptual change often tracks deeper shifts in scientific problems, practices, and forms of reasoning, and that key terms can carry multiple, sometimes conflicting meanings that shape how scientists interpret phenomena.

Recent work has renewed interest in conceptual history within the field (Schauz, 2015; Müller and Schmieder, 2016; Schauz, 2025), alongside explicit calls to connect conceptual analysis more systematically to quantitative and computational methods in conceptual history more broadly (Wevers and Koolen, 2020; Marjanen, 2023). In parallel, historians of science have increasingly pointed to the methodological potential of large-scale textual and bibliometric evidence and have called for computationally grounded histories of science (Laubichler, 2019; Lalli, 2023). Most recently, LLMs have intensified these discussions by making new forms and scales of contextual modelling available, prompting HPSS scholars to explore how these models can be integrated into concept historical research designs, as reflected by this volume and the contributions in this part of the book. For a more comprehensive survey of LLMs for HPSS, see our earlier work (Simons et al., 2026).

The article has two main parts. Chapter 2 reconstructs computational conceptual history before LLMs by bringing together the three strands of work: early digital methods in HPSS, distributional approaches from digital history and related research, and lexical semantic change detection. It then sums up the main challenges and opportunities, focusing on corpus construction, operationalization and modelling choices, and evaluation and interpretation. Chapter 3 turns to the era of LLMs. It starts with a short introduction to LLMs, reviews LLM-based work on lexical semantic change detection, and discusses relevant case studies in HPSS that make use of LLMs. It then revisits the earlier methodological questions, showing how issues like corpus construction, model choice and training data, operationalization trade-offs, and evaluation and interpretation play out in LLM-based workflows.

2. Computational conceptual history before LLMs

2.1 Early digital approaches to modeling conceptual change

2.1.1 Early digital methods in HPSS

Alongside the well-established qualitative tradition of conceptual history in HPSS, scholars, especially in the scientometrics community, have long explored digital methods for conceptual history and analysis. One of the most influential early contributions is Small's work on co-citation analysis (1973). Co-citation refers to the frequency with which two publications are cited together by later authors. Small links this technique to a broader claim about meaning in scientific literature. He argues that frequently cited papers can function as concept symbols (1978), serving as shorthand for ideas, methods, or findings that a research community recognizes as central. From this perspective, co-citation is useful because it traces how these symbolically important contributions are connected in practice and therefore reflects the conceptual associations that the wider population of researchers sees between key works. As these perceptions and citing habits shift, co-citation patterns change as well, offering a dynamic view of how a field's core ideas are linked and reorganized over time. If highly cited papers are taken to mark major concepts or methods, then co-citation analysis can map the evolving structure of these central elements with considerable detail (Small, 1999; Boyack et al., 2005; Boyack and Klavans, 2014).

Building on this foundation, Callon et al. (1983, 1986) introduce co-word analysis as a complementary computational strategy for mapping conceptual structures through language use in the scientific literature. Rather than inferring conceptual relations from shared references, this approach traces the co-occurrence of keywords and index terms to reveal how research communities connect problems, methods, and objects of inquiry within evolving discourses. As a result, co-word maps can highlight emerging themes, shifting alliances between topics, and the consolidation of new research areas, offering a vocabulary-centered view of conceptual change over time (Rip and Courtial, 1984; Ding et al., 2001).

A third pre-LLM computational approach to conceptual history in HPSS builds on topic modeling, in particular Latent Dirichlet Allocation (LDA) and its extensions (Blei et al., 2003; Blei and Lafferty, 2006). Topic models represent documents as mixtures of latent topics and topics as probability distributions over words, so that each topic can be read as a recurring constellation of terms that approximates a conceptual or thematic structure. By fitting such models to large scientific corpora and tracking how topics change in prevalence and internal composition over time, researchers have used LDA-based methods to investigate the emergence, transformation, and decline of research areas, thus providing a scalable way to study conceptual dynamics in science (Griffiths and Steyvers, 2004; Blei and Lafferty, 2007; Hall et al., 2008).

For critical comparisons between these different methods, see for example Boyack and Klavans (2010), Leydesdorff and Nerghes (2017), Benz et al. (2025), and Xie and Waltman (2025).

2.1.2 Digital Begriffsgeschichte

In parallel to these HPSS approaches, digital humanities and digital history have developed computational conceptual history in dialogue with Begriffsgeschichte (Koselleck, 2002; Koselleck and Richter, 2011), drawing on co-occurrence analysis and, later, distributional models to trace changing patterns of use. In Begriffsgeschichte, concepts are treated as historically layered semantic structures whose meanings accumulate, fracture, and compete over time, and which can be reconstructed to trace broader social and intellectual transformations. This perspective shares with conceptual history in HPSS the commitment to historical situatedness and contextual interpretation. Where Begriffsgeschichte tends to be more explicit, however, is in treating polysemy as a central methodological issue and in emphasizing the distinction between the words that appear in texts and the concepts behind them. This also brings with it a comparatively developed account of what concepts are and why the term–concept relation is methodologically fraught. By contrast, much pre-LLM computational work in HPSS has operationalized concepts through comparatively stable proxies (highly cited papers, index terms, topic labels) and mapped the resulting structures onto concepts in a more pragmatic, case-specific way, often without foregrounding the term–concept relation at the same level of theoretical depth. As a result, polysemy and sense competition tend to remain implicit or get smoothed over rather than becoming a first-order methodological concern.

Over the past decade, scholars in this field have increasingly tried to operationalise conceptual history with pre-LLM digital methods, especially through co-occurrence and early word embedding techniques to track shifts in contextual patterns and semantic neighborhoods across large corpora. Biemann and Friedrich (2016) highlight both the promise of large digitised corpora for mapping contexts of use and the difficulty of semantic ambiguity when meaning is inferred from distributional patterns. Schwandt (2018) similarly uses tools such as Voyant (Sinclair and Rockwell, 2016) to unsettle modern readings of historical concepts, while stressing how hard it is to compare uses across periods and genres. Gavin et al. (2019) explore a hybrid approach using vector semantics in computer-assisted close reading. Working with Word2Vec embeddings (Mikolov et al., 2013), they treat distributional representations not as stand-alone models of meaning but as tools that can support interpretation. In this way, embeddings serve to draw attention to certain patterns in the corpus rather than to replace close reading.

Where Gavin et al. highlight the interpretive potential of embeddings, Sommerauer and Fokkens (2019) focus on their methodological risks. Using "racism" as a case study, they show how distributional semantic models trained on the same corpus can yield divergent results depending on architecture, parameters, and corpus composition, demonstrating the need for controlled experimental design if such methods are to support historical claims. The most systematic pre-LLM account is Wevers and Koolen (2020), who use static embedding models such as Word2Vec (Mikolov et al., 2013) and GloVe (Pennington et al., 2014) on Dutch newspaper corpora to highlight opportunities to trace and measure semantic and conceptual change via shifts in vector positions and neighbourhoods. At the same time, they emphasise that embedding based results require interpretive grounding and validation, since the operationalization of concepts through lexical proxies, the size, quality, and bias of historical corpora, and the difficulties of evaluating model quality all require careful methodological reflection.

2.1.3 Pre-LLM approaches to lexical semantic change detection

Another tradition of studying shifting word and concept meanings with digital methods arises from research on lexical semantic change detection (LSCD). Emerging within computational linguistics, computer science, and natural language processing, LSCD uses many of the same distributional and embedding-based techniques as digital conceptual history, but typically focuses on technical implementation, the construction and evaluation of models and datasets, and systematic benchmarking rather than on substantive historical interpretation. Methodologically, its goal is to formalize and measure lexical semantic change in ways that allow different models and approaches to be compared. In recent years, however, LSCD techniques have increasingly been taken up and adapted within digital conceptual history, where they are combined with closer attention to historical context, sources, and interpretation.

Early LSCD approaches relied on methods such as dynamic topic modelling (Blei and Lafferty, 2006), statistical co-occurrence analysis (Gulordava and Baroni, 2011), and graph-based techniques (Mitra et al., 2014) to capture changes in the contexts in which words appear. With the rise of distributional semantics, vector-based methods became dominant. A central distinction is between static word embeddings, which represent each word type in a corpus as a single fixed vector (e.g., Word2Vec, GloVe), and more recent contextualized approaches, in which every word occurrence is represented by its own unique vector through the use of LLMs like BERT (Devlin et al., 2018) or GPT (Radford et al., 2018). Static embeddings make it comparatively straightforward to measure shifts in the predominant use of a word by comparing vectors across time slices, but they struggle with highly polysemous items, where multiple senses are collapsed into a single representation.

Within the pre-LLM, static embedding approach, Hamilton et al. (2016) propose influential methods for training separate Word2Vec models on different time slices, aligning the resulting spaces, and quantifying change by measuring cosine distance between a word's vectors across periods, which they use to suggest broader regularities linking semantic change to frequency and polysemy. Dubossarsky et al. (2017) subsequently caution that some such patterns may reflect artefacts of the models and evaluation procedures rather than genuine linguistic regularities, highlighting inherent biases in word representation models and the need for careful baselines and controls. Kutuzov et al. (2018) provide a comprehensive survey of pre-LLM SCD work, summarizing tasks, datasets, and methodological choices. Taken together, this tradition has strongly shaped the toolkits that conceptual historians now draw on, but as HPSS scholars we need to understand how these methods work, how to adapt them to our own questions and corpora, and how to combine them with the contextual interpretation and conceptual analysis at the core of our field.

2.2 Opportunities and challenges of pre-LLM computational conceptual history

Building on the three strands introduced in 2.1, this section draws together their key lessons to identify the main opportunities and recurring challenges of pre-LLM computational conceptual history, from the promise of scale to the constraints of corpus construction, concept operationalization, evaluation, and historically accountable interpretation.

2.2.1 Text as data and corpus construction

The availability of large digitized and, increasingly, born-digital text collections has often been framed as the decisive opportunity for computational approaches to conceptual history, because it allows scholars to study conceptual dynamics at scales of time and corpus size that were not feasible before (Biemann and Friedrich, 2016; Wevers and Koolen, 2020). Digitization makes sources easier to access and aggregate, but it also encourages historians to encounter their material through search, for example via keyword queries, indices, and frequency tools, so that individual word forms become common entry points into archives that were formerly approached through documents, genres, or debates. This invites new kinds of questions about how the relationship between words and concepts can be quantitatively operationalized and how linguistic change can be measured across large text corpora. As Marjanen (2023) notes in a wider discussion of quantitative approaches to conceptual history, this development is broadly good news for the field, but it also requires researchers to reflect carefully on what quantification captures and how large-scale linguistic patterns should be interpreted.

At the same time, what can be claimed from such large-scale analyses is constrained by decisions about corpus construction. What gets digitized, what is missing, how texts are segmented, and what metadata is available for filtering and comparison all shape what can reasonably be concluded from the analysis. Digitization is uneven across regions, institutions, and genres, and corpora are also uneven over time: recent decades often dominate, while earlier periods can be sparse. In addition, OCR quality, historical spelling variation, and corpus bias are primary methodological issues for embedding based historical analysis (Wevers and Koolen, 2020). Pre-LLM distributional approaches are also sensitive to preprocessing decisions, such as tokenization, lemmatization and stopword removal. HPSS corpora amplify these issues because scientific language is often highly domain-specific and heterogeneous (e.g., dense technical terms, abbreviations, and shifting terminological conventions). In addition, conceptual work is often distributed across different types of sources (e.g., articles, textbooks, grant proposals, review papers, lab manuals, etc.), and can also be encoded in non-textual elements (e.g., formulae, tables, figures, code, and diagrams) that are typically stripped during digitization and preprocessing when employing pre-LLM methods, further increasing the problem of what purely text-based models can measure.

2.2.2 Operationalization and modeling choices

One central methodological concern for computational conceptual history is the operationalization, or modeling, of concepts. To employ digital tools, researchers must translate historically layered and contested semantic structures into measurable proxies. This translation is never neutral and is further complicated by the fundamental word concept gap emphasized in conceptual history (de Bolla et al., 2019; Marjanen, 2023). As discussed in 2.1, pre-LLM approaches operate on words, citations, keywords, or topic distributions, and they infer conceptual meaning from these different surface forms. In practice, this often means relying on lexical proxies, for example seed terms, keyword sets, anchor papers, or topic word distributions, to stand in for broader conceptual structures. This can be productive, but it also makes results sensitive to the initial proxy choice and to whether the proxy actually captures the relevant uses, genres, and debates in the corpus.

In pre-LLM conceptual history, three recurring strands of operationalization can be observed, each with specific modeling trade-offs. Scientometric networks are closely tied to scholarly behavior and yield useful representations, but they primarily track patterns of association and community organization rather than conceptual content. Topic models enable scalable thematic mapping that is useful for exploration and periodization, but whether topics correspond to concepts remains contested (Chang et al., 2009). Early embedding methods align most closely with intuitions of conceptual meaning based on context and semantic neighborhoods, but they introduce difficulties with polysemy and aggregation. Pre-LLM approaches rely on static word embeddings that summarize all occurrences of a word into a single representation, which can flatten disputes and sense variations that are often central to conceptual historical analysis. A further limitation specific to HPSS is that scientific concepts are often considered material-semiotic: they stabilize not only through words, but also through instruments, experimental systems, and standardizing infrastructures, so conceptual change is not always legible as lexical change alone (Müller and Schmieder, 2018).

2.2.3 Evaluation and interpretation

Evaluation is a persistent challenge in pre-LLM computational conceptual history because most approaches lack straightforward ground truths and because in most humanities-aligned areas there are no widely-accepted evaluation datasets, an issue that is especially acute in HPSS (see the introduction to this volume, Simons et al., 2026). Topic models, for example, can yield substantially different topic structures depending on hyperparameters and preprocessing, raising the question of whether an inferred topic represents a stable conceptual formation or an artefact of the modeling setup. Static embedding workflows also depend heavily on corpus composition and parameter settings, so robust evaluation typically requires careful baselines, controlled comparisons, and triangulation of results (Dubossarsky et al., 2017; Sommerauer and Fokkens, 2019).

Across all these approaches, careful historical interpretation remains essential. In computational conceptual history, outputs such as frequency curves, co-citation clusters, co-word networks, or embedding shifts are rarely historical claims on their own. Instead, they point to patterns that have to be turned into arguments. Interpretation is this conversion step: deciding what a pattern can count as evidence for, which alternative explanations need to be checked and ruled out, and how the result connects to actors, instruments, practices, or institutions. LSCD often concentrates on measuring change and benchmarking models and typically leaves these interpretive steps outside its scope. For humanities and HPSS scholars, however, interpretation is the core of conceptual analysis, and pre-LLM work therefore kept interpretive responsibility firmly with the researcher, whether through a hybrid approach as in Gavin et al. (2019) or through embedding-based conceptual history that explicitly insists on interpretive validation as in Wevers and Koolen (2020).

3. LLMs for computational conceptual history

Building on the challenges and opportunities discussed in the previous chapter, the next two sections outline what LLMs can contribute to computational conceptual history, which problems they inherit, and which gaps they may help to close.

3.1 Recent advances and and HPSS case studies

3.1.1 Short primer on LLMs

As outlined in the introduction to this volume (Simons et al., 2026) and in our survey (Simons et al., 2026), we use LLMs to refer to neural language models built on the transformer architecture (Vaswani et al., 2017). Transformers build on and extend components introduced in earlier neural network models, most importantly attention mechanisms, encoder and decoder modules, and contextualized word representations, and integrate them into an architecture that scales effectively to very large datasets and parameter counts and produces rich representations of language.

Within this family, it is useful to distinguish two dominant architectural and training paradigms (see the Introduction to this volume for more details, Simons et al., 2026): Decoder-based, or generative models, often associated with chat-based systems such as ChatGPT, are trained with a next-token prediction objective, where given a text input sequence, the model learns to predict the most likely continuation (Radford et al., 2018). Encoder-based models, by contrast, are typically trained with masked language objectives, where the model learns to reconstruct tokens that have been hidden from an input sequence. The paradigmatic example here is BERT (Devlin et al., 2018), which is why they are also being referred to as BERT-like models. Because the encoder is oriented bidirectionally, it can use information from both the left and right context when representing a token, which makes it especially strong at modelling local meaning and sentence-level structure. Both model types produce context-sensitive vector representations for each token occurrence, so-called *contextualized* word embeddings (CWEs). Unlike *static* word embeddings, which assign a single vector to a word type across a whole corpus, CWEs are conditioned on the surrounding sentence or passage, so the same word form can have different representations in different contexts.

The distinction between decoder-based and encoder-based models matters for computational conceptual history because it tends to support different applications. Generative models are useful for text production and can, through prompt engineering, support structured extraction and forms of interpretive assistance. Encoder-based models, meanwhile, remain central for measurement oriented workflows, where CWEs serve as features for tasks such as lexical semantic change modelling, classification, or sequence labelling. One key advantage here is they can be efficiently adapted through task-specific fine tuning or domain adaptation on comparatively small datasets tailored to the historical or disciplinary question at hand. Generative models, by contrast, are often adapted at use time through in context learning, for example via few-shot or multi-shot prompting, or through retrieval-augmented generation (RAG), where relevant passages are retrieved from an external collection and included in the prompt to ground and influence the model's output (Gao et al., 2024).

3.1.2 Lexical semantic change detection (LSCD) using LLMs

LSCD with transformer models was initially driven mainly by encoder-based models, since their bidirectional training objective makes them well suited for detecting meaning shifts. A common workflow begins by extracting all occurrences of a target word from a diachronic corpus and representing each occurrence with a CWE extracted from a BERT-like model that is often domain adapted or further pretrained on the relevant historical material. These CWEs are then aggregated in two main ways: form-based approaches pool all occurrences into a single time-specific representation, while sense-based approaches use clustering methods to group the vectors into sense-specific groups. These time-specific representations can then be used to assess the amount of lexical change through distance-based measures such as cosine distance, or information-theoretic indicators such as Shannon entropy. This makes it possible to analyse different types of semantic change, most prominently shifts in a word's dominant meaning and changes in its degree of polysemy. Many recent studies both build on and further develop this pipeline (e.g., Martinc et al., 2020; Montariol et al., 2021; Kutuzov et al., 2022; Cassotti et al., 2023), and the survey by Periti and Montanelli (2024) offers the most systematic review of approaches, benchmarks, and model choices.

More recently, decoder-based or generative models have entered LSCD research along two somewhat different lines. First, they can be used directly for change detection through prompting, for example by asking the model for sense judgements, semantic relatedness decisions, or lexical substitutes. Periti et al. (2025) test a range of prompting strategies, including zero-shot and multi-shot setups with ChatGPT, on both long-term and short-term benchmarks for semantic change. They conclude that this direction is promising, but that encoder-based approaches with BERT-like models still achieve stronger performance, especially for fine-grained, short-term distinctions. Second, generative models are increasingly being used in hybrid approaches. Rather than replacing encoder-based measurement, they are applied to generate historically plausible, sense-specific example sentences that serve as synthetic, annotated diachronic datasets for training and evaluation, directly targeting the problem of sparse historical data. In these approaches, semantic change is still assessed with an encoder-based model. Cassotti and Tahmasebi (2025a), for instance, show how a Llama-based generative model can be used for sense-specific historical usage generation at scale, and how such data can be used for the standard downstream LSCD workflow. In a second paper that focuses on LSCD for highly polysemous target words (Cassotti and Tahmasebi, 2025b), they extend this idea by using decoder-based models not only to generate usages, but also to produce context- and time-specific sense definitions for the induced sense clusters, which are then used as anchors for the subsequent analysis.

3.1.3 HPSS case studies using LLMs

In recent years, HPSS scholars have increasingly adopted LLMs to computational analysis of scientific concepts. While these studies often draw on methods developed in LSCD, they also integrate other models of conceptual change and adapt them to the specific problems and questions raised by scientific concepts.

In this volume, several chapters show how HPSS scholars are adapting LLM-based methods to the analysis of scientific concepts. Malaterre and Lareau (2026) discuss how both encoder-based and generative models can be used to analyse the ways scientists deploy epistemic concepts in the biomedical domain. Focusing on epistemic markers such as "theory", "model", and "explanation", they propose a hybrid workflow in which LLMs are used to expand and refine lexicons of epistemic markers, classify statements in context, and map discipline-specific epistemic framings at scale. Ahmadi (2026) uses CWEs from BERT to compare how consistently terms are used in sociology versus astrophysics journal articles. She computes a field-level Semantic Uniformity Score by averaging cosine similarity across each term's own contextualized occurrences and aggregating across terms, interpreting higher similarity as more stable, standardized usage and lower similarity as more context-dependent use shaped by polysemy, topic, syntax, or genre. She finds astrophysics to be more semantically uniform than sociology. Aguilar Valdez et al. (2026) sketch how the generation of a graph structure of concepts, which could then be enriched with an LLM-based analysis, could take conceptual analysis to the next level, while also using their study to familiarise readers with key concepts for computational approaches to semantic change. Simons (2026) uses the term "Planck" as a deliberately polysemous test case with well-known senses for computational conceptual analysis in astrophysics, combining several complementary methods. Using a domain-adapted BERT model that is further pretrained on a large astrophysics and high energy physics corpus (Simons, 2024), he evaluates how well contextualized embeddings separate these senses through supervised sense prediction with sense prototypes, as well as through unsupervised clustering and cluster quality measures. Building on the resulting sense structure, he then traces lexical semantic change from 1990 to 2022 by tracking how the distribution of senses shifts over time and by combining sense based- and form-based indicators.

In an earlier study, Kleymann et al. (2022) approach the use of the concept of theory in digital humanities journals by combining a semasiological perspective with an onomasiological one. In their semasiological case study, they build dictionaries of theory-related references and use frequency based and co-occurrence analyses to map how specific theoretical frameworks are used across the corpus. In their onomasiological case study, they follow the LSCD pipeline, fine-tuning a BERT model on the journal corpus, extracting CWEs for all occurrences of "theory" and related epistemic terms such as "model" and "method", and then comparing period specific and aggregated representations via cosine similarity to trace how the concept's semantic neighbourhood, and with it its use and meaning, have changed over time.

Lastly, Zichert et al. (2025) trace the conceptual history of the virtual particle concept by using "virtual" as a linguistic marker and analysing its usage across a large physics corpus spanning 1924 to 2022. They domain adapt a pretrained BERT model on this corpus, extract CWEs for each occurrence of "virtual", and apply a range of form-based and sense-based LSCD methods to track both shifts in dominant meaning and changes in the degree of polysemy over time. The approach is complemented by dependency parsing, which identifies the nouns most often used with "virtual", and these syntactic patterns are then used as an interpretive cross check on the LLM-based methods. Across all of these HPSS case studies, the dominant methodological approach remains the use of CWEs generated by encoder-based LLMs, while decoder-based approaches do not yet play a comparable role.

3.2 Opportunities and challenges for the LLM-era

Building on the developments and case studies surveyed in 3.1, this section draws their key lessons together to update the main opportunities and recurring challenges of computational conceptual history in the LLM era, comparing them to the pre-LLM landscape and highlighting how encoder-based and decoder-based models, and the model choices they entail, reshape questions of corpus and training data, concept operationalization and modelling trade-offs, and the tasks of evaluation and interpretation.

3.2.1 Corpus construction and datasets

Corpus construction remains central in the LLM era, but it has become more layered. Many pre-LLM issues, like uneven digitization, missing genres, or sparse early periods, remain and still shape what can be measured in HPSS conceptual history. What changes is that LLM-based workflows also inherit the biases and blind spots of the datasets the models are trained on. This is especially true for large-scale generative models where analysis is influenced not only by the researcher-curated corpus but also by the often opaque pre-training datasets. As a result, questions of infrastructure, access, and power become more relevant: who controls model training and updates, who can afford computation, and how environmental and financial costs shape what can or should be researched (Lang, 2026). For conceptual history and HPSS more broadly, this increases the need for open-science infrastructures and transparent alternatives (Valleriani, 2025).

At the same time, LLMs also offer significant opportunities regarding text corpora: They are far more robust to raw text than earlier pipelines, reducing the need for heavy preprocessing that can erase rhetorically and historically meaningful signals. More importantly for HPSS, domain adaptation has become a practical and standard task. This is especially true for encoder-based models, which can be effectively further pretrained or fine-tuned on discipline-specific corpora to better capture the specialized vocabulary and shifting terminological conventions of scientific language (e.g., Simons, 2024; Zichert et al., 2025). Another promising direction regarding corpora is temporal adaptation. Because most current LLMs are not trained with historical questions in mind, they risk projecting contemporary meanings into earlier texts. Time-sensitive fine-tuning and explicit temporal modeling aim to counteract this tendency (Underwood, 2025; Büttner, 2026; Meding and Daugs, 2026). The problem of historical sparse data can also be addressed through the use of generative LLMs that can help support corpus work itself: assisting annotation and coding, improving search within conceptual fields, as well as generating synthetic data to support training or evaluation in sparse domains (Cassotti and Tahmasebi, 2025a; Danilova et al., 2026).

3.2.2 Operationalization

Operationalization remains a central concern in LLM based computational conceptual history, but the range of modelling strategies and outputs has expanded. The most important shift is from static to contextualized word embeddings, which most current case studies generate through encoder-based models (e.g., Kleymann et al., 2022; Zichert et al., 2025; Simons, 2026). CWEs represent individual occurrences and therefore handle polysemy more directly, which makes it possible to model time specific uses through clustering, prototype-based sense prediction, or related word-sense disambiguation workflows, and then assess change either at the level of dominant senses or at the level of shifting sense distributions, which can be used to approximate changes in polysemy over time. These contextualized approaches also introduce new modelling trade-offs: results depend on choices such as which model layers to use, how token representations are merged, and how corpora are divided into time periods, while sense modeling requires decisions about clustering methods, the number of senses, and how clusters are labelled (Periti and Montanelli, 2024). Since many workflows still aggregate representations over time slices or clusters, CWEs improve access to polysemy but do not by themselves determine what a historically meaningful sense is.

Decoder-based models, on the other hand, modify operationalization in a different way. Mostly employed in hybrid workflows, they can be used to extract candidate definitions or typical usages of a term through retrieval-augmented generation (Gao et al., 2024), to build and refine coding schemes (Dunivin, 2025), or generate summaries of how a concept is discussed within a time text or subcorpus (Zhang et al., 2024). In LSCD, this direction appears both in prompt engineering based approaches that ask the model to judge whether a term is used with the same meaning across different contexts, or to propose substitute words for a given usage (Periti et al., 2025), and in synthetic data generation strategies that produce sense and time specific example usages of a concept or term (Cassotti and Tahmasebi, 2025a). These workflows also introduce new modelling choices that require explicit documentation, including model use, prompt design, in context examples, and output schemas. Much of this work is still in an early stage, but it is likely to become more important as generative models improve and as scholars develop more reliable hybrid pipelines for historical analysis.

Despite these advances, the long-standing operationalization problem in computational conceptual history remains relevant: concepts are bound to words, but they are not reducible to them, and linguistic proxies only capture part of what conceptual historians ultimately care about (Koselleck, 1985; Marjanen, 2023; de Bolla et al., 2019). In HPSS, this point is especially important, since scientific concepts are often understood as material-semiotic constructs that are jointly shaped by language, scientific practices, and material arrangements (Müller and Schmieder, 2018). Rheinberger (1997), for example, argues that concept histories of epistemic objects cannot be reconstructed from texts alone, since diagrams, formulae, and the material organisation of research practice are part of how scientific meaning is produced and transformed. This is where the rise of multimodal LLMs (Wu et al., 2023; Yin et al., 2024) marks a potential new direction. By integrating non-textual data such as visual and oral communication, they may enable future work to connect these dimensions more systematically, for example by tracing how textual uses of a concept co-evolve with its diagrammatic and mathematical representations. At the same time, this is still early work, that, besides multimodal LLMs, requires suitable multimodal corpora, improved digitisation of figures and formulae, and reliable alignment between textual and non-textual elements.

3.2.3 Evaluation and interpretation

Evaluation remains an important issue for computational conceptual history using LLMs. Most of the workflows discussed in 3.1 already include some form of evaluation, whether through established LSCD benchmarks or through the use of auxiliary quantitative methods such as frequency and co-occurrence measurements or dependency parsing that can support or challenge embedding-based findings. This need is amplified by the fact that LLM-based embeddings are more opaque, more "black boxed", than most of the earlier quantitative methods in conceptual history, where assumptions and operationalization steps were often easier to inspect. As long as we have not established in detail how reliably these models capture conceptual dynamics, triangulation will remain the default evaluation strategy, and strong studies combine multiple quantitative lenses while pairing them with qualitative validation. As Marjanen (2023) argues, machine learning outputs only become meaningful when qualitative work clarifies what a measure is actually sensitive to (e.g., genre shifts, general linguistic trends, or changing publication volumes) and thereby helps distinguish conceptual dynamics from artefacts of the corpus or model (compare the discussion by Ahmadi, 2026, of whether high variability in embeddings signals semantic ambiguity, topical shifts, syntactic variation, or genre differences). This is especially important for HPSS, where we rarely have stable ground truths and where domain- and period-specific benchmark datasets are rare, and it is therefore unlikely that evaluation will center on a single standardized metric any time soon. For many (often commercial) generative models, this evaluation problem is sharpened by familiar reliability issues such as hallucinations, non-transparent model version updates and limited training-data transparency, that further complicate reproducibility (see Lang, 2026).

Interpretation remains the core operation of HPSS computational conceptual history, even in the LLM era. LLM-based workflows can support interpretive work in (at least) two complementary modes: first, they can test qualitative claims by empirically checking whether proposed shifts, stabilizations, or conceptual reorientations are visible across much larger corpora and longer timeframes than traditional case studies allow. Second, they can be used heuristically to discover interesting periods or candidate passages that call for closer analysis. As discussed earlier, LLMs in the form of generative models, can also support interpretation within hybrid workflows, for example by helping to build and refine coding schemes, extracting candidate definitions or typical uses of a term from retrieved contexts, and generating structured summaries of how a concept is framed within a time slice. At the same time, the increasing use of such systems raises a distinctive interpretive risk: even when models are used as assistants, their outputs can subtly shift interpretive control away from scholars (Khutsishvili, 2026). It remains to be seen how far such workflows can be pushed, since so far no case studies in computational conceptual history in HPSS have done so on a large scale.

4. Conclusion

In this contribution we have argued that LLM-based computational conceptual history in HPSS continues and reconfigures earlier digital approaches to modelling conceptual change, from co-citation methods to topic models and distributional semantics. LLMs expand what can be operationalised at scale by offering richer context-sensitive representations for tracing semantic and conceptual change through encoder-based models, and by enabling new hybrid workflows through the addition of generative models. At the same time, the core methodological questions remain, but LLMs shift how they have to be addressed, opening new possibilities while also adding new challenges. Corpus construction for analysis as well as model training, evaluation, concept operationalization, and domain as well as temporal adaptation are still decisive for what can be claimed. In practice, historically-accountable conceptual analysis continues to depend on triangulation, combining LLMs with established quantitative techniques and qualitative validation, with LLMs functioning as one component in a multi-method toolbox rather than as a self-sufficient solution.

Several avenues for future work in the field stand out. First, HPSS conceptual history needs more case studies that test generative models in hybrid workflows alongside encoder-based models, and that explore when, where, and for what tasks standalone generative approaches add value beyond supportive roles within conceptual history pipelines. Second, multimodal modeling opens a path toward operationalizations closer to how many HPSS scholars understand scientific concepts, as strongly connected with instruments, diagrams, formulae, and other material and representational practices, but this will require new corpora and reliable alignment between textual and non-textual elements. Third, progress across these lines of work depends on open and transparent infrastructures, open and HPSS-specific models and datasets, clear documentation standards, and shared evaluation resources that make results reproducible and meaningfully comparable across studies and over time.[1]

References

Aguilar-Valdez S, Phan-Tất B, Speelman D, et al. (2026) Discursive parallels of the chemical revolution. Topic modelling and distributional analysis. In: Simons A, Wüthrich A, Zichert M, et al. (eds) *Understanding Science with Large Language Models? Potentials for the History, Philosophy, and Sociology of Science*. Bielefeld: transcript, part-3.

Ahmadi ES (2026) Exploring disciplinary differences in semantic uniformity. A computational approach to codification. In: Simons A, Wüthrich A, Zichert M, et al. (eds) *Understanding Science with Large Language Models? Potentials for the History, Philosophy, and Sociology of Science*. Bielefeld: transcript, part-3.

Benz P, Pradier C, Kozlowski D, et al. (2025) Mapping the unseen in practice: comparing latent Dirichlet allocation and BERTopic for navigating topic spaces. *Scientometrics* 130(7): 3839–3870.

Biemann C and Friedrich A (2016) Digitale Begriffsgeschichte? Methodologische Überlegungen und exemplarische Versuche am Beispiel moderner Netzsemantik. *Forum Interdisziplinäre Begriffsgeschichte*. Epub ahead of print 1 January 2016.

Blei DM, Ng AY and Jordan MI (2003) Latent Dirichlet allocation. *Journal of machine Learning research* 3(Jan): 993–1022.

Blei DM and Lafferty JD (2006) Dynamic topic models. In: *Proceedings of the 23rd international conference on Machine learning*, New York, NY, USA, 25 June 2006, pp. 113–120. ICML '06. Association for Computing Machinery. Available at: https://doi.org/10.1145/1143844.1143859.

Blei DM and Lafferty JD (2007) A Correlated Topic Model of Science. *The Annals of Applied Statistics* 1(1). Institute of Mathematical Statistics: 17–35.

Boyack KW, Klavans R and Börner K (2005) Mapping the backbone of science. *Scientometrics* 64(3). Akadémiai Kiadó, co-published with Springer Science+Business Media B.V., Formerly Kluwer Academic Publishers B.V.: 351–374.

Boyack KW and Klavans R (2010) Co-citation analysis, bibliographic coupling, and direct citation: Which citation approach represents the research front most accurately? *Journal of the American Society for Information Science and Technology* 61(12): 2389–2404.

Boyack KW and Klavans R (2014) Creation of a highly detailed, dynamic, global model and map of science. *Journal of the Association for Information Science and Technology* 65(4): 670–685.

Büttner J (2026) Why pursue temporally-grounded AI for historical disciplines, and what makes it so challenging? In: Simons A, Wüthrich A, Zichert M, et al. (eds) *Understand-*

1 This chapter was written with support from large language models (LLMs). All model-generated text was reviewed and, where necessary, rewritten by the authors, who remain fully responsible for the final version. For details on the use of LLMs in this volume, see the statement in the volume's introduction.

ing Science with Large Language Models? Potentials for the History, Philosophy, and Sociology of Science. Bielefeld: transcript, part-2.

Cassotti P, Siciliani L, DeGemmis M, et al. (2023) XL-LEXEME: WiC Pretrained Model for Cross-Lingual LEXical sEMantic changE. In: *Proceedings of the 61st Annual Meeting of the Association for Computational Linguistics (Volume 2: Short Papers)* (eds A Rogers, J Boyd-Graber, and N Okazaki), Toronto, Canada, July 2023, pp. 1577–1585. Association for Computational Linguistics. Available at: https://aclanthology.org/2023.acl-short.135.

Cassotti P and Tahmasebi N (2025a) Sense-specific Historical Word Usage Generation. *Transactions of the Association for Computational Linguistics* 13: 690–708. Available at: https://doi.org/10.1162/tacl_a_00761.

Cassotti P and Tahmasebi N (2025) A Hypothesis-Driven Framework for Detecting Lexical Semantic Change. In: *Proceedings of the Eleventh Italian Conference on Computational Linguistics (CLiC-it 2025)* (eds C Bosco, E Jezek, M Polignano, et al.), Cagliari, Italy, 24 September 2025. CEUR Workshop Proceedings. CEUR. Available at: https://ceur-ws.org/Vol-4112/#18_main_long

Chang J, Gerrish S, Wang C, et al. (2009) Reading Tea Leaves: How Humans Interpret Topic Models. In: *Advances in Neural Information Processing Systems*, 2009. Curran Associates.

Callon M, Courtial J-P, Turner WA, et al. (1983) From translations to problematic networks: An introduction to co-word analysis. *Social Science Information* 22(2). SAGE Publications Ltd: 191–235.

Callon M, Law J and Rip A (1986) Qualitative Scientometrics. In: Callon M, Law J, and Rip A (eds) *Mapping the Dynamics of Science and Technology: Sociology of Science in the Real World*. London: Palgrave Macmillan UK, pp. 103–123. Available at: https://doi.org/10.1007/978-1-349-07408-2_7.

Danilova VV, Reed J, Burchell A, et al. (2026) Zero-shot generation of synthetic historical data with LLMs. In: Simons A, Wüthrich A, Zichert M, et al. (eds) *Understanding Science with Large Language Models? Potentials for the History, Philosophy, and Sociology of Science*. Bielefeld: transcript, part-2.

Daston L and Galison P (2007) *Objectivity*. New York: Zone Books.

de Bolla PD, Jones E, Nulty P, et al. (2019) Distributional Concept Analysis: A Computational Model for History of Concepts. *Contributions to the History of Concepts* 14(1). Berghahn Journals: 66–92.

Devlin J, Chang M-W, Lee K, et al. (2018) BERT: Pre-training of Deep Bidirectional Transformers for Language Understanding. *arXiv:1810.04805 [cs]*. Epub ahead of print 10 October 2018.

Ding Y, Chowdhury GG and Foo S (2001) Bibliometric cartography of information retrieval research by using co-word analysis. *Information Processing & Management* 37(6): 817–842.

Dubossarsky H, Weinshall D and Grossman E (2017) Outta Control: Laws of Semantic Change and Inherent Biases in Word Representation Models. In: *Proceedings of the 2017 Conference on Empirical Methods in Natural Language Processing* (eds M Palmer, R Hwa, and S Riedel), Copenhagen, Denmark, September 2017, pp. 1136–1145. Association for Computational Linguistics. Available at: https://aclanthology.org/D17-1118.

Dunivin ZO (2025) Scaling hermeneutics: a guide to qualitative coding with LLMs for reflexive content analysis. *EPJ Data Science* 14(1): 28.

Gao Y, Xiong Y, Gao X, et al. (2024) Retrieval-Augmented Generation for Large Language Models: A Survey. arXiv:2312.10997. arXiv. Available at: http://arxiv.org/abs/2312.10997.

Gavin M, Jennings C, Kersey L, et al. (2019) Spaces of Meaning: Conceptual History, Vector Semantics, and Close Reading. In: Gold MK and Klein LF (eds) *Debates in the Digital Humanities 2019*. Minneapolis: University of Minnesota Press, pp. 243–267. Available at: https://www.jstor.org/stable/10.5749/j.ctvg251hk.24.

Griffiths TL and Steyvers M (2004) Finding scientific topics. *Proceedings of the National Academy of Sciences* 101. Proceedings of the National Academy of Sciences: 5228–5235.

Gulordava K and Baroni M (2011) A distributional similarity approach to the detection of semantic change in the Google Books Ngram corpus. In: *Proceedings of the GEMS 2011 Workshop on GEometrical Models of Natural Language Semantics* (eds S Pado and Y Peirsman), Edinburgh, UK, July 2011, pp. 67–71. Association for Computational Linguistics. Available at: https://aclanthology.org/W11-2508.

Hacking I (1999) *The Social Construction of What?* Harvard University Press.

Hall D, Jurafsky D and Manning CD (2008) Studying the history of ideas using topic models. In: *Proceedings of the conference on empirical methods in natural language processing*, 2008, pp. 363–371. Association for Computational Linguistics. Available at: http://dl.acm.org/citation.cfm?id=1613763.

Hamilton WL, Leskovec J and Jurafsky D (2016) Diachronic Word Embeddings Reveal Statistical Laws of Semantic Change. In: *Proceedings of the 54th Annual Meeting of the Association for Computational Linguistics (Volume 1: Long Papers)*, Berlin, Germany, 2016, pp. 1489–1501. Association for Computational Linguistics. Available at: http://aclweb.org/anthology/P16-1141.

Kleymann R, Niekler A and Burghardt M (2022) Conceptual Forays: A Corpus-based Study of "Theory" in Digital Humanities Journals. *Journal of Cultural Analytics* 7(4).

Koselleck R (1985) *Futures Past: On the Semantics of Historical Time*. The MIT Press.

Koselleck R (2002) *The Practice of Conceptual History: Timing History, Spacing Concepts*. Stanford University Press. Available at: https://www.degruyterbrill.com/document/doi/10.1515/9781503619104/html.

Koselleck R and Richter M (2011) Introduction and Prefaces to the Geschichtliche Grundbegriffe: (Basic Concepts in History: A Historical Dictionary of Political and Social Language in Germany). *Contributions to the History of Concepts* 6(1). Berghahn Journals: 1–37.

Kuhn TS (1962) *The Structure of Scientific Revolutions*. The structure of scientific revolutions. University of Chicago Press: Chicago.

Kutuzov A, Øvrelid L, Szymanski T, et al. (2018) Diachronic word embeddings and semantic shifts: a survey. arXiv:1806.03537. arXiv. Available at: http://arxiv.org/abs/1806.03537

Kutuzov A, Velldal E and Øvrelid L (2022) Contextualized embeddings for semantic change detection: Lessons learned. *Northern European Journal of Language Technology* 8(1).

Fleck L (1979) *Genesis and Development of a Scientific Fact*. University of Chicago Press.

Khutsishvili K (2026) AI and the scientist. On the fracture of epistemic authority. In: Simons A, Wüthrich A, Zichert M, et al. (eds) *Understanding Science with Large Language Models? Potentials for the History, Philosophy, and Sociology of Science*. Bielefeld: transcript, part-1.

Lalli R, Laubichler M, Olbrich E, et al. (2023) Computational History: Challenges and Opportunities of Formal Approaches. *Journal of Social Computing* 4: 232–242.

Lang S (2026) Critical concerns for using LLMs in the (computational) humanities and beyond. In: Simons A, Wüthrich A, Zichert M, et al. (eds) *Understanding Science with Large Language Models? Potentials for the History, Philosophy, and Sociology of Science*. Bielefeld: transcript, part-1.

Laubichler MD, Maienschein J and Renn J (2019) Computational History of Knowledge: Challenges and Opportunities. *Isis* 110(3). The University of Chicago Press: 502–512.

Leydesdorff L and Nerghes A (2017) Co-word maps and topic modeling: A comparison using small and medium-sized corpora (N < 1,000). *Journal of the Association for Information Science and Technology* 68(4): 1024–1035.

Malaterre C and Lareau F (2026) Epistemic framings in science. Charting scientific knowledge with embeddings and LLMs. In: Simons A, Wüthrich A, Zichert M, et al. (eds) *Understanding Science with Large Language Models? Potentials for the History, Philosophy, and Sociology of Science*. Bielefeld: transcript, part-3.

Marjanen J (2023) Quantitative Conceptual History: On Agency, Reception, and Interpretation. *Contributions to the History of Concepts* 18(1). Berghahn Journals: 46–67.

Martinc M, Kralj Novak P and Pollak S (2020) Leveraging Contextual Embeddings for Detecting Diachronic Semantic Shift. In: *Proceedings of the Twelfth Language Resources and Evaluation Conference* (eds N Calzolari, F Béchet, P Blache, et al.), Marseille, France, May 2020, pp. 4811–4819. European Language Resources Association. Available at: https://aclanthology.org/2020.lrec-1.592/.

Meding H and Daugs A (2026) On the use and limitations of large language models in historical scholarship. In: Simons A, Wüthrich A, Zichert M, et al. (eds) *Understanding Science with Large Language Models? Potentials for the History, Philosophy, and Sociology of Science*. Bielefeld: transcript, part-1.

Mikolov T, Sutskever I, Chen K, et al. (2013) Distributed Representations of Words and Phrases and their Compositionality. arXiv:1310.4546. arXiv. Available at: http://arxiv.org/abs/1310.4546.

Mitra S, Mitra R, Riedl M, et al. (2014) That's sick dude!: Automatic identification of word sense change across different timescales. In: *Proceedings of the 52nd Annual Meeting of the Association for Computational Linguistics (Volume 1: Long Papers)* (eds K Toutanova and H Wu), Baltimore, Maryland, June 2014, pp. 1020–1029. Association for Computational Linguistics. Available at: https://aclanthology.org/P14-1096.

Montariol S, Martinc M and Pivovarova L (2021) Scalable and Interpretable Semantic Change Detection. In: *Proceedings of the 2021 Conference of the North American Chapter of the Association for Computational Linguistics: Human Language Technologies*, Online, 2021, pp. 4642–4652. Association for Computational Linguistics. Available at: https://aclanthology.org/2021.naacl-main.369.

Müller E and Schmieder F (2016) *Begriffsgeschichte Und Historische Semantik: Ein Kritisches Kompendium*. Originalausgabe, erste Auflage. Suhrkamp Taschenbuch Wissenschaft; 2117. Suhrkamp.

Müller E and Schmieder F (2018) Begriffsgeschichte und Wissenschaftsgeschichte: Bestandsaufnahme und Forschungsperspektiven. *Geschichte und Gesellschaft* 44(1): 79–106.

Pennington J, Socher R and Manning C (2014) GloVe: Global Vectors for Word Representation. In: *Proceedings of the 2014 Conference on Empirical Methods in Natural Language Processing (EMNLP)* (eds A Moschitti, B Pang, and W Daelemans), Doha, Qatar, October 2014, pp. 1532–1543. Association for Computational Linguistics. Available at: https://aclanthology.org/D14-1162.

Periti F, Dubossarsky H and Tahmasebi N (2024) (Chat)GPT v BERT Dawn of Justice for Semantic Change Detection. In: Findings of the Association for Computational Linguistics: EACL 2024 (S. 420–436). Association for Computational Linguistics. https://aclanthology.org/2024.findings-eacl.29.

Periti F and Montanelli S (2024) Lexical Semantic Change through Large Language Models: a Survey. *ACM Comput. Surv.* 56(11): 282:1-282:38.

Pickering A (1999) *Constructing Quarks: A Sociological History of Particle Physics*. University of Chicago Press.

Radford A, Narasimhan K, Salimans T, et al. (2018) Improving Language Understanding by Generative Pre-Training. OpenAI. Epub ahead of print 2018.

Rheinberger H-J (1997) *Toward a History of Epistemic Things: Synthesizing Proteins in the Test Tube*. Writing science. Stanford, Calif.: Stanford Univ. Press.

Rip A and Courtial JP (1984) Co-word maps of biotechnology: An example of cognitive scientometrics. *Scientometrics* 6(6): 381–400.

Schauz D (2015) Wissenschaftsgeschichte und das Revival der Begriffsgeschichte. *NTM Zeitschrift für Geschichte der Wissenschaften, Technik und Medizin* 23(1–2): 53–63.

Schauz D (2025) Contextualizing science and technology from the perspective of conceptual history. In: Ihalainen P and Marjanen J (eds) *Writing Conceptual Histories*. Bloomsbury Publishing, p. 93–118..

Schwandt S (2018) Digitale Methoden für die Historische Semantik. *Geschichte und Gesellschaft* 44(1). Vandenhoeck & Ruprecht: 107–134.

Simons A (2024) Astro-HEP-BERT: A bidirectional language model for studying the meanings of concepts in astrophysics and high energy physics. arXiv:2411.14877. arXiv. Available at: http://arxiv.org/abs/2411.14877.

Simons A (2026) Meaning at the Planck scale? Contextualized word embeddings for doing history, philosophy, and sociology of science. In: Simons A, Wüthrich A, Zichert M, et al. (eds) *Understanding Science with Large Language Models? Potentials for the History, Philosophy, and Sociology of Science*. Bielefeld: transcript, part-3.

Simons A, Wüthrich A and Zichert M (2026) Doing science studies with large language models. An introduction. In: Simons A, Wüthrich A, Zichert M, et al. (eds) *Understanding Science with Large Language Models? Potentials for the History, Philosophy, and Sociology of Science*. Bielefeld: transcript.

Simons A, Zichert M and Wüthrich A (2026) Large language models for history, philosophy, and sociology of science: Interpretive uses, methodological challenges, and critical perspectives. *Studies in History and Philosophy of Science* 117: 102151. https://doi.org/10.1016/j.shpsa.2026.102151.

Sinclair S and Rockwell G (2016) Voyant Tools. Web. Available at: http://voyant-tools.org/.

Small H (1973) Co-citation in the scientific literature: A new measure of the relationship between two documents. *Journal of the American Society for Information Science* 24(4): 265–269.

Small HG (1978) Cited Documents as Concept Symbols. *Social Studies of Science* 8(3). Sage Publications, Ltd.: 327–340.

Small H (1999) Visualizing science by citation mapping. *Journal of the American Society for Information Science* 50(9): 799–813.

Sommerauer P and Fokkens A (2019) Conceptual Change and Distributional Semantic Models: an Exploratory Study on Pitfalls and Possibilities. In: *Proceedings of the 1st International Workshop on Computational Approaches to Historical Language Change*, Florence, Italy, 2019, pp. 223–233. Association for Computational Linguistics. Available at: https://www.aclweb.org/anthology/W19-4728.

Underwood T, Nelson LK and Wilkens M (2025) Can Language Models Represent the Past without Anachronism? arXiv:2505.00030. arXiv. Available at: http://arxiv.org/abs/2505.00030.

Valleriani M (2025) Large language models that power AI should be publicly owned. *The Guardian*, 26 May. Available at: https://www.theguardian.com/technology/2025/may/26/large-language-models-that-power-ai-should-be-publicly-owned.

Vaswani A, Shazeer N, Parmar N, et al. (2017) Attention is all you need. *Advances in neural information processing systems* 30.

Wevers M and Koolen M (2020) Digital begriffsgeschichte: Tracing semantic change using word embeddings. *Historical Methods: A Journal of Quantitative and Interdisciplinary History* 53(4). Routledge: 226–243.

Wu J, Gan W, Chen Z, et al. (2023) Multimodal Large Language Models: A Survey. In: *2023 IEEE International Conference on Big Data (BigData)*, 1 December 2023, pp. 2247–2256. IEEE Computer Society. Available at: https://www.computer.org/csdl/proceedings-article/bigdata/2023/10386743/1TUOsrxd62A.

Xie Q and Waltman L (2025) A comparison of citation-based clustering and topic modeling for science mapping. *Scientometrics* 130(5): 2497–2522.

Yin S, Fu C, Zhao S, et al. (2024) A survey on multimodal large language models. *National Science Review* 11(12): nwae403.

Zhang H, Yu PS and Zhang J (2025) A Systematic Survey of Text Summarization: From Statistical Methods to Large Language Models. *ACM Comput. Surv.* 57(11): 277:1-277:41.

Zichert M, Simons A and Wüthrich A (2025) Expanding conceptual histories: using contextualized word embeddings for the history and philosophy of the virtual particle concept. *Computational Humanities Research* 1: e16.

Meaning at the Planck scale?

Contextualized word embeddings for doing history, philosophy, and sociology of science*

Arno Simons

1. Introduction

Scientific concepts are dynamic entities, continually shaped and reshaped by the socio-historical contexts from which they emerge. Their meanings are influenced by the intellectual, cultural, and material conditions of their time, as well as by the disciplinary frameworks that govern their usage. This dynamic nature is evident in medical concepts such as "syphilis" (Fleck, 1979) and "atherosclerosis' (Mol, 2002), physical concepts like "electricity" (Steinle, 2016) and "temperature" (Chang, 2007), as well as more abstract methodological notions like "probability" (Hacking, 1975) and "objectivity" (Daston and Galison, 2007).

From the perspective of the history, philosophy, and sociology of science (HPSS), understanding how scientific concepts acquire and stabilize their meanings is crucial for several reasons. Changes in scientific language reflect shifts in how researchers conceptualize their subject matter, and as concepts evolve, they reshape the frameworks through which reality is interpreted. By examining these processes, HPSS highlights how scientific knowledge is not merely discovered but constructed through complex social and intellectual negotiations, emphasizing the contingent and dynamic nature of scientific understanding (Fleck, 1979; Foucault, 1970; Hacking, 1999; Kuhn, 1962; Latour, 1987; Pickering, 1995). As Thomas Kuhn (1962: 128) famously remarked:

> "the Copernicans who denied its traditional title 'planet' to the sun were not only learning what 'planet' meant or what the sun was. Instead, they were changing the meaning of 'planet' so that it could continue to make useful distinctions in a world where all celestial bodies, not just the sun, were seen differently from the way they had been seen before".

---

* This chapter is almost identical to an earlier preprint (Simons, 2024b).

Transformations in the meanings of scientific concepts can also have significant societal repercussions. When concepts shift, they alter not only the internal structures of scientific knowledge but also the broader societal and cultural frameworks in which that knowledge is embedded. For example, historical changes in the meanings of terms like "nature" (Merchant, 1980) or "the normal and the pathological" (Canguilhem, 1991) have profoundly influenced education, policy, and social hierarchy. These shifts have reinforced or challenged societal norms and power dynamics. Redefining scientific concepts creates new frameworks of understanding that affect how institutions are organized, how resources are allocated, and how humans and non-human entities are categorized and treated (Bowker and Star, 1999; Clarke et al., 2010; Hacking, 1999).

To study the evolving meanings of scientific concepts across different socio-historical contexts, HPSS scholars have primarily relied on qualitative methods such as "close reading". While effective, these methods are often time-intensive, challenging to replicate, and limited in scalability. To complement these approaches, HPSS scholars have increasingly turned to "distant reading" techniques, such as co-word analysis (Callon et al., 1983), latent semantic analysis (LSA) (Deerwester et al., 1990), latent Dirichlet allocation (LDA) (Blei et al., 2003), and word vectors (Mikolov et al., 2013). These methods enable researchers to explore conceptual changes across large corpora, uncovering patterns in the evolution of individual terms or entire fields (Boyack et al., 2011; Chen et al., 2018; Courtial and Law, 1989; Lean et al., 2023; Leydesdorff and Rafols, 2009; Laubichler et al., 2019; Malaterre and Léonard, 2024; Overton, 2013; Pence and Ramsey, 2018; Rip and Courtial, 1984; Venturini et al., 2014; Wevers and Koolen, 2020). However, these approaches often struggle to capture context-dependent meanings of words and sentences.

This is where contextualized word embeddings (CWEs) provide a significant leap forward. CWEs rely on patterns of usage in large text corpora to encode words as points in a continuous vector space. In this space, semantic and syntactic relationships between words are represented through relative distances, with similar contexts grouped closely together. Unlike earlier word embeddings, which assign static representations to words, CWEs dynamically adjust a word's representation depending on its context. This adaptability enables researchers to analyze how meanings differ and evolve across time, disciplines, or other settings, surpassing the static associations revealed by traditional methods (Zichert and Simons, 2026).

To demonstrate the potential of CWEs for HPSS, I use "Planck" as a test case, a term with multiple distinct meanings in physics—such as the physicist Max Planck, Planck units, and the Planck mission—all well-defined and widely recognized within the physics community. This diversity of known interpretations offers a clear "ground truth" for evaluating how effectively CWEs can capture and disambiguate these meanings, while striking a balance between simplicity and complexity. While some senses, like "Max Planck Society" or "Fokker-Planck equation", can be disambiguated through straightforward co-occurrence patterns, others demand a deeper contextual understanding. For example, domain-adapted BERT models can discern synonymous meanings in contexts such as "a Planck function at that temperature" and "the spectral distribution for Planck's law", while also interpreting nuanced cases like "Planck's experiment yields Ps $\times$ 10−9 for the power of scalar modes at the end of inflation" as referring to the Planck space mission rather than Max Planck the physicist. The analysis focuses on five BERT-based models with varying domain specializations and training methodologies, including both general-purpose models and domain-specific models fine-tuned on scientific corpora. Among these is my own Astro-HEP-BERT (Simons, 2024a), a specialized model developed for conceptual analysis in astrophysics and high-energy physics (HEP), available on Hugging Face.[1]

As you may have already guessed, the title of this chapter is a play on words. In physics, the term "Planck scale" often refers to extremely small quantities of space and time, defined using natural units introduced by Max Planck in 1899. Here, "meaning at the Planck scale" serves as a metaphor for the linguistic task at hand: disambiguating the multiple senses of "Planck". Just as physics requires high resolution to measure reality at the smallest scales, understanding the contextual meanings of "Planck" requires CWEs to detect subtle semantic distinctions at the level of individual words and subwords.

The structure of this chapter is as follows. First, I introduce CWEs as a computational tool for HPSS, highlighting their advantages over traditional methods and the benefits of BERT-based architectures. Next, I describe the datasets and labels used, providing essential context for the analysis. This is followed by a detailed comparison of the five BERT-based models used in the empirical evaluation. The core of the paper consists of four empirical sections. The first evaluates CWEs' ability to disambiguate the term "Planck" through supervised word-sense prediction. The second assesses the quality of sense clusters formed in an unsupervised setting, measuring alignment with labeled data. The third examines the balance between cluster separation and internal cohesion. Finally, the fourth empirical section explores the diachronic evolution of "Planck", tracking shifts in the prominence of different senses over time. The paper concludes by synthesizing the findings and discussing their broader implications for HPSS research. It highlights how CWEs can illuminate the socio-historical evolution of scientific concepts, identifies their limitations, and proposes future directions for integrating CWEs into the study of scientific discourses.

2. CWEs as a new tool for HPSS

CWEs represent each word in a given sequence, such as a sentence or paragraph, as a unique numerical vector in a continuous embedding space. This space encodes relationships and patterns derived from large text corpora, where words with similar meanings or contexts are positioned closer together. By dynamically adjusting a word's representation based on its specific context, CWEs capture subtle differences in meaning, distinguishing between various senses of the same word or phrase. For example, the word "Planck" in the sentence "We used Planck's law to calculate the radiation emitted by a black body" will have a different vector representation than in "We used Planck's data to map the early universe", reflecting the distinct contexts and meanings of the word in each case.

The ability to generate context-aware embeddings is a hallmark of modern transformer-based architectures, particularly BERT (Bidirectional Encoder Representations

1 https://huggingface.co/arnosimons/astro-hep-bert

from Transformers) (Devlin et al., 2018). BERT's bidirectional design processes text by considering both the preceding and succeeding words in a sentence simultaneously. Through multiple layers of neural networks and attention mechanisms, it captures contextual relationships, making it especially well-suited for deep language understanding. This bidirectional approach gives BERT an edge over unidirectional models like GPT for analyzing complex linguistic contexts, though it is less effective for generative tasks.

CWEs lie at the heart of BERT's outstanding performance in natural language understanding tasks. As BERT is most often used for high-level applications like text classification, named entity recognition, or question answering, its embeddings normally remain "under the hood". However, CWEs can be accessed by extracting hidden states from BERT's intermediate or final layers. A common approach involves averaging the last four layers to create robust word embeddings (Devlin et al., 2018; Periti and Montanelli, 2024).

Though originally designed for higher-level tasks, CWEs are particularly well-suited for exploring the varying and evolving meanings of scientific concepts. Advances in word sense disambiguation (WSD) (Bevilacqua et al., 2021; Loureiro et al., 2020), word sense induction (WSI) (Periti and Montanelli, 2024; Sun and Platos, 2023), and lexical semantic change (LSC) analysis (Periti and Montanelli, 2024; Tahmasebi and Dubossarsky, 2023) suggest their capacity to uncover distinct meanings that scientific terms acquire across disciplines and historical contexts (Ahmadi, 2026; Kleymann et al., 2022; Zichert et al., 2025). WSD identifies the intended sense of a word in context, WSI clusters word occurrences to infer senses unsupervised, and LSC analyzes shifts in word meanings over time. LSC can be considered a temporal extension of both WSD and WSI when it uses information about a word's senses to measure semantic change, for example by tracking how changes in the relative prominence of these senses reflect changes in the word's overall meaning (Periti and Montanelli, 2024).

WSD, WSI, and LSC all use BERT's ability to map semantically similar words to similar vectors, enabling the modeling of semantic relationships. These tasks largely rely on three core operations:

Measuring distances between embeddings: This operation quantifies semantic similarity or divergence between terms, typically using cosine similarity or Euclidean distance. It can be used to compare either the same term in different contexts (as in WSD and WSI) or different terms within the same context. Additionally, average pairwise similarity or distance across a set of embeddings reveals broader properties, such as the degree of polysemy, by reflecting the dispersion of meanings.

Prototypical embedding creation: By averaging the embeddings of multiple instances of a term, researchers can derive a prototypical representation of its central or typical meaning. Depending on how instances are grouped (e.g. by predefined sense labels for WSD, discovered clusters for WSI, or temporal periods for LSC), prototypical embeddings can reveal patterns specific to these contexts.

Clustering embeddings: Clustering groups semantically similar terms or usages, making it valuable for WSI to uncover known or emerging senses, for WSD to generate prototypes and predict senses, and for LSC to track how senses evolve over time by grouping embeddings from different periods or contexts.

Effectively using CWEs for HPSS requires understanding domain adaptation, which customizes BERT models for specialized contexts. BERT's pretraining relies on masked language modeling, in which certain words in a sentence are hidden and predicted based on surrounding context. This process fine-tunes the model's internal parameters through extensive training on large, diverse corpora. However, while pretrained BERT models are robust general-purpose tools, they have found to underperform in the sciences, where specialized language dominates (Beltagy et al., 2019; Lee et al., 2020). Two main strategies address this limitation:

Training from scratch: A model is trained entirely on domain-specific data. This approach allows for a custom vocabulary tailored to the target domain, enhancing the model's ability to handle specialized terms. However, it requires substantial domain-specific data and significant computational resources (Grezes et al., 2022; Gu et al., 2021; Hellert et al., 2024).

Re-using weights from pretrained models: Pretrained models such as BERT-base are further trained on domain-specific corpora. This method efficiently builds on the general linguistic knowledge captured during pretraining, adapting the model to domain-specific terminology and contexts. Although this approach is resource-efficient, it typically reuses the original vocabulary, which may limit the model's ability to fully represent highly specialized terms. Subword tokenization helps mitigate this issue but cannot entirely address it (Devlin et al., 2018).

The choice between these approaches depends on the availability of data and computational resources. Additional training of a pretrained model is often sufficient for many applications—as demonstrated in this paper by the comparative performance of Astro-HEP-BERT against models like PhysBERT and astroBERT, which were trained from scratch—while training from scratch may be necessary for highly specialized tasks that demand more precise domain knowledge (Grezes et al., 2022, 2021; Hellert et al., 2024).

3. Datasets and labels

For this study, I use two custom-compiled datasets.

The Astro-HEP Corpus[2] consists of 21.84 million paragraphs extracted from 600,000 scientific articles (approximately 4.2 billion tokens) published on arXiv between 1986 and 2022, spanning the domains of astrophysics and high-energy physics (HEP). The para-

2 https://huggingface.co/datasets/arnosimons/astro-hep-corpus

graphs were extracted in plain text format from the original LaTeX files using Pandoc[3]. To ensure data quality, I replaced literature references with "[CIT]" and multiline formulas with "FORMULA". Additionally, paragraphs that were either unusually short or exhibited an abnormal frequency of whitespace characters were filtered out based on character frequency analysis.

The Wikipedia-Physics Corpus[4] contains 102,409 paragraphs extracted from 6,642 key physics-related Wikipedia articles. These articles were selected using the PetScan[5] tool, which generated a list of all pages categorized under "physics" or its immediate subcategories. Markup was removed and minimal cleaning applied to produce plain text paragraphs, and while references were removed, all formulas were retained.

To establish a ground truth for the WSD and LSC analyses of "Planck" (Sections 5–8), I annotated occurrences of the term in both corpora using 21 predefined labels, determined by theoretical insights, expert knowledge, and a qualitative review of the contexts. In the Astro-HEP Corpus, I manually labeled 2,932 occurrences from a random sample of 1,500 paragraphs, and this subset is referred to as the Astro-HEP-Planck Corpus[6]. In the Wikipedia-Physics Corpus, I labeled all 1,186 occurrences across 885 paragraphs. I identified occurrences using a case-insensitive regular expression to capture variations like "PLANCK" and "Planck(2015)" while excluding irrelevant forms like "planckian".

For further analysis, I focused on occurrences corresponding to the seven most common labels (see Table 1 and Figure 1). For each corpus, I created six subsets containing the two, three, four, five, six, and seven most common labels within the corpus, respectively. For each subset, I extracted all CWEs by averaging the model's final four hidden layers (Devlin et al., 2018; Periti and Montanelli, 2024).

*Figure 1: Distribution of labels and their cutoff thresholds for the word sense disambiguation and induction tasks in the Astro-HEP-Planck (**a**) and Wikipedia-Physics (**b**) corpora. The x-axis shows the labels ranked by frequency, while the y-axis represents the number of occurrences. Cutoffs indicate the number of labels included in each subset for subsequent analyses.*

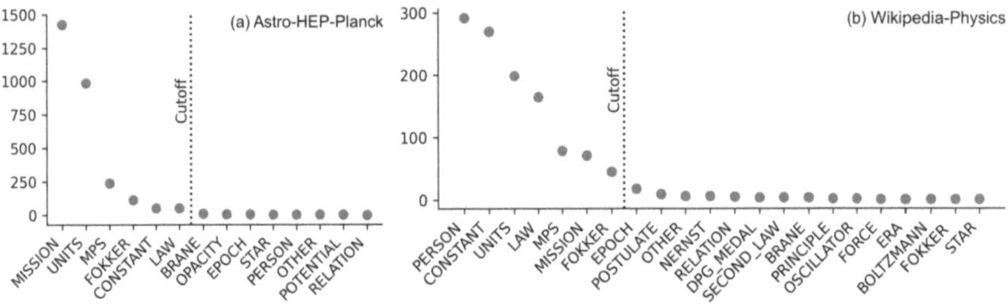

3 https://pandoc.org

4 https://huggingface.co/datasets/arnosimons/wikipedia-physics-corpus

5 https://petscan.wmcloud.org

6 https://huggingface.co/datasets/arnosimons/astro-hep-planck-corpus

Table 1: Description of the seven most common labels (in alphabetical order) used for word sense disambiguation of the term "Planck". Columns AHC and WPC depict the number of labeled occurrences in the Astro-HEP-Planck Corpus and the Wikipedia-Physics Corpus, respectively.

Label	AHC	WPC	Description
CONSTANT	52	270	Planck's constant h: A fundamental physical constant of foundational importance in quantum mechanics. A photon's energy is equal to its frequency multiplied by the Planck constant, and the wavelength of a matter wave equals the Planck constant divided by the associated particle momentum.
FOKKER	112	45	Fokker—Planck equation: A partial differential equation that describes the time evolution of the probability density function of the velocity of a particle under the influence of drag forces and random forces, as in Brownian motion.
LAW	52	165	Planck's law: A formula that describes the spectral density of electromagnetic radiation emitted by a black body in thermal equilibrium at a given temperature, when there is no net flow of matter or energy between the body and its environment.
MISSION	1426	71	Planck mission: A space observatory operated by the European Space Agency from 2009 to 2013. The project that mapped the anisotropies of the cosmic microwave background (CMB) at microwave and infrared frequencies, with high sensitivity and angular resolution.
MPS	239	79	The Max Planck Society for the Advancement of Science: A formally independent non-governmental and non-profit association of German research institutes
PERSON	4	292	Max Planck: The famous German theoretical physicist whose discovery of energy quanta won him the Nobel Prize in Physics in 1918.
UNITS	986	199	Planck units: A system of units of measurement defined exclusively in terms of four universal physical constants: c, G, $\hbar$, and kB. Expressing one of these physical constants in terms of Planck units yields a numerical value of 1.

4. Models

I compare the performance of five BERT-based models: The first model, BERT-base or simply BERT (Devlin et al., 2018), is a general-purpose model trained from scratch over 40 epochs on a dataset of Wikipedia articles and free books, totaling 3.3 billion tokens. It uses a 30,000-token uncased vocabulary and serves as the baseline model for this analysis. The second model, SciBERT-scivocab-uncased or simply SciBERT (Beltagy et al., 2019), was developed specifically for scientific text and trained from scratch on 1.14 million biomedical and computer science papers, with its uncased vocabulary learned directly from the corpus. The third model, astroBERT (Grezes et al., 2022), was trained on 400,000 astrophysics publications from the SAO/NASA Astrophysics Data System (ADS), specifically designed to identify, disambiguate, and tag entities in astrophysical literature. Its vocabulary is case-sensitive and was learned from the corpus. The fourth model, PhysBERT-uncased or simply PhysBERT (Hellert et al., 2024), was trained from scratch over 10 epochs on 1.2 million arXiv physics papers, with its uncased vocabulary learned directly from the corpus. The model was also fine-tuned to produce domain-adapted sentence embeddings.

The fifth model, Astro-HEP-BERT, is my own (Simons, 2024a). Instead of training it from scratch, as was done for the other four models, I reused BERT's learned linguistic patterns and vocabulary and retrained the model for three additional epochs using on my Astro-HEP Corpus. To enhance learning, I trained the model solely on entire paragraphs rather than packing in as many sentences as possible, as often suggested in BERT tutorials. This "full-paragraphs format" preserves sentences within their original context, which is especially meaningful in academic writing where paragraphs focus on one idea. Previous research shows that using multiple sentences from the same document improves results (Liu et al., 2019), so I expected this approach to help the model capture more semantic information.

5. Word sense prediction using sense prototypes

As a first WSD test I use a simple supervised 1-nearest neighbor (1NN) approach as described in (Loureiro et al., 2020). This test compares the models' ability to predict the correct sense of "Planck" occurrences by computing the cosine similarity between their CWEs and precomputed sense prototypes. The closest prototype is then used to assign the appropriate sense to the target occurrence. Failure to classify an occurrence correctly suggests that the CWEs for the senses are not well-separated. For each corpus, the sense prototypes are generated by averaging the CWEs of all labeled "Planck" occurrences corresponding to each sense.

Figure 2 compares the models' performance in predicting different senses of "Planck" using word sense prototypes across the two corpora. Each model was evaluated using six separate 1NN classifiers, with each classifier trained on a subset containing between two and seven of the most common labels in the corpus. Figures 2a and 2b present the weighted F-1 scores for the Astro-HEP-Planck and Wikipedia-Physics corpora, respectively. The x-axis represents the number of labels in the subset, while the y-axis shows the F-1 score for each model. This setup enables a direct comparison of how effectively the models distinguish between the senses of "Planck" as the classification task becomes more complex with the inclusion of additional senses (labels).

In both figures, we observe that the classifiers generally achieve higher F-1 scores when they have to predict fewer labels, with performance declining as the number of labels—and thus ambiguity—increases. This in itself should not be surprising because fewer labels make the classification task simpler, reducing ambiguity, whereas more labels increase complexity and the likelihood of misclassification. A minor exception occurs in the Wikipedia-Physics Corpus, where all models show a slight performance rebound after an initial decline when predicting the four most common labels. The initial decline in the four-label case is primarily due to misclassifying CWEs labeled as LAW—the newly introduced fourth most frequent label—as either CONSTANT or

PERSON, and vice versa. As the labels introduced in the five-, six-, and seven-label settings are less ambiguous, the overall F-1 scores improve slightly.

*Figure 2: Comparison of model performance in disambiguating the word "Planck" using 1-nearest neighbor (1NN) classifiers across subsets of labels from the Astro-HEP-Planck (**a**) and Wikipedia-Physics (**b**) corpora. The x-axis represents the number of labels in each subset, while the y-axis shows the weighted F-1 scores. Each model was evaluated on classifiers trained on CWEs extracted for two to six labels in the Astro-HEP-Planck Corpus and two to seven labels in the Wikipedia-Physics Corpus.*

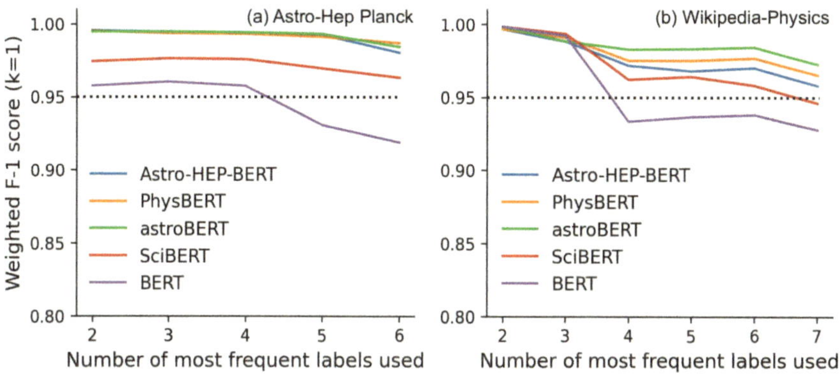

The results also highlight differences in how well the models handle the challenge of dealing with increasing numbers of labels and ambiguity: some maintain relatively high performance even as the number of labels grows, while others experience a steeper decline. Overall, PhysBERT, astroBERT, and Astro-HEP-BERT perform exceptionally well on the Astro-HEP-Planck Corpus, consistently achieving F-1 scores above 0.98, with most scores exceeding 0.99. SciBERT follows, with F-1 scores generally around 0.97. The weakest performer in the Astro-HEP-Planck Corpus is BERT, starting with an F-1 score near 0.96 and dropping to 0.92 in the six-label case. In the Wikipedia-Physics Corpus, astroBERT outperforms the other models, maintaining F-1 scores above 0.97. PhysBERT and Astro-HEP-BERT follow closely, with scores above 0.96 and 0.95, respectively. Although BERT starts strong with F-1 scores near 0.99 in the two-and three-label cases, its performance declines sharply, dropping to around 0.93 in the more complex label settings.

While BERT shows the weakest performance across both corpora, its F-1 score never falls below 0.92, reflecting the strong baseline performance of BERT models in word sense disambiguation (Wiedemann et al., 2019). In contrast, the stronger performance of astroBERT, PhysBERT, and Astro-HEP-BERT in disambiguating the term "Planck" may suggest the advantages of domain-specific pretraining.

The results of the word sense prediction task showed that domain-specific models outperform general-purpose ones in distinguishing between distinct senses of "Planck". However, sense disambiguation often requires unsupervised approaches, especially when sense labels are unknown. The next section evaluates the models' capacity to cluster senses effectively without supervision.

6. Cluster purity

While the previous section evaluated the models' ability to classify "Planck" occurrences into predefined senses, a classical WSD task, this section evaluates the models' ability to perform WSI by forming meaningful word sense clusters in an unsupervised manner employing the K-means clustering (cf. Periti and Montanelli, 2024; Sun and Platos, 2023). This algorithm partitions data into a predefined number of k clusters by initializing centroids (usually randomly) and iteratively refining them to minimize variance within each cluster, based on the average distance between data points and their nearest centroid. For each corpus, model, and subset of labeled CWEs, k is set to the number of labels l in the subset. For each value of k, the best clustering solution out of 100 random seed initializations is selected based on inertia—the sum of squared distances between CWEs and their nearest cluster center.

To evaluate how effectively a clustering solution s disambiguated the predefined senses of "Planck", I propose a purity indicator, using a coefficient of variation (CV) approach, where I compare the weighted average CV for the clusters against a theoretical maximum that can occur if clusters are perfectly pure. The CV is a standardized measure of dispersion of frequency distribution, defined as the ratio of the standard deviation σ to the mean μ. Building on this definition, my **purity indicator** is computed as:

$$\text{purity}_s = \frac{\text{waCV}_s}{\text{TM}_l}$$

where $waCV_s$ is the weighted average CV for the clusters in solution s, and TM_l the theoretical maximum value that any empirical waCV_s can attain, given l available labels. The waCV_s is computed as:

$$\text{waCV}_s = \sum_{i=1}^{k} w_i' \text{CV}_i$$

with k as the number of clusters, w_i' the normalized size of cluster i, and CV_i the CV of the label distribution in cluster i, defined as:

$$\text{CV}_i = \frac{\sigma_i}{\mu_i}$$

where σ_i is the standard deviation and μ_i the mean of the label distribution in cluster i. The theoretical maximum TM_l for waCV_s, given l labels, occurs when all clusters are "pure", meaning they contain only one type of label. Thus, we can simplify to a single-cluster-single-label scenario:

$$\text{TM}_l = \frac{\sigma_f((1,0^{l-1}))}{\mu_f((1,0^{l-1}))}$$

A high purity score indicates that clusters are mostly composed of instances carrying the same label, whereas a lower score suggests mixing of labels, indicating less effective disambiguation. Figures 3a and 3b display the purity scores of cluster solutions across models and subsets for the Astro-HEP-Planck and Wikipedia-Physics corpora. The x-axis represents the subset label count, while the y-axis shows the purity score. Each cluster solution is annotated with a permutation of dominant labels, reflecting their frequency in the corpus. For instance, "1234" represents clusters dominated by the top four labels in descending order of frequency. By contrast, "1243" (seen in BERT's four-label solution for the Astro-HEP-Planck Corpus) indicates that the third-largest cluster is dominated by the fourth label rather than the third, suggesting an artificial size increase due to "pollution" from other labels.

*Figure 3: Purity scores of clustering solutions across models and subsets of CWEs for the Astro-HEP-Planck (**a**) and Wikipedia-Physics corpora (**b**). The x-axis represents the number of predefined word sense labels in each subset, and the y-axis shows the purity scores for each model. Clustering solutions are annotated with permutations of dominant labels, reflecting the frequency distribution of word senses in the datasets. Detailed label mappings can be inferred from figures 1a and 1b for the Astro-HEP-Planck and Wikipedia-Physics corpora, respectively.*

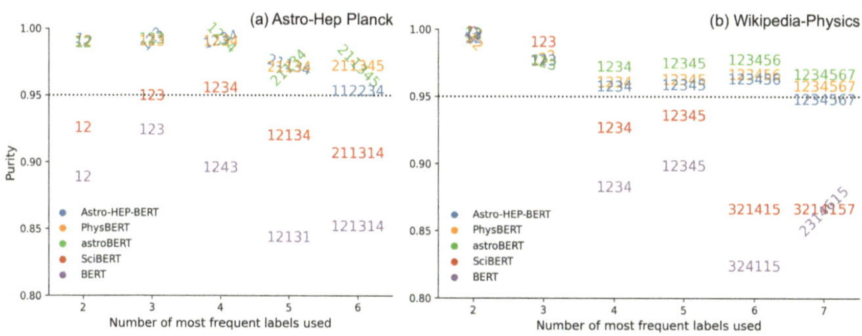

Results indicate distinct model performance patterns. In the Astro-HEP-Planck Corpus, Astro-HEP-BERT, PhysBERT, and astroBERT achieve the highest clustering quality, consistently above 0.95, and even exceed 0.99 with fewer labels. By contrast, SciBERT and BERT peak at 0.96 and 0.92, respectively, with performance declining faster as label complexity increases. Up to four labels, most models maintain the corpus's frequency order, with BERT as an exception. In the five- and six-label cases, however, all models deviate from frequency order, indicating challenges as complexity rises. Here, Astro-HEP-BERT, PhysBERT, and astroBERT tend to form multiple clusters dominated by labels like MISSION or UNITS, indicating potential context-specific subgroups within dominant senses. SciBERT and BERT also show overrepresentation, producing up to three MISSION-dominant clusters in the five- and six-label cases.

These trends may stem from domain-specific pretraining in Astro-HEP-BERT, PhysBERT, and astroBERT, enhancing their sensitivity to physics terms like "Planck". Conversely, SciBERT and BERT lack this focus and may thus merge distinct senses. Additionally, the K-Means algorithm, designed to minimize variance around centroids, may contribute to overrepresenting dominant senses by favoring larger clusters around common labels, often at the expense of rarer ones. Thus, beyond model-specific factors, the clustering method itself may introduce a bias toward frequent senses, influencing observed patterns

In the Wikipedia-Physics Corpus, astroBERT consistently outperforms, maintaining purity scores above 0.97, with PhysBERT and Astro-HEP-BERT closely following. SciBERT achieves the highest purity (0.99) in the three-label scenario, but its performance drops in six- and seven-label cases, while BERT shows inconsistent results. Models perform well in two- to five-label cases, aligning closely with corpus frequency order. Beyond five labels, deviations appear, particularly for SciBERT and BERT, which form multiple clusters dominated by the PERSON label in the six-and seven-label scenarios, highlighting challenges with less frequent senses.

Overall, astroBERT, PhysBERT, and Astro-HEP-BERT demonstrate b clustering quality, while BERT and SciBERT fall behind. Although the top three models maintain high purity scores across both corpora, they more accurately reproduce the label distribution in the Wikipedia-Physics Corpus, likely due to its more balanced label frequencies (see Figure 1), more conventional language, and clearer semantic boundaries. In contrast, the specialized Astro-HEP-Planck Corpus, with its technical language and niche contexts, seems to present challenges for all models and for K-Means in accurately distinguishing less frequent senses.

7. Cluster separation and cohesion

In addition to purity, I evaluate the balance between internal cluster cohesion and inter-cluster separation to better assess the quality of clustering solutions across different models, particularly in cases where purity scores were similar. To measure inter-cluster separation, I calculate the **average pairwise similarity (APS)** between CWEs from different clusters:

$$\mathrm{APS}(E_1,E_2) = \frac{1}{|E_1||E_2|} \sum_{e_{1,i}\in E_1, e_{2,j}\in E_2} \mathrm{CS}(e_{1,i},e_{2,j})$$

where E_1 and E_2 are the sets of CWEs in two different clusters, and CS is the cosine similarity function quantifying similarity between pairs of CWEs from different clusters.

For internal cluster cohesion, I compute the **average inner similarity (AIS)** within each cluster:

$$\mathrm{AIS}(E,E) = \frac{2}{|E|(|E|-1)} \sum_{e_i,e_j\in E, e_i\neq e_j} \mathrm{CS}(e_i,e_j)$$

where E represents the set of CWEs within a single cluster.

These metrics provide insight into how well-separated the clusters are and how cohesive each cluster is internally. Low APS scores suggest clear distinctions between clusters, while high AIS scores indicate strong internal coherence within clusters.

Figure 4 presents similarity heatmaps for the best clustering solutions (based on inertia) of Astro-HEP-BERT, astroBERT, PhysBERT, and BERT in the four-label case of the Astro-HEP Corpus. This configuration is ideal for a focused analysis of separation and cohesion across models because it combines balanced complexity—with clear, interpretable heatmaps that highlight trends across models—and similarly high purity scores of 0.99 for astroBERT, Astro-HEP-BERT, and PhysBERT, while SciBERT and BERT perform noticeably lower.

Figure 4: Heatmaps showing cluster cohesion and separation for each model's ***(a-e)*** *best four-label clustering solutions on the Astro-HEP-Planck Corpus. The diagonal cells display average inner similarity (AIS) scores for each cluster, reflecting internal cohesion, while the off-diagonal cells show average pairwise similarity (APS) between clusters, indicating separation. Cluster labels include the dominant label index, the six most frequent neighboring words within 10 tokens of "Planck", and the total number of embeddings in the cluster.*

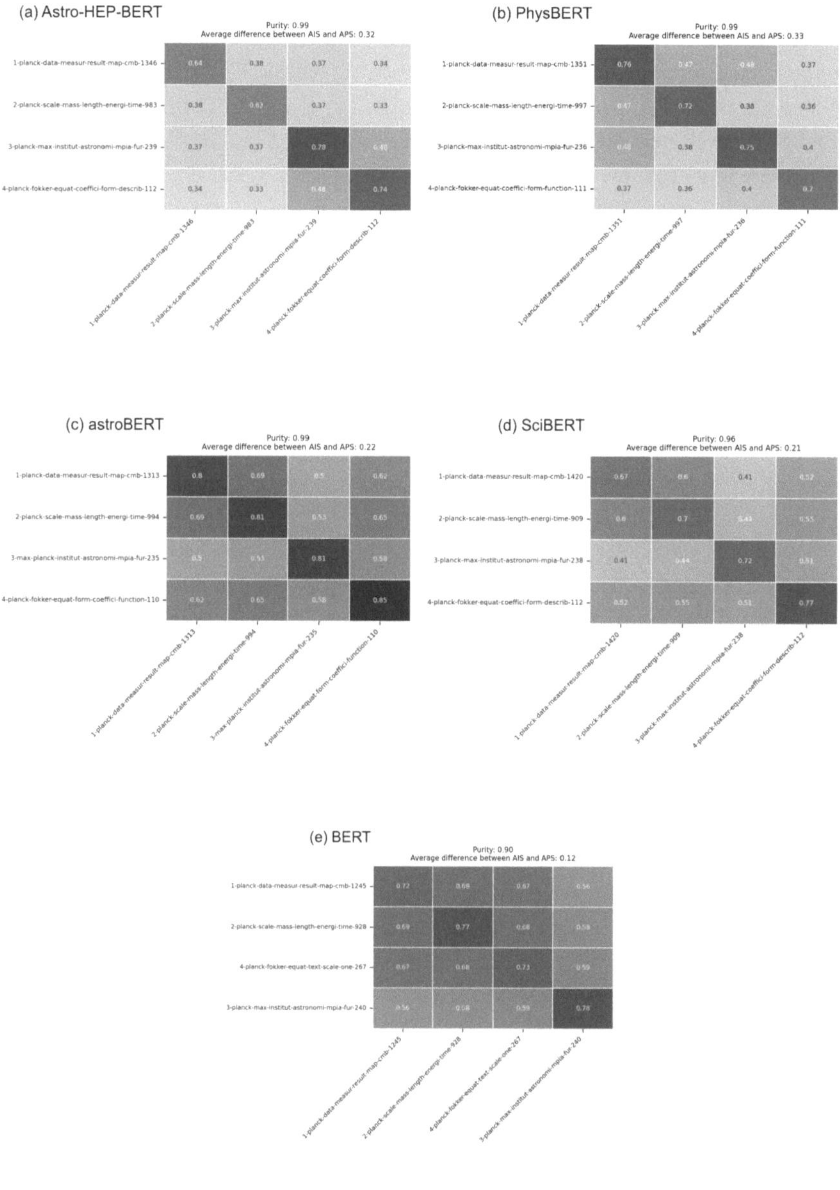

Each subfigure shows the mutual APS scores between the four clusters, with the AIS scores on the diagonal. For each clustering solution, the purity score as well as the average difference between AIS and APS scores are displayed at the top of the heatmap. Cluster names start with the index of the dominant label (as in Figure 3), followed by the six most frequent words within 10 tokens on either side of the "Planck" occurrences (stop words excluded), and the number of embeddings in the cluster. The heatmaps are sorted by cluster size, with the largest clusters positioned at the top and left.

Astro-HEP-BERT, PhysBERT, and astroBERT all achieve peak and matching purity scores, followed by SciBERT and then BERT, which performs significantly weaker. A comparison of the heatmaps reveals notable variations in APS and AIS scores across models, reflecting how well they balance internal cluster cohesion and separation. To evaluate this balance, we can examine the difference between a model's AIS and APS scores.

In the four-label case, PhysBERT and Astro-HEP-BERT achieve the best balance, with average differences of 0.33 and 0.32, respectively. PhysBERT demonstrates stronger coherence within its first two clusters, dominated by the MISSION and UNITS labels, though these clusters are less well-separated. In contrast, Astro-HEP-BERT has weaker coherence in its first two clusters but better separation, suggesting that they are more semantically diverse but also more distinct from each other. astroBERT achieves the highest AIS values overall (0.8 to 0.85), reflecting strong internal cohesion, but also has higher APS values, indicating less separation between clusters. For instance, clusters 1, 2, and 4 (recall that these indices reflect the actual frequency distribution of labels in the corpus rather than in the clustering solution) are much less distinct from one another in astroBERT compared to the other models. SciBERT shows very similar patterns, but performs worse than astroBERT. The weakest performer is clearly BERT, with smaller differences between AIS and APS, averaging 0.12. This indicates relatively poor separation between clusters, aligning with its lower purity scores and overall weaker performance in disambiguation.

As further detailed in the Supplement, the improved balance between AIS and APS observed in PhysBERT and Astro-HEP-BERT, compared to astroBERT and the other two models, may only partially result from differences in isotropy—the degree to which embeddings are evenly distributed within the vector space. While Astro-HEP-BERT demonstrates the highest global isotropy among all models, astroBERT lies at the opposite end, exhibiting the most anisotropic embedding space. Nevertheless, astroBERT outperforms both SciBERT and BERT on our WSD and WSI tasks, despite their relatively higher isotropy.

Having established the models' ability to distinguish the different senses of "Planck" effectively, we now turn to their application in a diachronic setting, tracking how the distribution of these senses evolves over time. This analysis bridges sense disambiguation with the broader task of understanding conceptual shifts in scientific language.

8. Lexical semantic change

Building on the models' tested performance in WSD and WSI from the previous sections, we can now explore their ability to measure semantic change for the term "Planck" in a real-world, large-corpus setting. By tracking the evolving distributions of CWEs across sense clusters, we examine changes over three decades in the unlabeled Astro-HEP Corpus, which includes over 300,000 instances of "Planck". A K-Means five-cluster solution, selected from 100 random seed initializations based on inertia, was applied to each model.

Figures 5a and 5b illustrate the relative prominence of these clusters (colored lines) over time for PhysBERT and BERT, respectively. The figures also show the overall relative frequency of "Planck" occurrences (dashed black line) and corpus growth (dashed green line). As before, cluster names were derived from the six most frequent words found in a 10-token window around "Planck" (excluding stop words), accompanied by the number of embeddings per cluster. The focus on PhysBERT (a domain-specific model) and BERT (a general-purpose model) reflects their contrasting design and the relative quality of their clustering solutions.

*Figure 5: Evolution of the relative frequency of "Planck" occurrences across five clusters (colored lines) and over time, as modeled using PhysBERT (**a**) and BERT (**b**). The x-axis shows the years from 1990 to 2022, and the y-axis represents the normalized frequencies of occurrences per year. The dashed black line indicates the overall relative frequency of "Planck", while the dashed green line represents overall corpus growth. Each cluster label includes the six most frequent neighboring words and the total number of embeddings assigned to the cluster.*

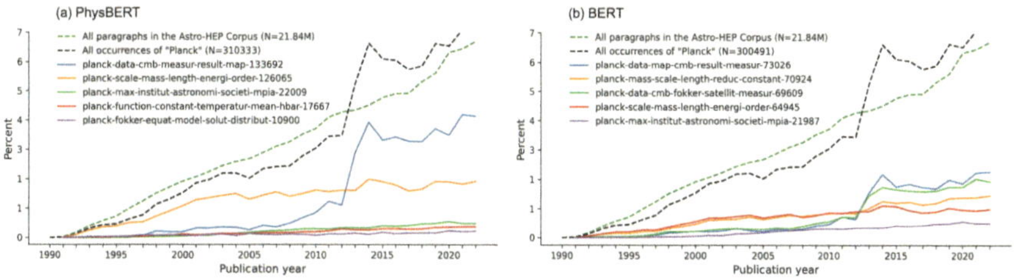

PhysBERT (5a) produced the most interpretable five-cluster breakdown, with clusters corresponding to prominent senses of "Planck": the Planck mission (blue), Planck units (orange), the Max Planck Society (green), Planck's law and constant (red), and the Fokker-Planck equation (purple). Notably, PhysBERT avoids splitting the mission and units senses into multiple clusters, yielding clearer and more distinct sense groupings. From 1996 to 1998, the (blue) mission cluster shows an initial rise, surpassing all clusters except the larger (orange) units cluster. Between 2007 and 2011, the mission cluster grows rapidly, overtaking the units cluster from 2012 to 2013. This growth peaks in 2014, after which the mission cluster stabilizes at a frequency roughly double that of the units cluster, maintaining this level through 2022.

In contrast, BERT (Figure 5b) disperses the mission and units senses across multiple clusters (blue/green for mission and orange/red for units), obscuring their relative growth. The model also blends the mission and Fokker-Planck equation senses in the green cluster, resulting in the least interpretable five-cluster breakdown. Still, the clusters referencing the mission sense (blue and green) surpass all others between 2012 and 2013.

These results align with milestones in the history of the Planck mission.[7] Originally proposed as the COBRAS/SAMBA mission in the early 1990s, it was approved by the European Space Agency (ESA) in 1996 and renamed the Planck mission in honor of Max Planck. Around this time, references to the mission began appearing in scientific literature.

Later, from 2011 to 2018, ESA released key data from the mission, including the 2013 first all-sky map of the cosmic microwave background (CMB), which may have contributed to the prominence of the mission-related sense during this period.

To quantify changes in the sense distribution of "Planck" and potentially the term's overall meaning, I applied two complementary indicators described in Periti and Montanelli (2024). The first, **Jensen-Shannon divergence (JSD)**, is a "sense-based" metric that captures shifts in the relative prominence of distinct senses over time by comparing their normalized frequency distributions. It is particularly sensitive to changes in the balance of senses, even if the dominant meaning remains stable. JSD is calculated as:

$$\mathrm{JSD}(d_1,d_2) = \frac{1}{2}(\mathrm{KL}(d_1,\mathrm{M}) + \mathrm{KL}(d_2,\mathrm{M})),$$

where d_1 and d_2 are the frequency distributions of clusters in two consecutive years, KL is the Kullback-Leibler divergence, and M is the average of the two distributions.

In contrast, the second indicator, **cosine distance between prototypes (CDPT)**, is a "form-based" metric that tracks changes in the overall meaning of a term by comparing the averaged word embeddings (prototypes) from consecutive years. CDPT is less affected by fluctuations in sense distributions and instead highlights broader shifts in dominant meaning. It is defined as:

$$\mathrm{CDPT}(\mathrm{PT}_1,\mathrm{PT}_2) = 1 - \mathrm{CS}(\mathrm{PT}_1,\mathrm{PT}_2),$$

where CS is the cosine similarity and PT_1 and PT_2 are the year-specific prototypes.

Figures 6a and 6b illustrate the results of JSD and CDPT, revealing significant semantic shifts in 1997 and 2013 that align with historical milestones of the Planck mission. In 1997, the renaming of the COBRAS/SAMBA mission to "Planck" coincides with a sharp spike in JSD, particularly in domain-specific models like Astro-HEP-BERT and PhysBERT, capturing the rise of the mission sense even as the dominant meaning remained tied to units. CDPT showed a smaller spike, suggesting limited impact on the overall dominant meaning. Similarly, in 2013, the release of key data from the Planck mission aligns with sharp increases in both indicators, reflecting a more cohesive shift in the dominant meaning of "Planck".

The complementary nature of JSD and CDPT is evident here: JSD is more sensitive to shifts in the balance of senses, such as the emergence of the mission sense in 1997, while CDPT highlights broader contextual changes, as seen in the major semantic shift in 2013. Post-2013, both indicators stabilize, but JSD captures ongoing smaller fluctuations in sense distributions, whereas CDPT reflects the settling of the dominant meaning.

---

7 https://www.esa.int/Science_Exploration/Space_Science/Planck_overview and https://www.cosmos.esa.int/web/planck/mission-history; https://en.wikipedia.org/wiki/Planck_(spacecraft)

Figure 6: Semantic change of the term "Planck" over time, measured using two indicators: Jensen-Shannon divergence (JSD) and cosine distance between prototypes (CDPT). JSD (a) captures shifts in the distributions of clusters, representing different senses, while CDPT (b) measures global shifts in the aggregated meaning of the term. Both plots track changes across the years from 1990 to 2022 for each model.

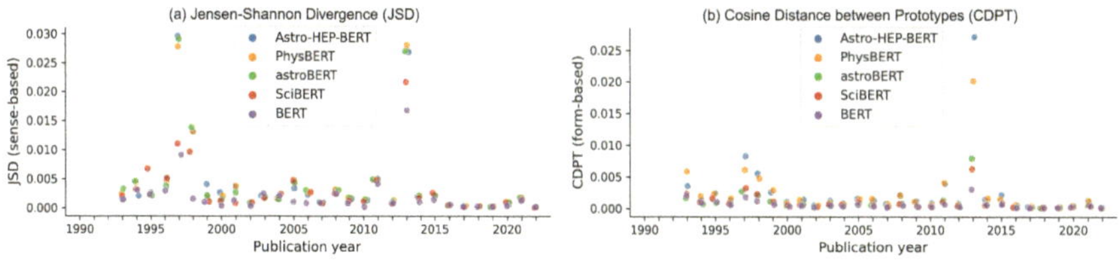

These findings demonstrate the value of combining sense-based and form-based LSC indicators to capture both fine-grained sense dynamics and broader shifts in dominant meaning, providing a comprehensive view of lexical semantic change for scientific concepts over time. Among the models, domain-specific ones like Astro-HEP-BERT and PhysBERT display greater sensitivity to both JSD and CDPT changes, reflecting their superior performance in WSD and WSI tasks.

9. Conclusion and discussion

This study demonstrates the value of contextualized word embeddings (CWEs) for the history, philosophy, and sociology of science (HPSS), providing a new tool for analyzing the meanings of scientific concepts. Using the term "Planck" as a test case, I evaluated five BERT-based models across tasks in word sense disambiguation, sense induction, clustering quality, and lexical semantic change. The findings show that CWEs effectively disambiguate scientific terms, uncover distinct meanings, and trace diachronic trends, revealing how concepts like "Planck" evolve alongside developments in science, technology, and society.

The study highlights the promise of domain-adapted CWEs for HPSS research, particularly when models are trained on specialized corpora tailored to specific scientific fields or historical periods. This approach maximizes contextual relevance, and while critics may fear that such specialization risks overfitting, we can instead view it as a form of context-sensitive tailoring that aligns a model's interpretive focus with the unique conceptual landscape of its target domain. By allowing models to learn the distinct characteristics of specific scientific discourses, this approach enhances performance in these contexts, even at the expense of broader generalizability. In this regard, the success of Astro-HEP-BERT serves as a proof-of-concept for developing specialized models for HPSS on a limited budget.

By reusing pretrained general-purpose weights, Astro-HEP-BERT achieved performance comparable to, or better than, more resource-intensive domain-adapted models like PhysBERT and astroBERT, despite using fewer training examples.

Beyond tracing "conceptual histories" (Wevers and Koolen, 2020; Zichert et al., 2024; Zichert and Simons, 2026), domain-adapted CWEs and their basic operations—such as measuring similarity, clustering, and generating prototypes—may offer broader applications for exploring other HPSS phenomena related to semantic similarity and divergence. These could include investigating lexical codification within research fields (Amahdi, 2026; Gläser et al., 2018; Teich et al., 2021) or addressing linguistic aspects of translating scientific concepts across paradigms (Kuhn, 2022) and within "trading zones" (Galison, 1997). For instance, semantic distance measures could reveal patterns of conceptual consistency, while clustering CWEs could uncover emergent terminologies or hybrid concepts bridging disciplines. Prototypical embeddings for terms from different paradigms could further illuminate semantic alignment or divergence, offering new and complementary methods for studying the evolution of scientific language. These possibilities remain speculative and sketchy, outlining potential approaches that must be explored further and specified through targeted studies and practical applications.

CWEs trained on scientific corpora can also enhance downstream tasks and advanced computational workflows, offering new opportunities for specialized applications in HPSS. For instance, they can improve domain-specific named entity recognition (NER) or the generation of specialized sentence embeddings, as demonstrated by recent studies (Hellert et al., 2024; Grezes et al., 2022). Beyond these tasks, domain-adapted CWEs could play a pivotal role in advanced computational workflows. A compelling example is retrieval-augmented generation (RAG), where embedding domain-specific queries and documents into a shared semantic space enhances retrieval accuracy and contextual relevance. This integration could enable more precise information retrieval, facilitate robust knowledge synthesis, and support dynamic interactions with complex scientific corpora.

For future research, scholars should consider developing HPSS-specific evaluation datasets tailored to a variety of computational tasks. Foundational tasks like WSD, WSI, LSC, and Word-in-Context (WiC) (Pilehvar and Camacho-Collados, 2019) remain essential. WiC, which tests whether a word maintains the same meaning across two different contexts, can also be extended to diachronic settings (Loureiro et al., 2022), providing insights into the evolution of scientific concepts over time. Beyond these, the scope of evaluation datasets could be broadened to include tasks such as domain-specific NER and the generation of sentence embeddings using models like SentenceBERT (Reimers and Gurevych, 2019). By creating tailored evaluation datasets for these and other tasks, researchers can address the unique challenges posed by specialized scientific languages, fostering more precise and context-aware computational analyses. My own two labeled datasets for the term "Planck," which provide sense annotations based on predefined categories and, in one case—the Astro-HEP-Planck Corpus—include the publication year, exemplify how HPSS-specific annotations can support the classical WSD task and prepare for both synchronic and diachronic analyses.

A key limitation is that BERT-based CWEs rely entirely on textual data, which restricts analyses to patterns and contexts within the analyzed corpus. Non-textual aspects of scientific practice—such as experimental settings, interpersonal communication, and institutional or cultural contexts—are often inaccessible or only indirectly reflected in texts. To address this, HPSS researchers can combine CWEs with qualitative methods and explore multimodal pipelines. Qualitative methods provide the interpretive depth needed to analyze complex senses and socio-historical factors, ensuring CWE-based analyses remain grounded in broader historical and disciplinary contexts. Multimodal pipelines, by integrating non-textual sources like figures, diagrams, metadata, and citation networks, might help bridge the gap between textual representations and the material and social dimensions of scientific activity. However, the feasibility of such approaches remains to be fully explored (Yin et al., 2024).

Staying informed about advancements in adjacent fields, such as computational humanities, computational linguistics, and machine learning, is essential for researchers employing CWEs in HPSS (cf. Simons et al., 2026). By engaging with developments in model architectures, semantic tools, and advanced pipelines, HPSS scholars can enhance the precision and interpretability of CWE-based analyses. This interdisciplinary approach will not only deepen our understanding of scientific concepts and their evolution but also open new avenues for studying the dynamics of scientific language and its broader cultural implications. By embracing CWEs and their extensions, HPSS researchers have an opportunity to lead a transformative shift in the computational analysis of scientific discourse.[8]

Acknowledgements

I am grateful to my colleagues Adrian Wüthrich and Michael Zichert for their insightful feedback on this draft, and for our collaborative exchange of ideas regarding the use of computational methods in HPSS. I also want to acknowledge funding by the European Union (ERC Consolidator Grant, Project No. 101044932). Views and opinions expressed are however those of the author only and do not necessarily reflect those of the European Union or the European Research Council. Neither the European Union nor the granting authority can be held responsible for them.

References

Ahmadi ES (2026) Exploring disciplinary differences in semantic uniformity. A computational approach to codification. In: Simons A, Wüthrich A, Zichert M, et al. (eds) *Understanding Science with Large Language Models? Potentials for the History, Philosophy, and Sociology of Science*. Bielefeld: transcript, part-3.

Ait-Saada M and Nadif M (2023) Is Anisotropy Truly Harmful? A Case Study on Text Clustering. In: *Proceedings of the 61st Annual Meeting of the Association for Computational Linguistics (Volume 2: Short Papers)* (eds A Rogers, J Boyd-Graber, and N Okazaki), Toronto, Canada, July 2023, pp. 1194–1203. Association for Computational Linguistics. Available at: https://aclanthology.org/2023.acl-short.103

8 This chapter was written with support from large language models (LLMs). All model-generated text was reviewed and, where necessary, rewritten by the authors, who remain fully responsible for the final version. For details on the use of LLMs in this volume, see the statement in the volume's introduction.

Beltagy I, Lo K and Cohan A (2019) SciBERT: A Pretrained Language Model for Scientific Text. http://arxiv.org/abs/1903.10676.

Bevilacqua M, Pasini T, Raganato A, et al. (2021) Recent Trends in Word Sense Disambiguation: A Survey. In: *Proceedings of the Thirtieth International Joint Conference on Artificial Intelligence*, Montreal, Canada, August 2021, pp. 4330–4338. International Joint Conferences on Artificial Intelligence Organization. Available at: https://www.ijcai.org/proceedings/2021/593

Biś D, Podkorytov M and Liu X (2021) Too Much in Common: Shifting of Embeddings in Transformer Language Models and its Implications. In: *Proceedings of the 2021 Conference of the North American Chapter of the Association for Computational Linguistics: Human Language Technologies* (eds K Toutanova, A Rumshisky, L Zettlemoyer, et al.), Online, June 2021, pp. 5117–5130. Association for Computational Linguistics. Available at: ht tps://aclanthology.org/2021.naacl-main.403.

Blei DM, Ng AY and Jordan MI (2003) Latent Dirichlet allocation. *Journal of machine Learning research* 3(Jan): 993–1022.

Bowker GC and Star SL (1999) *Sorting Things out: Classification and Its Consequences*. The MIT Press.

Boyack KW, Newman D, Duhon RJ, et al. (2011) Clustering more than two million biomedical publications: Comparing the accuracies of nine text-based similarity approaches. *PloS one* 6(3): e18029.

Cai X, Huang J, Bian Y, et al. (2020) Isotropy in the contextual embedding space: Clusters and manifolds. In: *International conference on learning representations*, 2020. Available at: https://openreview.net/forum?id=xYGNO86OWDH.

Callon M, Courtial J-P, Turner WA, et al. (1983) From translations to problematic networks: An introduction to co-word analysis. *Social Science Information* 22(2): 191–235.

Canguilhem G (1991) *The Normal and the Pathological* (trans. CR Fawcett). Revised edition. New York: MIT PR.

Chang H (2007) *Inventing Temperature: Measurement and Scientific Progress*. Oxford: Oxford University Press.

Clarke AE, Mamo L, Fosket JR, et al. (eds) (2010) *Biomedicalization: Technoscience, Health, and Illness in the U.S.* Duke University Press.

Courtial J-P and Law J (1989) A Co-Word Study of Artificial Intelligence. *Social Studies of Science* 19(2): 301–311.

Daston L and Galison P (2007) *Objectivity*. New York, NY: Zone Books.

Deerwester S, Dumals ST, Furnas GW, et al. (1990) Indexing by Latent Semantic Analysis. *Journal of the American Society for Information Science* 41(6): 391–407.

Devlin J, Chang M-W, Lee K, et al. (2018) BERT: Pre-training of Deep Bidirectional Transformers for Language Understanding. http://arxiv.org/abs/1810.04805.

Ding Y, Martinkus K, Pascual D, et al. (2022) On Isotropy Calibration of Transformer Models. In: *Proceedings of the Third Workshop on Insights from Negative Results in NLP* (eds S Tafreshi, J Sedoc, A Rogers, et al.), Dublin, Ireland, May 2022, pp. 1–9. Association for Computational Linguistics. Available at: https://aclanthology.org/2022.insights-1.1.

Ethayarajh K (2019) How Contextual are Contextualized Word Representations? Comparing the Geometry of BERT, ELMo, and GPT-2 Embeddings. http://arxiv.org/abs/1909.00512.

Fleck L (1979) *Genesis and Development of a Scientific Fact*. University of Chicago Press.

Foucault M (1970) *The Order of Things: An Archaeology of Human Sciences*. Random House.

Galison P (1997) *Image and Logic: A Material Culture of Microphysics*. illustrated edition. Univ of Chicago Pr.

Gläser J, Laudel G, Grieser C, et al. (2018) Scientific fields as epistemic regimes: new opportunities for comparative science studies. DEU. Epub ahead of print 2018.

Godey N, de la Clergerie É and Sagot B (2023) Is Anisotropy Inherent to Transformers? http://arxiv.org/abs/2306.07656.

Grezes F, Blanco-Cuaresma S, Accomazzi A, et al. (2021) Building astroBERT, a language model for Astronomy & Astrophysics. http://arxiv.org/abs/2112.00590.

Grezes F, Allen T, Blanco-Cuaresma S, et al. (2022) Improving astroBERT using Semantic Textual Similarity. http://arxiv.org/abs/2212.00744.

Gu Y, Tinn R, Cheng H, et al. (2021) Domain-specific language model pretraining for biomedical natural language processing. *ACM Transactions on Computing for Healthcare (HEALTH)* 3(1). ACM New York, NY: 1–23.

Hacking I (1975) *The Emergence of Probability: A Philosophical Study of Early Ideas about Probability, Induction and Statistical Inference*. Cambridge University Press.

Hacking I (1999) *The Social Construction of What?* Harvard University Press.

Hellert T, Montenegro J and Pollastro A (2024) PhysBERT: A Text Embedding Model for Physics Scientific Literature. http://arxiv.org/abs/2408.09574.

Kleymann R, Niekler A and Burghardt M (2022) Conceptual Forays: A Corpus-based Study of "Theory" in Digital Humanities Journals. *Journal of Cultural Analytics* 7(4).

Kuhn TS (1962) *The Structure of Scientific Revolutions*. Chicago: University of Chicago Press.

Kuhn TS (2022) *The Last Writings of Thomas S. Kuhn: Incommensurability in Science* (ed. B Mladenovic). First Edition. Chicago: University of Chicago Press.

Latour B (1987) *Science in Action: How to Follow Scientists and Engineers through Society*. Harvard University Press.

Laubichler MD, Maienschein J and Renn J (2019) Computational History of Knowledge: Challenges and Opportunities. *Isis* 110(3): 502–512.

Lean OM, Rivelli L and Pence CH (2023) Digital Literature Analysis for Empirical Philosophy of Science. *The British Journal for the Philosophy of Science* 74(4): 875–898.

Lee J, Yoon W, Kim Sungdong, et al. (2020) BioBERT: a pre-trained biomedical language representation model for biomedical text mining. *Bioinformatics* 36(4): 1234–1240.

Leydesdorff L and Rafols I (2009) A global map of science based on the ISI subject categories. *Journal of the American Society for Information Science and Technology* 60(2): 348–362.

Liu Y, Ott M, Goyal N, et al. (2019) RoBERTa: A Robustly Optimized BERT Pretraining Approach. http://arxiv.org/abs/1907.11692.

Loureiro D, Rezaee K, Pilehvar MT, et al. (2020) Language Models and Word Sense Disambiguation: An Overview and Analysis. Available at: https://www.semanticscholar.org/reader/a7cb7c67a181c35f2e2dd2767e21defc0af9c14a.

Loureiro D, D'Souza A, Muhajab AN, et al. (2022) TempoWiC: An Evaluation Benchmark for Detecting Meaning Shift in Social Media. http://arxiv.org/abs/2209.07216.

Malaterre C and Léonard M (2024) Epistemic Markers in the Scientific Discourse. *Philosophy of Science* 91(1): 151–174.

Merchant C (1980) *The Death of Nature: Women, Ecology, and the Scientific Revolution*. New York: Harper & Row.

Mickus T, Grönroos S-A and Attieh J (2024) Isotropy, Clusters, and Classifiers. http://arxiv.org/abs/2402.03191.

Mikolov T, Chen K, Corrado G, et al. (2013) Efficient Estimation of Word Representations in Vector Space. http://arxiv.org/abs/1301.3781.

Mol A (2002) *The Body Multiple: Ontology in Medical Practice*. Duke University Press.

Mu J, Bhat S and Viswanath P (2018) All-but-the-Top: Simple and Effective Postprocessing for Word Representations. http://arxiv.org/abs/1702.01417.

Overton JA (2013) "Explain" in scientific discourse. *Synthese* 190(8): 1383–1405.

Pence CH and Ramsey G (2018) How to do digital philosophy of science. *Philosophy of Science* 85(5). Cambridge University Press: 930–941.

Periti F and Montanelli S (2024) Lexical Semantic Change through Large Language Models: a Survey. *ACM Computing Surveys* 56(11): 282:1-282:38.

Pickering A (1995) *The Mangle of Practice: Time, Agency, and Science*. Chicago: University of Chicago Press.

Pilehvar MT and Camacho-Collados J (2019) WiC: the Word-in-Context Dataset for Evaluating Context-Sensitive Meaning Representations. In: *Proceedings of the 2019 Conference of the North American Chapter of the Association for Computational Linguistics: Human Language Technologies, Volume 1 (Long and Short Papers)* (eds J Burstein, C Doran, and T Solorio), Minneapolis, Minnesota, 2019, pp. 1267–1273. Association for Computational Linguistics. Available at: https://aclanthology.org/N19-1128.

Reimers N and Gurevych I (2019) Sentence-BERT: Sentence Embeddings using Siamese BERT-Networks. http://arxiv.org/abs/1908.10084.

Rip A and Courtial JP (1984) Co-word maps of biotechnology: An example of cognitive scientometrics. *Scientometrics* 6(6): 381–400.

Simons A (2024a) Astro-HEP-BERT: A bidirectional language model for studying the meanings of concepts in astrophysics and high energy physics. http://arxiv.org/abs/2411.14877.

Simons A (2024b) Meaning at the Planck scale? Contextualized word embeddings for doing history, philosophy, and sociology of science http://arxiv.org/abs/2411.14073.

Simons A, Zichert M and Wüthrich A (2026) Large language models for history, philosophy, and sociology of science: Interpretive uses, methodological challenges, and critical perspectives. *Studies in History and Philosophy of Science* 117: 102151. https://doi.org/10.1016/j.shpsa.2026.102151.

Steinle F (2016) *Exploratory Experiments: Ampère, Faraday, and the Origins of Electrodynamics*. University of Pittsburgh Press.

Sun Y and Platoš J (2023) A method for constructing word sense embeddings based on word sense induction. *Scientific Reports* 13(1): 12945.

Tahmasebi N and Dubossarsky H (2023) Computational modeling of semantic change. https://arxiv.org/abs/2304.06337.

Teich E, Fankhauser P, Degaetano-Ortlieb S, et al. (2021) Less is More/More Diverse: On The Communicative Utility of Linguistic Conventionalization. *Frontiers in Communication* 5.

Venturini T, Baya Laffite N, Cointet J-P, et al. (2014) Three maps and three misunderstandings: A digital mapping of climate diplomacy. *Big Data & Society* 1(2): 2053951714543804.

Wang L, Huang J, Huang K, et al. (2020) Improving Neural Language Generation with Spectrum Control. In: *International Conference on Learning Representations*, 2020. Available at: https://openreview.net/forum?id=ByxY8CNtvr.

Wevers M and Koolen M (2020) Digital begriffsgeschichte: Tracing semantic change using word embeddings. *Historical Methods: A Journal of Quantitative and Interdisciplinary History* 53(4): 226–243.

Wiedemann G, Remus S, Chawla A, et al. (2019) Does BERT Make Any Sense? Interpretable Word Sense Disambiguation with Contextualized Embeddings. http://arxiv.org/abs/1909.10430.

Zichert M and Simons A (2026) From early digital methods to LLMs. Computational conceptual history of scientific concepts. In: Simons A, Wüthrich A, Zichert M, et al. (eds) *Understanding Science with Large Language Models? Potentials for the History, Philosophy, and Sociology of Science*. Bielefeld: transcript, part-3.

Zichert M, Simons A and Wüthrich A (2025) Expanding Conceptual Histories: Using Contextualized Word Embeddings for the History and Philosophy of the Virtual Particle Concept. *Computational Humanities Research*: 1–22. Available at: https://doi.org/10.1017/chr.2025.10013

Supplement: isotropy analysis

In Section 7 I hypothesized that the improved balance between AIS and APS observed in PhysBERT and Astro-HEP-BERT, compared to astroBERT and the other two models may be partly due to a more isotropic em- bedding space, where embeddings are more evenly distributed, allowing for better cluster separation. To investigate this further, this section begins with a brief overview of the ongoing debate in the literature regarding the roles of anisotropy and isotropy in BERT models and their impact on model performance.

The discussion around anisotropy in BERT models originated with the surprising discovery by Ethayarajh (2019) that CWEs produced by BERT exhibit significant anisotropy—an uneven distribution of CWEs in the high-dimensional space. This finding contrasted with earlier research by Mu et al. (2018), which demonstrated that anisotropy in static word embeddings could lead to performance degradation. However, since BERT's contextualized representations have shown remarkable performance across numerous tasks, researchers began questioning whether anisotropy had the same negative effect on contextualized embeddings as it did on static ones.

Subsequent studies produced mixed results regarding the impact of reducing anisotropy in BERT and other Transformer models. Some researchers reported performance improvements by mitigating anisotropy, using methods such as post-processing techniques (Biś et al., 2021; Wang et al., 2020) or modifications to pretraining objectives

(Godey et al., 2023). However, others found no significant improvements or, in some cases, even performance declines when attempting to enforce isotropy (Ait-Saada and Nadif, 2023; Ding et al., 2022). This has led to varying opinions on whether reducing anisotropy consistently enhances the performance of Transformer models across different tasks.

A related discussion has emerged around the distinction between global anisotropy and local isotropy. Several authors found that while Transformer models, including BERT, display global anisotropy (an uneven embedding distribution across the entire space), they show local isotropy within specific subspaces or clusters (Ait-Saada and Nadif, 2023; Cai et al., 2020; Ding et al., 2022; Mickus et al., 2024). This local isotropy may explain BERT's high performance, as it allows for expressive representations in particular regions of the embedding space, despite the global anisotropy. Thus, the relationship between (an)isotropy and model performance, particularly for contextualized embeddings, remains complex and multifaceted.

The aim of this supplementary section is not to make a direct contribution to this broader debate, but rather to assess how our five models compare in terms of their global isotropy, which I calculated using the average cosine similarity (ACS) between the contextualized word embeddings (CWEs) of randomly sampled words—a standard approach in this research area. For random vectors with independent, identically distributed components, expected cosine similarity approaches zero as dimensionality increases. Thus, ACS values closer to zero suggest a more isotropic (and thus more balanced) embedding space.

For each model and corpus, 200,000 tokens were sampled, paired, and cosine similarity was calculated for 100,000 random pairs, with the resulting ACS providing a measure of global isotropy.

Results indicate that Astro-HEP-BERT has the most isotropic embedding space, with an ACS just below 0.1 in both corpora, followed by BERT and PhysBERT, which show ACS values slightly above 0.2 in the Astro-HEP Corpus. In contrast, PhysBERT shows a slightly higher ACS in the Wikipedia-Physics Corpus, nearing 0.3. SciBERT and astroBERT demonstrate higher anisotropy, with astroBERT reaching an ACS near 0.5 in the Wikipedia-Physics Corpus.

Certain models, like PhysBERT and SciBERT, display multiple peaks in their cosine similarity distributions, while Astro-HEP-BERT and BERT exhibit a more singular peak. This difference suggests that models with multiple peaks might have more compartmentalized embedding spaces, indicating anisotropic regions or clustering tendencies, whereas Astro-HEP-BERT's single, concentrated peak reflects a more isotropic, evenly distributed space.

These findings highlight two key observations: Astro-HEP-BERT's increased isotropy relative to models like PhysBERT and astroBERT may stem from its adaptation of BERT's pretrained embedding space rather than starting from scratch. Refining BERT's embeddings with astrophysics and HEP-specific data likely preserved and enhanced isotropy by organizing domain-specific terms within an existing, well-structured framework. In contrast, models trained from scratch on physics texts appear to have developed more clustered or anisotropic embedding spaces, potentially due to early overfitting on domain-specific patterns without a generalized framework.

Figures S1c and S1d show how isotropy evolved in Astro-HEP-BERT over three pretraining epochs. Notably, after the first epoch, ACS decreased by approximately 0.1, with smaller reductions in subsequent epochs, indicating progressive isotropy improvement as embeddings spread more uniformly.

Figure S1: Global isotropy analysis across models and corpora using average cosine similarity (ACS) and density distributions. (a) ACS and cosine similarity distribution for CWEs in the Astro-HEP Corpus, with each model's ACS represented by a colored vertical line. (b) ACS and cosine similarity distribution in the Wikipedia-Physics Corpus. (c, d) Evolution of ACS for Astro-HEP-BERT over three pretraining epochs, evaluated on the Astro-HEP (c) and Wikipedia-Physics (d) corpora, showing progressive improvement in isotropy. Lower ACS values indicate a more isotropic and balanced embedding space.

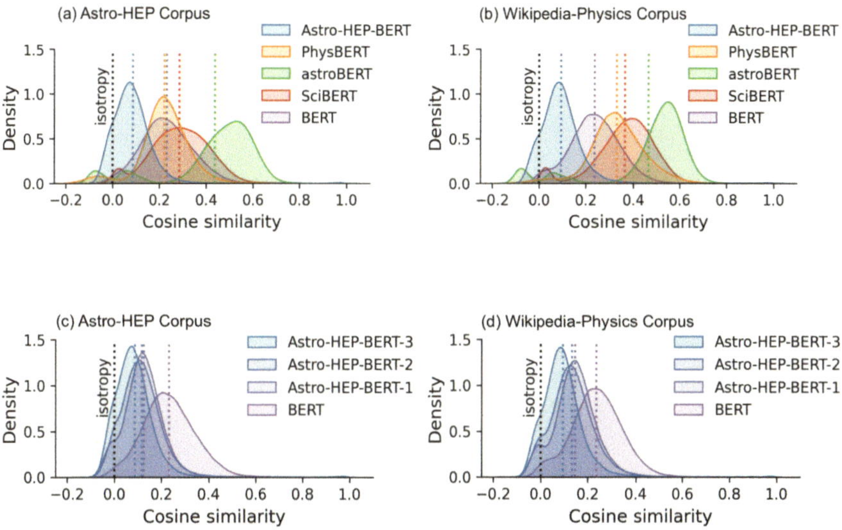

Exploring disciplinary differences in semantic uniformity

A computational approach to codification

Elaheh Sadat Ahmadi

1. Introduction

The study of disciplinary variation has long centered on the notion that fields differ not only in subject matter but also in how they organize knowledge and establish consensus. Zuckerman and Merton (1972: 507) introduced the concept of codification to capture these differences, defining it as "the consolidation of empirical knowledge into succinct and interdependent theoretical formulations." Highly codified disciplines such as physics and chemistry were seen as internally coherent, equipped with precise criteria for evaluating new contributions, and characterized by rapid citation turnover, what Derek Price called "citation immediacy." In these fields, the clarity of theoretical orientation allowed younger scholars to participate meaningfully at early stages in their careers. By contrast, in less codified disciplines such as zoology and botany, expertise depended on the accumulation of descriptive knowledge and loosely connected theories, making consensus harder to achieve and contributions slower to be recognized. Cole and Zuckerman (1975) extended this framework to their own field, arguing that sociology of science itself had, by the 1970s, achieved the level of codification necessary to become a recognized specialty. Their analyses of citation patterns, peer evaluations, and publication growth suggested that codification was linked to the consolidation of a specialty's identity and intellectual maturity. Yet subsequent work, including Cole, Cole and Dietrich (1978), complicated this picture. Although they sought to measure disciplinary consensus through indicators such as faculty ratings of scholars, reviewer agreement, and immediacy indices, the results were often inconsistent, statistically weak, or sensitive to the growth of literature. Stephen Cole (1983) went further, questioning the assumption that natural sciences inherently exhibit greater consensus and theoretical development than social sciences. His analyses showed that even in fields regarded as "hard sciences," disagreements at the research frontier were substantial. He concluded that disciplinary differences could not be neatly ordered along a fixed hierarchy, and that consensus was shaped as much by institutional and social mechanisms (peer review, funding structures, gatekeeping) as by cognitive organization alone.

Critics soon highlighted a deeper methodological limitation. Although the frameworks introduced by Zuckerman and Merton and later elaborated by subsequent scholars made important contributions by linking codification to observable scholarly behaviors, they fell short by relying on assumed field rankings and intuitive classifications, often grounded in Kuhnian authority or disciplinary tradition rather than empirical indicators (Cozzens, 1985). This lack of direct measurement rendered the interpretation of empirical findings ambiguous, as codification itself remained an unmeasured abstraction. Consequently, failures to confirm expected behavioral patterns could indicate that the codification hypothesis is incorrect, that commonsense rankings of disciplines along this dimension are flawed, or that codification is not a discipline-level phenomenon at all (Braxton and Hargens, 1996).

Addressing these debates and their methodological shortcomings, Gläser et al. (2024) reframe codification as a "global epistemic property" of research fields, one that cannot be attributed to individual research processes or their interrelations but exists only at the level of the scientific community. They contrast codification with analytical properties like "resource intensity" and structural ones like "epistemic diversity" of research processes and fields, emphasizing that codification reflects the holistic structure of a field's knowledge and language. Drawing on the foundational insights of Zuckerman and Merton, Gläser and colleagues define codification through two dimensions: theoretical organization and linguistic standardization. The former refers to how clearly concepts are defined, hierarchically structured, and theoretically integrated. The latter concerns the degree to which terminology is used consistently, avoiding homonyms and synonyms, and exemplified most extremely by mathematical language. To approximate codification empirically, Gläser et al. propose methods such as content analysis of scientific texts, bibliometric reconstruction of conceptual networks, and linguistic metrics. They particularly highlight the use of Contextualized Word Embeddings (CWEs) to detect semantic ambiguity and polysemy in field-specific language, a strategy this study adopts and explains in the following section.

The aim of this chapter is to report on an approach to a linguistically grounded measure of codification by examining the contextual stability of terminology in scientific discourse. Whereas previous research has largely relied on proxy indicators to approximate the degree of codification, this study builds on recent work that emphasizes the "standardization of a field's language" (Gläser et al., 2024). It introduces an embedding-based metric to capture how consistently key terms are used within disciplinary corpora, composed of journal articles in sociology and astrophysics, two disciplines conventionally positioned at opposite ends of the consensus spectrum. In doing so, the study investigates whether established differences in disciplinary consensus are reflected in their patterns of language use, particularly in their respective terminologies, and seeks both to operationalize the linguistic component of codification and to explore the use of large language models in the scalable comparative analysis of scientific fields.

This chapter is structured as follows: Section 2 outlines the methodological approach, including corpus construction, preprocessing, and the embedding-based procedure for computing Semantic Uniformity. Section 3 reports the empirical results, comparing sociology and astrophysics in terms of the contextual stability of their terminology. Section 4 addresses potential sources of error in corpus design and implementation, and Section

5 considers the broader conceptual and methodological implications of using contextual embeddings to operationalize codification.

2. Measuring codification with contextual embeddings

The experiment reported here focuses on one dimension of codification identified by Gläser et al. (2024: 12): the "standardization of a field's language". In their framework, a highly standardized language is one in which each term has a single, clearly defined meaning, and each concept is represented by only one term, minimizing ambiguity from synonyms and homonyms:

> "*The standardisation of a field's language* is high when among the concepts used there are few homonyms and synonyms, i.e. when only one word is used to represent a phenomenon, and when this word is used to represent only this particular phenomenon."

Building on this definition, standardization can be examined empirically through semantic uniformity, that is, by assessing whether a term is used consistently to refer to the same concept. A discipline's semantic uniformity score thus reflects the degree of consistency in how terms are employed within a given corpus of scientific writing. In highly codified fields, terms are expected to maintain stable, precisely defined meanings and therefore exhibit high semantic uniformity across contexts. In less codified disciplines, however, terms are more likely to carry multiple or even conflicting meanings, leading to greater polysemy and lower semantic uniformity.

2.1 Semantic Uniformity Score

Semantic Uniformity Score (SUS) can be built using contextual embeddings generated by a pre-trained BERT (Bidirectional Encoder Representations from Transformers) model (Devlin et al., 2019). BERT is a deep contextual language model based on Transformer architecture, pretrained on large-scale text corpora and its self-attention mechanism enables it to capture long-range dependencies and assign greater importance to contextually relevant words, producing embeddings that reflect a term's precise meaning in its specific context. BERT's effectiveness in detecting semantic ambiguity and polysemy is well established: it can distinguish word senses, with instances forming distinct clusters in the embedding space (Wiedemann et al., 2019) and can even predict a word's polysemy level from instances describing the same sense (Garí Soler and Apidianaki, 2021). These capabilities make BERT particularly well-suited for assessing the consistency of terminology in this experiment.

Leveraging these properties, BERT-generated contextual embeddings can be applied to quantify semantic uniformity across term occurrences in a corpus. Each occurrence of a word is represented as a high-dimensional vector $e_{w,i}$, informed by its co-occurring words and syntactic environment. For a given word w, appearing n_w times in the corpus D, all pairwise cosine similarities between its contextual embeddings are computed. The mean of these similarities, $Sim(w) = \frac{1}{\binom{n_w}{2}} \sum_{1 \leq i < j \leq n_w} \cos(e_{w,i}, e_{w,j})$, serves as an indicator of how stable the word's meaning is across contexts: tightly clustered embeddings indicate semantic consistency, while scattered vectors suggest contextual variability or polysemy. This type of calculation is mathematically equivalent to the *SelfSim* metric reported by Garí Soler and Apidianaki (2021), though here it is applied for a different analytical purpose by aggregating the word-level similarity scores into a discipline-level SUS. This aggregation can be performed either without weighting (treating all words equally) or with frequency-based weighting, which gives greater influence to commonly used terms. This is formalized as:

$$SUS_D^{(\alpha)} = \frac{\sum_{w \in W} \left[n_w^{\alpha} . sim\left(w\right) \right]}{\sum_{w \in W} n_w^{\alpha}}$$

where $\alpha = 0$ produces the unweighted SUS and $\alpha = 1$ the weighted variant. A higher SUS suggests that a field's language leans toward standardized, well-defined usage, while a lower SUS indicates that terms are deployed with more interpretive flexibility. Comparing weighted and unweighted scores can further show whether semantic consistency is concentrated in a high-frequency core vocabulary or distributed more evenly across the lexicon.

2.2 Data and method

To empirically examine disciplinary differences in semantic uniformity, the analysis focuses on two corpora, astrophysics and sociology. Using the Web of Science (WoS) categories *Astronomy & Astrophysics* and *Sociology* in combination with *Journal Citation Reports* (JCR), five high-impact journals were selected for each field based on disciplinary focus, original research content, consistent article length, and archival depth. From each journal, 100 highly cited research articles were sampled (25 from 1990, 2000, 2010, and 2020) resulting in 500 articles per discipline.

The raw PDF documents were first processed using the NOUGAT OCR model (Blecher et al., 2023), which extracts structured text directly into Markdown format while preserving mathematical notation and visual layout. This was followed by an extensive post-processing step, where abstract, reference and the appendix sections containing tables were removed, equations and variables were replaced with placeholder tokens ([EQUATION], [VARIABLE]), and errors such as misrecognized or missing content were manually corrected. In contrast to Storer (1967), who treated the number of tables as an indicator of codification, this study removed tables and figures from the corpus. While their captions were retained for their explanatory value, the tables themselves consist primarily of numerical data; since the present analysis focuses on linguistic units, they were excluded. Standard text normalization steps like stopword removal and lemmatization were deliberately avoided, as they risk removing syntactic and semantic cues essential for embedding quality (Alzahrani and Jololian 2021). Instead, efforts focused on normalizing Markdown artifacts (e.g., hyphens, asterisks, headers), and correcting formatting inconsistencies. Punctuation and capitalization were maintained, while noise reduction refined the corpus by removing extraneous fragments around placeholders (e.g., "m[VARIABLE]"), consolidating consecutive placeholders, and deleting common scholarly abbreviations (e.g., *et al.*, *cf.*) that added little value for embedding-based analysis. Finally, the text was segmented into blocks optimized for BERT's maximum input length of 512 tokens. Segmentation proceeded in stages: shorter blocks were merged to enhance coherence, oversized ones were split at sentence boundaries to prevent truncation, and residual undersized blocks were either expanded by borrowing sentences from adjacent segments or removed if they contained too little content. These procedures ensured that the resulting segments were structurally coherent, contextually rich, and of comparable length distributions across disciplines, thereby preparing the corpus for reliable embedding extraction.

To generate contextual embeddings for semantic analysis, the corpus was first initialized from preprocessed segmented text. Thresholds were applied to exclude low-frequency terms, defined as those appearing fewer than ten times in the corpus or in fewer than five documents (to reduce noise from typos or idiosyncratic vocabulary, ensure terms are representative of the broader discourse, and improve the stability of similarity estimates). Embedding extraction was performed using the *bert-base-uncased* implementation from the Hugging Face library, with a customized tokenizer that treated the predefined mathematical placeholder tokens as indivisible units. Since these tokens were absent from the original BERT vocabulary, they were assigned initial vectors by copying the pretrained embeddings for the base terms "variable" and "equation." This provided semantically grounded starting points instead of random vectors, reducing the risk of unpredictable contextual representations during inference. During inference, BERT's self-attention mechanism used these fixed embeddings as inputs and combined them with surrounding context to produce per-occurrence contextualized representations. Segments were tokenized in this way, padded, and passed through the BERT model, with final token embeddings computed by averaging the last four hidden layers. Subword units produced by WordPiece tokenization were reassembled and averaged to produce a single embedding per word.

After tokenization and embedding generation, a final vocabulary filter was applied before saving the terms for similarity analysis. This step did not alter the embeddings already produced, any token present in the segments, including noise items, had already contributed to the contextual encoding. The filter simply determined which tokens would be carried forward into the analysis. Non-alphabetic strings (e.g., numbers, mixed codes such as "fig_3"), domain-irrelevant artifacts from document extraction (e.g., "http", "html"), and very short tokens under three characters (e.g., "c4", "in") were removed, along with untagged mathematical expressions left over from incomplete NOUGAT recognition. Although such tokens remained in the original segments and influenced surrounding context during embedding generation, excluding them at this stage ensured that similarity calculations were based only on genuine lexical items, and that the Semantic Uniformity Score of the discipline was calculated from actual words rather than formatting artifacts or non-lexical symbols.

After that, cosine similarity scores were computed to assess how consistently each word was used across different contexts. To handle the large number of pairwise comparisons without exceeding memory limits, embeddings were first L2-normalized and then divided into fixed-size blocks. Each block was compared both internally (within-block) and externally (across-block) so that every unique pair of token embeddings was included exactly once. For each word type all occurrence-level scores were aggregated to produce a mean similarity value that reflects its overall contextual stability across the corpus and a standard deviation that captures how much its meaning varied across different uses.[1]

2.3 Results

The analysis reveals notable disciplinary differences in the consistency of language use. Comparing SUS values between corpora, astrophysics exhibits higher semantic uniformity than sociology, a distinction evident in both unweighted and frequency-weighted scores. While the gap between fields is relatively modest in absolute terms (around 0.04), its consistency across both weighted and unweighted measures lends it analytical significance. In the unweighted comparison, which treats all word types equally, astrophysics reaches a mean SUS of 0.71, compared to 0.67 in sociology. When weighting by word frequency, scores decline in both fields, reflecting the broader contextual variation of more commonly used terms, but the gap remains: astrophysics registers 0.64, while sociology falls to 0.61. The results indicate that language in astrophysics is more tightly codified, with terms tending to retain clearer and more stable meanings across contexts. In contrast, sociological language appears more interpretively flexible, particularly among frequently used terms, which exhibit greater contextual variability.

The distinctions between disciplines are evident not only in the average similarity scores but also in their distributional patterns. Kernel density estimates (KDE; Figure A, left plot) show that similarity values in astrophysics are more densely concentrated around higher scores (0.7 to 0.9), whereas sociology displays a broader and flatter distribution. A similar trend appears in the standard deviation of similarity scores, which serves here as a proxy for semantic volatility, with sociology exhibiting a slightly higher value (0.09) than astrophysics (0.08). This suggests more flexible and variable word usage in sociology. In contrast, astrophysics shows tighter semantic boundaries. These differences are visually reinforced in the KDE curves (right plot), where sociology reveals a wider spread and astrophysics a sharper peak. Beyond the aggregate corpus-level comparison, the analysis also examines whether the observed difference in semantic uniformity between the two disciplines persists across different parts of speech (POS). The lower panel in Figure A presents the SUS disaggregated by syntactic category, including nouns, verbs, adjectives, proper nouns, and stopwords, serving as illustrative examples. Although absolute SUS values vary across grammatical categories, the relative difference between disciplines remains stable, with astrophysics consistently exhibiting higher semantic uniformity than sociology. This pattern suggests that the observed contrast is neither confined to specific lexical classes such as technical nouns nor masked by high-frequency functional tokens like stopwords, which are not typically expected to show uniform usage. Rather, it reflects a broader tendency toward codified and consistent language use across grammatical roles. Notably, the higher SUS values in astrophysics per-

1 The implementation underlying the analyses reported in this chapter has been archived as part of a larger research project (Ahmadi 2025) and is publicly available via Zenodo: https://doi.org/10.5281/zenodo.18753591.

sist even in categories where greater contextual variability would normally be expected, such as adjectives and stopwords.

Figure A: Semantic uniformity across disciplines

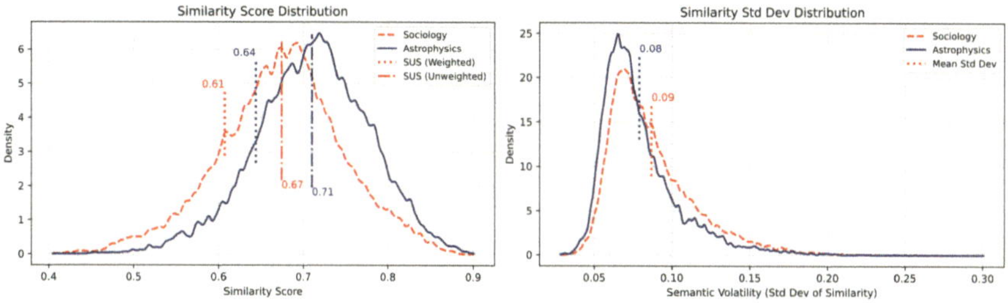

SUS by POS Group

Discipline	POS Group	Types (N)	Tokens (Σ freq)	SUS (Unweighted)	SUS (Weighted)
Sociology	Noun (NN+NNS)	5375	1278829	0.664	0.642
Sociology	Verb (all)	3358	424094	0.65	0.62
Sociology	Adjective (JJ)	2066	465541	0.679	0.641
Sociology	Proper Noun (NNP)	1925	122618	0.747	0.728
Sociology	Stopwords	108	1118273	0.56	0.535
Astrophysics	Noun (NN+NNS)	3321	1199085	0.695	0.677
Astrophysics	Verb (all)	2222	407587	0.687	0.666
Astrophysics	Adjective (JJ)	1274	425623	0.702	0.664
Astrophysics	Proper Noun (NNP)	2272	174721	0.769	0.751
Astrophysics	Stopwords	92	1143179	0.602	0.58

To examine the relationship between disciplines in greater detail with respect to frequency effects, an OLS regression model was estimated with word-level similarity as the dependent variable and log-transformed frequency, discipline, and their interaction as predictors. The results, presented in Figure B, confirm a general trend across both disciplines: semantic similarity tends to decrease as word frequency increases. However, the form and steepness of this decline differ notably.

Figure B: Model fit and residual diagnostics

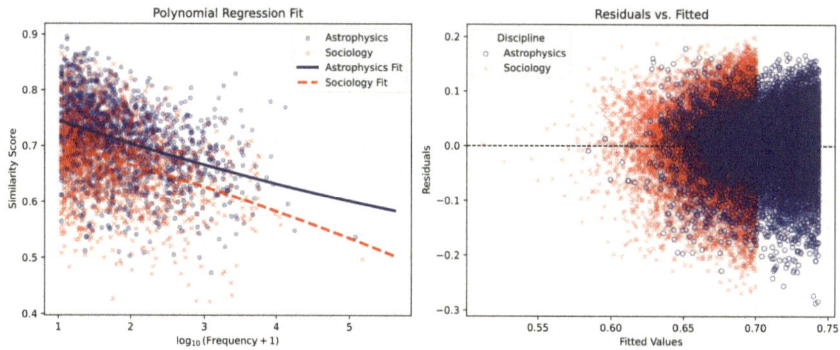

In astrophysics, the relationship is largely linear, reflecting a gradual decrease in similarity among more frequent terms. In sociology, by contrast, the interaction terms reveal a significant nonlinear effect, with a sharper drop in similarity at higher frequencies. This suggests that the core vocabulary of sociology (the most frequently used terms) is less semantically uniform than its periphery, whereas the gradient of variation in astrophysics is more restrained. Residual diagnostics, shown in the right panel of Figure B, support this interpretation: residuals in sociology are more widely dispersed, particularly at lower fitted values, while in astrophysics, they are more tightly clustered around zero, indicating a more consistent and predictable relationship between frequency and similarity.

Table A: OLS model summary with BP test

	Polynomial Interaction Model
Intercept	0.7112***
	(0.0007)
C(Discipline)[T.Sociology]	-0.0381***
	(0.0010)
log_count_c	-0.0401***
	(0.0013)
C(Discipline)[T.Sociology]:log_count_c	0.0036**
	(0.0018)
I(log_count_c ** 2)	0.0018
	(0.0012)
C(Discipline)[T.Sociology]:I(log_count_c ** 2)	-0.0039**
	(0.0017)
R-squared	0.1647
R-squared Adj.	0.1645
N	23372
R-squared	0.165
Adj. R-squared	0.165
F-statistic	935.688
Prob (F-stat)	0

Standard errors in parentheses.
* p<.1, ** p<.05, ***p<.01

Breusch-Pagan LM Stat: 285.9098
Breusch-Pagan LM p-value: 1.07e-59
Breusch-Pagan F Stat: 57.8753
Breusch-Pagan F p-value: 4.594e-60
Note: Standard Errors are heteroskedasticity robust (HC3)
log_count_c = log_count - mean(log_count)

Taken together, these results indicate a more semantically stable lexical structure in astrophysics, whereas sociology displays greater contextual flexibility in word usage. Although word frequency accounts for a meaningful share of the observed variation, it explains only approximately 16.5% of the variance in similarity scores. This implies that additional linguistic, contextual, or extra-contextual factors likely contribute to the disciplinary differences in semantic uniformity.

To evaluate the interpretive validity of the contextual embeddings as a proxy for codification, a qualitative validation was conducted using HDBSCAN. The aim was to test whether embeddings for key terms formed coherent clusters reflecting known or expected senses, thereby supporting the SUS metric derived from them. Consistent usage should produce tight clusters; polysemy should yield distinct groupings. This approach builds on Simons (2026), who showed that domain-tuned BERT models can recover multiple senses of terms like *Planck* in astrophysics. To examine both strengths and limitations, two terms from each field were analyzed: *order* and *paradigm* in sociology, *Planck* and *wave* in astrophysics.

The results present a mixed picture. Despite lacking domain-specific training, the general-purpose BERT model exhibited some promising capabilities. In sociology, the clustering of the term *order* (Sim(w) = 0.43, N = 2392) successfully differentiated several distinct meanings, including grammatical constructions (e.g., *in order to*), sociological concepts (*in-order-to motive*), ordinal scales (*first-order*), statistical modeling terminology, and macro-social structures (*Interaction Order*). These clusters demonstrated strong semantic coherence, suggesting that the embeddings effectively captured clear polysemy, though this may have been partially driven by syntactic cues. By contrast, the term *paradigm*, which showed moderate semantic consistency (Sim(w) = 0.66, N = 102), resisted clear semantic partitioning. Most instances referred to theoretical frameworks or research orientations, ranging from established analytic traditions (e.g., *network paradigm*, *race-comparative paradigm*) to disciplinary shifts described as *paradigm shifts*. Although many uses carried clear Kuhnian connotations, the term's meaning was often shaped more by pragmatic context than by discrete conceptual boundaries. The difficulty in forming distinct clusters and the frequent classification of plausible uses as noise highlight *paradigm*'s conceptual looseness and functional adaptability. Manual analysis confirmed that the term often serves less as a fixed label and more as a flexible framing device conveying intellectual orientation across diverse empirical contexts.

The astrophysics cases offered complementary insights. For the highly polysemous term *Planck* (N = 2991; Sim(w) = 0.71), clustering successfully identified three prominent senses: institutional references (e.g., Max Planck Institutes), the ESA satellite mission (*Planck Collaboration*), and the technical *Fokker-Planck equation*. However, the clustering also revealed notable limitations. Embeddings tended to overemphasize dominant senses, particularly the satellite mission, sometimes fragmenting conceptually unified instances into separate clusters. Conversely, certain semantically distinct senses, such as *Planck units* (marked by mathematical notation and identified in prior work, e.g., Simons, 2026), were not detected. This omission was likely a byproduct of preprocessing steps, especially the substitution of equations with placeholder tokens.

The term *wave* (N = 730; Sim(w) = 0.65) further illustrates the challenges of clustering semantically broad and polysemous terms using general-purpose language models. Although three primary clusters emerged (gravitational waves, shock or blast waves, and plasma waves) with reasonable internal coherence, the boundaries between them remained weak, and over half of the occurrences were classified as noise. This outcome reflects the conceptual and discursive overlap among different wave phenomena in astrophysics, where wave types frequently co-occur and are described using similar modeling frameworks. Overall, the clustering results suggest that the contextual embeddings used in this study are capable of capturing some meaningful distinctions, but lack the granularity required to reliably separate closely related technical senses, particularly when terms are both frequent and broadly applied.

Across both disciplines, clustering analyses surfaced key methodological challenges stemming from the limited sensitivity of general-purpose BERT embeddings to subtle contextual variation (cf. Simons, 2026). In specialized scientific language, conceptually coherent usages were sometimes split due to superficial lexical differences, while distinct meanings blurred when surface cues aligned too closely. The clustering algorithm, though well-suited for irregular data, struggled at times to balance noise filtering with semantic granularity, either fragmenting dominant senses or merging subtly distinct ones. These effects were amplified by frequency disparities and the lack of a gold-standard sense inventory, highlighting both the interpretive demands and the technical indeterminacy of unsupervised clustering in scientific discourse.

3. Limitations

The corpus design inevitably introduces potential sources of error that may influence the absolute values of the SUS. These can be disentangled into several categories.

Genres: This study draws exclusively on journal articles, which, while offering structural comparability, omit other genres such as books, conference papers, grant proposals, or letters. Codification may differ across these contexts, particularly between formal publications and more exploratory forms of communication. For the observed relationship to reverse (i.e., sociology appearing more codified than astrophysics), one would have to assume that language in sociology's excluded genres is *more* semantically uniform than in its journal articles, while language in astrophysics' excluded genres is *less* uniform than in its journal articles. This seems theoretically unlikely: it would require that sociological monographs differ substantially from research articles in ways that increase codification, while astrophysics books are less semantically uniform than astrophysics articles. In such a case, including all genres, or only books, would show the opposite relationship between disciplines. Nonetheless, this assumption could be tested directly by varying the corpus composition to include additional genres.

Field delineation: Field delineation also matters for interpreting the results in light of earlier arguments about codification. Cole and Zuckerman (1975) proposed that codification becomes most visible when a research area consolidates into a specialty with shared evaluative standards and a stable vocabulary. The present study, however, relies on WoS subject categories that define disciplines at a broad level, combining heterogeneous subfields with distinct terminologies. Such breadth introduces semantic diversity that can lower SUS values and makes codification harder to detect at the disciplinary scale. A narrower delineation at the specialty level would align more closely with the conditions under which Cole and Zuckerman argued codification becomes identifiable, and would likely increase absolute SUS scores in both sociology and astrophysics. Yet even under this assumption, a reversal of the observed disciplinary ordering would require that sociological specialties be systematically more codified than those in astrophysics, for which there is no empirical basis.

Temporal coverage: The corpus includes articles from 1990, 2000, 2010, and 2020. Spanning multiple decades tends to increase semantic diversity, as vocabularies and meanings change over time. For the observed relationship to reverse under a different sampling window, for example, using only articles from 2020–2025, astrophysics would need to have become substantially less uniform after 2020, to the point that sociology's SUS would overtake it. Conversely, for the reversal to occur by extending coverage backwards, sociology would need to have been significantly more codified before 1990 than it is in later decades. Both scenarios seem improbable given what is known about the historical development of the two fields, although this assumption requires empirical validation through time-specific corpora.

Journal quality: Restricting the sample to journals with high JCR metrics reduces the inclusion of material less likely to contribute to the body of knowledge, concepts, and methods that the research community actively uses and builds upon. For the observed relationship to reverse, sociology's lower-JCR literature would have to exhibit more semantically uniform language use than its higher-JCR literature, while astrophysics' lower-JCR literature would have to exhibit less, a questionable pattern. Although JCR rankings are not a perfect proxy for cognitive core membership, higher-ranked journals often have more standardized editorial practices, which can encourage (but do not guarantee) terminological consistency. Nonetheless, this could be tested by including mid- and low-JCR journals and comparing the results.

The overall measurement pipeline also remains sensitive to several implementation choices, each of which can influence the absolute values of the SUS.

Frequency thresholds: Removing infrequent terms is an effective way to reduce noise, yet it can also disproportionately exclude highly specialized terms that are consistently used within subfields, potentially lowering the SUS for fields with more niche vocabularies. This effect can be tested by recomputing the SUS under different cut-off points to verify whether the relative disciplinary ordering remains stable.

Segmentation: Segmentation strategy presents another source of variation which directly affects which words are included and how *context* is defined, ultimately influencing the generated embeddings. The size of the textual context fed into BERT affects the amount of co-text available to each term; shorter segments (e.g., 100 tokens) limit context and may increase variability in embeddings, whereas longer segments (e.g., 500 tokens) capture broader linguistic environments and may result in more stable representations. Robustness here can be assessed by recalculating the SUS with alternative segment lengths and comparing the results.

Placeholders: A third consideration concerns the treatment of mathematical notation in astrophysics. In the present corpus, equations and variables were replaced with placeholders to facilitate processing, yet these expressions are among the most standardized elements in the field. If considered part of the language, their omission likely underestimates astrophysics' semantic uniformity and may contribute to the relatively small observed difference between the two disciplines. This can be tested by reprocessing the astrophysics corpus with a math-aware BERT model capable of embedding equations directly, then comparing the resulting SUS with the placeholder-based approach.

Taken together, these considerations underline that no corpus design or modeling pipeline can entirely eliminate uncertainty. However, the potential biases identified here are unlikely to systematically favor one discipline over the other in ways that would invert the observed pattern. The numerical results reported here are specific to the analyzed disciplinary corpora and should thus be understood as relative measures of semantic uniformity and the relative degree of language standardization, rather than as absolute indicators of codification. Within this scope, the relative difference seems to represent a robust feature of the two fields, though its precise magnitude may vary under alternative design choices. This study therefore can be further refined through targeted variations in corpus construction and processing.

4. Conclusion

This chapter has explored how computational models of meaning can be leveraged to operationalize the linguistic dimension of codification in scientific discourse. By applying the Semantic Uniformity Score to corpora from astrophysics and sociology, it provides an empirical lens on long-standing claims about disciplinary differences in cognitive consensus. The results suggest a visible contrast: astrophysics tends toward more semantically uniform usage of terms, whereas sociology exhibits greater contextual variability, particularly among frequently used vocabulary. While these patterns are suggestive, their interpretation must remain exploratory rather than definitive.

The attempt to capture codification through an embedding-based semantic uniformity metric necessarily entails reducing the complexity of meaning-making in language. The model employed here, pretrained on general-purpose corpora, represents meaning through statistical regularities in language use, specifically patterns of word co-occurrence and attention distributions. Ontologically, this reflects the distributional hypothesis: concepts are treated as emergent from patterns of use rather than grounded in stable, context-independent definitions. Yet this assumption is marked by theoretical ambiguity. While some perspectives hold that words possess relatively stable core senses modulated by context, others argue that meaning is wholly relational and constructed within situated usage (Ravin and Leacock, 2000). Contextual embeddings further inherit the inherent ambiguity of both natural language and the training corpora on which they are based. Consequently, high variability in embeddings does not necessarily signal genuine semantic ambiguity (polysemy); it may instead arise from topic shifts, syntactic variation, or genre differences.

In this sense, the Semantic Uniformity Score should be understood as a partial indicator of codification. It captures regularities in how terms are distributed across contexts, but it does not fully reflect the conceptual clarity or epistemic consensus such terms may embody. Complementary approaches, such as clustering contextual embeddings, inspecting the sentences from which they are drawn, or qualitative content analysis, can help validate SUS and mitigate its limitations by moving beyond static textual windows and incorporating broader interpretive and paratextual elements. Ultimately, interpreting meaning requires sensitivity to the conceptual traditions, interpretive frameworks, and historical developments that shape how terms like *structure*, *agency*, or *order* are deployed. In fields like sociology, where such terms span multiple schools of thought, disentangling meanings often demands knowledge that exceeds what statistical modeling can extract from local linguistic context alone.

Beyond the general limitations of distributional semantic models, which affect all research using contextual embeddings for semantic variation or shift detection, this study also faced constraints specific to corpus construction and preprocessing. Such design choices can influence the absolute values of the SUS, but they are unlikely to overturn the relative disciplinary differences observed here, since reversing them would require empirically implausible conditions. Within these boundaries, the approach nevertheless illustrates how semantic features of disciplinary language can be examined systematically at scale, in ways that complement more established qualitative and bibliometric perspectives. Rather than presenting a definitive tool, it adds to the repertoire of methods available for studying how knowledge is articulated through shared and stabilized terms. Future work could refine this approach by incorporating expert annotation, testing models better suited to specialized language, and focusing more closely on specialty domains where conceptual alignment is often clearer. Expanding the analysis to multiword expressions or terminological units may also capture more stable conceptual constructs. Ultimately, integrating computational indicators like SUS with interpretive analysis can offer a way to trace how concepts are negotiated into shared meanings, situated within broader theoretical frameworks, and stabilized within scientific communities.[2]

References

Ahmadi ES (2025). Measuring the codification of scientific knowledge with large language models. Discussion Paper, Social Studies of Science and Technology, Institute of History and Philosophy of Science, Technology, and Literature, Technische Universität Berlin.

Alzahrani E and Jololian L (2021) How Different Text-preprocessing Techniques Using The BERT Model Affect The Gender Profiling of Authors. arXiv:2109.13890. arXiv.

Blecher L, Cucurull G, Scialom T, et al. (2023) Nougat: Neural Optical Understanding for Academic Documents. arXiv:2308.13418. arXiv.

Braxton J and Hargens L (1996) Variation among academic disciplines: Analytical frameworks and research. In: *Reprinted from: Higher Education: Handbook of Theory and Research*. Agathon Press. New York.

Cole JR and Zuckerman H (1975) The Emergence of a Scientific Specialty: The Self-Exemplifying Case of the Sociology of Science. In: Coser LA (ed.) *The Idea of Social Structure. Papers in Honor of Robert K. Merton*. New York, Chicago, San Francisco, Atlanta: Harcourt Brace Janovich, Inc., pp. 139–174.

Cole S (1983) The hierarchy of the sciences? *American Journal of Sociology* 89: 111–139.

Cole S, Cole JR and Dietrich L (1978) Measuring the cognitive state of scientific disciplines. In: Elkana Y, Lederberg J, Merton RK, et al. (eds) *Toward a Metric of Science*. New York: Wiley, pp. 209–251.

2 This chapter was written with support from large language models (LLMs). All model-generated text was reviewed and, where necessary, rewritten by the authors, who remain fully responsible for the final version. For details on the use of LLMs in this volume, see the statement in the volume's introduction.

Cozzens SE (1985) Comparing the Sciences: Citation Context Analysis of Papers from Neuropharmacology and the Sociology of Science. *Social Studies of Science* 15(1): 127–153.

Devlin J, Chang M-W, Lee K, et al. (2019) BERT: Pre-training of deep bidirectional Transformers for language understanding. In: *Proceedings of the 2019 Conference of the North American Chapter of the Association for Computational Linguistics: Human Language Technologies* (eds J Burstein, T Solorio, and C Doran), Minneapolis, Minnesota, 2019, pp. 4171–4186. Association for Computational Linguistics.

Garí Soler A and Apidianaki M (2021) Let's Play Mono-Poly: BERT Can Reveal Words' Polysemy Level and Partitionability into Senses. *Transactions of the Association for Computational Linguistics* Roark B and Nenkova A (eds) 9. Cambridge, MA: MIT Press: 825–844.

Gläser J, Hoffmann M, Laudel G, et al. (2024) The empirical investigation of epistemic properties of research processes and fields. Berlin.

Ravin Y and Leacock C (2000) Polysemy: An Overview. In: *Polysemy: Theoretical and Computational Approaches*. Oxford, New York: OUP Oxford, pp. 1–29.

Simons A (2026) Meaning at the Planck scale? Contextualized word embeddings for doing history, philosophy, and sociology of science. In: Simons A, Wüthrich A, Zichert M, et al. (eds) *Understanding Science with Large Language Models? Potentials for the History, Philosophy, and Sociology of Science*. Bielefeld: transcript, part-3.

Storer NW (1967) The Hard Sciences and the Soft: some sociological observations. *Bulletin of the Medical Library Association*. 55(1): 75–84.

Wiedemann G, Biemann C, Chawla A, et al. (2019) Does BERT Make Any Sense? Interpretable Word Sense Disambiguation with Contextualized Embeddings. arXiv:1909.10430, Preprint.

Zuckerman H and Merton RK (1972) Age, aging, and age structure in science. In: Riley MW, Johnson M, and Foner A (eds) *Aging and Society, Volume 3: A Sociology of Age Stratification*. New York: Russell Sage Foundation, pp. 292–356.

Use of large language models in the classification of scientific texts into disciplines

Diego Alves, Sergei Bagdasarov, Badr Mohammed Abdullah, and Stefania Degaetano-Ortlieb

1. Introduction

The classification of scientific publications into disciplines is a fundamental task for organizing and understanding the architecture of knowledge. Yet, this task is complicated by the fact that disciplinary boundaries are not static; they are fluid, historically contingent, and shaped by institutional practices, funding priorities, and evolving research agendas (Abbott, 2001; Frickel and Gross, 2005). This fluidity is evident in the proliferation of interdisciplinary fields and the continuous emergence, differentiation, and reconfiguration of subdisciplines.

In practice, there is no universal consensus on what constitutes a discipline, a fact reflected in the divergent classification systems used across academia and research analytics. Frameworks such as the OECD's Fields of Research and Development (OECD, 2015), the QS Subject Areas[1], and Elsevier's All Science Journal Classification[2] (ASJC) employ varying levels of granularity and organizational logic (Alves et al., 2024). For instance, as summarized in Table 1, these systems differ significantly in the number of categories they define and their use of hierarchical grouping or "other" labels. This lack of a consistent framework underscores a fundamental challenge: any attempt to classify scientific texts must first define a suitable disciplinary taxonomy, as the categories themselves are moving targets.

In linguistic terms, the evolution of disciplines also manifests in lexical and semantic change, as specialized vocabularies develop in response to emerging knowledge domains (Halliday and Martin, 2003; Ure, 1982). These sublanguages or registers are closely tied to socio-cultural and cognitive developments in the respective scientific communities. Against this backdrop, computational tools offer promising avenues for tracing how such categories evolve over time. In particular, the rise of large language models (LLMs) has

1 https://support.qs.com/hc/en-gb/articles/4539968720924-QS-Subject-Categorisation

2 https://support.qs.com/hc/en-gb/articles/4406036892562-All-Science-Journal-Classifications-ASJC-Codes

introduced new possibilities for the automated classification of scientific texts and for mapping disciplinary development on a large scale.

Table 1: Overview of subject area counts in major classification frameworks

	Broad area	Subject
QS	5	56
ASCJ	-	388
FORD	6	42

This study investigates whether and how LLMs can be applied to classify scientific texts into disciplines, using the Royal Society Corpus (RSC; Fischer et al., 2020), a diachronic corpus spanning over 300 years of scientific publication. By employing a fine-tuned LLM (Hermes-2-Pro-Llama-3-8B), we assess the model's capacity to assign discipline labels to historical scientific articles from 1665 to 1996, based on a predefined taxonomy. Building on previous research on topic modeling in historical corpora (Bizzoni et al., 2020; Menzel et al., 2021) and guided by the evolving structure of the Royal Society's own disciplinary divisions, we aim to inspect diachronic patterns of specialization and interdisciplinarity in scientific discourse.

In doing so, the paper contributes to two lines of inquiry. First, it provides a computationally grounded method for the historiography of science, enabling the large-scale tracking of how the prevalence and relationships of disciplinary identities shift over time. Second, it evaluates the reliability and limitations of using a fine-tuned LLM for historical text classification, a task complicated by lexical and semantic change (Halliday and Martin, 2003). By clearly defining our taxonomic framework and applying it consistently, we aim to shed light on the dynamic architecture of scientific knowledge while also highlighting the critical methodological considerations required when using AI tools in historical research.

2. Disciplines in the Royal Society Philosophical Transactions and Proceedings

Brought to existence in 1665, the *Philosophical Transactions* of the Royal Society of London, joined two centuries later by the *Proceedings*, was the world's earliest periodical dedicated explicitly to the dissemination of high-quality scientific knowledge. In its early stages, the *Transactions* consisted of letter-like observations, travel accounts, experiment reports, and book reviews and read more like a patchwork of intellectual curiosities with no clear disciplinary boundaries than a modern scientific journal. Astronomy, medicine, biology, chemistry, physics, and an eclectic array of other subjects coexisted on its pages.

Throughout the following decades and centuries, the *Philosophical Transactions* experienced substantial changes in its organization, procedure for paper selection and paper format, yet these changes had little to no influence on the notion of discipline in the journal. Well into the first half of the 19th century, it remained a smorgasbord of topics, where works on optics or anatomy of the eye were directly followed by an "Account of a Monstrous Lamb" or an "Account of an Elephant's Tusk, in Which the Iron Head of a Spear was Found Imbedded".

However, the growing specialization of scientific knowledge, coupled with the rapid development of new techniques and methods throughout the 19th century, gradually rendered the journal's generalist approach untenable. Adding to this pressure was the emergence of newer, field-specific periodicals, which began to compete with the *Philosophical Transactions and Proceedings* both for audience and contributors. A natural, if somewhat delayed, response to this new reality in the scientific world was the eventual division of the journal into Sections A and B, devoted respectively to the mathematical & physical sciences and biological sciences.

The branching of scientific inquiry found further expression in the establishment of six Sectional Committees, each dedicated to a major domain: Mathematics, Botany, Zoology, Physiology, Chemistry & Physics, and Geology. These committees assumed the task of reviewing incoming submissions, embodying the growing need for disciplinary expertise. Today, the Royal Society recognizes eleven such Sectional Committees, spanning Computer Sciences, Mathematics, Astronomy and Physics, Chemistry, Engineering and Materials Science, Earth and Environmental Sciences, Molecules of Life, Cell Biology, Multicellular Organisms, Patterns in Populations, and Health and Human Sciences. While these committees no longer oversee peer review, they nonetheless offer a revealing insight into how the Royal Society continues to structure and conceptualize disciplinary boundaries—at least at their broadest contours.

Turning again to publishing practices, it is interesting how the continued evolution of science in the 20th and 21st centuries has left the overall structure of the *Philosophical Transactions and Proceedings* largely untouched. Their bisectional division has persisted into the present day, with new disciplines and emerging fields of study absorbed into one section or the other. On its website, the Royal Society draws a distinction between Disciplines and Subjects within the two sections: the former denote broad, overarching domains of knowledge, while the latter serve as more granular, often interdisciplinary tags used to describe individual papers.

The disciplines in both sections are summarized in Table 2. The inclusion of Biology and the creation of a separate "discipline" for cross-disciplinary research within Section A as well as the appearance of Mathematics in Section B are particularly striking, illustrating the difficulties inherent in assigning a single disciplinary label to research that might be actually equally rooted in multiple scientific domains and highlighting potential challenges related to automatic classification of disciplines.

However, providing information on document domain is crucial for large collections of data. An attempt to accomplish this task for Royal Society texts was made by Menzel et al. (2021) who trained a probabilistic topic model using the Royal Society Corpus (RSC; Fischer et al., 2020). The model used an unsupervised machine-learning approach that scanned the documents in the corpus and identified different topics based on words occurring together in similar contexts. Then, labels were assigned manually after the examination of the most distinctive words of each topic. This procedure resulted in a hierarchically organized system of topics illustrated in Figure 1.

Figure 1: Topics in the Royal Society Corpus as identified by an unsupervised topic-modelling approach

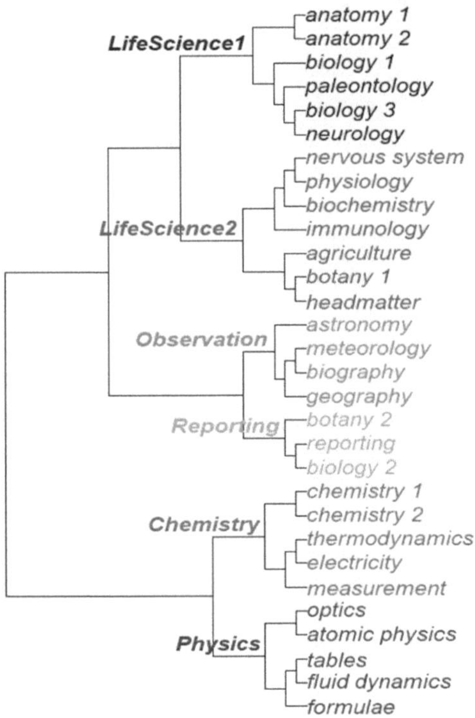

Figure taken from Menzel et al. (2021)

Table 2: Disciplines in Sections A and B of the Philosophical Transactions and Proceedings according to the Royal Society website.

Section	Disciplines	Number of Papers (Transactions + Proceedings)
Section A	Astronomy	503
	Biology	15
	Chemistry	946
	Computer Science	1,246
	Cross-Disciplinary Studies	13
	Earth Sciences	1,164

Section	Disciplines	Number of Papers (Transactions + Proceedings)
	Engineering and Technology	2,595
	Mathematics	3,153
	Physics	3,903
Section B	Biology	15,568
	Mathematics	1

The appearance and rapid development of LLMs in recent years offers researchers and the DH community in general new opportunities in many NLP tasks. Evidently, this applies to metadata generation and text classification as well. We explored the potential of LLMs in this field in a case study described in the next section.

3. Case study: using LLM to classify texts into scientific disciplines

In this case study, we aimed to analyze the ability of a large language model to classify scientific texts spanning from 1665 to 1996 according to a predefined list of disciplines. The task is challenging because the concept of discipline is fluid and undergoes constant change over time. However, this type of classification can still be useful for studies in the history of science and in other research areas, for example, in linguistics, to analyse differences within the scientific register.

Our aim is to provide an overview of the distribution of disciplines over time and to identify potential areas of interdisciplinarity.

3.1 Methodology

In our study, we use the Royal Society Corpus (RSC; Fischer et al., 2020), a diachronic corpus of scientific English spanning the period from 1665 to 1996. The corpus is based on the Philosophical Transactions and Proceedings of the Royal Society of London, and comprises 47,837 texts (295,895,749 tokens). The table in Figure 2 shows the overall distribution of texts across different periods of the RSC.

Regarding the LLM, we used Hermes-2-Pro-Llama-3-8B[3], a fine-tuned version of the Llama 3 language model family. Based on the 8B-parameter version of Llama 3, this model has been instruction-tuned and further optimized for producing structured outputs, such as JSON and YAML. It outperforms earlier models like Llama 2 and Mistral, particularly in tasks requiring clear, machine-readable responses. Its enhanced ability to

3 https://huggingface.co/NousResearch/Hermes-2-Pro-Llama-3-8B

follow instructions and generate well-formatted outputs makes it especially suitable for classification tasks.

Figure 2: Number of Texts and Tokens per Period in the RSC.

Years	# Texts	# Tokens
1665–1699	1 325	2 582 856
1700–1749	1 686	3 414 795
1750–1799	1 819	6 342 489
1800–1849	2 774	9 112 274
1850–1899	6 754	36 993 412
1900–1949	10 011	65 431 384
1950–1996	23 468	172 018 539

As is common in prompt-based workflows, we defined a system prompt to describe the task, while the user prompt contained the information and data from the texts to be classified. Although our main focus in this study is the classification of RSC texts by discipline, the prompt used included four tasks: (a) suggesting an alternative title; (b) providing a TL;DR[4] summary; (c) identifying the main topics of the paper; and (d) determining the primary discipline (from a predefined list) along with a suitable sub-discipline.

The system prompt is structured as follows:

1. Role: Act as a librarian
2. Objectives: Read, analyze, and organize
3. Input: Texts from the RSC
4. Tasks:
   a. Provide an alternative title
   b. Generate a TL;DR summary
   c. Identify main topics
   d. Determine primary discipline and sub-discipline
5. Examples: Sample input and output

The user prompt, on the other hand, follows an iterative process, providing information about the text to be analyzed:

1. Article ID
2. Author(s)
3. Journal
4. Year
5. Title
6. Text snippet (i.e., the first 1,000 words of the paper)

4 "Too Long; Didn't Read"

The detailed scripts, including prompt configurations, are available on GitHub[5] for reference.

Given the ever-evolving nature of scientific disciplines, using a predefined list helps avoid an overly fine-grained classification. Therefore, based on the information described in Section 2 about the RSC (i.e., its history and the topic-modeling classification by Menzel et al. (2021)), we defined the following list:

1. Physics
2. Chemistry
3. Environmental and Earth Sciences
4. Astronomy
5. Biology and Life Sciences
6. Medicine and Health Sciences
7. Mathematics
8. Engineering and Technology
9. Social Sciences, Art, and Humanities

Requesting the LLM to provide the main topics and a subdiscipline categorization for each text is also intended to enhance the classification, as some subdisciplines in the corpus may be highly recurrent and could help refine and expand the predefined list.

3.2 Evaluation

The evaluation of the LLM's output shows a high level of reliability, with 99.42% (47,558 out of 47,837) of the generated responses being valid YAML formats, indicating strong respect of the required structured output.

Regarding the classification accuracy, 90.95% (43,257 out of 47,558) of the predicted scientific disciplines matched those in the predefined list, demonstrating the model's effectiveness in assigning appropriate categories. However, some discrepancies were observed, such as labeling *Earth Sciences* instead of the broader *Environmental & Earth Sciences* and the occasional generation of novel categories, like *Music*. These findings suggest that while the model performs well overall, there is room for refining category definitions and handling outliers.

The *Music* category, for example, was used to classify seven articles: one in the subdiscipline *Acoustics* and the other six in *Theory*. Two papers are from the 18th century (one from the beginning and one from the end), while the rest are from the second half of the 19th century. The themes vary. One article presents a method for tuning musical instruments using precise mathematical proportions to achieve perfect harmonies. Another examines the conditions, extent, and realization of a perfect musical scale on instruments with fixed tones. There is also a discussion of the Hindoo division of the octave and its implications for music theory. Thus, while the LLM's reasoning was accurate, it considered the topics too specific to classify these papers within the broader category of *Social Sciences, Arts, and Humanities*.

5 sci_scope/llm_code at main · badrex/sci_scope · GitHub

Figure 3 shows the distribution by decade of RSC articles, indicating the percentage of articles published in each decade that were categorized under disciplines not included in our predefined list.

Figure 3: Proportion of articles published in each decade of the RSC classified under disciplines not covered by the predefined categories.

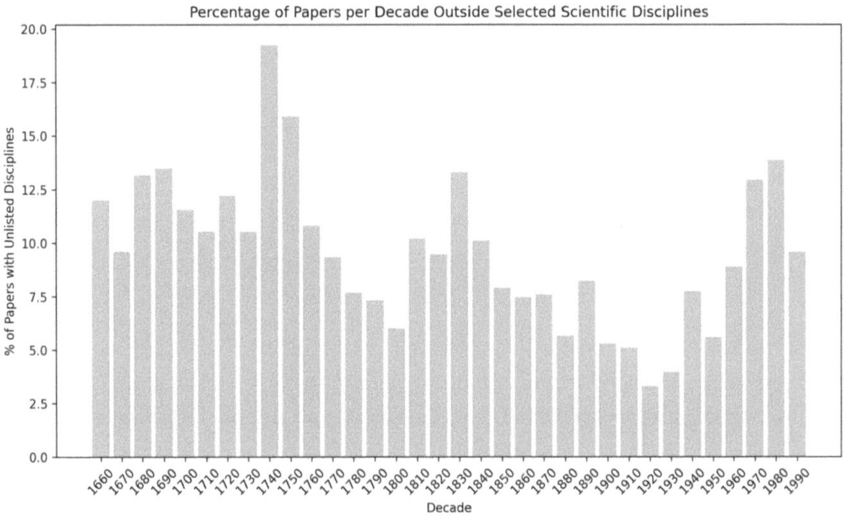

It is possible to observe better performance by the LLM in the second half of the 18th century, which may be attributed to the clearer delineation of disciplines following the changes brought about by the scientific revolutions. After a slight increase in hallucinations during the first half of the 19th century, performance improves again. However, in the second half of the 20th century, the number of papers classified under disciplines not included in the predefined list increases once more. These peaks may also reflect periods of greater interdisciplinarity, possibly associated with the emergence of new disciplines.

Regarding the texts classified into the nine predefined categories, we also conducted an accuracy evaluation of the LLM's classification. First, we randomly selected 15 texts from each discipline category identified by the LLM, ensuring that at least two texts from each 50-year period of the RSC were included. These texts were then labeled according to the list of disciplines by two German annotators, both university students in the humanities and fluent in English.

To evaluate the reliability of the manual annotations, we calculated inter-annotator agreement using Cohen's Kappa coefficient, a statistical measure that accounts for the degree of agreement between two annotators while correcting for chance agreement. The annotation campaign resulted in a Cohen's Kappa of 0.72, which indicates substantial agreement, indicating a high level of consistency between annotators. Figure 4 presents the confusion matrix indicating the disagreements between the 2 annotators.

Some expected discrepancies between annotators can be observed. For instance, four papers labeled as *Physics* by Annotator 1 were classified as *Engineering and Technology* by another annotator. Differences in annotation are also noticeable in the fields of *Biology, Medical and Health Sciences*, and *Chemistry*.

Figure 4: Confusion matrix showing the agreements and disagreements between the two annotators across the assigned labels.

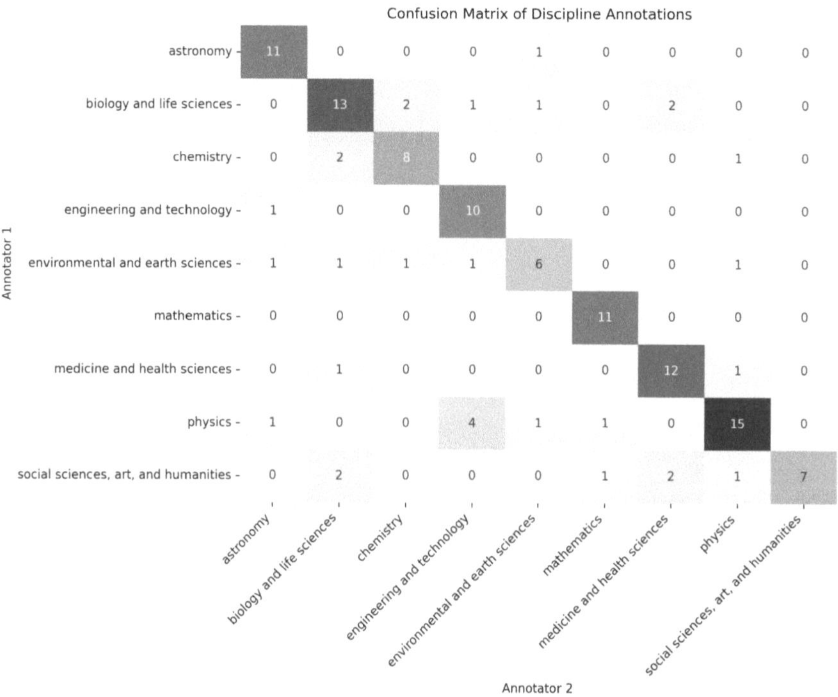

To evaluate the consistency of discipline classification by the LLM, we compared the output from our two annotators against labels provided automatically. First, we calculated the accuracy by considering a classification correct if the reference label matched the label assigned by at least one of the two annotators, resulting in an accuracy of 82.22%. Additionally, we computed accuracy only on the subset of cases where both annotators independently agreed on the discipline, resulting in a slightly higher accuracy of 83.87%. These results demonstrate the potential of using LLMs for discipline categorization of texts, even with a diachronic dataset spanning three centuries of scientific publications.

Figure 5 presents the confusion matrix comparing the LLM labels to the categories defined by the annotators, considering only cases where both agree.

The disciplines that appear most challenging for the LLM are Chemistry, Engineering and Technology, and Physics. This may be due to the application of some concepts from the two more theoretical disciplines within practical and technological contexts in some papers.

Figure 5: Confusion matrix showing the agreements and disagreements between the LLM (Predicted label) and the annotators (True label).

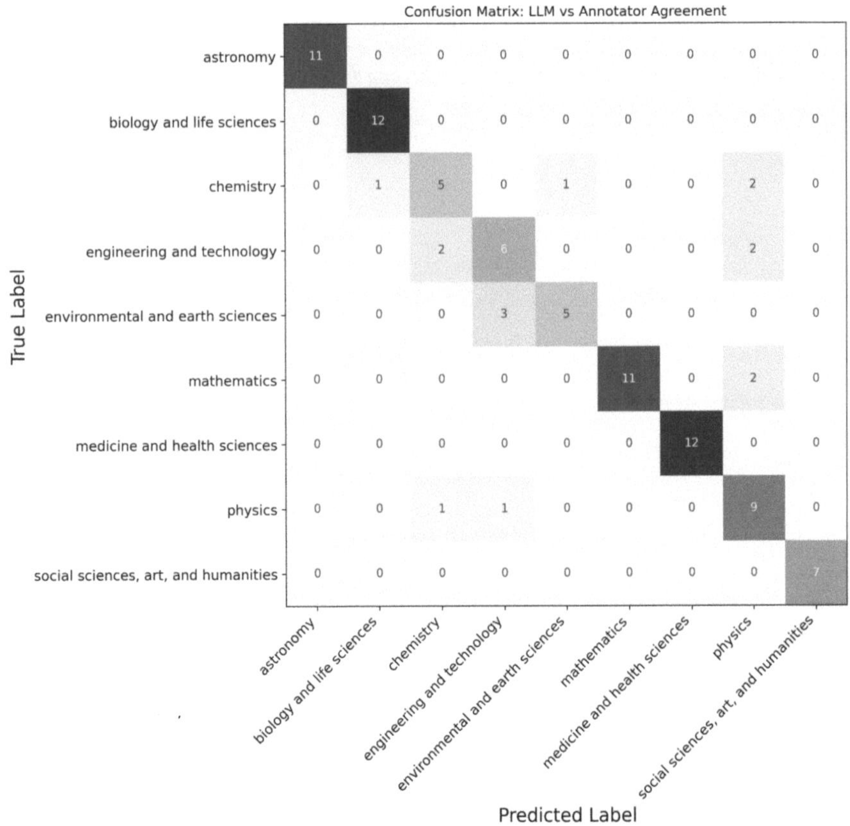

3.3 Analyzing interdisciplinarity

Figure 6 presents the distribution of the number of RSC papers per discipline, grouped by 20-year periods.

The graph in Figure 6 indicates a greater diversity of disciplines represented in publications from the earlier periods of the RSC (up to the beginning of the 19th century). *Environmental and Earth Sciences* and *Medicine and Health Sciences* were relatively well represented during this time, but their presence diminished over the years. However, *Environmental and Earth Sciences* shows a slight increase in the final two 20-year periods, possibly linked to growing scientific awareness of environmental issues and ecology-related research. The impact of the Chemical Revolution (late 18th century) is evident in the increased number of chemistry-related publications from that period onward. Moreover, an increase in physics publications is also noticeable in the early 20th century, reflecting the discipline's major transformations with the development of quantum theory and Einstein's theory of relativity.

Figure 6: Distribution of RSC Papers by Discipline (20-Year Periods)

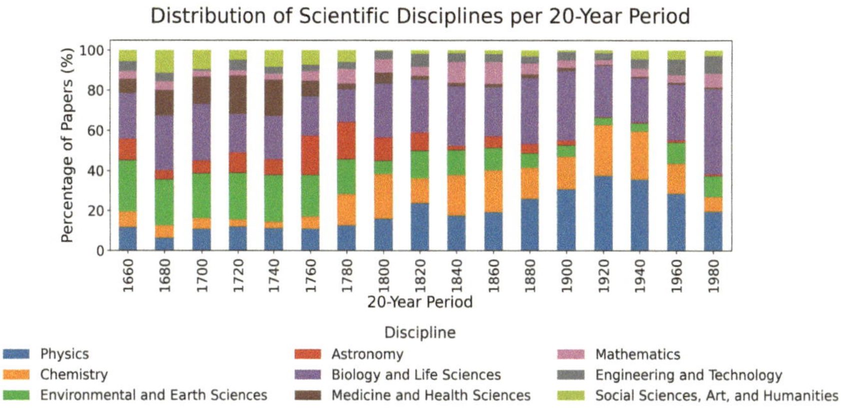

To analyze interdisciplinarity, we used the TL;DR summaries generated by the LLM to create text vectors, using the FastEmbed library dictionary[6] with the BAAI/bge-small-en-v1.5 model, for all RSC papers classified under one of the nine predefined disciplines. These summaries were chosen to minimise the influence of text length and linguistic features specific to different historical periods. We then applied a t-SNE projection to plot the resulting vectors in the two-dimensional graph shown in Figure 7.

Although discipline clusters can be identified, their boundaries are not clearly defined. *Astronomy* and *Mathematics* appear more distinct, though signs of interdisciplinarity are visible, particularly between *Astronomy* and *Physics*, and between *Mathematics* and *Earth Sciences*. *Chemistry* is positioned between *Physics* and *Biology*, showing a broad area of overlap. As expected, *Biology* is also closely interconnected with *Medicine and Health Sciences*, and *Physics* with *Engineering*.

This large overlap between Chemistry and Biology suggests the potential need to add Biochemistry as a tenth discipline in the list. However, Biochemistry was not recognized as a distinct scientific discipline until the late 19th to early 20th century. Thus, depending on the research question, this could introduce a bias in the analysis if texts are taken from a limited period onward.

These results demonstrate that using LLMs to classify scientific publications into predefined disciplines can be a valuable tool for analyzing interdisciplinarity. This approach can be extended to a more detailed diachronic analysis, for example by comparing different periods, revealing how disciplines and their intersections evolve over time and providing a foundation for the emergence of new disciplines. Moreover, the method of Aguilar-Valdez et al. (2026), originally applied to the entire RSC, can also be used to trace conceptual shifts within individual disciplines.

6 https://github.com/qdrant/fastembed

Figure 7: t-SNE projection of RSC texts based on vectors generated from TL;DR summaries. Each color represents one of the nine predefined disciplines.

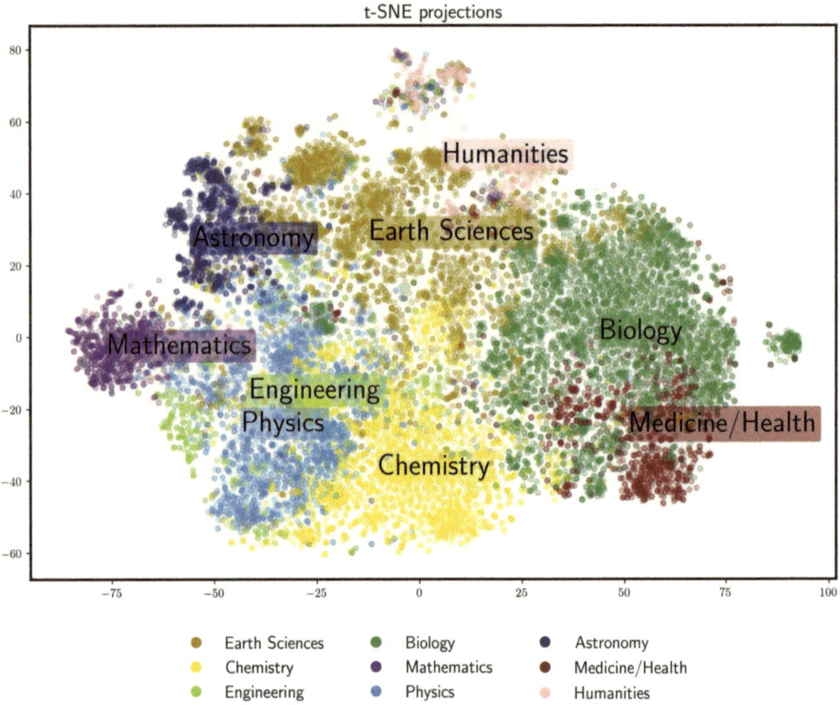

4. Conclusions

The classification of scientific disciplines is a complex and dynamic process, shaped by historical developments, institutional structures, and evolving research paradigms. This study shows that the boundaries between disciplines are fluid, with fields continuously fragmenting, merging, and redefining themselves in response to new knowledge and societal needs. The case study of the Philosophical Transactions and Proceedings of the Royal Society illustrates how scientific publishing has adapted to increasing specialization.

Using large language models (LLMs) to classify historical scientific texts presents both opportunities and challenges. While the model in this study achieved high accuracy in assigning predefined discipline labels, discrepancies arose due to the inherent ambiguity in disciplinary boundaries and the emergence of new fields over time. The analysis also revealed significant interdisciplinary overlaps, particularly between chemistry, biology, and physics, suggesting that rigid classifications may not always capture the true nature of scientific inquiry.

These findings highlight the importance of carefully defining disciplinary frameworks when conducting digital humanities research. Researchers must consider the historical context, the granularity of classification, and the potential biases introduced by automated systems. Future work includes testing other LLMs, refining discipline lists (with different predefined lists across various time periods), incorporating subdisciplinary distinctions, and exploring diachronic trends in interdisciplinarity.

Ultimately, this study underscores the value of computational methods for tracing the evolution of scientific knowledge while emphasizing the need for critical, context-aware approaches to classification.

5. Limitations

Historical texts might be underrepresented in our model's pre-training data, which might have negatively impacted the classification results for earlier periods. Moreover, while the model was fine-tuned for structured output generation and instruction following, it was not fine-tuned for discipline classification. Although we prompted the model in a few-shot setting, providing relevant examples, an additional task-specific fine-tuning might improve the classification results.

Because our prompt required both classification and generation, we opted for an encoder-decoder model as a unified solution. However, for classification alone, a fine-tuned encoder-only model might offer better results.

We defined our list of disciplines based on the historical evolution of the Transactions and Proceedings of the Royal Society and on the previous topic modelling results. However, our labels fail sometimes to account for highly interdisciplinary texts, resulting in misclassifications and model hallucinations. Moreover, applying a modern disciplinary framework to historical texts can introduce bias, as many current fields were not conceived as independent disciplines in the past.[7]

References

Abbott A (2001) *Chaos of Disciplines*. Chicago: University of Chicago Press.

Aguilar-Valdez S, Phan-Tất B, Speelman D, et al. (2026) Discursive parallels of the chemical revolution. Topic modelling and distributional analysis. In: Simons A, Wüthrich A, Zichert M, et al. (eds) *Understanding Science with Large Language Models? Potentials for the History, Philosophy, and Sociology of Science*. Bielefeld: transcript, part-3.

Alves D, Tadić M and Rehm G (2024) Which Domains, Tasks and languages are in the focus of NLP research on the languages of Europe? In: *Proceedings of the Second International Workshop Towards Digital Language Equality (TDLE): Focusing on Sustainability @ LREC-COLING 2024* (eds F Gaspari, J Moorkens, I Albade et al.), Torino, Italia, 25 May 2024, pp. 18–32. Totina: ELRA and ICCL.

7 This chapter was written with support from large language models (LLMs). All model-generated text was reviewed and, where necessary, rewritten by the authors, who remain fully responsible for the final version. For details on the use of LLMs in this volume, see the statement in the volume's introduction.

Fischer S, Knappen J, Menzel K, et al. (2020) The Royal Society Corpus 6.0: providing 300+ years of scientific writing for humanistic study. In: *Proceedings of the twelfth language resources and evaluation conference* (eds. N Calzolari, F Béchet, P Blache et al.), Marseille, France, 11–16 May 2020, pp. 794–802. Marseille: European Language Resources Association (ELRA).

Frickel S and Gross N (2005) A general theory of scientific/intellectual movements. *American Sociological Review* 70(2): 204–232.

Halliday MAK and Martin JR (2003) *Writing Science: Literacy and Discursive Power*. London: Routledge.

Menzel K, Knappen J and Teich E (2021) Generating linguistically relevant metadata for the Royal Society Corpus. *Research in Corpus Linguistics* 9(1): 1–18.

OECD (2015) *Frascati Manual 2015: Guidelines for Collecting and Reporting Data on Research and Experimental Development*. Report, 8 October. Paris: OECD Publishing.

Ure J (1982) Introduction: approaches to the study of register range. *International Journal of the Sociology of Language* 1982(35): 5–24.

Discursive parallels of the chemical revolution

Topic modelling and distributional analysis

Sofía Aguilar-Valdez, Bách Phan-Tất, Dirk Speelman, Dirk Geeraerts, and Stefania Degaetano-Ortlieb

1. Introduction

How do conceptual changes show up in scientific writing? This paper addresses this question by examining both the topics that authors foreground and the immediate linguistic contexts in which key terms are embedded. Focusing on the late phase of the Chemical Revolution (1750–1800), we analyze a subset of the Royal Society Corpus (Kermes et al., 2016; Fischer et al., 2020). We adopt a two-level approach that connects broad thematic trends at the document level with local linguistic evidence at the word level, allowing us to link where change appears in the discourse to how it is linguistically realized.

At the document level, we trace how topics emerge, stabilize or decline across decades. At the word level, we focus on the concept of *air*, a central term in late eighteenth-century chemistry, and investigate how it undergoes conceptual change by analyzing the company it keeps, following Firth's (1957) distributional insight. While our approach is grounded in distributional semantics, a perspective that also underlies LLMs, we deliberately rely on more established methods of topic modelling based on Latent Dirichlet Allocation (LDA; Blei et al., 2003) combined with interpretable information-theoretic measures. This choice is not merely pragmatic. For questions in the field of HPSS, it is essential to make explicit how epistemic knowledge is structured, contested, and consolidated in discourse. Probabilistic topic models yield explicit probability distributions over themes, which can then be analyzed using information-theoretic measures based on entropy or divergence. In this framework, entropy measures the diversity of topics within a period. Rising entropy signals periods in which multiple thematic perspectives coexist, corresponding to phases of epistemic competition, while declining entropy indicates consolidation around fewer, more dominant themes. Although LLMs and embedding-based approaches can also quantify semantic shifts, such divergence is defined over opaque latent representations that encompass multiple underlying sources of variation (lexical, semantic, grammatical, etc.), which cannot be easily disentangled, making the interpretability considerably more difficult

(Tahmasebi et al., 2021). In particular, these approaches do not provide direct access to the grammatical configurations in which a concept is embedded, nor do they allow changes in syntactic function and lexical realization to be inspected independently. Data sparsity in historical corpora further reinforces this methodological choice (see below and cf. Lareau and Malaterre, 2026). We do, however, also sketch ideas for taking the analysis presented here to the next level, which would include the generation of a graph structure of concepts which could then be enriched with an LLM-based analysis. By way of presenting our analysis, we also intend to familiarize the reader with key concepts for the computational analysis of semantic change—be it by LLMs or other means.

Our document-level analysis asks a simple question: which topics are most characteristic of each decade, and how much do those decade profiles differ from one another? The results reveal a recognizable sequence tied to oxygen-centered chemistry, with decades in which themes around *water*, then *air*, and later *acid* gain comparative prominence.

Our word-level view zooms in on *air*, and quantifies its changing use by examining the words that most often appear in its immediate context—especially adjectives such as *fixed*, *dephlogisticated*, and *vital*. By grouping surrounding words into syntactic "slots" (for example, adjectives that modify *air*), we track how the distribution of slot fillers changes over time. This makes recontextualization visible as a measurable shift from general or environmental descriptors toward chemically specific ones.

Taken together, these two perspectives—topic trends across documents and local context around air—converge on the same picture: a gradual reorganization of discourse rather than a single abrupt break. This interpretation is compatible with standard histories of the oxygen program, while remaining grounded in observable textual patterns (cf. Holmes, 1994; Thagard, 1990; and Bensaude-Vincent and Abbri, 1995 from a discursive perspective). Building on earlier work showing that lexical and grammatical distributions mark periods of scientific reorientation (Degaetano-Ortlieb and Teich, 2018), our finer-grained slot-filler analysis complements document-level trends by identifying the local linguistic contexts in which conceptual change is realized. This aligns with usage-based and variationist perspectives in lexical semantics, where changes in distribution are meaningful for how concepts are encoded (Geeraerts, 2009; Geeraerts, 1997), and ties directly to our empirical observations on *air* in the late eighteenth century.

Finally, the two-level signals (document-level topics and local linguistic evidence) identified here will serve as input in future graph-based modeling of conceptual change.

2. Related work

From a cognitive linguistic perspective, Geeraerts (2009) and Geeraerts et al. (2012) demonstrate how meanings evolve around culturally salient prototypes that shift over time. For instance, in the category bird, a robin or sparrow is often considered more prototypical than a penguin or ostrich, because it more closely matches the characteristics people associate with birds, such as flying or singing. However, these prototypes can shift over time and across communities. Geeraerts et al. (2012) further highlight how different communities may use the same words differently, depending on their social or disciplinary context. The idea that changes in term usage or terminology often reflect shifts in conceptual framing has also been explored in diachronic corpus linguistics. Before a shift, competing perspectives often surface and circulate within a scientific community. Once the shift occurs, familiar terms may acquire new meanings, or entirely new terminology may emerge to capture novel concepts. Degaetano-Ortlieb and Teich (2019) show that during revolutionary periods, the lexicon exhibits a wave-like tendency in which new terms are coined and old terms acquire new meanings or are used in different linguistic environments (e.g. in different syntactic positions, i.e. as a subject or object). These shifts often precede or co-occur with major scientific reorientations. After a shift the diverging peaks within the lexicon stabilize showing periods of consolidation.

An illustrative example is the chemical revolution of the late 18th century, which centered around the topic of combustion. At the time, the dominant explanation for combustion was the phlogiston theory, which stated that a fire-like substance called phlogiston was released during burning. However, increasingly detailed experimental work, especially by Lavoisier and Priestley, led to the discovery of oxygen, a new interpretation of combustion as a process of oxidation. This finding posed a fundamental anomaly that the phlogiston paradigm could not explain. The in-depth experimental work on combustion within the phlogiston paradigm led to the oxygen discovery, which eventually became the established paradigm (Kuhn, 1962, pp. 66–76; 148–150). In this case, a new framework replaced an older one, and the scientific field moved toward a paradigm that offered a different perspective on both old and new problems.

Recent work in computational semantics has greatly advanced the analysis of semantic change. Surveys by Tahmasebi et al. (2021) and Cassotti et al. (2024) categorize methods into frequency-based, context-based, and embedding-based approaches, emphasizing challenges in aligning models across time periods. Periti et al. (2022) address this challenge through incremental updating of word embeddings, which facilitates continuous tracking of word usage shifts without the need for post hoc alignment. Their approach has been applied to diachronic corpora in both scientific and political domains.

In their recent review, Periti and Montanelli (2024) discuss how large language models (LLMs) can be leveraged for semantic shift detection. While LLMs capture nuanced contextual meanings, their lack of transparency limits interpretability. Periti and Tahmasebi (2024) introduce evaluation protocols for comparing LLM-based shift detection with classical distributional approaches, showing that LLMs outperform in capturing nuanced sense distinctions but tend to overfit when applied to sparse historical data, which is exactly the case in our study.

In this study we prioritized interpretability with a framework based on topic modeling and information-theoretic metrics, aiming to explore the conceptual shifts around the chemical revolution from two linguistic dimensions: semantics, expecting distinct themes over time and drastic change in their dynamics after the discovery of oxygen, and lexico-grammar, where syntactic relationships (slots) and their lexical content (slot fillers) would reflect change that would be consistent with the literature (Degaetano-Ortlieb and Teich, 2019; Chang, 2011). In terms of methods, we combine document-level topic modeling with interpretable syntactic slot-filler analysis. While most studies focus on either word-level or topic-level analysis, we integrate both to bridge the gap between lexical semantic drift and conceptual reorganization. We test this approach envisioning it to offer a richer, more interpretable model of how linguistic traces reflect underlying changes during periods of change.

3. Methods

This section describes the dataset and why it is well-suited for our task, and the methods used to represent and analyze the linguistic dimensions of a conceptual shift. The code[1], data samples, and full-texts[2] are available. As a case study, our aim is to analyze two competing theories during the period of the Chemical Revolution: phlogiston theory (e.g., "a lighter burns because it contains phlogiston") and oxygen theory (e.g., "a lighter burns because some substances ignite when in contact with oxygen"). We seek to explore the decline and rise of phlogiston and oxygen respectively through two dimensions: at the **document level**, we estimate conceptual properties of the drift, at the **word level**, we aim to represent linguistic properties of the drift. By combining these approaches, our goal is to explore to what extent these reflect the conceptual change produced by the discovery of oxygen.

3.1. Data

We draw on the Royal Society Corpus (RSC), a diachronic corpus of English scientific writing covering publications in the Philosophical Transactions and Proceedings of the Royal Society of London from 1665 to 1996 (Kermes et al., 2016; Fischer et al., 2020). The RSC is designed to support research on the development of scientific English, amounting at 47 837 texts. It provides metadata such as publication year, author, and document type, allowing for fine-grained analyses across linguistic and extralinguistic variables (Fischer et al., 2020; Menzel et al., 2021). We focus on the period between 1750 and 1800, which corresponds to the period before and after the Chemical Revolution. This period is particularly rich in documents addressing combustion, gases, and chemical reactions, key themes in the shift from phlogiston to oxygen theory. Preprocessing involved tokenization, lemmatization, and sentence splitting using spaCy. Lemmatized tokens were used throughout to enhance consistency and reduce lexical sparsity in our analyses.

3.2 Document level

This approach consisted of three stages. First, to create a representative corpus, we filtered publications given specified decades and oxygen terminology. Next, we estimated the most distinctive topics over time using topic modeling. Finally, to approach measurement of the shift assessment, we used the topic entropy trajectories to identify key topics and calculated the Jenson-Shannon (JS) distance between them.

---

1 https://github.com/sa-aguilarv/llms-hpss

2 https://fedora.clarin-d.uni-saarland.de/rsc_v6/

I. Data sampling We retrieved texts published in the period, then filtered those with oxygen terms derived from previous studies: Bizzoni et al. (2021), which used Kullback-Leibler Divergence (KLD) to identify distinctive terms during the Chemical Revolution; and Chang (2011), who explains why the term "oxygen" persisted despite theoretical revisions, arguing that its operational meaning remained stable while phlogiston theory lost operational grounding. Lastly, to test prior text influence on downstream analyses, we created cumulative and non-cumulative datasets: in cumulative sampling each decade includes the indicated decade and all previous decades' documents, in non-cumulative sampling each decade includes only documents from that period. This is to test the influence of prior textual data on downstream analyses, i.e. how utilizing or not documents from the past would impact subsequent analyses.

II. Topic modeling Document-topic distributions offer insight into how specific themes are distributed across time, see e.g. Griffiths and Steyvers (2004) who applied this to find scientific disciplines and based on these evaluated collaborations between researchers. Our topic modeling workflow consisted of constructing a document-term matrix (DTM) from the preprocessed corpus, counting the occurrences of each lemmatized term per document. We evaluated model performance and topic interpretability using log-likelihood (Bengio et al., 2003) and topic coherence (Mimno et al., 2011), to mitigate the often manual and arbitrary parameter selection in LDA (Schmidt, 2012; Mohr and Bogdanov, 2013). Following best practices in interpretability, we generated topic labels using LDAvis (Sievert and Shirley, 2014), which are determined based on relevance scores that are more meaningful compared to word rankings (i.e., if *air* is the top-5 word yet its relevance score is higher compared to other terms, this will be the topic label).[3] We modeled topics separately for each sampling strategy (cumulative and non-cumulative) to evaluate the robustness of the trajectories. We found 2,337 documents after sampling, and that a model of 6 topics seems adequate as it achieved low perplexity and high coherence across all decades.

III. Shift assessment. Entropy, originally defined by Shannon (1948), quantifies the degree of unpredictability or diversity in a distribution. In our context, it measures the variability of each topic's probability distribution over time, indicating how conceptually stable or unstable a topic is. This approach has been used to capture thematic volatility in scientific discourse (Hall et al., 2008; Luhmann and Burghardt, 2022). We focused on topics whose top terms included one or more oxygen-related keywords and whose entropy trajectories showed a sequential activation across the decades. This allowed us to identify topics that plausibly represent distinct stages in the development of oxygen theory.

3 Note that, while current state-of-the-art approaches for scientific document classification rely on contrastive learning over citation graphs (Singh et al., 2023), this was not feasible for our study due to the RSC's sparse and non-standard citation formatting. We thus selected LDA as an interpretable and well-established method for estimating thematic shifts.

3.3 Word level

We used data from 1750 to 1799, filtered by the topic labels *Biochemistry*, *Chemistry 1*, and *Chemistry 2*, which were derived from an earlier LDA-based topic modeling study on the Royal Society Corpus (Fankhauser et al., 2016).

To analyze linguistic change at the micro-level, we investigated the lexeme *air*, a term central to both phlogiston and oxygen theories. The corpus was split into five-year time bins to capture fine-grained temporal resolution. All data were part-of-speech and dependency parsed using the Stanza library (Qi et al., 2020).

To examine syntactic and semantic shifts of *air*, we employed SynFlow (Phan-Tất, 2025), a custom Python package designed for extracting syntactic co-occurences of words from dependency-parsed corpora for collocational analyses. SynFlow builds a syntactic tree based on CONLLU format annotations and retrieves (and counts) the syntactic slots and slot fillers for a given target word or part of speech. For example, given the target *air* as a *NOUN*, some extracted slots include chi_det (determiner, as in *the air*) and pa_nsubj (subject, as in *the air heats up*). SynFlow uses the prefix chi (children) when the target is the head in a dependency relation and pa (parent) when it is the dependent. This distinguishes the directionality of syntactic relations. It also enables the reconstruction of syntactic slot-filler distributions. Thus, it was chosen for its ability to systematically retrieve and quantify the frequencies of both syntactic slots (e.g., chi_amod) and their fillers (e.g., adjectives like "pure" or "vital") across historical stages. In our workflow, we first used SynFlow to retrieve the frequencies of all syntactic slots of *air* in each five-year time bin (e.g., 1750–1755, 1755–1760) before calculating Jensen-Shannon Divergence (JSD) between slot distributions to measure distributional shifts over time. JSD is a symmetric and bounded measure that quantifies the difference between two probability distributions.

4. Results and discussion

We begin this section with a macro-analytical view by modeling the document level, analyzing topic drifts, and then move on to a micro-analysis at the word level by analyzing linguistic properties of the changes observed.

4.1 Topic drifts at document level

4.1.1 Topic trajectories across decades under cumulative and non-cumulative modeling

Table 1 shows the 6 topics and 3 top-words of every decade based on the cumulative strategy (decades and past decades), and the results suggest this strategy provides stable topics over time, with recurring labels such as *author*, *year*, *observation*, *distance*, *plant*, *water*, *air*, *wind*, *foot*, and *blood*. Note that topics sharing a column with different labels are not equivalent; the alignment is arbitrary.

Table 2 presents a seemingly more diverse set of topics through time. This is based on the non-cumulative strategy (only documents from that decade), and the results hint toward more fine-grained representations of the themes discussed. Here, the term *air* does not appear at all in the 1750s; it enters the topic of *water* in the 1760s, becomes a distinct topic label in the 1770s, and transitions into a top term under the *acid* topic in the 1790s. By the 1800s, *air* disappears as a salient term, leaving *acid* as the dominant theme.

Table 1: Topic labels of the 6 topics with their 3 top-words over time based on the cumulative strategy

	1		2		3		4		5		6	
		year		time		plant		water		water		blood
		author		observatio		fig		experimen		time		time
1750	**author**	time	**observation**	motion	**plant**	tree	**water**	air	**wind**	place	**blood**	body
		year		time		plant		water		time		blood
		author		observatio		fig		experimen		water		time
1760	**author**	number	**observation**	sun	**plant**	animal	**water**	air	**wind**	place	**blood**	body
		year		time		plant		water		wind		blood
		author		observatio		animal		air		water		time
1770	**year**	number	**observation**	distance	**plant**	fig	**water**	experimen	**wind**	time	**blood**	body
		year		time		plant		time		air		blood
		author		observatio		animal		water		water		body
1780	**year**	number	**observation**	distance	**plant**	fig	**air**	wind	**wind**	experimen	**blood**	time
		year		time		plant		air		time		blood
		author		observatio		fig		water		day		body
1790	**year**	number	**observation**	distance	**plant**	tree	**air**	experimen	**wind**	water	**blood**	animal
		year		time		plant		air		time		animal
		author		observatio		tree		water		foot		blood
1800	**year**	number	**distance**	distance	**plant**	fig	**air**	experimen	**foot**	water	**blood**	body

Table 2: Topic labels of the 6 topics with their 3 top-words over time based on the non-cumulative strategy.

	1		2		3		4		5		6	
		water		day		plant		time		number		foot
		experiment		time		animal		water		year		inscription
1750	**experiment**	glass	**day**	year	**plant**	body	**earthquake**	foot	**number**	time	**inscription**	word
		point		animal		time		wind		year		water
		line		body		observation		day		inscription		air
1760	**point**	ratio	**animal**	part	**observation**	sun	**wind**	foot	**year**	letter	**water**	experiment
		experiment		year		bird		observation		number		air
		body		animal		country		wind		height		water
1770	**ball**	inch	**animal**	tree	**bird**	water	**wind**	time	**number**	difference	**air**	experiment
		body		animal		star		day		degree		air
		inch		part		distance		time		thermometer		water
1780	**body**	velocity	**animal**	time	**star**	observation	**day**	foot	**thermometer**	experiment	**air**	acid
		experiment		eye		point		day		observation		acid
		inch		animal		line		time		star		air
1790	**inch**	water	**eye**	experiment	**gravity**	gravity	**day**	year	**observation**	time	**acid**	water
		experiment		animal		star		ray		iron		acid
		body		stomach		motion		glass		gold		water
1800	**vessel**	vessel	**animal**	portion	**star**	observation	**ray**	experiment	**gold**	copper	**acid**	gas

4.1.2 Distinctiveness of topics across decades

To assess the distinctiveness of topics across decades, we computed the JS distance between them, validating whether each decade had estimated distinct themes. Figure 1 shows a comparison of the 1800s: compared to the non-cumulative (decade without past decades) approach, the cumulative (decade including past decades) approach estimates topics that have lower scores which hints toward higher similarity. These results were consistent in all decades.

Fig. 1: JS distance scores for the 6 topics from the 1800s. From left to right, it shows the cumulative and non-cumulative strategies.

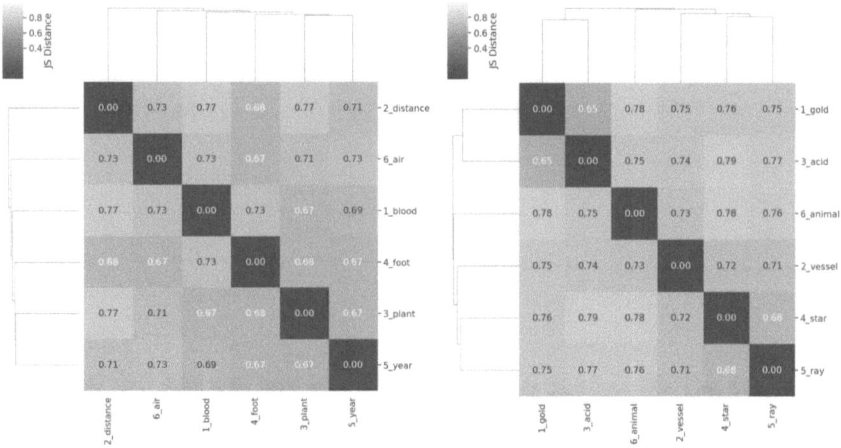

4.1.3 Assessing shifts in topics variability

To assess shifts in the variability of each topic's probability distribution over time, we used entropy as an indicator of how conceptually stable or unstable a topic is. We found higher values in the cumulative strategy (decade with past decades). Figure 2 shows topics with oxygen terms; notably, *wind* and *water* seem to behave similarly from the 1750s to the 1770s, while *air* seems to be the continuation of *water*. This is consistent with Tables 1 and 2, where the *air* term often appears with the water label and vice versa, supporting the view that *air* and *water* are thematically linked across decades. Based on the non-cumulative strategy (decade without the past), Figure 3 hints the trajectory of *wind* is continued by *air*, and from there *acid* follows. Overall, while the cumulative strategy shows rising entropy across oxygen-related topics, the non-cumulative declines then rises post-1774 oxygen discovery. This decline of diversity pre-1780 could be interpreted as the resistance of phlogistonists to decenter *air* in their findings (e.g., Priestley's *dephlogisticated air*), while the rise of diversity post-1780 could be attributed to Lavoisier's coining the term *oxygen*. These observations support Thagard's (1990) on why Lavoisier made the conceptual shift instead of Priestley. However, to clarify this, future studies will have to incorporate Lavoisier's articles in French, which are absent from our corpus.

Lastly, based on the observed patterns, we selected representative topics given the following: (1) the non-cumulative strategy (decade without past) offers more fine-grained topics, (2) the topic of *water* seems to develop into the topic of *air*, and (3) the topic of *air* drifts to become the topic of *acid*. Figure 4 shows the JS distance between these topics, with two maximum values obtained between the 1760s-1800s (*water-acid*), and the 1780s-1800s (*air-acid*). Interestingly, these values seem to match the conceptual shift if we assume the *water* topic as reflecting phlogiston theory, *air* as a transition phase, and *acid* as representing the oxygen paradigm. In the 1770s, Priestley contested phlogiston theory by demonstrating the existence of what he called *dephlogisticated air*, a phenomenon that Lavoisier later conceptualized coining the term of oxygen (Chang, 2011). Hence, the key patterns of text reuse and rhetoric related to the competing oxygen and phlogiston concepts may be contained in the documents that are members of the topics *acid* (1800s) and *water* (1760s), respectively, while *air* (1780s) could be studied as a transition stage between both.

Fig. 2: Entropy values over time based on the cumulative strategy for oxygen-related terms.

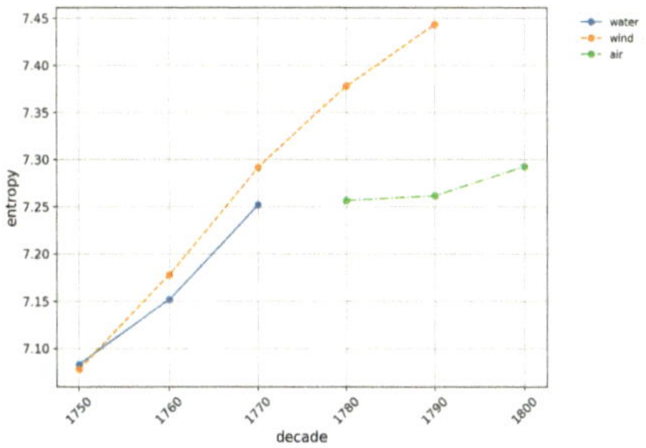

Fig. 3: Entropy values over time based on the non-cumulative strategy for oxygen-related terms.

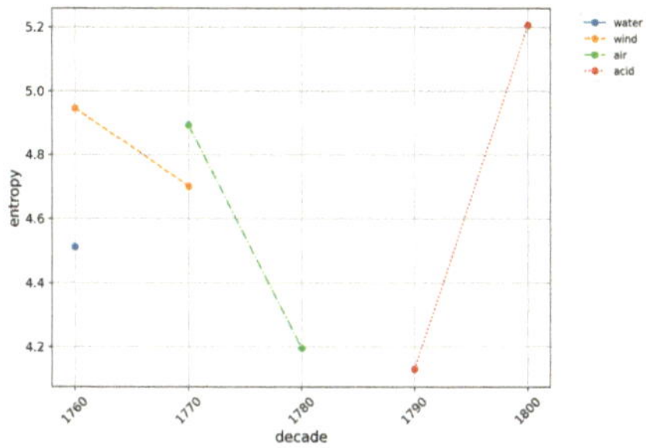

Fig. 4: JS distance across selected topics from the non-cumulative strategy: water (1760), air (1770), air (1780), acid (1790), and acid (1800).

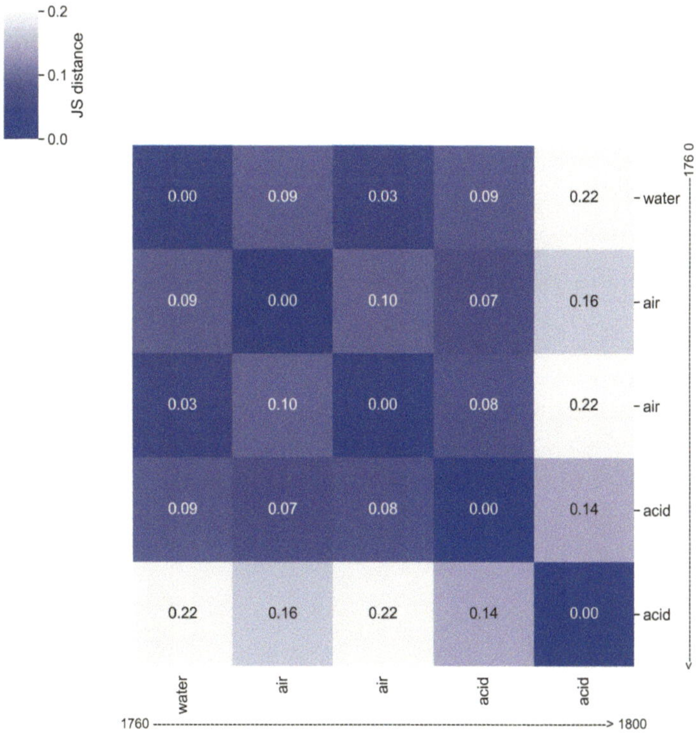

4.2 Word level

To investigate change at the word level, we focused on the transition phase and analyzed the term *air*, which plays a central role in the transition from phlogiston to oxygen theory as shown in our document level analysis. Using SynFlow, a syntactic analysis tool for dependency-parsed corpora, we extracted and analyzed the slot-filler distributions of *air* across five-year time bins between 1750 and 1800 as described in our method section. Our goal was to track how the linguistic environments of *air* evolved over time, as a proxy for changing conceptual and epistemic roles.

4.2.1 Diachronic distribution of syntactic slots associated with *air*

Figure 5 presents the relative frequencies of all syntactic slots associated with *air* over time. *Air* was syntactically stable as the relative frequency of the syntactic slots did not fluctuate much. Slots such as chi_obj (object) and pa_nsubj (nominal subject) remained stable in relative frequency throughout the period, indicating persistent grammatical roles. In contrast, the relative frequency of the chi_amod (adjectival modifier) and chi_det (determiner) slots showed quite some variation. The frequency of *amod* was stable at first, then suddenly increased during the first half of the 1760s, then further increased from the latter half of the 1760s to early 1790s then suddenly decreased. The frequency of *det* fluctuated at first then decreased gradually. This suggests a shift in how *air* was conceptualized. In earlier decades, *air* was frequently introduced as a known, observable entity (*the air*), consistent with a referential and phenomenological view. Over time, the decline of this pattern may reflect a growing interest in the internal properties of *air*, i.e. its quality, typology, and theoretical framing, rather than its immediate physical instantiation. This change aligns with the transition from descriptive to theoretical discourse in chemistry and also explains the increased use paired with a premodifying adjective (amod).

Fig. 5: Frequency of top-7 slots between periods

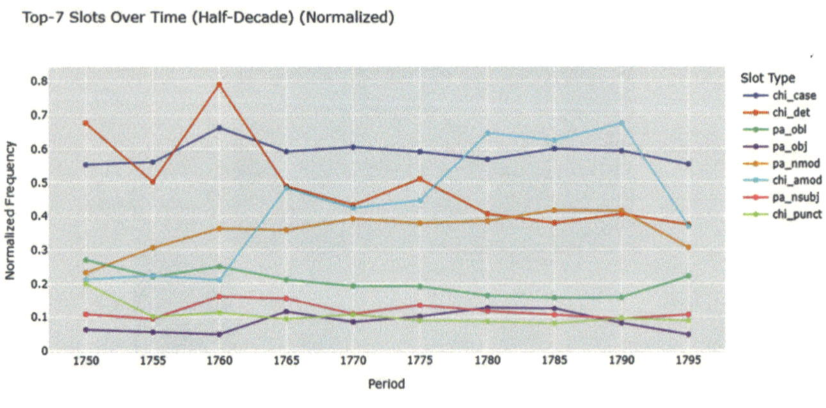

4.2.2 Syntactic signatures of the conceptual change of *air*

Figure 6: Jensen-Shannon Divergence of all slots

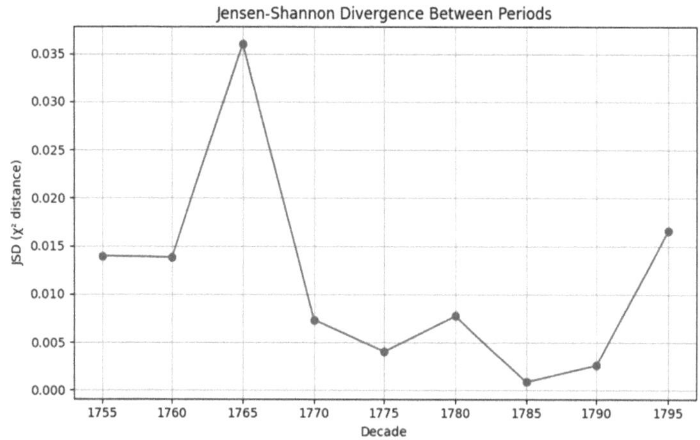

Figure 6 shows the Jensen-Shannon Divergence (JSD) values across successive five-year time bins from 1750 to 1800, measuring the distributional change in syntactic slots associated with *air*. The most noticeable pattern is a sharp increase in divergence around the 1760–1765 period, peaking just above 0.035. This suggests a substantial shift in the way *air* was used syntactically and coincides with the slot-filler adjective modifier as shown in the analysis of the diachronic distribution, reflecting the emergence of new descriptive terminology (e.g., *fixed air, dephlogisticated air*). Following this peak, JSD drops steeply from 1765 to 1770 and continues to decline to its lowest value around 1785. This period can be interpreted as a phase of relative stabilization, in how *air* was syntactically described, showing a phase of terminological consolidation as certain descriptions gained acceptance in the community. After 1785, divergence gradually increases again, with a marked increase between 1790 and 1795, likely corresponding to the transition toward oxygen-centered theory, as Lavoisier's ideas gained traction. Together, these fluctuations trace the linguistic signature of conceptual instability and eventual transformation.

However, note that the absolute value of the JSD was extremely small, which means that generally, the distributions were very stable. When visualizing the contributions of different slots, the picture becomes clearer. All the big increases in the JDS pattern were mostly contributed by the *amod* slot (Figure 7).

During the 1750s and early 1760s, *amod* contributed insignificantly. However, it contributed significantly to the big divergence increment in the late 1760s. For the next 30 years, the contribution became minimal again, except for the early 1780s. And suddenly, in another rise in divergence at the end of the period, its contribution rose again.

Figure 7: Contributions of different slots in the JSD

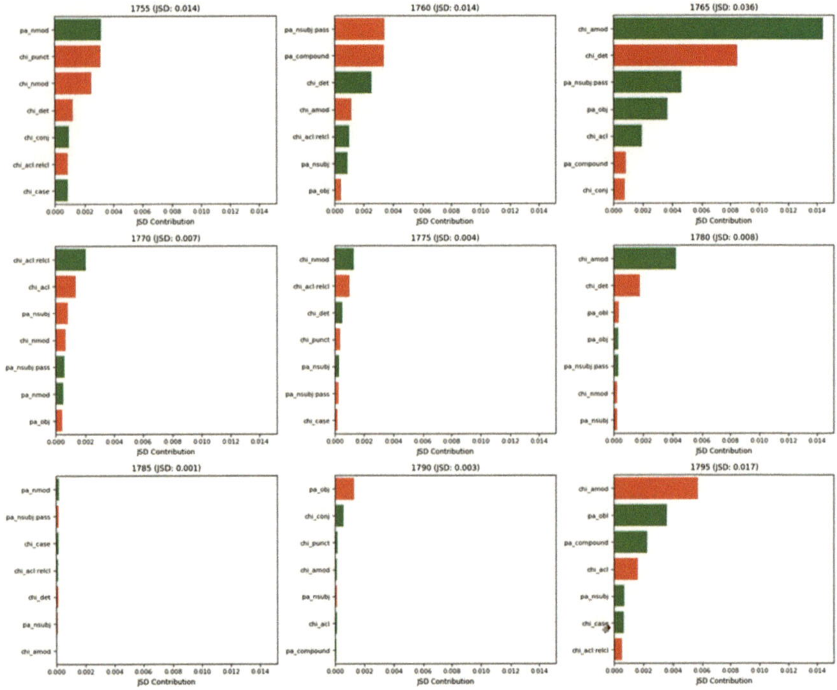

4.2.3 Assessing shifts in adjectival slot-filler distributions

We applied the same workflow with the adjective slot. First, we visualized the shift in relative frequencies of the adjectives and found that the discovery of oxygen changed the way scientists described *air*, causing great fluctuations of *technical* adjectives (e.g., *phlogisticated*, *fixed*) throughout the period (Figure 9). The frequencies of other *non-technical* adjectives (e.g., *fresh*, *open*) decreased and stayed low most of the time (Figure 8).

Figure 8: Relative frequency shifts of the adjectives filling the amod slot

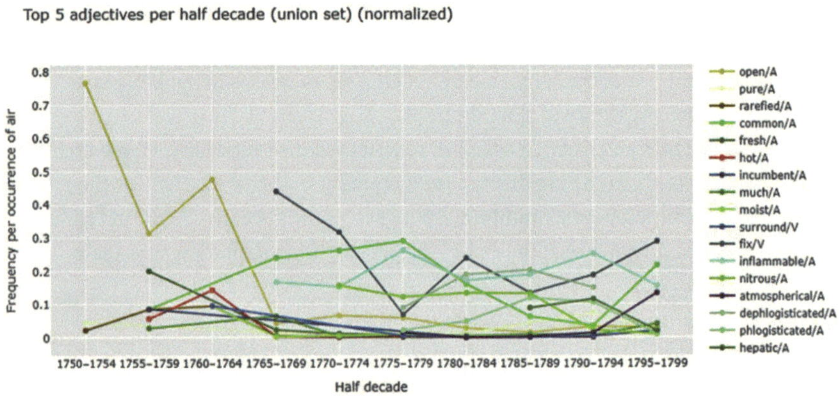

Timewise, we can see some similarities in terms of the small periods with that of the above figures.

After computing the JSD score for the distributions of the adjectives, we found that it matches perfectly with that of the slots' JSD (Figure 10). The consistent pattern is: stability in the first decade, then a big rise followed by a stable period and an increase at the end.

Figure 9: The frequency shift of technical adjectives

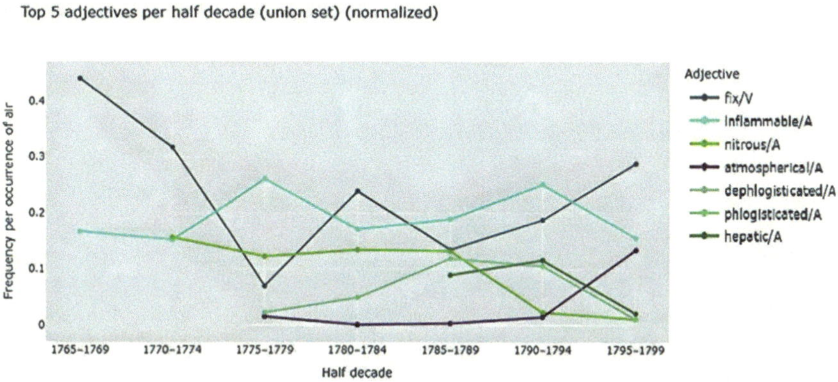

Figure 10: JSD of the adjectives

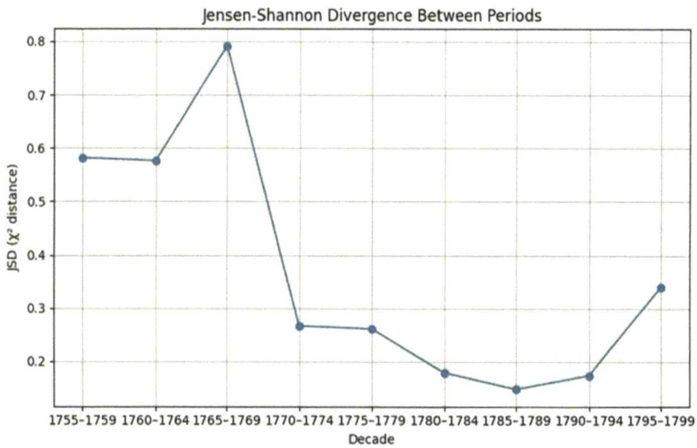

Figure 11 presents the JSD contributions of different adjectives modifying the noun *air* across 5-year intervals from 1755 to 1799. Each subplot represents a time bin and shows which adjectives (slot-fillers of the amod slot) contributed most to the distributional shift in that period. This fine-grained analysis allows us to trace how conceptual associations with *air* changed during the Chemical Revolution. The results converge with and help to quantify the linguistic aspect of the established narrative findings from the history and philosophy of science (Chang, 2011; Kuhn, 2012; McEvoy, 1978;

Stewart, 2012). Early adjectival modifiers reflect physical or environmental properties, while mid-century modifiers indicate taxonomization of different air types (e.g., *fixed, mephitic*), consistent with the exploratory period of phlogiston theory. In the 1770s and 1780s, the divergence score decreased considerably, indicating stabilisation around main competing descriptors (e.g., *(de-)phlogisticated, fixed*) at the height of debate. By the 1790s, phlogiston-related terms disappeared while compositional and functional terms dominated, such as *respirable* and *uncombined*, causing the divergence to rise again. This aligns with the defeat of phlogiston theory and the rise of oxygen theory. These changing slot-filler patterns offer linguistic evidence for the conceptual restructuring that defined the Chemical Revolution.

Figure 11: Contributions of different adjectives to the JSD

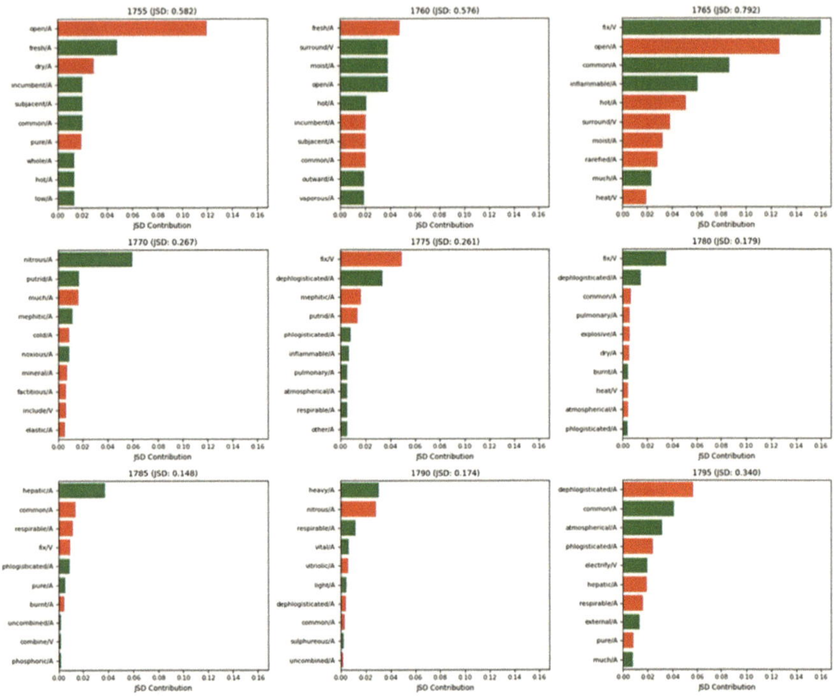

4.2.4 Exploratory comparison with LLM-based interpretations

Based on the findings, we explored whether commercial LLMs (ChatGPT and Grok), trained on large amounts of secondary scientific and historical material, would be able to provide similar results in terms of a comparable high-level narrative. To this end, we conducted a brief, illustrative comparison by prompting ChatGPT and Grok in a fresh session to avoid contamination from prior interactions. The convergence between the LLMs' responses and our finding lies in the identification of three broad periods characterized by epistemic struggle rather than abrupt replacement, as well as the prominence of key terms such as *(de)phlogisticated, common, fixed*. However, their responses only track macro-level conceptual change and largely reflect established philosophical and historical narratives, whereas our analysis provides a fine-grained, quantitative account grounded in tractable and transparent distributional and syntactic evidence. In this sense, LLMs appear well suited for exploratory purposes, but they do not (yet) replace the need for interpretable analyses when the goal is to trace how conceptual change is realized through language and to relate quantitative findings to a known and controllable data source.

4.3 Toward graph-based modeling of conceptual changes

The present study provides a first step towards modeling the multi-layered nature of conceptual changes in an interpretable, data-driven fashion. By combining topic modeling and syntactic slot-filler analysis across time, we have extracted core elements that can be operationalized as nodes and relationships in dynamic knowledge graphs. We envision these graphs to be able to represent the trajectories, tensions, and reconfigurations that constitute conceptual change in science as shown in recent studies (Kaye et al., 2024).

At the document level, our topic modeling reveals a sequential progression across three topics: *water* (1760s), *air* (1780s), and *acid* (1800s). These topics correspond to key stages in the theoretical shift from phlogiston to oxygen theory. In a graph-based framework, each of these topics can be modeled as a node anchored in time. The edges between them can encode their pairwise similarity using Jensen-Shannon Divergence (JSD) scores calculated from topic distributions. For instance, the edge between *water* (1760s) and *air* (1780s) would be weighted by their JSD, capturing the degree of topical shift. The same applies to the edge between *air* and *acid* (1800s), representing the continued drift in conceptual content (i.e. the meaning of *air* moving conceptually from *water* to *acid*). More importantly, the *air* node is not just one point on a timeline. It functions as a transitional hub. Based on our findings, *air* appears linguistically and conceptually between the decline of phlogiston (*water*) and the rise of oxygen (*acid*), both temporally and in terms of content similarity. The graph should reflect this mediating function explicitly: *air* is connected to both *water* and *acid*, and these connections are directional and weighted, signifying that *air* inherited semantic traits from one and contributed to the emergence of the other.

From the word-level analysis using SynFlow, we derive additional relational data that can be layered into the graph. For instance, the slot-filler analysis revealed that modifiers of *air* changed significantly over time. In the earlier period, *air* was often preceded by spatial or environmental adjectives (e.g., *open*, *fresh*), while in the later period, modifiers became chemically loaded (e.g., *vital*, *fixed*, *dephlogisticated*). This shift in the amod (adjective modifier) slot can be modeled as a series of edges connecting the *air* node to adjective nodes representing its contextually salient attributes. These attributes themselves can evolve and be clustered, representing the change in the conceptual schema surrounding *air*. Furthermore, we observed a decline in the relative frequency of the determiner slot (det) for *air*, suggesting a shift from referential usage (*the air*) to more abstract, substance-like usage without the determiner. In graph terms, this could be represented as a change in node type or ontological status, i.e. a transition from a generic, external referent (i.e. *the air*) to a theoretically posited scientific entity. This supports modeling *air* not just in semantic or distributional terms but also as a concept that changes category within the scientific ontology, something graph-based ontologies can represent explicitly.

Entropy values from the topic analysis add another dimension. High entropy in a topic suggests conceptual plurality or competition, while declining entropy reflects conceptual consolidation. Thus, entropy values can be used as node-level attributes or as weights modulating the confidence or centrality of a node at a given time. For example, the high entropy of the *air* topic during the 1780s may indicate a state of conceptual fluidity, just before *acid* (oxygen theory) became dominant in the 1800s.

All of these elements (topic trajectories, JSD values, syntactic environments, slot-filler shifts, determiner frequency changes, and entropy dynamics) can be encoded as graph elements. Topics such as *air*, *water*, and *acid*, as well as key lexical terms like *air*, can serve as nodes. These nodes may be enriched with associated properties derived from modifier terms extracted via syntactic analysis, such as *vital*, *fixed*, or *dephlogisticated*, which reflect the evolving semantic framing of these concepts. Additionally, rhetorical functions such as *observation* or *substance* identified in topic contexts could themselves be modeled as nodes, enabling the tracking of shifts in conceptual roles over time. Connections between these elements are encoded as edges. Topic-to-topic edges, for example, can be weighted using the JS distance to quantify thematic divergence across time. Term-to-term edges may be derived from co-modification patterns (e.g., when *air* and *gas* share similar adjectives), while term-to-property edges arise from slot-filler patterns (e.g., *air* modified by *vital* in the amod slot). Edges can also capture temporal progression, allowing chronological anchoring of conceptual developments. Each node can carry specific attributes that reflect the linguistic and conceptual state of the entity it represents. Entropy scores can capture topic diversity at a given time; relative frequency can indicate prominence or decline; grammatical role distributions (e.g., the presence or absence of determiners) can reflect shifts in ontological framing. The types of edges themselves can be semantically rich: they may represent semantic similarity (as computed from word embeddings or co-occurrence), rhetorical continuity (when one topic or term extends the discourse function of another), syntactic modification (capturing grammatical patterns), or ontological transitions (e.g., when *air* shifts from being treated as an environmental referent to a chemical entity). This formalization provides the structural backbone for graph-based modeling of conceptual shifts, allowing us to integrate linguistic observations into a dynamic and interpretable network of conceptual change. This graph could be dynamically constructed over time slices and queried to trace how concepts evolve, converge, or disappear.

Ultimately, the analyses presented here serve as a scaffolding: they identify the elements and dynamics that need to be encoded in future graph-based models. They also offer quantitative backing for where conceptual transitions occur and how they manifest in language. By translating these observations into graph structures, we aim to model conceptual shifts not as isolated changes but as dynamic, interconnected processes of semantic and epistemic evolution.

This view is similar to Vogl et al.'s (2026) where LLMs are described "*as a compass rather than a quick fix*". Although our tasks have different objects of study, the overarching goal of detecting shifts in discourse is the same. Their approach used topic modeling and LLMs combined to generate interpretable labels for a knowledge graph, and despite the challenges of characterizing the driving forces of the narratives or the participants, we are excited to explore akin implementations in future works.

5. Conclusion

This paper examined how conceptual change during the late phase of the Chemical Revolution is reflected in scientific writing by combining document-level topic modeling and word-level syntactic slot-filler analysis with information-theoretic measures to interpret change through linguistic insight. The results show a gradual reorganization with phases of thematic competition and consolidation at both levels. In particular, we observed the transition from phlogiston to oxygen theory, with topic shifts (*air* to *acid*) accompanied by higher entropy and slot-fillers with rising divergence during that period.

Beyond the historical case study, the contribution of this work lies in showing the value of interpretable, distribution-based methods for linguistically HPSS-oriented questions such as conceptual changes. Topic entropy and divergence make explicit how thematic diversity and stabilization unfold over time, while slot-filler patterns allow conceptual change to be traced through observable grammatical configurations. Compared to LLM-based approaches, which recover high-level narratives but conflate lexical, semantic, and grammatical information and do not offer direct access to the linguistic configurations through which change is enacted, our approach provides a transparent link between quantitative signals and the linguistic realization of epistemic change in a clearly defined primary corpus.

By translating these interpretable linguistic features into nodes, edges, and attributes, future work can integrate the strengths of distributional analysis with more expressive representational frameworks, potentially complementary with LLM-based methods. In this sense, the analyses presented here are not an endpoint, but a scaffolding for modeling conceptual change as a dynamic, multi-layered process grounded in observable language use.[4]

Acknowledgements

Funded by the European Union. Views and opinions ex-pressed are however those of the author(s) only and do not necessarily reflect those of the European Union. Neither the European Union nor the granting authority can be held re-sponsible for them.

4 This chapter was written with support from large language models (LLMs). All model-generated text was reviewed and, where necessary, rewritten by the authors, who remain fully responsible for the final version. For details on the use of LLMs in this volume, see the statement in the volume's introduction.

References

Bengio Y, Ducharme R, Vincent P, et al. (2003) A neural probabilistic language model. J. Mach. Learn. Res. 3(null): 1137–1155.

Bensaude-Vincent B and Abbri F (1995) Lavoisier in European Context: Negotiating a New Language for Chemistry. Science History Publications/USA, Canton, Massachusetts.

Bizzoni Y, Degaetano-Ortlieb S, Menzel K, et al. (2021) The diffusion of scientific terms – tracing individuals' influence in the history of science for English. In: Proceedings of the 5th Joint SIGHUM Workshop on Computational Linguistics for Cultural Heritage, Social Sciences, Humanities and Literature (eds S Degaetano-Ortlieb, A Kazantseva, N Reiter, et al.), Punta Cana, Dominican Republic (online), November 2021, pp. 120–127. Association for Computational Linguistics. Available at: https://aclanthology.org/2021.latechclfl-1.14

Blei DM, Ng AY and Jordan MI (2003) Latent dirichlet allocation. J. Mach. Learn. Res. 3(null): 993–1022.

Cassotti P, Periti F, De Pascale S, et al. (2024) Computational modeling of semantic change. In: Proceedings of the 18th Conference of the European Chapter of the Association for Computational Linguistics: Tutorial Abstracts (eds M Mesgar and S Loáiciga), St. Julian's, Malta, March 2024, pp. 1–8. Association for Computational Linguistics. Available at: https://aclanthology.org/2024.eacl-tutorials.1/

Chang H (2011) The Persistence of Epistemic Objects Through Scientific Change. Erkenntnis 75(3): 413–429.

Degaetano-Ortlieb S and Teich E (2018) Using relative entropy for detection and analysis of periods of diachronic linguistic change. In: Proceedings of the Second Joint SIGHUM Workshop on Computational Linguistics for Cultural Heritage, Social Sciences, Humanities and Literature (eds B Alex, S Degaetano-Ortlieb, A Feldman, et al.), Santa Fe, New Mexico, August 2018, pp. 22–33. Association for Computational Linguistics. Available at: https://aclanthology.org/W18-4503

Degaetano-Ortlieb S and Teich E (2019) Toward an optimal code for communication: The case of scientific English. Corpus Linguistics and Linguistic Theory 18(1): 175–207.

Fankhauser P, Knappen J and Teich E (2016) Topical Diversification Over Time In The Royal Society Corpus. In: ADHO 2016 – Kraków, 2016. Available at: https://dh-abstracts.library.cmu.edu/works/2457

Firth JR (1957) A Synopsis of Linguistic Theory, 1930–1955. In: Studies in Linguistic Analysis. Oxford: Basil Blackwell. Available at: https://cs.brown.edu/courses/csci2952d/readings/lecture1-firth.pdf

Fischer S, Knappen J, Menzel K, et al. (2020) The Royal Society Corpus 6.0: Providing 300+ Years of Scientific Writing for Humanistic Study. In: Proceedings of the Twelfth Language Resources and Evaluation Conference (eds N Calzolari, F Béchet, P Blache, et al.), Marseille, France, May 2020, pp. 794–802. European Language Resources Association. Available at: https://aclanthology.org/2020.lrec-1.99/

Geeraerts D (1997) Diachronic Prototype Semantics: A Contribution Historical Lexicology. Oxford University Press. Available at: https://doi.org/10.1093/oso/9780198236528.001.0001

Geeraerts D (2009) Theories of Lexical Semantics. Oxford University Press. Available at: https://doi.org/10.1093/acprof:oso/9780198700302.001.0001

Geeraerts D, Grondelaers S and Bakema P (2012) The Structure of Lexical Variation: Meaning, Naming, and Context. De Gruyter Mouton. Available at: https://www.degruyterbrill.com/document/doi/10.1515/9783110873061/html?lang=en

Griffiths TL and Steyvers M (2004) Finding scientific topics. Proceedings of the National Academy of Sciences 101(suppl_1). Proceedings of the National Academy of Sciences: 5228–5235.

Hall D, Jurafsky D and Manning CD (2008) Studying the history of ideas using topic models. In: Proceedings of the Conference on Empirical Methods in Natural Language Processing – EMNLP '08, Honolulu, Hawaii, 2008, p. 363. Association for Computational Linguistics. Available at: http://portal.acm.org/citation.cfm?doid=1613715.1613763

Holmes FL (1994) What was the chemical revolution about. In: Abstracts of Papers of the American Chemical Society, USA, 1994, pp. 13–HIST. American Chemical Society.

Kaye A, Schlattmann R, Vogl M, et al. (2024) Socio-Epistemic Networks: A Framework for History of Knowledge. In: Lerg CA, Östling J, Weiß J, et al. (eds) History of Intellectual Culture 3/2024: Experimental Spaces: Knowledge Production and Its Environments in the Long Nineteenth Century. De Gruyter Oldenbourg, pp. 253–274. Available at: https://www.degruyterbrill.com/document/doi/10.1515/9783111291383-012/html

Kermes H, Degaetano-Ortlieb S, Khamis A, et al. (2016) The Royal Society Corpus: From Uncharted Data to Corpus. In: Proceedings of the Tenth International Conference on Language Resources and Evaluation (LREC'16) (eds N Calzolari, K Choukri, T Declerck, et al.), Portorož, Slovenia, May 2016, pp. 1928–1931. European Language Resources Association (ELRA). Available at: https://aclanthology.org/L16-1305

Kuhn TS (1962) The Structure of Scientific Revolutions. University of Chicago Press: Chicago.

Kuhn TS (2012) The Structure of Scientific Revolutions. 50th Anniversary Edition. University of Chicago Press: Chicago.

Lareau F and Malaterre C (2026) When less is more? Topic model performance across title, abstract, and full-text corpora. In: Simons A, Wüthrich A, Zichert M, et al. (eds) *Understanding Science with Large Language Models? Potentials for the History, Philosophy, and Sociology of Science*. Bielefeld: transcript, part-4.

Luhmann J and Burghardt M (2022) Digital humanities—A discipline in its own right? An analysis of the role and position of digital humanities in the academic landscape. Journal of the Association for Information Science and Technology 73(2): 148–171.

McEvoy JG (1978) Joseph Priestley, 'Aerial philosopher': Metaphysics and methodology in Priestley's chemical thought, from 1772 to 1781, part I. Ambix 25(1): 1–55.

Menzel K, Knappen J, and Teich E (2021) Generating linguistically relevant metadata for the Royal Society Corpus. Research in Corpus Linguistics, 9(1). Available at: https://doi.org/10.22028/D291-38086

Mimno D, Wallach HM, Talley E, et al. (2011) Optimizing semantic coherence in topic models. In: Proceedings of the Conference on Empirical Methods in Natural Language Processing, USA, 27 July 2011, pp. 262–272. EMNLP '11. Association for Computational Linguistics.

Mohr JW and Bogdanov P (2013) Introduction—Topic models: What they are and why they matter. Poetics 41(6). Topic Models and the Cultural Sciences: 545–569.

Periti F and Montanelli S (2024) Lexical Semantic Change through Large Language Models: a Survey. ACM Comput. Surv. 56(11): 282:1-282:38.

Periti F and Tahmasebi N (2024) A Systematic Comparison of Contextualized Word Embeddings for Lexical Semantic Change. In: Proceedings of the 2024 Conference of the North American Chapter of the Association for Computational Linguistics: Human Language Technologies (Volume 1: Long Papers) (eds K Duh, H Gomez, and S Bethard), Mexico City, Mexico, June 2024, pp. 4262–4282. Association for Computational Linguistics. Available at: https://aclanthology.org/2024.naacl-long.240/

Periti F, Ferrara A, Montanelli S, et al. (2022) What is Done is Done: an Incremental Approach to Semantic Shift Detection. In: Proceedings of the 3rd Workshop on Computational Approaches to Historical Language Change (eds N Tahmasebi, S Montariol, A Kutuzov, et al.), Dublin, Ireland, May 2022, pp. 33–43. Association for Computational Linguistics. Available at: https://aclanthology.org/2022.lchange-1.4/

Phan-Tất B (2025) phantatbach/SynFlow: 20250739. Zenodo. Available at: https://zenodo.org/records/16610427 (accessed 21 October 2025).

Qi P, Zhang Yuhao, Zhang Yuhui, et al. (2020) Stanza: A Python Natural Language Processing Toolkit for Many Human Languages. In: Proceedings of the 58th Annual Meeting of the Association for Computational Linguistics: System Demonstrations, Online, 2020, pp. 101–108. Association for Computational Linguistics. Available at: https://www.aclweb.org/anthology/2020.acl-demos.14

Schmidt BM (2012) Words Alone: Dismantling Topic Models in the Humanities. Journal of Digital Humanities 2(1): 49–65.

Shannon CE (1948) A Mathematical Theory of Communication. Bell System Technical Journal 27(3): 379–423.

Sievert C and Shirley K (2014) LDAvis: A method for visualizing and interpreting topics. In: Proceedings of the workshop on interactive language learning, visualization, and interfaces, 2014, pp. 63–70.

Singh A, D'Arcy M, Cohan A, et al. (2023) SciRepEval: A Multi-Format Benchmark for Scientific Document Representations. In: Proceedings of the 2023 Conference on Empirical Methods in Natural Language Processing (eds H Bouamor, J Pino, and K Bali), Singapore, December 2023, pp. 5548–5566. Association for Computational Linguistics. Available at: https://aclanthology.org/2023.emnlp-main.338/

Stewart J (2012) The Reality of Phlogiston in Great Britain. Hyle 18(2): 175–194.

Tahmasebi N, Borin L, Jatowt A, et al. (2021) Computational Approaches to Semantic Change. Language Science Press. Available at: https://langsci-press.org/catalog/book/303

Thagard P (1990) The Conceptual Structure of the Chemical Revolution. Philosophy of Science 57(2). [The University of Chicago Press, Philosophy of Science Association]: 183–209.

Vogl M, von Schwerin A and Kirschke S (2026) Large language models in interdisciplinary research settings. A reflection. In: Simons A, Wüthrich A, Zichert M, et al. (eds) *Understanding Science with Large Language Models? Potentials for the History, Philosophy, and Sociology of Science*. Bielefeld: transcript, part-4.

Part 4: Knowledge graphs and topic modeling

The potential of LLMs for constructing a socio-legal knowledge graph

Christian Boulanger

1. Knowledge graphs as a tool to study disciplinary knowledge production

Why and how do academic disciplines and research fields emerge and evolve? This is one question that the field of the history and sociology of science is concerned with (Stichweh, 1992). When one surveys the literature, it becomes apparent that traditionally, a major focus has been on the natural sciences, and it is only recently that the social sciences or humanities are receiving more attention (Kuznetsov, 2019). I am looking into the question of disciplinary development in the context of my research on the post-war history of the German-language sociology of law, which was unable to institutionalize itself as an academic (sub-)field in a way similar to the U.S. or the UK, despite Germany's rich pre-1945 socio-legal intellectual history (Machura, 2020).

The traditional way to study the history of an academic discipline has been the "close reading" of the available literature, the study of the biographies of scientists and scholars, archival work on the actors and the institutions that have been crucial in the development of the discipline. Hermeneutic methods remain essential in grasping the complex history of a research field. But in recent years, the use of computational methods has been on the rise not only being increasingly used in the fields of the digital humanities (Luhmann and Burghardt, 2022), digital history (Cohen et al., 2008), or digital legal research (Livermore, 2019), but also in the field of the history of science (Gibson and Ermus, 2019). Digital Humanities methods thus provide an additional perspective on disciplinary history. Their promise is to discover and reliably document structures and relationships that only appear when algorithmically analyzing a large body of data (Jänicke et al., 2015), often referred to as "distant reading" (Underwood, 2017). They allow us to model and visualize these structures and develop metrics for numerical results and comparisons that can be reproduced by others. This is in contrast to hermeneutic studies where we need to trust the plausibility of the author's interpretation and have no access to the body of data on which she based this interpretation on, beyond those sources that are quoted in the scholarly work.

To be sure, this does not mean that computational methods by definition provide more "objective" results. As with any quantitative studies, the selection of the data that is included involves (unacknowledged) value judgements. This is not only limited to data collection and selection. It also affects preprocessing, processing, and, in particular, visualization. As Gibson and Emus express it, rather than bringing research closer to objectivity, "each step of the workflow adds another layer of subjectivity" (Gibson and Ermus, 2019: 555). This has to be kept in mind when interpreting the results of our computational methods.

For the purpose of a Digital Humanities approach, I understand the history of a research field as the history of knowledge production in this field. In contrast to an understanding of scientific knowledge as the process of progressively eliminating false believes by empirical evidence or logical reasoning and replacing them by (temporary) "true" beliefs, the focus of knowledge production centers on the social processes involving human actors in tempo-spacial contexts that influences the form and content of the knowledge they produce in textual form. These processes can be modelled by knowledge graphs, which store information on actors, their products, and the context as a network of entities and the relationships between them (Haslhofer et al., 2018). To be sure, this is an incomplete picture of knowledge production. These methods cannot tell us anything about the context in which these publications have been produced, nor can they capture the real-life processes of knowledge production. Any insights to be gained by taking a bird's-eye view always have to be checked against studies that rely on "close reading".

A knowledge graph always needs a data model which describes what entities and relationships the model contains. This data model is an extremely reduced representation of the empirical reality which gave rise to the research question, and, at a minimum, the model must contain the entities that are necessary to answer it. However, not all the entities or relationships that one would like to include in the model can be populated with existing data easily. There is a tradeoff between the complexity of the model and the costs (in terms of time and money) necessary for data acquisition. In my case, the most interesting data points are scholars, works, venues and institutions.

Fig. 1 presents a possible data model for the graph. The entity types "scholars" and "institutions" are self-explanatory. "Works" refer to scholarly publications such as articles, chapters, or books; "venues" refer to containers in which such works are included or of which they are a part, such as academic journals, edited volumes, or book series. The relationship "EDITOR_OF" can express the editorship of books, but an additional purpose is to represent the involvement in the production of serial publications which are represented by the "venue" entity type.

Finally, works can cite other works. This relationship looks inconspicuous in the data model, but is this edge in the graph which contains most of the information. A distinctive characteristic of scholarly knowledge production is the use of references and citations, i.e. the rigorous statement of the literature used, sometimes with the information on the exact location within this literature where an idea or quote has been taken from. In the data model, this information is represented by the CITES relationship between Work nodes. In fact, this is one of the most interesting parts of our data model in terms of the research question. If we know which work cited which other work, it is possible to reconstruct relationships that exist between these works. Needless to say, the fact that a citation exists does not tell us – by itself – the exact nature of this relationship, let alone answer the question whether one work was influenced by another. There are certain questions that could only be answered by the inclusion, into the data, of the citation context/intent, which tell us, for example, if a reference was cited approvingly or in opposition, for what purpose it was cited, how often or at which particular page (On this, see Small, 2011; Berrebbi et al., 2022. See also Liesegang and Gläser (2026) , Simons (2026), and Simons et al. (2026). For the purpose of the current study I am interested only in the fact whether a reference can be proven to exist.

Fig. 1: The data model of the (socio-)legal theory knowledge graph

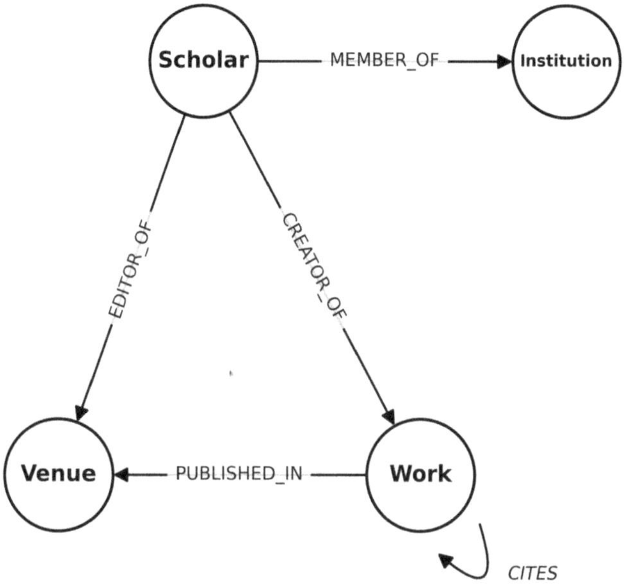

What is missing in the visualization of the data model is the temporal aspect. For example, scholars belong to one or more institutions for a specific time period. Thus, for example, the "MEMBER_OF" relationship between "scholar" and "institution" nodes should have a temporal property that contains the information from which start date to which end date the affiliation existed. While this is completely intuitive for humans, the modelling, representation and analysis of temporal relationships is computationally not straightforward and the subject of research on temporal knowledge graphs (Gottschalk and Demidova, 2018; Zhang et al., 2021).[1]

1 There are many more data points that could be added to this model. For example, it would be fascinating to include information on academic conferences which scholars attended. Unfortunately, no databases exist in which information on academic conferences could be simply retrieved from. Instead, getting this data would require text-mining of conference web pages or printed conference proceedings.

2. Challenges for populating the knowledge graph: lack of data and tools

Although this seems like a relatively simple model, populating the model with data encounters many non-trivial challenges. Theoretically, a data source exists for each of the data points in the model. For example, the Research Organization Registry (ROR) contains metadata on institutions (https://ror.org/). Metadata on German Scholars can be retrieved from places like the German Integrated Authority File (GND) (https://www.dnb.de) through a linked data interface (https://lobid.org), which also gives access to bibliographic data on published books. For recent publications (roughly the last two decades), data on journal articles can be found in bibliometric databases such as the Web of Science, Scopus, or OpenAlex.

In the context of a knowledge graph, one major challenge is to disambiguate and link datasource entries reliably. How do we know, for example, if a certain "John Smith" has published an article, which of the many persons of that name is the given author? Specialized databases such as orcid.org which provide unique ids are of very recent origin and only cover a miniscule subset of authors.[2]

The disambiguation and linking problem becomes only relevant once one has large datasets exported from unrelated data sources which need to be consolidated in the knowledge graph. The main challenge the current project faces, in contrast, is the *lack* of data for this particular domain of knowledge. Bibliographic databases such as the one mentioned mainly contain data on contemporary, English-language literature from the life sciences and biomedicine, the physical sciences, business and technology. In contrast, the coverage for non-English language Humanities, law and social sciences has been quite limited (Hammarfelt, 2016; Gläser and Oltersdorf, 2019; Vera-Baceta, Thelwall and Kousha, 2019; Boulanger, Fejzo and Rimmert, 2025). In addition, the time period covered by the research concerns decades of knowledge production prior to the digital age or the assignment of DOIs. This means that metadata coverage of older literature from the target domain is usually very poor. This is confirmed by testing the mere availability of citation data of the Zeitschrift für Rechtssoziologie in the traditional bibliometric data sources. Fig. 2 shows almost no citation data before the turn of the century, and very little afterwards, as compared, for example, to the Journal of Law and Society (UK).

There is another probable reason for the lack of bibliographic data on German-language law, humanities and social sciences in bibliometric databases: a large part of the journals in this area use footnotes (or endnotes) for references. We don't know the technology that the commercial database vendors use for extracting citations from the journals they index. However, existing open source reference extraction software might offer another clue to why there is such little data. The existing software works well for extracting references from separate, well-formed bibliographic sections (Cioffi and Peroni, 2022). In contrast, it is still an unsolved problem to extract dispersed citation data, by which I mean information on references that is not centrally organized in a bibliography, but which might occur anywhere in the text of a scholarly work. For the purpose of the

2 In particular, ORCID is a self-maintained service and therefore cannot be used, for example, to retroactively identify deceased authors.

project, this mainly involves bibliographic information which is embedded in footnotes, often incomplete, mixed with commentary, and using heavily abbreviated terms. Fig 3 and 4 show examples of footnotes which demonstrate how different citation practices in the Humanities and Law are as compared to the highly structured reference section that can be found, for example, at the end of this chapter.

Fig. 2: Citation data on ZfRsoz in bibliometric data sources, as compared to the Journal of Law and Society

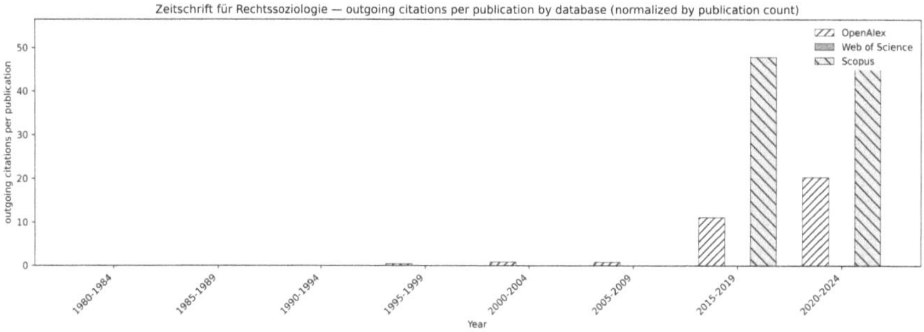

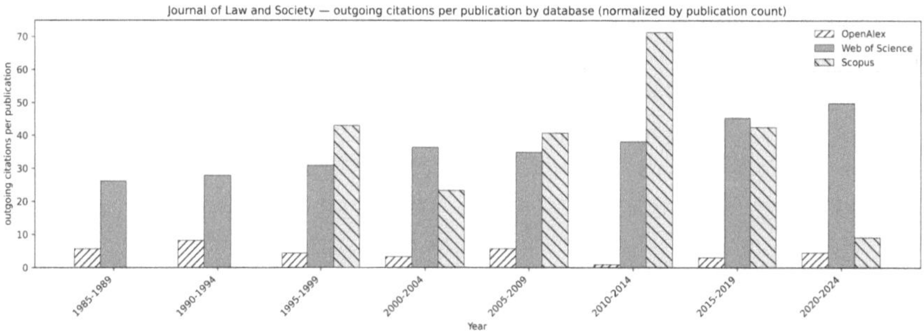

Fig. 3: Footnotes from a German historical article.

1 LUDWIG WITTGENSTEIN, Tractatus logico-philosophicus, London 1922, 1.1.

2 Vgl. die Artikel »Fact« im Oxford English Dictionary und »Thatsache« in Grimms Wörterbuch. Das Wort »T[h]atsache« scheint erstmals ins Deutsche eingeführt worden zu sein durch J. J. Spaldings Übersetzung von Bischof Joseph Butlers Schrift The Analogy of Religion, Natural and Revealed, to the Constitution and Course of Nature, 1736: W. HALBFASS, Tatsache I, in: Historisches Wörterbuch der Philosophie, hg. von JOACHIM RITTER und KARLFRIED GRÜNDER, 10 Bde., Darmstadt 1971 ff., Bd. 10, St–T, Sp. 909–914.

3 HALBFASS, Tatsache (Fn. 2), Sp. 911.

4 Die Beziehungen zwischen den juristischen und den naturphilosophischen Schriften von Francis Bacon werden in der wissenschaftlichen Literatur immer stärker diskutiert. Zwei jüngere Studien, auf die ich mich stark gestützt habe, sind JULIAN MARTIN, Francis Bacon, the State and the Reform of Natural Philosophy, Cambridge 1992, und BARBARA J. SHAPIRO, A Culture of Fact: England, 1550–1720, Ithaca, London 2000.

DOI: 10.12946/rg01/036-055

Fig 4: Footnotes from a German law journal article.

27 Vgl. den Beschluss der Kultusministerkonferenz: „Ausweitung der Nutzungsmöglichkeiten der Hochschulbibliotheken, ihre notwendige Ausstattung und verbesserte Verwendung ihrer Ressourcen" vom 27. Januar 1995, abrufbar unter: http://www.kmk.org/fileadmin/veroeffen tlichungen_beschluesse/1995/1995_01_27-Nutzungsmoeglichkeiten-Hochschulbibliotheken. pdf, zuletzt abgerufen am 6.1.2020, S. 4: „Neue elektronische Formen der Fernleihe sind zunächst auf regionaler und nationaler Ebene zu entwickeln, dabei sind die Dienstleistungen zu beschleunigen und zu rationalisieren.".

28 Zu den verfassungsrechtlichen Hintergründen grundrechtsbezogener Optimierungsgebote, die hier nicht vertieft werden können, vgl.: *Böckenförde*, Der Staat 29 (1990), S. 21 f. et passim.

29 Dieser Sachverhalt wurde jüngst bestätigt in: EuGH, Urt. v. 19.12.2019 – C-263/18, ECLI:EU:C:2019:1111 = BeckRS 2019, 32135 – *Tom Kabinet*.

DOI: 10.5771/2699-1284-2020-1-16

Footnotes may contain several citations, often in one line. Also, many disciplines, such as law, heavily rely on abbreviated journal names, court cases, and shorthands. The challenge for extraction algorithms therefore is to find information embedded in a lot of noise. The existing tools, based on traditional machine-learning algorithms are not optimized for this kind of data and perform quite badly when faced with footnoted literature (Rodrigues Alves et al., 2018; Boulanger and Iurshina, 2022).

3. From structured prediction to large language models

This was the situation when I started work on the Legal Theory Knowledge Graph project in mid-2021.[3] The first goal was the implementation of a pipeline that would be able to process the pilot data: the PDFs of the "Zeitschrift für Rechtssoziologie" and "Journal of Law and Society". As no citation data exists for the "Zeitschrift", it quickly became apparent that the extraction of citations would be the main problem to solve, before any other part of the knowledge graph could be worked on.

At that point in time, LLMs were known only to a specialized audience and were not generally available. I experimented with several tools available at the time, which were all based on traditional machine learning techniques. GROBID (Testori, 2020) is a popular and well-maintained tool for the extraction of bibliographic data from PDFs. It is extremely well architected, using layered models that iteratively segment the document in order to extract structured bibliographic data. Ad-hoc tests showed, however, that it was not able to parse my source material correctly. Grobid provides ways of retraining it with domain-specific data. Unfortunately, the hierarchical layering of models that distinguishes Grobid from other tools also means that it requires highly complex training data. GROBID uses the TEI-XML format and a layered set of models and corresponding training datasets. Producing this data would have meant an investment of more time and resources than I could afford at the time.

3 https://www.lhlt.mpg.de/2514927

After experimenting with ExCite (Hosseini et al., 2019; Boulanger and Iurshina, 2022), I switched to using the AnyStyle citation extraction engine[4], which, like ExCite, relies on predictive machine learning and had shown to be, on average, the best performing tool in a comparison (Cioffi and Peroni, 2022). Similar to Grobid, the AnyStyle's default model had been trained with publications with structured bibliographies. AnyStyle's training data format is much simpler than Grobid's: AnyStyle uses only two models, one for extracting whole lines containing reference strings from text, and one for segmenting the references into its constituent parts.[5] Using a dedicated web application[6], and with the help of students, I produced a dataset of over a hundred XML annotations of PDF documents containing footnotes. Of these, 69 document annotations were used for training the extraction model and 882 fully annotated references for the segmentation model.[7] Using AnyStyle's own evaluation algorithm, we saw the following performance improvements:[8]

Fig. 5: Results of the evaluation using AnyStyle's check command

	Number of sequences	Sequence errors	Sequence error rate	Number of tokens	Token errors	Token error rate
Finder Model						
Default	69			57895	7033	**12.15%**
Custom	69			57895	1292	**2.23%**
Parser Model						
Default	882	734	**83.22%**	19428	7984	**41.10%**
Custom	882	170	**19.27%**	19428	1269	**6.53%**

Even with the quite limited number of annotations, we see a substantial decrease from a 12.15% error rate for the default extraction model that has been trained with non-footnoted literature to 2.23% for our custom dataset. For the segmentation model, the

4 https://github.com/inukshuk/anystyle.

5 The simplicity of AnyStyle comes with a cost: in contrast to GROBID, AnyStyle cannot profit from implicit semantic information provided by the coordinates of the page content: the placement helps to identify titles, authors, abstracts, or the footnotes, and differentiate these elements from other information.

6 In order to be able to involve non-technical support staff, I developed a web application that allows to visually annotate documents and produce training data in the AnyStyle format: https://github.com/cboulanger/citext. The application also allows using the currently trained model to automatically annotate the text loaded into the editor. The predicted annotations can be corrected and saved, to then re-train the model. This allows an iterative process with the aim to have a well-performing model that can deal with the complexity of footnote citations.

7 For this, the standard AnyStyle reference annotation label set was extended. In particular, the default segmentation model had problems with the fact that footnotes very often do not just contain reference data, but also "signal" words and phrases such as "see", "cf.", etc. and the author's commentary ("See, for example, the excellent overview by...") or other phrases that used to introduce sources ("As argued by ...")(On this, see Vogel, 2012). Both types of content were annotated as "signal". This significantly improved the accuracy of reference segmentation.

8 The limitation of this algorithm is that it is testing the model against its gold standard data instead of cross-validating it with a split dataset. However, since the dataset is fairly heterogeneous, the results are meaningful for measuring the model's performance. Also, the check counts exact predictions only.

improvement is even more pronounced: the error rate goes down from 41.1% to 6.53%. However, even with this massive improvement, the extraction still did not yield results that matched our expectations. The resulting data was good enough to show some general trends, for example, to do a bibliometric analysis of a socio-legal journal (Boulanger, 2023; Boulanger, Creutzfeldt and Hendry, 2024). For these kinds of analyses, the errors in the data were acceptable, as we could expect them to be distributed over the feature space and therefore not to distort the results. However, for the purpose of populating a knowledge graph, the quality of the data was not sufficient.[9]

I had initially intended to publish the dataset[10] after adding more annotations and doing some quality control. However, after the "LLM revolution" at the end of 2022, it quickly became apparent that investing further time into training data for AnyStyle did not make sense. Ad-hoc experiments showed that even very early models such as OpenAI's "text-davinci-003" were outperforming the ML models without any training.[11] Therefore, I restarted the reference extraction project from scratch, relying on the ability of LLMs to infer the semantics of a reference instead of the "simple" pattern recognition afforded by traditional ML.

4. LLamore: Large LAnguage MOdels for Reference Extraction

In collaboration with David Carreto Fidalgo (mpcdf) and Andreas Wagner (mpilhlt), a new attempt was made to set up a workflow to extract citation data from the domain of (socio-)legal theory scholarship. Given the propensity of LLM models to hallucinate, it was clear that we needed a robust testing and evaluation solution before we could apply LLM at scale to analyse our data.

A cornerstone of this effort is the development of a high-quality "Gold Standard" dataset specifically tailored to Law and Humanities scholarship. In the previous attempts, the annotations were done on the source material that was of interest for the research question. The advantage was that the annotation efforts produced data that could directly be used for domain-specific analyses. A severe limitation of these documents was that they are not Open Access and therefore the source data, i.e. the PDFs, could not be distributed along with the Open Source code. We realized that if we wanted to advance the technology, in particular in collaboration with others, it was paramount to use Open Access journals to create a dataset. This would allow us to publish complete datasets including the source PDFs.

Therefore, we abandoned our previous data, and started again with an initial set of articles from Open Access journals. This dataset is currently being annotated using the

9 In particular, adding more training data did not seem to remove certain classes of parsing errors, such as the constant misclassification of "See ... " as an author name instead of a signal word.

10 As the PDFs I was working with were not Open Access, for legal reasons, the dataset would have consisted only of raw footnote strings and the segmentation of these footnotes into references and the reference elements.

11 See https://pad.gwdg.de/s/LWUjPOi1F.

TEI XML standard.[12] TEI (Text Encoding Initiative) was chosen for its well-established, comprehensive framework for text interchange, which supports detailed markup beyond simple reference management, encompassing citations, cross-references, and contextual elements. While TEI offers robustness, it does not entirely resolve all conceptual or technical complexities inherent in representing these diverse citation practices.[13]

To leverage the capabilities of LLMs and evaluate their performance, we are using a lightweight Python package named "Llamore" (https://github.com/mpilhlt/llamore). Llamore facilitates two key functions: on the one hand, it can extract citation data from raw text or PDF inputs using various LLM APIs, which can be remote or locally hosted models, outputting structured data suitable for further processing. On the other hand, it provides methods for evaluating the accuracy of this extraction. Evaluation involves comparing the extracted references against the TEI-annotated Gold Standard, computing metrics such as F1-score by aligning the predicted elements with the gold elements.

Initial evaluation results comparing Llamore (using Gemini 2.0 Flash) with the established tool Grobid demonstrate the efficacy of the LLM-based approach for this specific domain, at least when using Grobid's default model. While Grobid and Llamore exhibit comparable performance on a standard dataset (F1 scores around 0.61-0.62)[14], Llamore significantly outperforms Grobid on the project's specialized dataset of footnoted SSH literature, achieving an F1 score of 0.45 compared to Grobid's 0.14. This represents approximately a threefold improvement in performance for handling complex footnotes.

To be sure, Grobid's model has not been trained with domain-specific material. In a cooperation with the current maintainer of Grobid, Luca Foppiano, we are annotating the same dataset for use by the Grobid models. Initial tests have shown that this has improved its performance, however, the production of a complete and big enough training dataset will take some time.[15] So the jury is still out as to whether LLMs or specialized tools will provide better performance, in particular since Grobid is much faster and consumes much less computing power. On the other hand, as we do not have a sufficiently large dataset yet, we have not been able to evaluate the performance of fine-tuned small LLMs.

5. Future work and outlook

These findings suggest that LLMs offer a viable path for extracting citation data from law and Humanities texts where traditional tools struggle. While Grobid remains a faster

12 We are using an online editor especially developed for this purpose. See https://github.com/mpilhlt/pdf-tei-editor.

13 See https://github.com/mpilhlt/bibliographic-tei/ for the documentation of cases where the TEI annotation guidelines do not provide clear rules for the markup of bibliographic information.

14 We used the "plos_1000" dataset provided by Grobid, published at https://zenodo.org/records/7708580.

15 See https://grobid.readthedocs.io/en/latest/Training-the-models-of-Grobid and https://grobid.readthedocs.io/en/latest/Grobid-specialized-processes/. In a cooperation with the University of Stuttgart, we are working on a set of high-quality annotations for training and evaluation.

and less resource-intensive option for the literature it was trained on, LLamore's performance on complex footnotes highlights the potential of modern language models. Future work will focus on expanding the Gold Standard dataset, refining evaluation metrics, and exploring the effectiveness of smaller, local, open-source models for this task. At the same time, the potential of solutions that do not rely on generative ai should be actively explored.[16]

Significant progress can only be made via open and collaborative efforts. First of all, the lack of training material for the citation practices in Law and the Humanities is one of the main reasons why current models perform badly. Since the production of high-quality training and evaluation data is notoriously difficult and time-consuming, we need a community effort to produce such material in amounts that will allow us to fine-tune and evaluate specialized models. In addition, the potential of synthetic datasets for finetuning needs to be explored.

Another stumbling block for collaboration concerns the lack of established standards for data interchange. We have decided to use TEI, which provides a syntax for expressing bibliographic information both as structured data and as annotated text, unlike other serialization formats such as BibTeX, MODS or CSL-JSON.[17] TEI provides a forward-compatible data format which allows to encode additional semantics such as the citation context or citation intent in future work.

In the context of knowledge graphs, a push towards more standardization, or at least for better interoperability, might come from the research on ontologies. Ontologies have been developed for bibliographic and citation data (Peroni and Shotton, 2018) or for data on scholars (Pertsas and Constantopoulos, 2017; Nguyen et al., 2020).[18] These ontologies might be useful for the eventual publication and long-term storage of the graph data produced in the project presented here.

In any case, much remains to be done before we have a sufficiently large, open knowledge graph that allows us to monitor and explore the evolution of a research field using data on knowledge production in the field of (socio-)legal theory. [19]

Acknowledgements

The data for fig. 2 were provided by the German Comptence Network for Bibliometrics funded by the Federal Ministry of Education and Research (Grant: 16WIK2101A)

---

16 See Wagner and Hermes (2026).

17 The corresponding TEI elements are <biblStruct> and <bibl>, respectively.

18 See https://opencitations.net/publications (last accessed 12.11.2025).

19 This chapter was written with support from large language models (LLMs). All model-generated text was reviewed and, where necessary, rewritten by the authors, who remain fully responsible for the final version. For details on the use of LLMs in this volume, see the statement in the volume's introduction.

References

Berrebbi D, Huynh N and Balalau O (2022) GraphCite: Citation Intent Classification in Scientific Publications via Graph Embeddings. In: 2nd International Workshop on Scientific Knowledge: Representation, Discovery, and Assessment. Available at: https://inria.hal.science/hal-03648498 (accessed 14 July 2025).

Boulanger C (2023) Citation graph of the Journal of Law and Society (1974–2022). DOI: 10.5281/ZENODO.8389925.

Boulanger C, Creutzfeldt N and Hendry J (2024) The Journal of Law and Society in context: a bibliometric analysis. Journal of Law and Society 51(1): 1–25. DOI: 10.1111/jols.12465.

Boulanger C, Fejzo, D and Rimmert, C (2025) Law Doesn't Count? Measuring Bibliometric Coverage of German Law Journals. Max Planck Institute for Legal History and Legal Theory Research Paper Series No. 2025–10. Available at http://dx.doi.org/10.2139/ssrn.5350481.

Boulanger C and Iurshina A (2022) Extracting Bibliographic References from Footnotes with Excite-Docker. In: Backes T, Iurshina A and Mayr P (eds) Proceedings of the Workshop on Understanding Literature References in Academic Full Text. Cologne: CEUR Workshop Proceedings, pp.26–33. Available at: http://ceur-ws.org/Vol-3220/#paper3 (accessed 14 July 2025).

Cioffi A and Peroni S (2022) Structured references from PDF articles: assessing the tools for bibliographic reference extraction and parsing. ArXiv. DOI: 10.48550/arXiv.2205.14677.

Cohan A et al. (2019) Structural Scaffolds for Citation Intent Classification in Scientific Publications. arXiv. DOI: 10.48550/arXiv.1904.01608.

Cohen DJ et al. (2008) Interchange: The Promise of Digital History. The Journal of American History 95(2): 452–491. DOI: 10.2307/25095630.

Gibson A and Ermus C (2019) The History of Science and the Science of History: Computational Methods, Algorithms, and the Future of the Field. Isis 110(3): 555–566. DOI: 10.1086/705543.

Gläser J and Oltersdorf J (2019) Persistent Problems for a Bibliometrics of Social Sciences and Humanities and How to Overcome Them. In: Catalano G et al. (eds) Proceedings of the 17th International Conference on Scientometrics & Informetrics. Rome: Edizioni Efesto, pp.1056–1567. Available at: https://www.issi-society.org/publications/issi-conference-proceedings/proceedings-of-issi-2019 (Accessed: accessed 14 July 2025).

Gottschalk S and Demidova E (2018) EventKG: A Multilingual Event-Centric Temporal Knowledge Graph. In: Gangemi A et al. (eds) The Semantic Web. Cham: Springer International Publishing, pp.272–287. DOI: 10.1007/978-3-319-93417-4_18.

Hammarfelt B (2016) Beyond Coverage: Toward a Bibliometrics for the Humanities. In: Ochsner M, Hug SE and Daniel H-D (eds) Research Assessment in the Humanities: Towards Criteria and Procedures. Cham: Springer International Publishing, pp.115–131. DOI: 10.1007/978-3-319-29016-4_10.

Haslhofer B, Isaac A and Simon R (2018) Knowledge Graphs in the Libraries and Digital Humanities Domain. In: Sakr S and Zomaya A (eds) Encyclopedia of Big Data Technologies. Cham: Springer International Publishing, pp.1–8. DOI: 10.1007/978-3-319-63962-8_291-1.

Hosseini A et al. (2019) EXCITE – A Toolchain to Extract, Match and Publish Open Literature References. In: 2019 ACM/IEEE Joint Conference on Digital Libraries (JCDL). Champaign: IEEE, pp.432–433. DOI: 10.1109/JCDL.2019.00105.

Jänicke S et al. (2015) On Close and Distant Reading in Digital Humanities: A Survey and Future Challenges. Eurographics Conference on Visualization (EuroVis) – STARs. DOI: 10.2312/EUROVISSTAR.20151113.

Jha R et al. (2017) NLP-driven citation analysis for scientometrics. Natural Language Engineering 23(1): 93–130. DOI: 10.1017/S1351324915000443.

Kuznetsov A (2019) Changed but Undescribed? What STS Could Say on the Research Practices of Social Sciences. EASST Review 38(1): 4–5. Available at: https://easst.net/easst-review/381/changed-but-undescribed-what-sts-could-say-on-the-research-practices-of-social-sciences/ (Accessed: 14 July 2025).

Liesegang L and Gläser J (2026) Supporting citation context analysis with large language models raises questions that should have been asked 40 years ago. In: Simons A, Wüthrich A, Zichert M, et al. (eds) *Understanding Science with Large Language Models? Potentials for the History, Philosophy, and Sociology of Science*. Bielefeld: transcript, part-6.

Lima JF, Amaral CMG and Molinaro LFR (2010) Ontology: An Analysis of the Literature. In: Varajão JEQ et al. (eds) ENTERprise Information Systems. Berlin: Springer, pp.426–435. DOI: 10.1007/978-3-642-16419-4_44.

Livermore MA and Rockmore DN (eds) (2019) Law as Data: Computation, Text, and the Future of Legal Analysis. Santa Fe: SFI Press.

Luhmann J and Burghardt M (2022) Digital humanities—A discipline in its own right? An analysis of the role and position of digital humanities in the academic landscape. Journal of the Association for Information Science and Technology 73(2): 148–171. DOI: 10.1002/asi.24533.

Machura S (2020) Milestones and Directions: Socio-Legal Studies in Germany and the United Kingdom. German Law Journal 21(7): 1318–1331. DOI: 10.1017/glj.2020.81.

Nguyen VB et al. (2020) Ontologies Supporting Research-related Information Foraging Using Knowledge Graphs: Literature Survey and Holistic Model Mapping. In: Keet CM and Dumontier M (eds) Knowledge Engineering and Knowledge Management, Springer International Publishing, 2020. DOI: 10.1007/978-3-030-61244-3_6.

Peroni S and Shotton D (2018) The SPAR Ontologies. In: Vrandečić D et al. (eds) The Semantic Web – ISWC 2018. Cham: Springer International Publishing, pp.119–136. DOI: 10.1007/978-3-030-00668-6_8.

Pertsas V and Constantopoulos P (2017) Scholarly Ontology: modelling scholarly practices. International Journal on Digital Libraries 18(3): 173–190. DOI: 10.1007/s00799-016-0169-3.

Rodrigues Alves D, Colavizza G and Kaplan F (2018) Deep Reference Mining From Scholarly Literature in the Arts and Humanities. Frontiers in Research Metrics and Analytics 3: 21. DOI: 10.3389/frma.2018.00021.

Simons A (2026) Scaling In, Not Up? Testing Thick Citation Context Analysis with GPT-5 and Fragile Prompts. In: Simons A, Wüthrich A, Zichert M, et al. (eds) *Understanding*

Science with Large Language Models? Potentials for the History, Philosophy, and Sociology of Science. Bielefeld: transcript, part-6.

Simons A, Arnaout H and Gurevych I (2026) Reconstructive citation context analysis using large language models. A roadmap. In: Simons A, Wüthrich A, Zichert M, et al. (eds) *Understanding Science with Large Language Models? Potentials for the History, Philosophy, and Sociology of Science*. Bielefeld: transcript, part-6.

Small H (2011) Interpreting maps of science using citation context sentiments: a preliminary investigation. Scientometrics 87(2): 373–388. DOI: 10.1007/s11192-011-0349-2.

Stichweh R (1992) 'The Sociology of Scientific Disciplines: On the Genesis and Stability of the Disciplinary Structure of Modern Science. Science in Context 5(1): 3–15. DOI: 10.1017/S0269889700001071.

Testori M (2020) GROBID: when data extraction becomes a suite. In: OpenMethods. Available at: https://openmethods.dariah.eu/2020/09/09/grobid-when-data-extraction-becomes-a-suite/ (accessed 14 July 2025).

Underwood T (2017) A Genealogy of Distant Reading. Digital Humanities Quarterly 11(2). Available at: http://digitalhumanities.org:8081/dhq/vol/11/2/000317/000317.html (accessed 14 July 2025).

Vera-Baceta MA, Thelwall M and Kousha K (2019) Web of Science and Scopus language coverage. Scientometrics 121(3): 1803–1813. DOI: 10.1007/s11192-019-03264-z.

Vogel R (2012) Verbs for referring to sources in humanities and social sciences: Grammatical and lexical analysis of their distribution. Discourse and Interaction 5(1): 63. DOI:10.5817/DI2012-1-63

Wagner A and Hermes J (2026) Encoded humanities, or: not everything has to be generative. A dialogue on AI tasks and roles. In: Simons A, Wüthrich A, Zichert M, et al. (eds) *Understanding Science with Large Language Models? Potentials for the History, Philosophy, and Sociology of Science*. Bielefeld: transcript, part-1.

Zhang F et al. (2021) RDF for temporal data management – a survey. Earth Science Informatics 14(2): 563–599. DOI: 10.1007/s12145-021-00574-w.

From source to structure

Extracting knowledge graphs with LLMs

Raphael Schlattmann, Aleksandra Kaye, and Malte Vogl

Introduction

Extracting structured information, such as knowledge graphs (KGs), from unstructured historical texts presents significant challenges for researchers in the History, Philosophy and Sociology of Science (HPSS). While manual annotation is often considered precise, it scales poorly. Recent advancements in Large Language Models (LLMs) offer potential for automating parts of this process, yet require careful integration to handle the often hidden nuances of historical sources and research questions (see also Boulanger, 2026; Meding and Daugs, 2026).

To address this challenge, we developed a computational pipeline[1] aimed at extracting KGs from historical data. It aims to connect large, semi-standardized, but previously isolated sources like biographical lexicons and make them structurally analysable. The goal is to supplement classical historical analysis, with a structural perspective that focuses on relations. Such an approach makes it possible to ask new kinds of questions from these sources and, at the same time, forces a new engagement with their nature (e.g., Weiß et al., 2025). The pipeline employs a two-stage architecture grounded in the principles of Task Decomposition, Human-in-the-Loop (HiL) validation, and a research question & data-driven approach.

Stage 1 implements ontology-agnostic Open Information Extraction (OIE), extracting factual triples (subject-predicate-object) without explicitly defining an ontological structure. The process begins with data preprocessing, loading, and chunking. Already at this stage, we utilize multimodal capabilities of LLMs to extract textual input data for later KG construction directly from initial sources, often characterized by complex layouts. Raw SPO triples (plus time as a property) are then extracted from these text segments using a first LLM. These triples undergo ensemble validation, with a second LLM

1 The code will be made available under https://github.com/raphschlatt/Knowledge_Graph_Extraction

antagonistically scoring, refining, and normalizing them ("LLM-as-a-judge"). Stage 1 includes systematic evaluation against manually created gold-standard samples to ensure extraction quality meets research-question-specific thresholds. This ontology-agnostic approach enables the discovery of unexpected relationships, mitigating premature constraints based on researcher assumptions about the data, while complementing the research-question-driven focus of the second stage.

Stage 2 focuses on ontology-driven knowledge graph construction. The process begins with an LLM analyzing a representative sample of validated triples from Stage 1 to draft potential Competency Questions (CQs) – research questions that the final knowledge graph should be able to answer. These machine-generated, data-oriented questions are refined, discarded, or expanded by a human expert, ensuring alignment with the actual research questions. The resulting expert-validated CQs and the same sample of triples then guide an LLM in drafting a domain-specific ontology designed to address these questions. After manual ontology review in another HiL step, validated OIE triples are mapped onto this ontology, and an LLM drafts SHACL shapes based on the ontology and validated triples, which are again subsequently manually refined. Entity linking and disambiguation then resolves textual entity labels to persistent identifiers from authority files through a multi-source strategy: asynchronous searches against Wikidata with VIAF as fallback, followed by LLM-based contextual disambiguation of entity candidates. Entities without matches undergo character-embedding-based candidate blocking, again followed by LLM-based contextual disambiguation. Successfully matched entities receive authoritative labels, and for locations, geographic coordinates. The pipeline concludes with RDF-star output generation and SHACL validation.

Human-in-the-Loop integration occurs at multiple points: gold-standard creation for OIE evaluation, competency-question refinement, ontology and SHACL shape review and correction, manual disambiguation verification, and quality assessment at each pipeline stage. This tries to balance LLM-driven automation for speed and scalability with expert validation for accuracy and domain appropriateness. The modular design facilitates component replacement as technology evolves, while comprehensive metadata logging at each input/output step (model details, parameters, prompts, timestamps, costs, providers, etc.) ensures partial reproducibility, given the non-deterministic nature of LLM outputs and transparency. This human oversight is crucial both for validating the accuracy of extracted relations and for iteratively refining the ontology based on encountered complexities.

Rather than presenting the intricacies of a finalized tool, this contribution centers on the process of its creation and adaptation. The theoretical framework for the utility of knowledge graphs or networks in general is discussed elsewhere (see e.g., Kaye et al., 2024). Likewise, the specific application of the KG for historical analysis is not the subject of this article. The adopted presentation format here is that of an expert interview, presenting an edited interview transcript (Part 1) featuring the research team. This serves as a research vignette tracing the pipeline's evolution. The interview explores the rationale behind the pipeline's design, discusses concrete problems encountered (e.g., optimizing OIE prompts, designing effective validation steps, balancing automation and control), and details the adaptive strategies employed.

Following this detailed account (Part 1), the paper includes a machine feedback component (Part 2). LLMs were tasked to analyse the introduction and interview transcript and provide critical commentary from two distinct viewpoints: that of a HPSS researcher considering epistemological and methodological implications and that of a Computer Scientist assessing technical choices and implementation.

By combining detailed process documentation via an interview with AI-generated commentary, this work aims to contribute practical insights into the challenges and strategies involved in leveraging LLM capabilities within toolbuilding workflows for HPSS, fostering methodological discussion on the responsible and effective integration of such technologies in humanities research.

The choice of this unconventional, two-part format – an expert interview complemented by LLM-generated commentary – is deliberate. We believe it offers a transparent and reflexive lens through which to examine the often "messy", iterative, and collaborative realities of developing and integrating these new tools within humanities research. Instead of presenting a final result, we aim for a process documentation, revealing the complex decision-making, unforeseen challenges, and adaptive strategies that characterise the actual research process. By contrast, the interview format allows us to capture these nuances, offering a more authentic research vignette and to reflect on our own choices and decisions. Furthermore, juxtaposing the human narrative of the development process with LLM feedback, aims to provoke deeper methodological reflection on the nature of such human-AI collaborations and the epistemological implications of using these new tools in fields like HPSS not only for automatable tasks but also processes of critical interaction.

Part 1: Expert Interview

Notes on format

Why a full-length interview? We present the interview answers in full because Part 2's LLM commentaries are derived from and should be read against this complete transcript. Trimming would firstly compromise this comparative analysis and make some LLM remarks unintelligible. Secondly, as this contribution serves as a research vignette it is meant to narratively capture human experiences in real life settings, giving "the reader a sense of being there in the scene" (Erickson, 1985: 150), in this case the development process.

How was the Interview conducted? The interview questions were collaboratively developed through iterative discussion and redaction. We internally reviewed and refined the questions, then answered them individually in written, asynchronous rounds, iterating until all authors confirmed clarity and completeness. The transcript presented here is slightly edited (light copy-editing for readability, without substantive changes).

Questions and answers

Motivation and challenges

Question: *Let's start at the very beginning. What was the initial problem that you aimed to address with an LLM-based pipeline for knowledge graph extraction, and what did your first conceptual sketch of such a pipeline look like?*

RS: I've been working with automated network data creation for a while now, and in 2023 I participated in a workshop on Heterogeneous Referencing where I had already started using LLMs experimentally to map person names referenced in publications (implicit citations) to explicit citations. That worked "reasonably" well with GPT-3.5 back then, and since then there have been several ideas for using the contextual "understanding" of LLMs to extract otherwise difficult-to-extract semantic information. In my work on the history of science of gravitational research in the GDR, for example, I repeatedly consulted the biographical database *Who was who in the GDR? A lexicon of East German biographies*[2], and each time I had to think about how much information is contained in this lexicon, while being completely disconnected internally (it is a classic lexicon, originally printed, then got digitized but only text-based, so you have individual entries without linking). I always thought there's so much structural information in there, that could be utilized additionally to the way it was used already. So to be able to combine micro and macro perspectives. That was the initial idea: to automatically extract networks from it, before the next step became the idea of using knowledge graphs. The first version was also pretty rudimentary and straightforward: Define a few entity classes with a few relations and then write a prompt. And because this simple experiment already seemed to work so "well", it became a side project and then later eventually a larger project that also took quality control and source criticism in historical research more seriously and tried to mitigate the known problems of LLMs.

AK: For a while now, I've been interested in better understanding the role of migrants in the transnational circulation of scientific information and methods in the nineteenth century. What knowledge and in what formats did migrants contribute towards production and exchange? What underlying conditions were key for facilitating knowledge circulation? I have been researching this in the context of exchanges between various Latin American countries and the territories of the partitioned Polish Lithuanian Commonwealth. It was in relation to this work that I came across a particularly interesting source: Stanisław Zieliński's "Mały słownik pionierów polskich kolonialnych i morskich"

2 "Who was who in the GDR? A lexicon of East German biographies" is part of the biographical databases of the Federal Foundation for the Study of the Communist Dictatorship in Germany. Together with the "Biographical Handbook of German Communists", the database contains around 5,400 structured short biographies of central actors from the GDR and the communist movement. The entries offer fact-based life stories with information on origin, function(s), party affiliation, and professional career (Müller-Enbergs et al., 2010; Weber and Herbst, 2013). Available online at: https://www.bundesstiftung-aufarbeitung.de/de/recherche/kataloge-datenbanken/biographische-datenbanken.

[The Little Dictionary of Polish Colonial and Maritime Pioneers][3], where the author listed over 1070 Polish 'travellers, discoverers, achievers, researchers, explorers, and migrant diarists, activists and writers' (Zieliński, 1933: 1). If the information Zieliński included in his work could be extracted and structured, then digital humanities methods could be applied, allowing us to reconstruct the social networks of Poles working abroad in the nineteenth century, not just in Latin America, but globally. Such an approach could enrich the research I was already conducting in relation to historical Polish migration, which, among other things, involved careful appraisal of "The Little Dictionary" as a historical source. At that point, Raphael had already been applying LLMs in his work for OCR correction with good results, and it seemed like these models could also be useful for retrieving and processing information from "The Little Dictionary".

MV: I had already worked with structured sources previously in my role as a research software engineer, e.g. a list of all ancient Greek settlements in the Mediterranean sea or a three-volumed dictionary of analysed handwritings of medieval manuscripts written all over Europe. For these sources I had built regular expression search methods that gave sufficient results, but required a lot of tweaking and inclusion of edge cases. In the end the actual historical research with these newly organized and more accessible sources was not possible because the time to build the tool to work with the sources was taking so long. The additional aim of extracting the content of a source as a knowledge graph was inspired by another project where colleagues had constructed such a resource after literally years of discussion on the ontology. In the end their knowledge graph was great to describe a specific source but not easily adjustable to other historical questions or source material. My hope with this LLM-based approach was to get results of better quality and shorten the amount of time necessary to verify and clean the results to have more time for the actual historical work with this type of source material.

Question: *When you began implementing these initial ideas, what were the most significant and perhaps unexpected challenges you encountered, particularly concerning the nature of historical (biographical) sources and the then-current capabilities of LLMs for extraction tasks?*

RS: The biggest and initially underestimated challenges were actually three things in retrospect: **First**, precisely operationalizing what we actually want. Because biographical encyclopedia entries contain more information than you might initially think. Take the sentence: "X was a professor of theoretical physics at University Y from 1950–1960." Depending on how explicitly or implicitly you ask and define it, many statements are contained in that one sentence. Just a few examples. Explicitly, for instance:

1. Person X existed (at least between 1950–1960).
2. University Y existed (at least during that time).
3. X held a professorship (at least during that time).
4. X specialized in *theoretical physics* (at least during that time).

---

3 Available online at: https://polona.pl/preview/ed8bca19-0311-43db-b535-f8c160a449b2

Implicitly (via logic, world knowledge, or domain expertise):

1. X was alive and professionally active (at least between 1950–1960).
2. X taught and/or conducted research at University Y (at least during that time).
3. University Y had a theoretical physics department (at least during that time).

And that's just a small selection of possible inferences. Since we want to ask structural questions, we need standardization/operationalization for that. So: which statements like the ones above actually interest us, and which are actually present in the data? That usually happens through ontologies, and creating them took quite some time, as it does in many KG projects. Later, we reversed the process and began by formulating the questions we wanted the final KG to answer. **Second**, the precision of extracting those statements. At first glance, the results looked surprisingly good. But as so often in digital humanities, the devil is in the details. The more data we tested and the closer we looked, the clearer it became that we needed to proceed very granularly to achieve the desired quality, including quality checks at multiple stages. **Third**, the merging and disambiguation of "identical" entities. That turned out to be one of the hardest problems. Sometimes the same entity is written very differently, abbreviated or spelled out, like "HU Berlin" vs. "Humboldt University Berlin". Sometimes entirely different things have the same name, like two people or places. Or there are OCR/transcription errors. You have to resolve all that if you really want only the "correct" connections in your graph. And of course, across all three tasks, you can observe large differences between time points and model versions in LLM development. The large frontier models are naturally better at extraction with naive prompting and zero-shot approaches. But if you take the right model, prompt cleverly and/or combine techniques, the open, smaller, faster models are often good enough.

AK: Like Raphael already pointed out, the disambiguation challenge stemming from the historical sources themselves was particularly troublesome. In the Little Dictionary, instances are common of the same publications, individuals or institutions being referred to in multiple ways across different dictionary entries, making it difficult to determine whether or not the extracted names all refer to the same person or entity. Additional manual and thus time-consuming checking and matching before further analysis can be conducted was necessary. A separate underestimated challenge relates to the fact that it is difficult to know what the LLMs have been trained on. Likely, it was not nineteenth-century documents in Polish. A slight inconvenience, hence, was ensuring that when LLMs were used for the extraction of information or OCR correction, the spelling and grammar were not 'corrected' to the present-day format. What I also found concerning at the beginning and to some extent continue to do now, is that it is so difficult to ascertain the level of trust we can have in the accuracy of the results we obtain when employing LLMs for information extraction.

MV: I was actually surprised by the quality of the results already at the start. But it was also clear that open-source LLMs did not reach the same quality as SOTA for-pay services. I am still struggling with the idea that paid-for services should be part of a scientific software pipeline. On the one hand because I like to make software reusable and findable but also because a piece of closed-source software will make the research less reproducible. One other thing that is still challenging to me is the design of the actual prompt. To me the results seemed very sensitive to the concrete language and structure of the prompt. In the beginning of the implementation of this idea it was not clear to me how to deal with this sensitivity. How could one be sure that the current prompt gives the correct results? I accept this part of the development work as more of a craft instead of a science at the moment. Maybe I will change my position on this later.

Question: *Could you describe a specific early-stage component or process that didn't work as anticipated? What were the failure points, and what did you learn from that setback regarding the complexity of the task?*

RS: In the beginning, we first defined an ontology and then extracted statements based on its structure. But it quickly became clear that some types of subject-predicate-object (SPO) triples were extracted much more reliably than others. For instance, relations like Person ->has_role -> Role worked very well, almost no relevant cases were missed. Others, like Person -> achieved -> Achievement, performed not as good or inconsistently, which is of course not surprising. Roles are often short, well-defined terms with clear semantics – "physicist", "mayor", "professor" – and are in our sources usually prominently mentioned, often in the first sentence. Achievements, by contrast, are more complex and diffuse, often embedded in longer, narrative-style sentences like: "Through groundbreaking work on gauge symmetry, X helped redefine approaches to unifying gravity and quantum theory." There's no simple verb like "received" here, and the contribution is conceptual, not tied to a concrete title or award. But this also reflects a more fundamental limitation of the graph approach itself: when structuring unstructured text, I trade narrative nuance for operationalization. This inevitably reduces interpretive flexibility, but in return, I gain scaled comparability and the ability to detect structural patterns. These patterns have to be treated with care and rather often point to tendencies, offering interpretation at a different, macro layer. But that's why combining micro- and macro-level interpretation is so crucial. Meaningful analysis requires both. This also means, when designing an ontology, distinctions between classes and relations must be as clear and non-overlapping as possible. The more precise the concept of a statement, the better the extraction, but this applies to humans and machine extractors alike. Also, extraction works better when the area of attention is smaller. Sentence-level extraction, but with the whole paragraph as context is significantly more accurate than directly paragraph-level extraction. This again is no surprise, but the combination of a narrow focus and a broad (or broader) context is an important quality factor. And an insight, that is also common knowledge: LLM ensembling proved highly effective. If one LLM extracts a triple and a second LLM reviews both the sentence and the proposed output of the first, the quality improves noticeably.

AK: One process that did not work as well as anticipated was simply switching between the two biographical dictionaries during testing, from "Who was who in GDR" to "The Little Dictionary". Initially, we were getting much fewer and less accurate results from the Polish source than from the German source, so we had to adapt and adjust the process to the different input dictionaries, for example, by providing context-specific prompts and examples. As Raphael mentioned, another issue we encountered in the early runs was that certain entities were frequently miscategorised; in both cases, it was the persons' achievements that proved particularly difficult to get right.

MV: The earliest version of the approach gave a list of JSON files that included co-mentioned names in each entry. Aleks then had to provide a list of names that were not persons, or differently spelled names and so on. This required several cycles of data cleaning. In later variants we therefore changed the prompt for the LLM to e.g. expand abbreviations or correct capitalizations and so on. This made things easier. Another aspect of solving this was the inclusion of disambiguation parts in the pipeline. Again all of this is just a way to solve the data generation and leave more time for the scientific work, which of course contains a critical reflection of the way the data was generated.

Adaptive Solutions

Question: *The introduction highlights a two-stage architecture and, more generally, Task Decomposition. How did you arrive at this specific design? Was it a gradual evolution or a more distinct pivot based on the early challenges?*

MV: In my memory the two-stage architecture emerged gradually as a result of the data quality. Even with quite good initial results, it was always necessary to reconcile the data e.g. concerning name disambiguations. To me the current architecture is also a result of the question of how to judge the sensitivity of a result to the structure of the prompt. By having several stages with clear target goals, it becomes possible to have an answer to the question of what is the best prompt. We just compare it to the hand-created small datasets for each step.

RS: I would agree with that. It was a gradual process where more and more intermediate steps were added. The more precisely we tried to extract certain things, the more obvious the problems became. And the more granular we made the sequence of steps, the clearer it also became which points in this ladder were problematic. And as said before, current models are usually particularly good when the task is clearly defined. If you want to simultaneously extract statements regarding an ontology and return the extracted things in original Polish form and normalized (e.g., directly in English and spelled out), that's usually worse than first extracting in Polish and then normalizing in a further run. The more granular, the better you usually can formulate few-shot examples, for instance what you exactly mean by normalization, since you only provide examples for very specific tasks, which is more unambiguous. Quality control is easier as well, if you have more steps. Also in terms of modularity and the choice of special models for special tasks (also regarding the future), it became increasingly clear that radical decomposition makes sense.

AK: Quite early on it became apparent to us that to limit the black-box effect and increase our understanding of the overall process, it would be helpful to break it down into smaller tasks and thoroughly interrogate the impact of the individual steps. As Malte and Raphael said, this realization came about as a result of the unexpected peculiarities we were observing in the extracted data.

Question: *Human-in-the-Loop (HiL) is also central to your approach. Could you elaborate on the different HiL stages? For instance, how does the HiL for validating LLM-extracted triples differ from the HiL involved in refining the domain ontology, and what considerations were driving the process?*

AK: We keep tweaking the different steps of the pipeline, so I would say Human-in-the-Loop is omnipresent. Pretty early on in our journey of utilising LLMs for extracting information from historical sources, we had manually created detailed and accurate examples of triples, in the form of subject-predicate-object expressions, based on excerpts from our source material. Based on the close reading of the biographical dictionaries, we also came up with a series of competency questions. Deciding on these questions helped us both to think through the types of information we can actually expect to find in these specific sources, and later to assess the accuracy of the triples extracted using LLMs. The HiL validation of these triples is pretty straightforward and involves returning to the source material and cross-referencing to determine whether the triples are correct or not. We are not expecting complete accuracy, so the difficulty lies rather in deciding what level of accuracy is sufficient and what requires further refining of the obtaining process. Refining the domain ontology is a more involved process, as the complexities of the base language really come to the fore.

RS: There are several points in the current pipeline where HiL is central. These are actually either quality control, i.e., human feedback loops, or questions that are crucial for later analysis. The first central HiL point is the Competency Questions generation step. These are questions that the final Knowledge Graph (KG) or the ontology should be able to answer when properly designed and filled with the appropriate data. In our design, these points are mostly co-working tasks, so in this step, for example, a SOTA reasoning model designs about 25 questions based on the OIE-extracted triples that can be asked of the available data at all.[4] The domain expert then goes through these questions, corrects them, discards them, or adds their own, and this iteratively, until it's clear: these are (1) the questions I want to be able to answer and (2) they are actually askable of the available data via KG. The second important HiL step is the ontology creation. This works basically similar to creating the CQs. A reasoning model makes a proposal based on the already approved CQs and the OIE-extracted triples. So we take on one side what statements are really present, and on the other, what our ontology should be able to answer, and have a first draft for an ontology created from that. A domain expert then goes through this proposal again and revises until the ontology fits the questions and the real data (or sometimes you have to go back to the questions). There are other interfaces, but these two are probably the most relevant. Because that's where it's decided what and how the KG contains information.

4 This idea is based on (Kommineni et al., 2024a, 2024b).

MV: I would say that the LLMs are just tools that help human researchers to perform certain tasks more quickly. So actually I could maybe call this LLMs-in-the-Loop. Similar to traditional research methods the results of LLMs in each stage have to reach a quality that is good enough for the current research question and source material. That is to me the core principle: Each new source material and research question concerned with it requires a new tweaking of the method. There may not be one LLM prompt or one ontology that is useful for all questions. The process of the triple validation was interesting to me since it requires a careful reflection of what implicit statements are contained in everyday language. To give a good example we need to express these, which was not always straightforward to me. While the first step requires careful reading of the sources, the second step requires reflection again but this time on the current research question. To be able to validate the usefulness of an ontology, we had to ask ourselves what information we would need to be able to answer the question. Since I am interested in applying methods from historical network research, I had to go even one step further and consider the potential relation that I would like to query based on the future knowledge graph. This added a level of complexity to me. All of these are examples of the hermeneutic circle of doing research. LLMs are a new tool that requires new ways of reflecting. I guess this proposed pipeline is one way of making these reflections possible.

Question: *Focusing on Stage 1, the LLM-driven Open Information Extraction: Firstly, why did you choose to use OIE? And secondly: What were some of the key learnings in prompt engineering or model selection to get "useful" factual triples from your historical texts, and how iterative was this process?*

RS: We only introduced OIE relatively late, when it became increasingly clear that we wanted to make KG construction seriously data- and research-question-driven, not ontology-driven. There are tons of comprehensive ontologies or even specialized ones for person data, for example. In our experience, the problem was always: if we focus directly on the ontology, we lose sight of what questions we actually want to answer and also what the data potentially provides at all. Moreover, it was clear from the beginning that we wanted to extend the project to other data and not just use it for the two case studies. So at some point we stumbled upon OIE, because it's about structuring all statements contained in text without a prefixed structure. If you then apply this to a representative sample of the original dataset, for example, it often only becomes apparent which statements are actually present or relevant. So before you start constructing a sophisticated ontology, it might become apparent in advance that the structure you wanted to ask about or model is only covered by a minimal fraction of the statements. For instance, out of 40,000 statements, there are actually only 8 that represent a certain relation. That can of course also be interesting and produce a historically meaningful narrative, but to track structural change over time, that's probably too little and likely makes the ontology unnecessarily complex. Regarding prompt engineering, I had already said something about this earlier. The most important learning in my opinion is: the more specific and clearly delineated or defined the task is, the better the output. Good few-shot examples make a huge difference. And model selection should actually be empirically accompanied from the get-go, even though we unfortunately didn't do that consistently. Because sometimes small changes already make a big difference. So classically iterative, always optimizing quality/cost/speed and open-source vs. proprietary. In the newer versions of the pipeline, we consistently include metadata so that every output logs all relevant parameters (model, model parameters, prompt, etc.), so we can systematise what config gives the best output.

MV: I can not say much about the prompt engineering since Raphael did a brilliant job with that. However, the core reason for me to introduce this step was really the point that it became possible to validate the usefulness of each prompt. There was a hugely reduced space left to hallucinations since text units were smaller and it became easier to validate the output quality.

AK: I think a learning outcome regarding prompt engineering that is worth mentioning relates to the fact that we have been working multilingually on this computational pipeline for knowledge graph extraction. The sources I study here are all in Polish, while the sources Raphael has been working with are in German. Our common language is English, and so this is the language we communicate in and discuss the research challenges we face. While Polish, German and English are all Indo-European languages, they differ in their grammars, structure and vocabulary, so necessarily the HiL interventions needed to be specific to the language of the source. This, on the one hand, makes things more difficult, as we can't just freely apply the same fixes across contexts, but on the other hand, it is helpful as it encourages us to interrogate more thoroughly the possible difficulties.

Question: *Regarding Stage 2, structure imposition and validation: How did you balance the need for a domain ontology with the complexities of the source texts? How did the HiL feedback loop contribute to evolving the ontology itself?*

MV: To me the crucial shift in perspective was to get away from the idea of a central knowledge resource that is gradually filled by the triples generated in diverse research projects. Coming from this old perspective I first had in mind to build and extend an ontology based on CIDOC-CRM, since I had used it in previous projects. However, the event-based approach made the resulting structure quite complex, which of course reflected in the necessary SPARQL queries to actually do any analysis. To my mind, by speeding-up the process of triple extraction and ontology building by LLMs we now have the chance to see each ontology and each source as unique, while keeping the option to link to existing knowledge through Linked-Open-Data paradigms. This to me is giving the benefit of structured output while staying flexible to source material and research questions. The validation by SHACL constraints is again an elegant and useful way of limiting the potential of the LLMs to hallucinate. The circular nature of our approach seems natural to the hermeneutic reasoning of traditional research. The full approach is based on the source material and the research questions concerned with it. If the ontology is not able to provide answers to central aspects of the research, we have to adapt the prompts. This process gives opportunity to reflect and explicate, which is very useful in interdisciplinary teams.

RS: I agree with Malte. The goal is a combination of standardization and source focus. I had previously worked with other ontologies and my association was often: modeling data with CIDOC-CRM, for example, uses enormous resources and often seems to be the main task compared to analysing the data modelled that way. But coming from network research, we were actually interested in the analysis. And for that, it's also helpful if the whole process, data creation and analysis, can run iteratively. The potential scale effects of LLMs allow us to proceed more experimentally (without downplaying the necessary historiographical rigour or the significance of environmental impacts). With our current case studies, for example: running the Polish data completely through the current pipeline (with all steps, including linking and disambiguation) takes about 2 hours and costs around $ 8–10. That means if you have to, you can also adjust something afterwards if necessary and do it again. That's not the goal and we of course always test extensively and evaluate the intermediate steps, but what I want to express with this is: you don't need to design a perfect ontology for 2 years that covers all eventualities if you only want to answer three specific questions for the current project.

AK: There is a limit to what we can learn from any given historical source. If the ontology is good enough, it will help the researcher learn not only what is in the source but also identify what is missing. Navigating between the source and developing a domain ontology that connects the information contained in the source with a broader context (through, for example, manually matching entities to their Wikidata item identifiers) can be viewed as having a function of meaning-making, as we make sense of the content in this reiterative process. Through the HiL feedback loop, we have been taking stock of what we can learn from the two biographical sources alone and have identified directions for the necessary further research to be conducted as we proceed to analyse the obtained structured information.

Decision-Making & Trade-offs

Question: *Looking back, what were one or two of the most critical design decisions that significantly shaped the pipeline's current form? What alternatives did you consider, and what were the primary factors or trade-offs (e.g., accuracy vs. scalability, cost vs. control, automation vs. interpretability) that tipped the scales?*

AK: All these factors (accuracy vs. scalability, cost vs. control, automation vs. interpretability) merit careful consideration, and indeed they all played into the decisions we made. For example, it was important for us to keep the costs, both monetary and environmental, as low as possible, but in some cases, the more 'expensive' models worked better for our needs. It was a process of experimentation and balancing different criteria. I would say, however, that perhaps the aspect that often ultimately decided for us was time – whether we had sufficient time to properly implement the design decisions we were considering. We'd ask, given the time constraints we face, what are the most efficient changes and improvements that we can make?

MV: The design of the pipeline in different stages allowed us to select different LLM models depending on the task at hand, some of them quite small-scale. This is crucial to keep the costs low, which would otherwise be a big problem concerning the reproducibility of the results. I would say we gradually got more away from the idea to automate everything and reintroduced the human component more and more. Of course this also means that I can not simply analyse a corpus without the domain knowledge providing the human in the loop aspect. Like I said earlier this also limits the scalability of the approach, but I do not consider that an issue any more.

RS: As I mentioned in other questions: I do think that the three core principles that crystallized, Research-Driven & Data-Oriented, Human-in-the-Loop, and Task Decomposition, were the critical design decisions. So the iterative discovery and then making it a guiding principle, that was critical for the current form. The good thing is that this allows everything to be built modularly, and if there's a better/cheaper/faster model for one step, like OIE extraction, you can simply integrate it there and all other steps aren't affected by it. Alternatively, we thought at the beginning that we might be able to use something out of the box or adapt it slightly, because there are theoretically some packages or other projects that supposedly do something similar, like the Neo4j GraphBuilder (neo4j-labs/llm-graph-builder 2025), but they either do something completely different or have very limited possibilities to control the result, which speaks rather against them in a scientific environment, which ideally incorporates transparency and traceability.

Question: *The project involves team members with different expertise/domain knowledge. How did this interdisciplinary collaboration influence the problem-solving process and the design choices? Were there specific instances where these different perspectives led to a particularly innovative solution or a necessary compromise?*

RS: Interdisciplinary collaboration has definitely influenced how the pipeline looks now and will look in the future. The whole project was designed from the beginning to bring together different domain experts with similar source types (Polish biographies and GDR biographies). And also that it must be extensible to other sources. I think the multilingualism of the sources was decisive in some design decisions, for example. For instance, the separation of extraction and normalization (e.g. from "Biblj. Uniw. Warsz." becomes "University of Warsaw Library") might not have been emphasized so much otherwise, as well as the approach to entity linking and disambiguation (taking a lot of context from the original and the normalized language and emphasizing, for example, diacritics or other language peculiarities for disambiguation). But also the emphasis on what's important or which relationships had to be modelled emerged essentially collaboratively.

AK: I am grateful to have been able to collaborate on this and other projects with Malte and Raphael, because I think together we come up with interesting research directions and creative solutions to the questions we seek to answer. Combining expertise from different domains enables us to identify viable avenues more quickly, thoroughly and effectively than if we were to work on a similar project alone. For example, we can determine together how much pre-processing work would be necessary to attempt to use digital methods for analysis and whether the effort would be worthwhile given the likely output. We can also discuss which questions are well-suited to be answered, and what sorts of arguments we can substantiate, using the historical sources and methods we plan to employ. That is to say that we can work out the necessary compromises early on in the research process. Due to the division in expertise, it does mean that we rely on one another to take the lead in different sections of the project – good teamwork is key. Regarding our current project in relation to deriving knowledge graphs from biographical dictionaries, I believe that our approach is fresh and innovative. I have worked with similar historical documents for more than fifteen years, consulting individual entries, but with the approach we are implementing, I will for the first time be able to use such sources more holistically, and observe the bigger picture they paint.

MV: I find this hard to answer, since interdisciplinarity was such an essential part of the whole process. I am sure, if I would have done something similar alone I would have taken several shortcuts and just looked at the source material as is. The collaboration forced me to become more reflective in my work, which makes things harder but leads to more substantial results in the long run. So I could say the compromise on my side was to take the sources seriously.

Question: *Reflecting on the entire development journey, what advice would you give to other humanities researchers or teams looking to integrate LLMs for complex analytical tasks with historical or nuanced textual data? What are the key "lessons learned" about the "messiness" and iterative nature of such projects?*

RS: I've kind of answered this in other questions: Especially while working with LLMs, proceed iteratively and experimentally. Take the model of the hermeneutic figure eight seriously, that is, the image often used in digital humanities to visually represent the iterative dynamic loop between parts and wholes, data and context, as well as analysis and self-reflection (even if it is sometimes partly generated, see Part 2: LLM Commentary). And only use LLMs in your research, if you are also interested in reflecting their strengths and weaknesses.

AK: My advice would be to remember that while it might be comfortable to think of LLMs as something separate from our messy humanness, they are, after all, fundamentally built on human languages, and humanities scholars, historians included, have much experience in the domain of studying human expression. As Nick Petrić Howe put it, "the 'language' part of the large language model, fundamentally impacts how they can function, because language is intrinsically linked with culture, history, belief, identity, wellbeing, inequity and progress". Moreover, I also think those looking to use LLMs for complex analytical tasks with historical or nuanced textual data, and to bridge the gap between human and computer language, would do well not to neglect the earlier approaches to language analysis from the humanities and Natural Language Processing.

MV: Similar to Aleks, I would say that LLMs should be treated with a grain of salt all the time. Similar to the critical reading of a source, the answer of an LLM requires reflection. And since the LLMs language can be quite convincing, I think we have to be even more alert in this reflection. The basic idea of the pipeline to break down a large task into much smaller tasks to make them understandable and verifiable is key to me. Reflecting on the academic journey that led to this current project as a whole, I want to highlight that although our disciplinary backgrounds and fields of expertise differ, we share an interest in finding ways to address research questions related to knowledge evolution, and I would say that some sort of question or problem that interests everyone involved in a project is vital to the success of a project such as ours. Coming from different disciplines can result in a different vocabulary and focus, so what is also necessary is the willingness to communicate, listen, learn, and teach one another.

Question: *Thinking about the process of developing this pipeline, how did the choices you made in terms of what data to extract, how to structure it, and where to implement HiL, reflect or even shape your own epistemological assumptions about constructing historical knowledge from these biographical sources?*

RS: I would say that, as with other projects, I naturally went into this process with a certain idea of why the data might be interesting and why connecting it could add value. Behind this lies, perhaps, an internalized structuralist epistemology, the assumption that structures matter for historiographical interpretation. That historical persons or phenomena can be better understood, if you look at them both individually, on a micro level, but also on the meso/macro level through their positions in graphs, for instance. The whole project is guided by this idea: that creating connections reveals structures that can help answer structural questions. The problematic thing about the whole matter is somewhat that you potentially reproduce or perpetuate biases (such as collection, gender, training data or authority file bias etc.) because you're investigating structures that are based on a biased data basis or a biased model/algorithm. Sometimes, what you uncover tells you more about editorial practices, like how biographical entries were written, than about the historical setting of the described persons. It's a kind of structural enrichment of the original layer of production, but one that can offer valuable historical insights.

AK: Similarly when evaluating sources as historical evidence — considering their context, purpose, significance, biases, strengths, and limitations — thinking about the process of developing this pipeline has encouraged me to reflect on my role and positionality within the research process. I am keenly aware that the choices we made about what data to extract and how to structure it will contribute to the sort of results we observe. As we ask questions, explain, and argue, our choices contribute to shaping and creating new knowledge. The outputs that we produce will need to be scrutinised by future researchers before they choose to use the results we produce or to otherwise follow in our footsteps. To help with that, it is vital that we not only observe the FAIR principles but also be transparent about our decision-making that went into developing the pipeline. I hope this interview is a step towards that.

MV: One thing that comes to mind is the writing of an entry in the source and the splitting in small facts in the pipeline. How far did the author of e.g. the dictionary consider what information is entailed in a sentence? Could we then potentially enrich the sentence beyond what was intended by the author and would that be problematic? Or are we just making use of the general structure of language to extract new knowledge that the author of the dictionary did not even have? One central point is in addition the fact that we can look at structures of entries. On the other hand if I think e.g. of the index and abbreviations introduced by the author, it seems to reflect a networked knowledge by the author of the considered person. However these details get somehow lost to the readers of a dictionary. In that sense our approach reflects the process of writing the dictionary. Another aspect is our background in research on historical epistemology and what is called socio-epistemic networks, i.e. that knowledge is carried by people, expressed with time-dependent semantics and also has material conditions. I would say this background has strongly influenced our implementation.

Question: *Beyond the specifics of this pipeline, what broader methodological questions or potentials do you see for HPSS in adopting such LLM-driven, HiL-based approaches to knowledge extraction? How might this change the way researchers in HPSS interact with, analyze, or even conceptualize their source material in the future?*

AK: I see in LLM-driven, HiL-based approaches to knowledge extraction a potential for pre-processing information contained in the historical documents. If we develop robust, reliable methods of extracting knowledge graphs we would be able to use them to quickly identify sections of the documents, or indeed entire big corpora that contain the information we are interested in, that we can then proceed to interrogate using our chosen methods of analysis, whether that is network analysis, close reading, or some other well-established historical research method. Essentially, I see in these approaches a way of structuring information contained in historical sources that allows for further analysis using digital methods.

MV: Maybe my answer is a bit on the pessimistic side, but I think there is not really a fundamental change in how HPSS will work with source material. All the practices of critical reflection of source material are still necessary and structural questions have been asked on more limited sets of sources in the past. There might be a real shift in scale however and maybe a chance to include less spoken of actors like lab assistants or collectors of data. That would be a real benefit of such methods.

RS: I agree with the others in part, but would go a bit further. I think scale is essential here. These kinds of extractions and structurings can “unlock” corpora on a structural level much faster, cheaper, and at larger scale than ever before, including previously untapped sources, which also potentially changes the types of questions we ask. For example, with RAG techniques you could locate very specific passages for close reading across many more documents than was previously feasible, while the kind of data creation we do here also makes longue durée, data-driven distant reading analyses on large datasets more accessible. This scaling alone already brings in new methodological approaches from other fields like Complexity Sciences or Computational Linguistics into HPSS. Take what Aleks mentioned: with these techniques, one could for example analyse how certain text components, like introductions or methods sections and so on, change across thousands or millions of scientific publications in a certain field over extended periods. The same applies to other specific text sections across large representative samples of any other text category. So yes, I think LLMs and other ML techniques will and already are changing both how we interact with sources and how we conceptualize research questions and programs around the central question of "what to do with which sources".

Part 2: LLM Commentary

In **Part 1** we documented the development of a two-stage pipeline for knowledge-graph extraction. **Part 2** now utilizes LLMs as *role specific, digital sparring partners* to analyse that process from two standpoints that we think are relevant for (HPSS) humanities tool-building: (a) epistemological adequacy within History, Philosophy & Sociology of Science (HPSS) and (b) technical rigour within Computer Science & AI/ML engineering. The following two sub-sections present a commentary generated by different LLMs after being provided with the complete transcript of the interview and the introduction above.

Drawing on the concept of *AI-as-Sparring-Partner* (SP), we treat these commentaries not as decorative "add-ons," but as dialogical instruments designed to challenge and extend our own analytical stance. A related discussion of LLMs as "thinking partners" in humanities research is offered by Wagner and Hermes (2026). Herrmann (2025: 170) defines an AI sparring partner as an artefact that

> interacts with users in a combination of a cooperative as well as a challenging, competitive mode, [...] which either expands the range of ideas and viewpoints considered [...] or fosters discussion and critical thinking about and reflecting on [...] the quality of ideas or problem solving [...], the skills and knowledge of the user [...] or the performance of the AI system itself.

Such a dual mode of cooperation and challenge aligns with Human-Centered AI's call for keeping the "human-in-the-loop" while still provoking cognitive stretch (Shneiderman, 2022; Dellermann et al., 2019). A well-designed sparring session can offer several potential benefits (Herrmann, 2025: 166): (1) perspective broadening: exposing blind spots by injecting counter-arguments or alternative framings (2) critical-thinking: obliging the human team to reflect and/or justify theoretical or design decisions, echoing work on generative AI and higher-order reasoning (Kasneci et al., 2023) and (3) reflexive benchmarking of one's own ideas and methods, testing whether the pipeline's narrative can withstand critique from systems that were not involved in its construction, thus functioning as an externalised assessment loop.

However, these same benefits can constitute fundamental limitations. LLM-generated commentary often lacks broader situated or very specific "understanding" of research contexts, cannot draw on embodied experience, may be factually incorrect and may reproduce biases embedded in their training data while appearing neutral.[5] For this reason, we do not outsource final evaluation or conceptual decision-making to machines. Rather, we solicit structured, externalised critiques that can expose blind spots in the messy, iterative, and collaborative realities of integrating AI into humanities scholarship.

On the implementation side we created an OpenRouter[6] chatroom querying eight state-of-the-art LLMs in parallel with synchronised prompts and sampling parameters. All logs (model versions, system prompts, model parameters, and full chat logs) are accessible unedited via Zenodo[7], alongside their summaries for both disciplinary roles.

HPSS research perspective analysis

The first commentary prompt tasked LLMs to provide a critical analysis from the perspective of a senior HPSS researcher. This persona was chosen to critically interrogate the project's theoretical assumptions and methodological choices, handling of source nuance, and underlying epistemological assumptions, thereby challenging the core of the historical research design. The initial prompt given to each of these models is shown below. Subsequently, a summary meta-commentary was created from the eight resulting analyses, for the purpose of reducing text length and to highlight consensus and unique standpoints of the different models.

Prompt Used: HPSS Perspective Analysis

Critically analyse the preceding interview transcript and introduction from the perspective of a senior researcher in the History, Philosophy, and Sociology of Science (HPSS). Focus on the following aspects:

- Implications for historical methodology.
- How the team handled source nuance, ambiguity, and potential biases.
- The epistemological assumptions seemingly underlying the project.
- The alignment of the project's goals and methods with core concerns in HPSS.
- How the described process reflects the "messiness" of integrating computational tools in humanities research.

Be critical. Be concise. Don't overhype. No bloat.

From the eight resulting analyses, a summary/meta-commentary was generated using the synthesis prompt below:

5 For a broader discussion of the ethical and epistemic implications of using LLMs in humanities and HPSS research, including issues of bias, data labour, and the situatedness of machine "understanding", see Lang (2026).

6 OpenRouter is an AI model router that provides an API endpoint to access a diverse array of models from various providers, and also features a chat interface for simultaneous, comparative prompting of multiple models using consistent parameters. See: https://openrouter.ai/chat.

7 You can find the materials under https://doi.org/10.5281/zenodo.16737269

Prompt Used: Synthesis Prompt

Instructions and Rules:

Role: You are an AI assistant specialised in synthesising multiple complex text sources into a coherent meta-commentary.
Context: Your sources are the eight available LLM responses that provide a critical analysis of an interview from the perspective of an HPSS researcher.
Task: Produce a thematically ordered meta-commentary that summarises the core statements, points of consensus, and unique critiques across the eight responses. The text must emulate their critical, analytical tone and use a clear, paragraph-based structure.

1. Headings and Structure: Use the following headings **verbatim and in this order**.

- Implications for Historical Methodology
- Handling Source Nuance, Ambiguity, and Bias
- Epistemological Assumptions
- Alignment with Core Concerns in HPSS
- The "Messiness" of Integrating Computational Tools
- Brief Summary

2. Content Synthesis:

- For each thematic heading (the first five), provide a substantive synthesis in prose (< 200 words recommended). In your paragraph(s), first identify Consensus points, then highlight Distinctive perspectives.
- For the final "Brief Summary" heading, provide a single, more detailed paragraph distilling the core critique (< 100 words).
- Include short illustrative quotations (< 25 words) where helpful.

3. Citation Requirements (Strict):

- Every claim—quoted or paraphrased—must carry an in-text citation in this form: (Model-Name).
- For multiple models, separate names with semicolons, e.g., (Grok-4; O3-Pro; DeepSeek-RI).
- Use these keys exactly: Gemini-2.5-Pro, Claude-Opus-4, Grok-4, DeepSeek-R1, O3-Pro, Qwen3-235B, Magistral-Medium, Llama-4-Maverick.

4. Style and Limits:

- Write in a scientific style adhering to the principle: "Be critical. Be concise. No over-hype. No bloat."
- Avoid first-person narration.
- Do not introduce information or opinions not present in the sources.

Keep the total length under ∼1100 words.

LLM-Response:

Summary

Implications for Historical Methodology

A broad consensus across the analyses posits that the pipeline's core methodological implication is a trade-off that exchanges "narrative nuance for operationalization" (Gemini-2.5-Pro; Claude-Opus-4). This process of "methodological flattening" (DeepSeek-R1; Qwen3-235B) inherently privileges quantifiable, relational facts over ambiguous or contextual meaning, risking the reinforcement of a positivist bias in historical inquiry (Grok-4; Qwen3-235B). The models agree that while the shift to a data-driven, ontology-last approach is an intelligent adaptation, the very act of structuring sources into subject-predicate-object triples imposes a framework that may not align with the historical context (Claude-Opus-4).

Distinctive critiques further refine this point. The primary scholarly act is seen as shifting from interpretive reading to the design and validation of an "extraction and structuring apparatus" (Gemini-2.5-Pro). This introduces a new methodological risk: the research agenda can be subtly shifted from the historian to the model, particularly when using LLM-generated "competency questions" to guide the inquiry, which may naturalise accidental patterns in the training data (O3-Pro). The loss of nuance is thus not merely an acceptable trade-off but a "historical problem to be theorised" in its own right (O3-Pro).

Handling Source Nuance, Ambiguity, and Bias

The analyses concur that while the project team demonstrates awareness of surface-level ambiguities like name variants and OCR errors, its handling of deeper, systemic biases is underdeveloped (Grok-4; DeepSeek-R1). A central point of consensus is that the pipeline risks laundering or reifying the inherent biases of the source material—such as editorial selection or gender representation—by converting them into structured, seemingly objective data (Gemini-2.5-Pro; Qwen3-235B). The strategy of linking to authority files like Wikidata is identified as particularly problematic, as it imports contemporary, often Western-centric, categorisations and their own latent biases into historical contexts (DeepSeek-R1; Grok-4).

Unique perspectives sharpen this critique. One analysis argues that nuance, such as the "diffuse" nature of an achievement, is treated as "noise to suppress" rather than as historically significant data (Qwen3-235B). Another notes that ambiguity is not embraced as a feature of the source but is instead "forced to be determinate" by the extraction process, which atomises sentences and strips them of their rhetorical function (O3-Pro). The most critical missing step is identified as a robust analysis of how source biases and model biases interact and amplify each other, a dynamic that the current technical mitigations do not adequately address (Gemini-2.5-Pro).

Epistemological Assumptions

There is a strong consensus that the project is grounded in a "structuralist epistemology" (Gemini-2.5-Pro; Grok-4) or a form of "structuralist positivism" (Qwen3-235B). This stance assumes that historical knowledge is structurally extractable and that relational patterns revealed in graphs constitute meaningful insight. This approach is seen to carry a "latent positivism" (Claude-Opus-4) and a degree of "techno-optimism" (Grok-4), where language is treated primarily as a carrier of information rather than as a form of social action (O3-Pro). The pipeline's design, therefore, implicitly de-prioritises forms of knowledge that are not easily structured or quantified (Gemini-2.5-Pro).

Distinctive critiques elaborate on this foundation. One analysis frames the core epistemological problem as a risk that "what can be easily structured" becomes a proxy for "what is important to know" (Gemini-2.5-Pro). Another challenges the iterative refinement process as implying a "quasi-asymptotic notion of truth", which is at odds with constructivist understandings of how facts are produced within situated practices (O3-Pro). The claim of a "data-driven" ontology is also challenged as illusory, since the LLM prompts used in the initial extraction stage already impose implicit, non-neutral categories on the source material (Qwen3-235B).

Alignment with Core Concerns in HPSS

The analyses generally agree that the project's stated goals—exploring socio-epistemic networks and knowledge circulation—align with established HPSS interests (Grok-4; Claude-Opus-4). The reflexive documentation of the research process is also seen as a positive contribution to the field (O3-Pro). However, this alignment is consistently described as partial or incomplete. The consensus is that the project's methods do not sufficiently engage with critical HPSS concerns regarding power asymmetries, the sociology of knowledge tools, or post-colonial critiques of data and archives (Grok-4; DeepSeek-R1; O3-Pro). Divergent points specify this gap. One critique warns that the focus on formal, documented affiliations may inadvertently reinforce "great-man narratives" by overlooking the informal networks and practices central to STS scholarship (Claude-Opus-4). Another argues that the project reduces the practice of source criticism to "accuracy validation", failing to interrogate how the sources themselves function as ideological or knowledge-making artifacts (Qwen3-235B). A more profound HPSS framing, it is suggested, would use the pipeline not merely to extract data from sources but to ask critical questions about the construction of those sources (Gemini-2.5-Pro).

The "Messiness" of Integrating Computational Tools

All analyses commend the interview format for successfully capturing the technical "messiness" of the research process, including the iterative design pivots, the "craft" of prompt engineering, and the pragmatic negotiation of trade-offs between cost, accuracy, and control (Gemini-2.5-Pro; Grok-4). This transparent documentation of a non-linear workflow is considered a valuable contribution to the digital humanities community (Gemini-2.5-Pro; O3-Pro).

However, several of the more critical analyses argue that this documented messiness is largely confined to technical implementation, while the deeper epistemological messiness is "somewhat sanitized" or glossed over (Grok-4; Qwen3-235B). A distinctive critique is that the narrative presents an underlying belief in technical solutions for what are "fundamentally interpretive problems" (Claude-Opus-4). The pipeline's modularity is also seen as masking how each stage embeds its own non-neutral assumptions, which are not subject to the same level of critical reflection as the final output (Qwen3-235B). The real messiness—how computational constraints actively reshape the historical questions being asked—remains under-examined (Qwen3-235B).

Brief Summary

The collective critique finds the project to be a methodologically transparent and technically sophisticated effort that nonetheless remains epistemologically shallow. Its core weakness, identified across the analyses, is the treatment of fundamental historiographical challenges as technical problems amenable to computational solutions. The pipeline operationalises historical inquiry by trading narrative depth for quantifiable structure, a process that risks reifying source biases and flattening interpretive complexity. While the Human-in-the-Loop framework is praised for managing surface-level errors, it is seen as insufficient for confronting how the computational apparatus itself co-produces historical meaning. The project thus serves as a valuable case study, but one that highlights the persistent gap between computational capability and deep historical critique.

Technical engineering perspective analysis

In the second round of the sparring process, the LLMs were prompted to take the role of a senior AI/ML engineer. This perspective shifts the focus of the critique towards the pipeline's technical architecture, scalability, and maintainability. The aim here is to "stress-test" the functionality and design choices from a practical implementation standpoint, providing a complementary, engineering-focused evaluation. As with the HPSS perspective, the outputs from multiple models were synthesized into a meta-commentary.

Prompt Used: Technical Engineering Perspective

Critically analyse the preceding interview transcript and introduction from the perspective of a Senior AI/ML Engineer specializing in Natural Language Processing and Digital Humanities. Focus on the following aspects:

- Evaluation of the technical design choices mentioned. Potential optimisations or alternative approaches.
- Assessment of the task decomposition strategy.
- Effectiveness and implementation details of the Human-in-the-Loop (HiL) components from a systems view.
- Considerations regarding scalability, modularity, and maintainability.
- How the described process aligns with the known capabilities and challenges of LLMs in scholarly workflows.

Be critical. Be concise. Don't overhype. No bloat.

LLM-Response:

Summary

Evaluation of the technical design choices mentioned. Potential optimisations or alternative approaches.

There is a broad consensus that the two-stage architecture—commencing with ontology-agnostic Open Information Extraction (OIE) before imposing a schema—is a methodologically sound and pragmatic design choice (Gemini-2.5-Pro; Grok-4; O3-Pro; Claude-Opus-4). This approach grounds the final knowledge graph in the source material's content rather than researcher preconceptions (Gemini-2.5-Pro). Similarly, the use of an "LLM-as-a-judge" for ensemble validation is recognised as a common pattern for improving quality, though it is criticised for increasing cost and complexity (Gemini-2.5-Pro; Grok-4; DeepSeek-R1; Qwen3-235B). The reliance on proprietary models is consistently identified as a primary weakness that compromises reproducibility and inflates costs (Grok-4; O3-Pro; Claude-Opus-4; Qwen3-235B).

Distinctive critiques and optimisations were offered. Several models suggest that fine-tuning smaller, open-source models on domain-specific data would be more efficient and cost-effective than relying on prompt engineering with large, general-purpose APIs (Grok-4; O3-Pro). One analysis proposes that modern "closed-IE" models could potentially render the second mapping stage redundant by extracting ontology-ready triples directly (O3-Pro). The entity linking strategy was also scrutinised, with one critique dismissing the character-embedding-based blocking as "over-engineered" for historical text and suggesting sourcespecific gazetteers as a more practical alternative (Qwen3-235B). Finally, the claim of using "multimodal capabilities" was deemed an overstatement for what is essentially a standard document processing task (Gemini-2.5-Pro).

Assessment of the task decomposition strategy.

The "radical decomposition" of the workflow into granular, single-purpose steps is praised across the board as an exemplary design principle (Gemini-2.5-Pro; Grok-4; Llama-4-Maverick; Qwen3-235B). This modularity is seen as a core strength that enhances debuggability, allows for targeted model selection, and represents a logical evolution from the initial, less successful monolithic approach (Grok-4; Claude-Opus-4). The strategy of focusing on "sentence-level extraction with paragraph context" is noted as a fundamental and effective NLP practice (Claude-Opus-4).

However, this decomposition is not without its flaws. A distinctive critique identifies a "critical flaw" of circular validation, where Competency Questions (CQs) derived from OIE output may fail to reflect genuine research needs if the initial extraction was incomplete (Qwen3-235B). Other analyses warn of the high risk of error propagation and increased latency across the long chain of tasks (O3-Pro; Claude-Opus-4). From a systems perspective, the decomposition lacks a formal orchestration layer (e.g., Airflow) to manage dependencies (Grok-4; O3-Pro), and one model warns of accruing "integration debt" without a strict, version-controlled data schema to define the contracts between the discrete stages (Gemini-2.5-Pro).

Effectiveness and implementation details of the Human-in-the-Loop (HiL) components from a systems view.

The models concur that the integration of Human-in-the-Loop (HiL) is essential and well-conceived, correctly focusing expert validation on high-leverage, epistemologically critical tasks such as refining the ontology and Competency Questions (Gemini- 2.5-Pro; Grok-4; Llama-4-Maverick; O3-Pro). Despite this conceptual strength, the implementation is uniformly assessed as ad-hoc, synchronous, and a significant scalability bottleneck that relies on unstructured expert availability (Gemini-2.5-Pro; Grok-4; DeepSeek-R1; Qwen3-235B).

A strong consensus emerged around a key distinctive critique: the failure to implement active learning loops. Such loops would prioritise human review on the most uncertain or ambiguous LLM outputs, thereby systematically reducing the expert workload (DeepSeek-R1; Claude-Opus-4; O3-Pro; Qwen3-235B; Grok-4). Another pointed critique is the missed opportunity to use the valuable human feedback to fine-tune the models, which means the system augments the human but does not "learn from them systemically" (Gemini-2.5-Pro). The lack of structured protocols, such as inter-annotator agreement metrics or dedicated annotation UIs, was also highlighted (Grok-4). One analysis aptly re-frames the process as "LLM-in-the-Loop", a term that better captures how human expertise remains the ultimate driver of the workflow (Grok-4).

Considerations regarding scalability, modularity, and maintainability.

There is a clear consensus that while modularity is the pipeline's greatest strength, scalability is its most significant weakness (Gemini-2.5-Pro; Grok-4; Llama-4-Maverick; Magistral-Medium). The modular design allows for individual components to be updated or replaced as technology evolves (Gemini-2.5-Pro). However, the current implementation is deemed unscalable due to the combination of high proprietary API costs, long processing times, and the non-scalable dependency on expert time for HiL validation (Gemini-2.5-Pro; Grok-4; Claude-Opus-4; Qwen3-235B). The practice of logging metadata is considered a necessary but insufficient step toward true maintainability and reproducibility (Gemini-2.5-Pro; Grok-4; Claude-Opus-4).

Distinctive critiques focus on the lack of MLOps rigor. The description of prompt engineering as a "craft" is heavily criticised as unmaintainable, with multiple models calling for systematic prompt versioning and automated testing frameworks (Gemini-2.5-Pro; Qwen3-235B; O3-Pro). The claim of achieving "partial reproducibility" is pointedly dismissed as "inadequate" and "reproducibility theater" given the use of non-deterministic, closed-source models (Claude-Opus-4). To achieve genuine modularity, a formal contract, such as a versioned I/O schema, is required between each component (O3-Pro).

How the described process aligns with the known capabilities and challenges of LLMs in scholarly workflows.

The analyses agree that the project aligns well with emerging best practices for using LLMs in the humanities: as a tool for augmentation and exploration, not blind automation (Gemini-2.5-Pro; Grok-4; Llama-4-Maverick). The team's thoughtful reflection on the trade-off between "narrative nuance" and "operationalization" is praised as astute and central to digital humanities discourse (Gemini-2.5-Pro). The pipeline's architecture correctly acknowledges and attempts to mitigate known LLM challenges like hallucination through its extensive HiL components (Grok-4; Magistral-Medium).

However, several distinctive critiques argue the approach underestimates key LLM limitations. One sharp analysis contends that the "ontology-agnostic" OIE stage is misleading because it inevitably "embeds the prompt designer's ontology" (Qwen3-235B). Others note that while the team reflects on source and model bias, the pipeline itself lacks technical components to actively measure, audit, or mitigate it (Gemini-2.5-Pro; O3-Pro). A "critical gap" is the absence of confidence scoring for LLM outputs, which would provide more transparent error analysis (DeepSeek-R1). Finally, the risk of LLMs "correcting" or modernizing historical language is highlighted as a significant challenge that erases crucial period-specific nuance (Grok-4; Qwen3-235B).

Brief Summary

The collective analysis assesses the described pipeline as a methodologically thoughtful and pragmatically designed research prototype that successfully integrates scholarly rigour with LLM-driven automation. Its core strengths lie in its modular, two-stage architecture and its principled centering of human expertise. However, from a systems engineering perspective, the project is critically flawed by significant operational weaknesses. The ad-hoc implementation of Human-in-the-Loop components creates an unscalable bottleneck, while the heavy reliance on costly, non-deterministic proprietary APIs undermines reproducibility and long-term viability. The pipeline lacks standard MLOps practices such as active learning, prompt versioning, and automated regression testing. Consequently, it functions more as a well-documented proof-of-concept for an exploratory process than as a robust, scalable, or maintainable piece of scholarly infrastructure.

Conclusion

This contribution sought to document the development of a knowledge graph extraction pipeline through an experimental format: an expert interview complemented by LLM-generated commentary. The approach highlights tensions that are fundamental to computational humanities work in general.

Our pipeline emerged from a straightforward goal, namely to unlock the structural information latent in biographical dictionaries. The resulting two-stage architecture attempts to balance data-driven discovery with research-question-driven structuring, incorporating Human-in-the-Loop steps that reflect our conviction that scholarly decision-making requires human supervision and/or intervention. The LLM commentaries highlighted fundamental tensions embedded within the proposed pipeline, such as narrative nuance vs. operationalization, interpretive flexibility vs. scalable comparability, and reproducibility vs. practical feasibility, among others.

In this sense, the experimental use of LLMs as disciplinary "sparring partners" proved fruitful. Their critiques targeted several supposedly undervalued/underdeveloped points or blind spots. A Whiggish tendency in our technological framing that implied inevitable progress through a focus on structure and automation, the circularity between structuring data and presuming a structuralist epistemology, or the risk of naturalizing design decisions that are, in fact, contingent. They also questioned the privileging of disambiguation over ambiguity and how our claims to transparency clashed with the black-box nature of the AI systems used. However, these analyses also exposed the models' limitations, specifically their context quality-dependence. For instance, the epistemic tension between narrative nuance and operationalization, a well known debate in Historical Network Research that we hinted at in the interview, was treated by the models as a blind spot rather than a deliberate design choice. A human reviewer might have weighted this differently while possibly overlooking other issues. This mirrors the potential and risk of these tools: they sometimes can augment scholarly analysis well while subtly reshaping the focus of our critique in other cases.

Three insights, while not new, were highlighted by this exercise. First, the "messiness" of integrating computational tools in humanities research is not merely technical but fundamentally epistemological, each design decision embeds assumptions about what historical knowledge is and how it can be represented. Second, transparency about this process matters since documenting not only what worked but also what failed, as well as the potential solution and its iterations, reveals the interpretive work that computational methods often obscure. Third, the structuralist epistemology inherent in knowledge graph approaches is not a bug but a feature, one that demands theoretical framing rather than technical "workarounds".

This experiment also suggests that the focus of tool-building in computational humanities lies not only in mitigating the tension between technical capabilities and interpretive nuance, but in rendering that tension productive by making the underlying negotiation processes transparent. The pipeline extracts structures that enable new forms of analysis, but only by imposing frameworks that themselves require critical reflection. In this sense, our tools and their theoretical framing are active participants in the construction of historical knowledge, a role that, as this experimental documentation might show, benefits from both human reflection and machinic provocation.[8]

References

Dellermann D, Calma A, Lipusch N et al. (2019) The future of human-AI collaboration: A taxonomy of design knowledge for hybrid intelligence systems. In: *Proceedings (HICSS 2019)*. DOI: 10.24251/HICSS.2019.034.

Boulanger C (2026) The potential of LLMs for constructing a socio-legal knowledge graph. In: Simons A, Wüthrich A, Zichert M, et al. (eds) *Understanding Science with Large Language Models? Potentials for the History, Philosophy, and Sociology of Science*. Bielefeld: transcript, part-4.

Erickson F (1985) *Qualitative Methods in Research on Teaching*. East Lansing, MI: Institute for Research on Teaching, Michigan State University. (Occasional Paper, Institute for Research on Teaching, Michigan State University).

Herrmann T (2025) AI as a sparring partner – An HCAI approach to promote human capabilities. In: Degen H and Ntoa S (eds) *Artificial Intelligence in HCI*. Cham: Springer Nature Switzerland, pp. 162–176. DOI: 10.1007/978-3-031-93415-5_10.

Kasneci E, Sessler K, Küchemann S et al. (2023) ChatGPT for good? On opportunities and challenges of large language models for education. *Learning and Individual Differences* 103: 102274. DOI: 10.1016/j.lindif.2023.102274.

Kaye A, Schlattmann R, Vogl M, et al. (2024) Socio-epistemic networks: A framework for history of knowledge. In: Lerg C et al. (eds) *Experimental spaces: knowledge production and its environments in the long nineteenth century*. Berlin: De Gruyter Oldenbourg. pp. 253–274. DOI: 10.1515/9783111291383-012.

8 This chapter was written with support from large language models (LLMs). All model-generated text was reviewed and, where necessary, rewritten by the authors, who remain fully responsible for the final version. For details on the use of LLMs in this volume, see the statement in the volume's introduction.

Kommineni VK, König-Ries B and Samuel S (2024a) From human experts to machines: An LLM-supported approach to ontology and knowledge graph construction. Preprint. DOI: 10.48550/arXiv.2403.08345.

Kommineni VK, König-Ries B and Samuel S (2024b) Towards the automation of knowledge graph construction using large language models. In: *Proceedings of the 3rd NLP4KGC International Workshop on Natural Language Processing for Knowledge Graph Creation*, pp. 19–34. Available at: https://ceur-ws.org/Vol-3874/paper2.pdf (accessed 29 October 2025).

Lang S (2026) Critical concerns for using LLMs in the (computational) humanities and beyond. In: Simons A, Wüthrich A, Zichert M, et al. (eds) *Understanding Science with Large Language Models? Potentials for the History, Philosophy, and Sociology of Science*. Bielefeld: transcript, part-1.

Meding H and Daugs A (2026) On the use and limitations of large language models in historical scholarship. In: Simons A, Wüthrich A, Zichert M, et al. (eds) *Understanding Science with Large Language Models? Potentials for the History, Philosophy, and Sociology of Science*. Bielefeld: transcript, part-1.

Müller-Enbergs H, Wielgohs J, Hoffmann D et al. (2010) *Wer war wer in der DDR? Ein Lexikon ostdeutscher Biographien*. 5th ed. Berlin: Links.

Neo4j Labs (2025) *LLM Graph Builder*. Available at: https://github.com/neo4j-labs/llm-graph-builder (accessed 29 October 2025).

Shneiderman B (2022) *Human-Centered AI*. Oxford and New York: Oxford University Press.

Wagner A and Hermes J (2026) Encoded humanities, or: not everything has to be generative. A dialogue on AI tasks and roles. In: Simons A, Wüthrich A, Zichert M, et al. (eds) *Understanding Science with Large Language Models? Potentials for the History, Philosophy, and Sociology of Science*. Bielefeld: transcript, part-1.

Weber H and Herbst A (2013) *Deutsche Kommunisten: biographisches Handbuch 1918 bis 1945*. Berlin: Dietz.

Weiß L, Von Welczeck L, Vogl M et al. (2025) Past, present, and future of HNR: Reflections on the practices and methods in historical network research based on a quantitative survey. *Journal of Historical Network Research* 11(1). DOI: 10.25517/JHNR.V11I1.103.

Zieliński S (1933) *Mały słownik pionierów polskich kolonialnych i morskich: podróżnicy, odkrywcy, zdobywcy, badacze, eksploratorzy, emigranci – pamiętnikarze, działacze i pisarze migracyjni*. Warszawa: Instytut Wydawniczy Ligi Morskiej i Kolonialnej.

Large language models in interdisciplinary research settings

A reflection

Malte Vogl, Alexander von Schwerin, and Sabrina Kirschke

1. Introduction

In the research project "Analysis of narratives in the genetic engineering discourse"[1], an interdisciplinary group of researchers with backgrounds in history (of science), political science, and digital humanities jointly analyzed the narratives of stakeholders in the German discourse on genetic engineering over a period of forty years. The challenge of the research project was to extend the established method of narrative analysis to a long period of time and to compare several points in time. For this aim, a large collection of documents, such as statements, press releases, and position papers, was collected and cataloged. The cataloged documents were analyzed using qualitative methods of empirical social research. In order to complement the qualitative analysis, a topic analysis workflow was developed that included a Large Language Model (LLM) to produce "meaningful" names for the detected topics. In this contribution, we want to reflect our experiences – the necessary translation processes, constructive friction, and critical points from the different backgrounds – in the usage of that method. We see the LLM as a connecting center for our interdisciplinary research, which led to new insights in the corresponding disciplines. However, we also challenge the "black box" that is introduced by the usage of the LLM.

Genetic engineering (GE) is an increasingly prevalent, yet controversial topic. From a social and political science perspective, there is an increasing amount of literature on the discourse of GE in different countries. However, temporal comparison may be key in understanding the role of narratives and respective contextual factors in policy-related discourse and the development of GE policies. Against this background, our study provided a historical narrative approach and analyzed how different stakeholder groups

1 https://www.museumfuernaturkunde.berlin/de/forschung/analyse-von-narrativen-im-gentechnikdiskurs

have used narratives at different times between 1985 and 2024 in Germany. We implemented the narrative analysis using a multi-step qualitative text analysis based on the Narrative Policy Framework, which is common in political science. Based on the current discourse, we have identified eight narratives. Of these, we selected one narrative and examined the extent to which the various stakeholders used this narrative and how it changed or did not change over time, focusing on three different time phases in the overall study period. The results show that certain stakeholder groups used the narrative over the entire period, but that certain features of the narrative changed in the process.[2] We assume that the value and goal framework of the stakeholders is expressed in the continuity and that changes in the context cause changes in the narrative. However, these assumptions have not yet been investigated further.

We developed the topic analysis in parallel to the main qualitative work in order to test the extent to which a computer-based topic analysis can support a conventional text evaluation or can be seen as an alternative research approach. The need to develop alternative methods arises in particular from the fact that the comparison of different points in time and stakeholder groups greatly increase the complexity of conventional text analysis.

Our article focuses on the development and application of the topic analysis in this research context, referring to the interaction between developer and users. By "developer", we refer to MV as the DH specialist in our project; by "users", we refer to AS (historian) and SK (political scientist), who expected topic modeling to assist them in narrative analysis.

2. Data and computation

As stated above, the source material for the research project was heterogeneous. AS collected data from online sources and archives into a Zotero database. Approximately 1200 documents of varying lengths (from several lines to several pages) have been collected in this way. One entry corresponded to one document. Each entry got tagged with a controlled vocabulary concerning the group that authored the material (e.g., governmental, industry and business, science, and non-governmental organization), which general values the group represented (e.g., ecologically aligned, industry-related, technological-innovative etc.) and also the position towards genetic research and its applications. Importantly, each entry also had a date and we had access to the full text of the document. With the help of Aron Marquart from University of Leipzig a pipeline that runs optical character recognition on all collected sources was developed. To get an overview picture of the material, MV developed a pipeline to run topic modelling that made use of the BERTopic package written in Python, specifically the module TextGeneration (Grootendorst 2025).

In the pipeline, several steps are based on machine learning (ML) models, e.g., the embedding of sentences. However, in its standard configuration the pipeline returned topics simply as lists of a fixed number of words. Depending on the selected parameters

2 See (Von Schwerin et al 2025) for details of the research context.

the algorithm would potentially output more than a hundred of such lists. To verify the validity of the result, an expert of the source material would need to look over the lists and give an interpretation of what they would entail. This exact step was then replaced by the LLM in a subsequent version of the software.

The prompt that is provided by the BERTopic tool gathers the mentioned lists of words with a random selection of the source text material together with an example of such a question and the desired answer. With the aim of reproducible research a number of open LLMs were tested with larger models strongly improving the results. For testing, LLAMA 3–8 was sufficient. To produce research output, the pipeline was run with LLAMA 3–70B on a local workstation. The Python code closely follows an example by the BERTopic package (Vogl 2024 , Grootendorst 2025a). The relevant prompt[3] is constructed for each topic by randomly selecting 10 texts from the found topic together with the returned words describing the topic. The LLM is tasked with a system prompt to be an "honest assistant for labeling topics" and the task is described in an example prompt. The final part of the prompt is then the task to create a short label of the topic in a selected language and return only this label. Since the locally available GPU only provides 48GB of RAM, the process is effectively running on CPU and 512GB of RAM. The number of topics that has to be processed by the LLM depends on parameters of the BERTopic pipeline, which greatly influences the runtime. For fine-grained visualizations, the computation can be up to 12 hours and more.

One challenge in the adaptation of this process with our source material lies in the relative length of the material. The prompt provided by the package apparently was designed for rather short text snippets. The material for this research project ranges from 0,5 to more than 30 pages and has diverse levels of structuring. To overcome this, a simple cutting approach was chosen, that clipped all text into pieces of a fixed word length. While this potentially cuts the texts in the middle of an argumentative structure, each text still has potentially several topics assigned to it, as is usually the case in topic modeling. The rationale behind this approach is that the large number of used text parts will still allow a meaningful picture of the evolving debates in the corpus. An additional challenge could have been the language of the corpus which is mostly in German. However, the different ML parts of the BERTopic package provide multilingual versions and the prompt was slightly reformulated to include the information which language the source material has and the returned topics should have.

3. (Limits of) interdisciplinary translation work

The discussion between the developer (MV) and the users (AS, SK) led to a number of modifications to the original computer-based approach. Initially, the changes consisted of the basic adaptation of the computational tools to the research project and required some development work, as described in section 2. In addition, problems with respect to the LLM data processing arose during the joint exchange that were less easy to solve.

3 The prompt for the LLAMA LLM is based on https://maartengr.github.io/BERTopic/getting_started/representation/llm#prompt-engineering

After an initial development phase, the software gave results that seemed convincing to the developer MV. However, the communication about the gathered results opened another challenge to the team. How could one talk about the resulting document classifications and relations in the corpus? Luckily, an additional external software package, datamapplot (McInnes 2025), allowed the use of BERTopic output for interactive visualizations, see Figure 1. In the first adaptation, the text snippets (publications or parts of publications) were visualized in the form of dots with a color that corresponds to the found topic and the LLM-generated topic name was displayed.

Figure 1: Screenshot of the topic modeling result with LLM topic names.

The figure shows the first adaptation (visualization of topics and topic labels) in a zoomed out view for text parts selected between 2018 and 2024 (color online).

While maybe visually pleasing, the analysis from the user's perspective was still challenging for several reasons: How could one compare the identified topics and labels with the source material? How could one identify the temporal aspects of the topic without checking all text snippets (dots)? What actors in the debate used which topics as part of their narrative? This feedback led to a second adaptation of the interactive visualization which incorporated a histogram of the times for which documents exist, a hover menu with the classification and text snippet of the text document, a selection menu to limit the visualization to any time period, and a selection menu by actor group and attitudes, see Figure 2.

Figure 2: Screenshots of the topic modeling result with LLM topic labels.

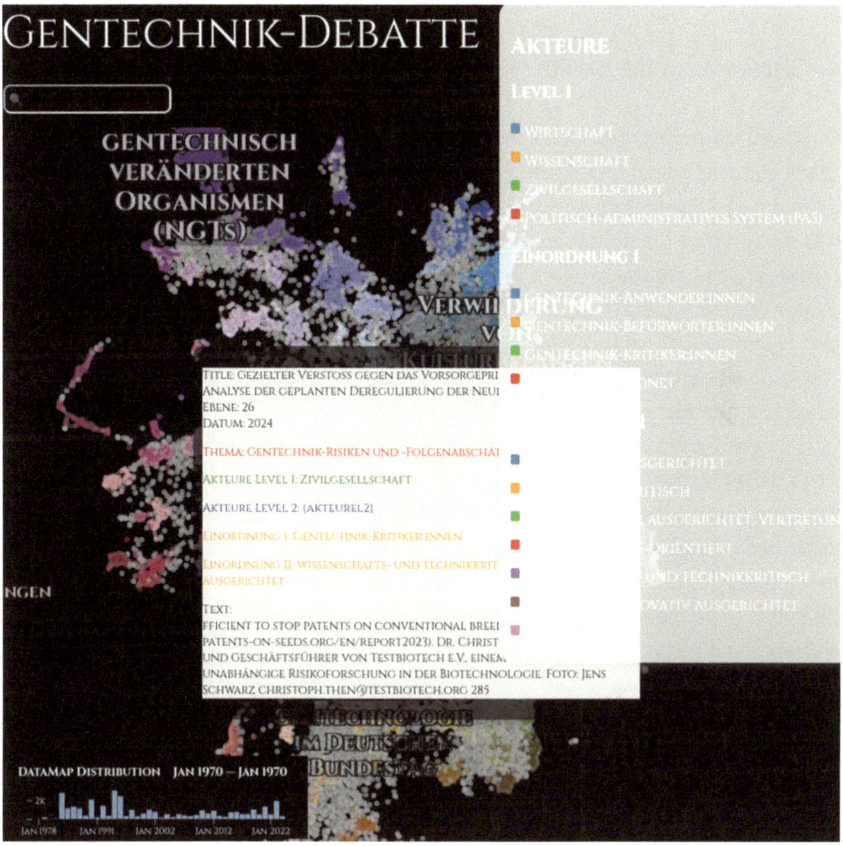

The figure shows the second adaptation (integration of time and document text information and selection modi) displaying information about a certain text, the timeline overview and selection menu and the selection menus for the whole corpus (color online).

Once the software tool had been developed, the users (AS, SK) had further questions. Their wishes were to adapt the results of the topic analysis and the topic labels even more closely to the requirements of the narrative analysis and to manipulate the calculation methods implemented in the LLM accordingly. For example, they noted that the topic names were too general; instead, they wanted the topic names to express the content preferences and values of the actors even more strongly. The developer then explained to the users the various settings available to him to change the results of the LLM and the prompt in the visualization.

During the joint discussion, it became clear that it was very difficult to explain the LLM calculation methods to the users. However, it also became clear that neither the developer nor the users were able to clarify exactly what changing the adjusting screws meant for the validity of the results. It also remained unclear how far one could go to ultimately make the results fit in the perspective of the users, i.e. where the limit of inadmissible manipulation of the data lay.

4. Discussion of the results

Our research process revealed some mixed experiences regarding the combination of LLMs and traditional qualitative research. On the one hand, our collaboration highlighted well-known challenges of interdisciplinary cooperation. In our case, the focus was on applying the digital humanities (DH) to a historical-political science issue, by translating the requirements of a methodology in the historical and political science (historical policy narrative analysis) into a DH methodology. If the relevant competencies are not combined in a single person, there is automatically an increased need for communication between the researchers in order to establish a mutual understanding of approaches, methods, their possibilities and limitations.

However, given the high amount of trust between the researchers, it was not necessarily a disadvantage that the DH methodology was a black box for the users, i.e., that the users only understood the DH methodology to a limited extent. Also, the researchers dedicated enough time to discuss open questions. In fact, the adaptation of the DH methodology to the historical-political science requirements took place in several steps. And even after the successful joint development of the software tool in its individual elements and processing steps, there was a need for further adaptation on the part of the users.

These post-development adjustments described above can be categorized as a normal procedure when handling statistical procedures that can be classified as a form of calibration. Processes of empirical social research often involve such interactive and, in principle, incomplete adaptation of tools to the research question. One example is the calibration of the calculation methodology for cluster analyses.

On the other hand, we faced limitations in our collaborative research process that go beyond these well-known interdisciplinary challenges. Calibrating the finalized computational tool to the requirements of the historical and political science research question would have required an interdisciplinary understanding of the exact relationship between the DH computational method, outputs, and the respective research questions. In other words, the black box would have had to be opened up. However, a mutual understanding could not be easily established in the exchange because such an understanding required a deeper competence in the other discipline – in our case, an understanding of the users of the structure of LLMs.

Further, and contrary to more established statistical or DH methods, the use of LLMs in interdisciplinary research settings leads to three main LLM-specific challenges.

First, most disciplines in the humanities and social sciences have not yet established what should be considered best-practice in the use of LLMs, leading to a certain lack of trust in such methods. Second, and related to the first point, few if any universities offer introductory courses on the statistical background of LLMs and ML in general. Contrasting this, e.g., with courses such as quantitative sociology or foundations of digital humanities, shows again the lack of established usage patterns. The third and final challenge is the more fundamental question of reproducibility. On the one hand, LLMs are designed to give statistical answers, i.e. not necessarily repeat the same answer to the same question. On the other hand, slight changes in the prompt formulation can give significantly different results. It remains unclear how one should weigh the error of different topic representations or what error bars should follow for different formulations of a prompt. Together, these challenges make using LLMs in an interdisciplinary setting an "extra-dark" black box – for both, the users AND the developers.

5. Conclusion for future cooperation

It is important to manage expectations as to what the tool can and cannot achieve: The identification of topics is exciting, but is limited in its informative value. In our case, the tool was very helpful in identifying which topics are taken up by which actor. However, when it came to questions of attitudes of the actors and the identification of narratives, the informative value was limited. For example, a specific term could be taken up but evaluated very differently by different types of actors.

The use of LLMs is associated with a subject-specific language that is not immediately comprehensible to qualitative researchers. This leads to communicative challenges in the research process, which can also be addressed if sufficient time is provided. This also includes building trust in the methods and in the researchers applying the respective methods.

As an outsider, it is not possible to understand how the tools come to specific results. This is also a problem in qualitative analysis given the subjectivity of the coder. However, in qualitative research, there are clearly defined processes for increasing the intersubjectivity of the results such as double coding of a minimum of two researchers (Mayring 2014), which are not applicable here, although similar standards exist with respect to LLMs such as pair programming or code review through third parties. The question is therefore whether the results can be recognized as scientifically valid and under what conditions this is the case.

In sum, LLMs can be considered as a helpful compass rather than a quick fix in our field of research related to the identification of narratives of different actor groups and in different moments in time, see also (Purificato et al 2025). A sufficient amount of time is needed for expectation management, traceability of results and communication as it is common in interdisciplinary work, more generally. Future research also has to put special emphasis on developing standards for cross-checks that are acceptable for both the LLM and qualitative researchers involved.

References

Grootendorst M (2025) BERTopic. Available at: https://maartengr.github.io/BERTopic/getting_started/representation/llm (accessed 6 January 2025).

Grootendorst M (2025a) Topic modelling with LLAMA. Available at: https://colab.research.google.com/drive/1QCERSMUjqGetGGujdrvv_6_EeoIcd_9M (accessed 16 September 2025).

Mayring P (2014) Qualitative content analysis: Theoretical foundation, basic procedures and software solution. Klagenfurt: Beltz. Available at: https://nbn-resolving.org/urn:nbn:de:0168-ssoar-395173 (accessed 6 November 2025).

McInnes L (2025) DataMapPlot. Available at: https://datamapplot.readthedocs.io/ (accessed 6 November 2025).

META-LLAMA (2025) Llama 3. Available at: https://huggingface.co/meta-llama/Meta-Llama-3-8B (accessed 16 September 2025).

Purificato E, Bili D, Jungnickel R, Ruiz Serra V, Fabiani J et al. (2025) The Role of Artificial Intelligence in Scientific Research – A Science for Policy, European Perspective. *Publications Office of the European Union.* DOI: 10.2760/7217497.

Vogl M (2024) SemanticLayerTools (1.0.0). Zenodo. DOI: 10.5281/zenodo.14190505.

Von Schwerin A, Kirschke S, Marquardt A, and Vogl M (2025) The narrative of (technological) progress in the German genetic engineering discourse. A temporal comparison approach. *(submitted)*.

When less is more?
Topic model performance across title, abstract, and full-text corpora

Francis Lareau and Christophe Malaterre

1. Introduction

Topic modeling has emerged as an indispensable computational tool for analyzing large volumes of scientific literature, particularly in fields such as history, philosophy, and sociology of science. By automatically identifying thematic patterns within textual data, topic modeling enables researchers to track research trends, identify paradigm shifts, analyze the evolution of scientific vocabulary, and map the interrelations of scientific themes. However, a crucial methodological question remains largely unexplored in the case of article-based corpora: what level of textual granularity—titles, abstracts, or full texts—provides the optimal balance between computational efficiency and analytical depth?

This question has significant practical implications for mining corpora of academic articles. Full-text analysis demands considerable computational resources for acquisition, preprocessing, and analysis, while title or abstract analysis offers a more economical alternative. Yet, the potential information loss when using only titles or abstracts must be weighed against these efficiency gains. Despite the importance of this methodological consideration, systematic comparisons across text structures remain scarce in the literature.

This contribution examines the performance of two prominent topic modeling approaches—Latent Dirichlet Allocation (LDA) and BERTopic—across three text structures (titles, abstracts, and full texts). Through a set of quantitative evaluations, we seek to determine whether the additional resources required for full-text analysis are justified by proportional analytical gains, or if more parsimonious approaches using titles or abstracts might suffice for certain research objectives.

2. Methodological considerations

The comparison centers on two fundamentally different topic modeling approaches: LDA, a classical statistical technique based on word-count vectors and Dirichlet distributions (Blei et al., 2003), and BERTopic, a more recent approach leveraging large language model embeddings (Grootendorst, 2022).

LDA represents documents as bags-of-words and models topics as latent variables following a Dirichlet distribution. Its relatively straightforward implementation has made it the de facto standard in computational text analysis for over a decade. LDA can handle documents of varying lengths, making it suitable for all three text structures under investigation.

BERTopic, conversely, represents a newer generation of topic modeling that capitalizes on contextual embeddings from large language models. Rather than treating topics as latent variables, BERTopic identifies topics through document clustering in the embedding space. Until recently, BERTopic was limited in its ability to handle long texts due to token limitations in transformer models. However, recent innovations in long-text embeddings have expanded BERTopic's applicability to full-text corpora, making this comparative analysis particularly timely.

This study used a corpus of 3,698 full-text articles from three leading astrobiology journals (1974–2023), curated by Malaterre and Lareau (2023). Only documents that include a title, abstract, and full-text were retained, totaling 3,542 documents. After standard preprocessing—tokenization, POS tagging, and lemmatization— stopwords were excluded (Manning, Raghavan and Schütze, 2008). Only nouns, verbs, adverbs, and adjectives were kept, and documents were vectorized into term-document matrices for LDA and BERTopic. LDA modeling was performed using a Python API with a word frequency TDM. For BERTopic, document embeddings were generated from full texts using the high-scoring stella model (stella_en_1.5B_v5) based on Alibaba-NLP (Zhang et al., 2025). The BERTopic pipeline was executed via Python API, utilizing the TDM for word ranking and outlier reassignment. Note that the same protocol was used for titles and abstracts.

3. Comparative analysis framework

To systematically compare model performance across text structures, a multi-faceted evaluation framework is essential. Here, as a start, we examine performance with three complementary metrics:

Topic Diversity: Assesses whether topics within a model are represented by distinct word sets. Since this is the percentage of distinct words among all the top words representing topics, greater diversity suggests more differentiated topics with less semantic overlap.

Joint Recall (mJIR): Measures how effectively a topic's representative words collectively recall the documents assigned to that topic (Lamirel et al., 2025). Higher recall indicates better alignment between topic keywords and document content:

$$mJIR = \frac{1}{|D|}\sum_{c=1}^{K} |\{d \in c | \exists i, i \in Top_W[c] | d[i] \neq 0\}|$$

where W is the number of top words chosen as description of any cluster c, $|D|$ is the number of documents in the corpus, $Top_W[c]$ is the set of the top W words describing topic c, $d[i]$ represents the presence/absence of word i in the document d.

Topic Coherence: Evaluates whether the words describing each topic semantically belong together in a meaningful way. Higher coherence suggests more interpretable topics. We used the coherence CV of Röder et al. (2015).

As a complementary measure, we also calculated the standard deviation of document counts per topic to capture how evenly documents are distributed. Lower values indicate balanced topic representation, while higher values suggest dominance by a few topics. For LDA, although each document has a full topic distribution, we assigned documents to their dominant topic to enable comparability with crisp clustering models such as BERTopic.

These metrics, when considered collectively, provide a first framework for evaluating model performance across different text structures and algorithm choices. Of course, the models should also be evaluated along other dimensions, notably qualitatively (in terms of topic interpretability) but also in how document assignment is balanced across topics. This is work under progress.

4. Preliminary results

The framework consisting of the different metrics was applied to 5 models of approximately 10, 20, 30, 40 and 50 topics, and this for each type of model, either BERTopic or LDA trained on titles, abstracts or full text:

Topic Diversity: Preliminary diversity results show that all models offer a minimum diversity of at least 60%. The best-performing models being BERTtopic trained on titles or full text, offering a diversity of more than 83% (Fig. 1).

Fig. 1: Topic diversity performance comparisons between topic models (10 top-words by topic).

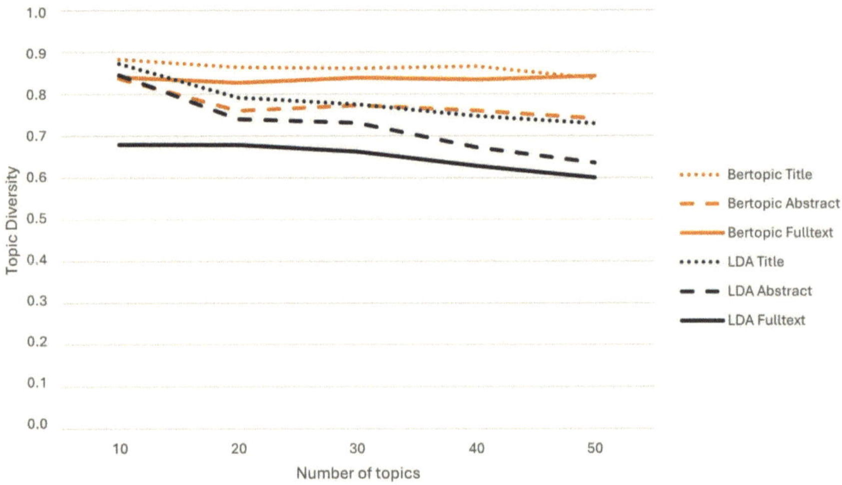

Joint Recall: The preliminary results of joint recall show two clear trends: that LDA models outperform BERTopic models when considering the same document type, and that full text outperforms abstracts, which in turn outperform titles, when considering the same type of algorithm (Fig. 2).

Fig. 2: Joint recall performance comparisons between topic models (10 top-words by topic).

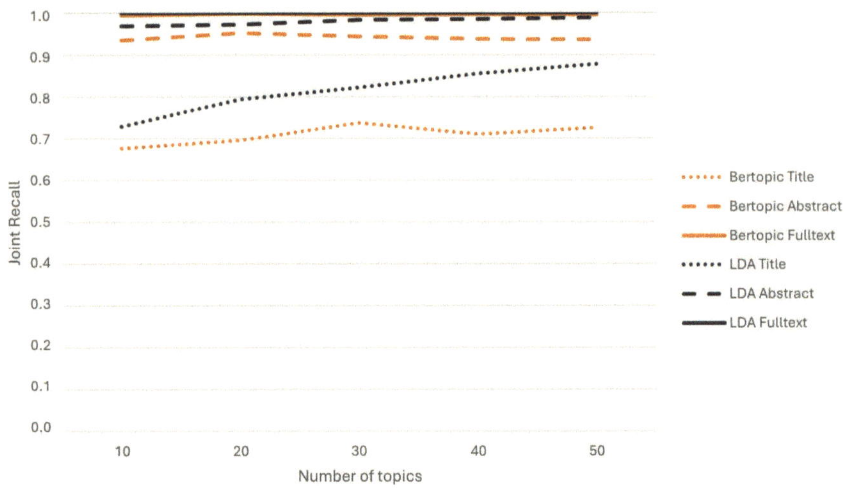

Topic Coherence: Preliminary coherence results show that BERTopic models trained on abstracts and full text perform best, followed by LDA models on the same document structures. Models trained on titles perform worst (Fig. 3).

Fig. 3: Coherence CV performance comparisons between topic models (10 top-words by topic).

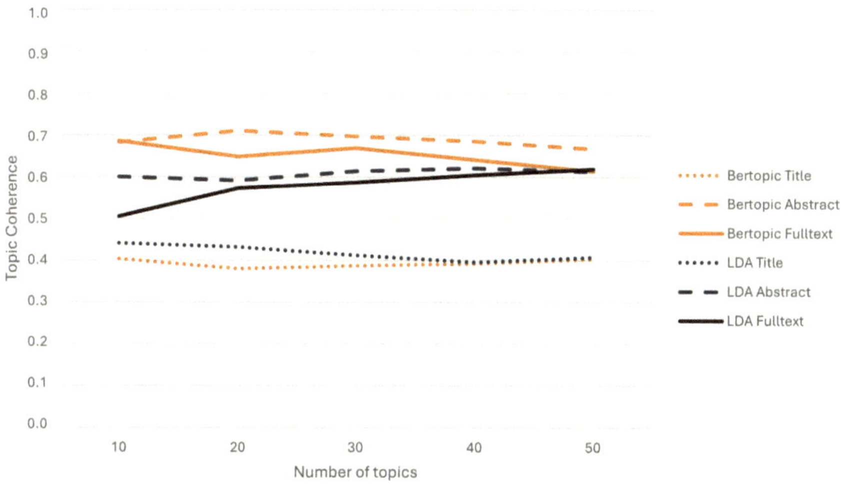

Standard deviation (SD): The dispersion of the documents in the topics varied substantially by model and document type when comparing the *N*=25 topics models. LDA applied to full texts yielded a standard deviation of approximately 52.2, closely aligned with the value obtained from abstracts (SD ≈ 53.4), suggesting; by contrast, LDA on titles produced the lowest dispersion (SD ≈ 28.4). In comparison, BERTopic exhibited markedly higher variability: full texts produced the greatest spread (SD ≈ 116.9), abstracts showed intermediate dispersion (SD ≈ 57.0), and titles yielded unexpectedly high variability (SD ≈ 96.8). This indicates that LDA produces more homogeneous document allocations to topics regardless of text length, and more so than BERTopic which tends to generate less balanced topics in terms of number of documents assigned to them.

5. Key findings

5.1 Title-based models: information scarcity challenge

Topic models based solely on titles consistently underperform across most metrics. Although the models trained on the titles tend to generate topics with high diversity, especially the BERTopic model (Fig. 1), they significantly lag behind their abstract and full-text counterparts for the other two metrics, namely topic coherence and most notably document recall (Fig. 2 and 3). This means that title-based models do not manage to properly cover the entire spectrum of documents.

The primary challenge appears to be information scarcity. Titles, by design, provide only the most succinct representation of document content, often emphasizing novelty or appeal over comprehensive thematic coverage. This limitation manifests in poor coherence scores and problematic document-topic assignments. However, it is noteworthy that despite these limitations, BERTopic title models can still identify topics scoring high on diversity and covering about 70 to 80% of documents (as measured by recall).

5.2 Abstract models: the Goldilocks zone

Abstract-based models consistently demonstrate strong performance across evaluation metrics. The BERTopic abstract model, in particular, achieves the highest coherence scores (Fig. 3), though its performance in terms of diversity and recall is not as good as the BERTopic full-text model (Fig. 1 and 2). LDA abstracts models consistently offer better coherence and diversity scores compared to their full-text counterparts, though again recall is less satisfactory.

This overall excellent performance may be attributed to the nature of abstracts themselves. As condensed summaries designed to highlight key findings and themes, abstracts offer a focused representation of document content without the noise and methodological details often present in full texts. This balance between information density and signal clarity positions abstracts in a "Goldilocks zone" for topic modeling—providing sufficient information for robust topic identification without the computational and analytical complexities of full-text processing.

5.3 Full-text models: information overload effects

Full-text models demonstrate complex performance patterns. While they achieve strong joint recall scores (Fig. 2)—indicating excellent coverage of document content by topic keywords—they exhibit lower performance than their abstract-based counterparts, especially in terms of topic coherence (for both LDA and BERTopic), but also diversity (for LDA, though not BERTopic).

One possible explanation is the presence of methodological aspects alongside conceptual content in full-text documents: this may indeed blur both the consistency of topics by mixing methodological terms with substantive ones (thereby lowering topic coherence measures) and their diversity (at least in the case of the LDA model), possibly by adding cross-thematic methodological terms. This may suggest an "information overload" effect, where the wealth of details in full texts can complicate the identification of clear thematic boundaries. However, full-text models may be useful to capture nuanced themes that might be absent from title- and abstract-based models, particularly regarding methodological approaches and secondary research themes. Such qualitative aspects need to be further examined.

6. Practical implications

These findings have several practical implications for researchers employing topic modeling in scientific literature analysis:

Research-Goal Alignment: The choice of text structure should align with specific research objectives. For broad thematic mapping, abstract models offer an excellent balance of performance and efficiency. For fine-grained analysis of methodological approaches or secondary themes, full-text models may provide crucial insights despite their computational costs.

Resource Optimization: The strong performance of abstract models suggests that in many cases, researchers can achieve robust results without the substantial computational and preprocessing requirements of full-text analysis. This is particularly valuable for large-scale studies or when working with limited computational resources.

Multi-Level Modeling: Rather than viewing text structures as competing alternatives, researchers might benefit from a multi-level approach that leverages the complementary strengths of different text structures. For instance, abstract models might serve for initial thematic mapping, with full-text analysis reserved for deeper exploration of specific areas of interest.

Model Selection: BERTopic generally outperforms LDA across text structures, particularly for coherence and diversity metrics (though not recall nor standard deviation). This suggests that despite their greater computational requirements, embedding-based approaches offer tangible analytical benefits that may justify their adoption.

Conclusion

The question of whether less is more in topic modeling depends fundamentally on research objectives and resource constraints. Abstract-based models emerge as particularly strong performers, offering a compelling middle ground that captures most major thematic structures without the computational burden of full-text analysis. Title-based models, while computationally lightweight, sacrifice too much information for most analytical purposes. In any case, the different models need to be submitted to further complementary evaluations, notably in terms of topic interpretability (for instance, by examining top-words and top-documents assigned to the models, depending on textual structure), but also in terms of how document assignment is representatively balanced across topics (notably to avoid large generalist topics as well as extremely small overfitted topics). These are aspects that we are currently investigating.

These findings suggest that the field might benefit from developing hybrid approaches that intelligently combine information from different text structures, potentially extracting the focused signal from abstracts while selectively incorporating the richer detail available in full texts. As topic modeling continues to evolve with advances in language models and computational techniques, such multi-level approaches may represent the next frontier in scientific literature analysis.[1]

1 This chapter was written with support from large language models (LLMs). All model-generated text was reviewed and, where necessary, rewritten by the authors, who remain fully responsible for the final version. For details on the use of LLMs in this volume, see the statement in the volume's introduction.

Acknowledgments

F.L. acknowledges funding from Canada Social Sciences and Humanities Research Council (Postdoctoral Fellowships 756–2024-0557, Grant 430–2018-00899). C.M. acknowledges funding from Canada Social Sciences and Humanities Research Council (Grant 430–2018-00899) and Canada Research Chairs (CRC-950-230795). The authors thank Arno Simons, Adrian Wüthrich, Michael Zichert and Gerd Graßhoff for organizing the *LLMs in HPS* workshop (TU Berlin, 2025) and inviting them to contribute to this volume.

References

Blei, D M, Ng, A Y and Jordan, M I (2003) Latent dirichlet allocation. *Journal of Machine Learning Research*, 3(Jan): 993–1022.

Grootendorst, M (2022) BERTopic: Neural topic modeling with a class-based TF-IDF procedure. *arXiv preprint* arXiv:2203.05794.

Lamirel, J-C, Lareau, F and Malaterre, C (2025) Old but Not Obsolete: Bag-of-Words vs. Embeddings in Topic Modeling. *Proceedings of the 20th International Conference on Scientometrics and Informetrics*, Jun 2025, Yerevan, Armenia.

Malaterre, C and Lareau, F (2023) The Emergence of Astrobiology: A Topic-Modeling Perspective. *Astrobiology*, 23(5): 496–512. https://doi.org/10.1089/ast.2022.0122

Manning, C D, Raghavan, P and Schütze, H (2008) *Introduction to Information Retrieval*. Cambridge: Cambridge University Press.

Röder, M, Both, A, and Hinneburg, A (2015) Exploring the Space of Topic Coherence Measures, *Proceedings of the 8th ACM International Conference on Web Search and Data Mining*. New York: ACM, 399–408.

Zhang, D, Li, J, Zeng, Z and Fulong, W (2025) Jasper and Stella: distillation of SOTA embedding models. *arXiv preprint* arXiv:2412.19048.

Part 5: Retrieval-augmented generation (RAG)

From RAGs to rich responses

Enhancing LLM reliability through retrieval-augmented generation

Holle Meding and Aurel Daugs

1. Introduction

Retrieval-Augmented Generation (RAG) offers a technical response to a key problem in the use of Large Language Models (LLM) across the humanities: their tendency to hallucinate facts in the absence of verified sources. By supplementing the generative process with targeted retrieval from external resources such as archival corpora, knowledge graphs and document databases, RAG shifts the epistemic burden from the model's internal weights to verifiable information. This article reports on the early-stage implementation and critical evaluation of a RAG pipeline developed at the Chair for Digital History at Humboldt-Universität zu Berlin for historical text analysis using a broad *DER SPIEGEL* magazine corpus containing more than 100,000 articles. While this collection covers a wide variety of themes, we illustrate its use here via a case study on decolonisation in Asia and Africa (1940s–1970s) (Hiltmann et al., 2025).

The pipeline comprises four key stages: (1) *Chunking*, which segments documents into coherent units that preserve semantic meaning within token limits; (2) *Embedding*, using a pre-selected *Sentence Transformer* model to encode each *chunk* into high-dimensional vector representations, which were then stored in *ChromaDB*; (3) *Retrieval*, using cosine-similarity-based matching to identify and return thematically relevant text fragments in response to user queries; and (4) *Generation*, in which the retrieved content was inserted into the prompt of an LLM (HU-LLM-3 and GPT-4o) to produce contextually grounded responses.

The RAG pipeline was developed in a collaborative research project with students aimed at exploring the potential and limitations of LLMs for historical research, while being designed with broader applicability in mind. Its modular design, corpus-driven architecture and transparent evaluation, with all the challenges that entails, allow it to be adapted for a wide range of use cases across the humanities and social sciences, wherever source-critical text generation is required.

The article addresses an interdisciplinary readership, including historians who are beginning to integrate LLMs into their research workflows, as well as scholars with more advanced experience in computational methods. The subject is complex, and we have sought to provide a transparent account of RAG's retrieval and generation workflow, while maintaining accessibility for readers with varying degrees of technical proficiency. Explanatory boxes clarify key terms to support those less familiar with the field without oversimplifying the challenges.

This contribution offers an overview and evaluation of RAG's practical utility in historical scholarship, while also addressing challenges common to its wider use across the humanities and social sciences. We argue that its effectiveness depends on informed prompt engineering based on high-quality retrieval sources and ongoing domain-specific evaluation.

2. Retrieval-augmented generation (RAG)

Retrieval-augmented generation, or RAG for short, refers to an architecture that combines LLMs with external knowledge bases, for example, digitised archives, institutional records or curated corpora, to produce more context-sensitive responses. Rather than relying solely on the implicit knowledge encoded in the LLM's training, a RAG system retrieves relevant texts in real time and uses them to contextually ground its outputs. The core objective of RAG is to reduce hallucinations and enhance the factual reliability of generated content by introducing a retrieval mechanism, without needing to retrain the base LLM from scratch (Lewis et al., 2020; Schuster et al., 2021).

Most commonly, RAG systems employ either sparse or dense retrieval methods. The first is keyword-based and operates strictly at the lexical level, while the second embeds both the user query and segments of the knowledge base into a shared semantic vector space, retrieving relevant passages based on similarity metrics (e.g., cosine similarity). Because both approaches have distinct strengths and limitations, hybrid retrieval, which combines both methods, is often used to balance performance, improving precision on rare or exact terms while maintaining strong recall for semantically similar content (Fan et al., 2024: 6492f.).

RAG has been successfully applied across multiple domains and shows particularly promising results in healthcare. For instance, MMed-RAG (Xia et al., 2024) introduced a versatile multimodal RAG system for medical vision LLMs, achieving an average improvement of 43.8% in factual accuracy compared to baseline generation. In a real-world deployment, Nguyen-Duc et al. (2025) present a multi-agent, retrieval-augmented chatbot system designed for university admissions counseling in Vietnam (MARAUS). Their system demonstrates high accuracy, averaging 92% across more than 6,000 user interactions, while reducing hallucination rates from 15% to just 1.45%.

In the field of history, philosophy and sociology of science (HPSS), however, RAG remains a relatively recent development with only a handful of projects exploring its potential. One such example is the work by Sergeev et al. (2025), who developed a RAG-based chatbot for searching and summarising large, heterogeneous humanities datasets. Through a hybrid strategy combining full-text and semantic retrieval, their findings demonstrate promising results in improving access to archival materials such as the Prozhito diary entries from 1900–1916, though the work lacks user-centered evaluation by historians or archivists. Similarly, Hill (2026) introduces the 'data interview', a hybrid methodological approach that integrates RAG into qualitative workflows. Drawing on a corpus of 2.5 million anonymised Facebook comments, Hill explores how LLMs can be integrated into reflexive, interpretative research, thereby offering a critical template for integrating automation into social inquiry. Her approach both leverages RAG to enhance qualitative analysis and contextualises LLMs within the broader trajectory of knowledge production.

Several recent studies, while demonstrating significant improvements in factual accuracy compared to baseline model outputs, identify the retrieval phase as a critical area for further development. As RAG systems remain vulnerable to noisy or fabricated content in the retrieved documents, they can, in some cases, introduce additional hallucinations based on irrelevant or misleading retrieval results (Fan et al., 2024: 6498; Yang et al., 2024; Asai et al., 2024). Taking a more critical stance, Chatzikyriakidis (2025) examines the presumed generalisability of RAG systems and their claimed performance gains in the domain of semantic historical event extraction. Using Thucydides' *History of the Peloponnesian War* as a controlled test case, he shows that purely inferential generation not only outperforms enhanced RAG configurations in extensively documented domains overall but also delivers markedly higher accuracy. Indeed, RAG can sometimes diminish output quality by distorting the semantic relationships between historical actors and events (8).

In summary, these concerns underscore the need for a robust evaluation step, not only for the specific implementation of our SPIEGEL-RAG pipeline, but more broadly as a response to the epistemic opacity of such systems. Consistent with the principles of the historiography and HPSS, this evaluation should elucidate both the technical components (e.g. retrieval algorithms and similarity metrics) and the interpretative frameworks embedded in the system prompts, thereby keeping methodological transparency central to historical research (Simons et al., 2026; Oberbichler and Petz, 2025).

In its basic form, the framework follows a multi-stage processing pipeline:

1. **Database setup**: Relevant documents are first indexed within a structured database.
2. **Retrieval phase**: Using dense retrieval, embedded document representations are then semantically fetched based on the user's query.
3. **Prompt construction**: An input prompt is crafted that incorporates both the user's original question and the retrieved content from the knowledge base.

This sequential design allows RAG systems to selectively search targeted collections and feed relevant excerpts into the language model, resulting in responses grounded in verifiable context. Additionally, it enables source attribution by linking generated outputs to specific retrieved documents, allowing users to trace the origins of the information and verify its accuracy.

3. Practical applications of RAG in the historical sciences

As part of the LLM project at Humboldt-Universität zu Berlin involving researchers and graduate students, we developed a RAG pipeline to support the analysis of historical texts from the West German news magazine *DER SPIEGEL*. Thematically, the project aimed to trace how the process of decolonisation in Asia and Africa was represented in *SPIEGEL*'s reporting between the late 1940s and the 1970s. Through individual case studies, students explored both the potential and the limitations of integrating LLMs into the historical research process.

One of the key challenges in this historical analysis was the implicit nature of references to decolonisation in the analysed articles. Rather than using the term 'Dekolonisation' ('decolonisation') directly, SPIEGEL authors discussed the topic through contemporary language such as 'Befreiungsbewegung' ('liberation movement'), 'Dritte-Welt-Bewegung' ('Third World movement'), or 'Entwicklungspolitik' ('development policy').

This diversity in terminology poses serious difficulties for traditional, rule-based text-mining approaches which rely on keyword matches. Conversely, this challenge highlights the particular strength of LLMs when used in conjunction with a RAG pipeline: In this setup, historical terms can be embedded into a high-dimensional semantic space, allowing the system to retrieve text passages that are not lexically identical but semantically similar. The LLM can then analyse these retrieved texts through what Simons et al. (2026: 5) term 'interpretive sparring', helping to identify subtle conceptual overlaps and latent discursive patterns, potentially revealing new perspectives on how decolonisation was framed or narrated, even when the term itself was not explicitly mentioned (McGillivray et al., 2024).

In our configuration, retrieval is not primarily a fact-checking or citation mechanism, but a tool for exploratory discovery. This design offers two key advantages. First, semantic retrieval expands the scope of inquiry by surfacing relevant sources that conventional keyword searches would miss, thereby drawing attention to documents that engage with related ideas without using expected terminology. Second, by making the connection between retrieved texts and generative output explicit, the system takes a necessary step towards transparency in alignment with academic integrity. Researchers can verify the sources underlying the model's interpretations, ensuring that computational insights remain accountable to the historical record. In this way, the pipeline supports large-scale interpretative analysis while preserving the core scholarly principle of source-based verification.

4. Understanding the RAG architecture: chunking, embeddings and retrieval

To lay the groundwork for analysing and implementing a RAG pipeline, an external knowledge base is required. In our case, this took the form of a large corpus of 105,257 articles from the *DER SPIEGEL* online archive, extracted via web scraping and converted into a structured CSV database. The corpus comprises approximately 84 million *tokens*, an extensive body of historical material that would be too large to analyse thoroughly without computational methods.

However, this dataset also presents a number of qualitative limitations that must be made explicit. As with most digitisation projects, a degree of information loss is unavoidable (Porter, 2021). Original images have been removed and the physical layout of the articles is no longer preserved, making the visual experience that a contemporary reader would have had impossible to reconstruct. In addition, the text contains occasional OCR errors, especially in the rendering of foreign names with diacritics or non-German characters such as ç, ğ, ł, or ñ, which are often misrecognised or replaced with incorrect symbols. No official documentation has been released by *SPIEGEL* detailing the digitisation process undertaken in 2007.

While this lack of transparency complicates assessments of data provenance and technical procedures, it is not an isolated case. On the contrary, it reflects a broader challenge faced by historians working with digitised source material. The absence of metadata or processing protocols is symptomatic of the infrastructural realities that shape much of today's digital source landscape. That such a large corpus has been made available at all should not be taken for granted. It is the outcome of a major archival effort from which historical research continues to benefit, even as we, as researchers, grapple with its limitations.

Despite these challenges, the corpus provided a workable and sufficiently rich basis for building and testing the RAG pipeline. Before the texts could be processed by the language model, however, they needed to be adapted to the technical requirements of semantic retrieval and generation. We first segmented the texts into smaller, yet semantically coherent units – commonly referred to as *chunks* – in order to remain within the context window limits of the models. This step was essential not only for technical compatibility but also for structuring the data in a way that preserves interpretative coherence. To that end, we employed a *context-overlap* strategy, that is, the final portion of each *chunk* was repeated at the beginning of the subsequent one. This approach preserves continuity between sections, ensuring that key semantic relationships were not lost during processing.

A *token* represents the smallest unit of processing in Natural Language Processing (NLP). Depending on the *tokenisation* method used, it may correspond to an entire word, a word fragment, or a punctuation mark.

For the RAG pipeline, we applied *recursive chunking* with a *chunk size* of 500 *tokens* and an *overlap* of 50 tokens.

Chunking is a method of progressively breaking down text, where a text is divided into smaller segments according to a fixed number of characters or by using delimiters (e.g., paragraphs, sentences). In recursive chunking, the text is not simply divided once into fixed sections (e.g., paragraphs), but if the initial division does not yield the desired segment size or structure, smaller units based on delimiter criteria (e.g., sentences) are used step by step. The aim is to produce semantically coherent and roughly equally sized *chunks* without imposing rigid boundaries, thereby preserving the context within the segments as effectively as possible.

Subsequently, we converted each of these *chunks* into *sentence embeddings* using a *Sentence Transformer model* (*all-MiniLM-L6-v2*) and stored them in the open-source vector database

ChromaDB. Instead of representing words merely as sequences of characters, the model captures relationships between words and sentences and expresses them in a numerical format.

> *Embeddings* are high-dimensional numerical vector representations of text that capture semantic meaning relative to context. Because language models do not 'understand' text as humans do, they instead rely on *embeddings* to translate language into a mathematical form suitable for computation. *Embeddings* are essential for semantic search, enabling systems to retrieve conceptually similar text even without shared keywords.

The retrieval module identified the most semantically relevant *chunks* in response to each user query. We converted the query into *embeddings* and compared them with the stored vectors using *cosine similarity*, allowing the module to fetch not only exact matches but also conceptually related passages. These retrieved *chunks* ultimately served as the contextual basis for generating the model's response.

> *Cosine similarity* is a metric used to measure the semantic proximity between two vector representations by calculating the cosine of the angle between them. The result always falls between –1 (vectors pointing in exactly opposite directions) and +1 (identical direction), with 0 indicating orthogonality. Higher scores (closer to +1) correspond to stronger semantic similarity, while values towards –1 indicate increasing dissimilarity.

5. Towards reliable LLM output: system prompts for uncertainty and source traceability

Ensuring the reliability of LLM-generated responses requires careful prompt design (Chen et al., 2025). We implemented a *system prompt* that grounds the model's answers strictly in the documents retrieved during the search phase (Box 1). To reduce hallucinations, the prompt explicitly instructs the model to acknowledge uncertainty. When no relevant information is found, it responds with 'I cannot answer that', rather than speculating. In an ideal scenario, this approach maintains the integrity of the output and reinforces source traceability within the workflow. In addition, we employed role-based prompting, in which the model was instructed to act as a historian with specialised expertise in the thematic case study under investigation and to contextualise its responses with reference to the relevant time frame (1940s–1970s). This framing is intended to guide the model towards historically grounded, context-sensitive responses.

> A *system prompt* is a high-level instruction that defines fundamental rules for an LLM such as its behaviour, tone, and scope of responses. It is included with every API call alongside the user prompt and any prior messages, ensuring that its influence persists throughout the interaction.
>
> In contrast, a *user prompt* is formulated individually for each query. While the system prompt governs the model's overall behaviour, the user prompt determines the specific content of a given interaction.

Box 1 **Sample system prompt used in the RAG pipeline, own translation (see Appendix A, Box 1 for the original system prompt).**

System prompt (English translation): You are a historian with specialised expertise in the Algerian War and the role of Charles de Gaulle. Your task is to analyse articles from the West German news magazine *DER SPIEGEL*, published between the 1940s and the 1970s, and to answer the following question as accurately and factually as possible. Your response must rely exclusively on the content of the retrieved *SPIEGEL* articles. Do not draw on external sources or make inferences beyond what is stated in the texts. If the retrieved material does not provide sufficient or clear information, make this explicit and respond with: 'I cannot answer that.'

Fig. 1: Meding and Daugs: Early-stage RAG Pipeline for the Analysis of Historical Texts from the SPIEGEL Online-Archive (1940s–1970s).

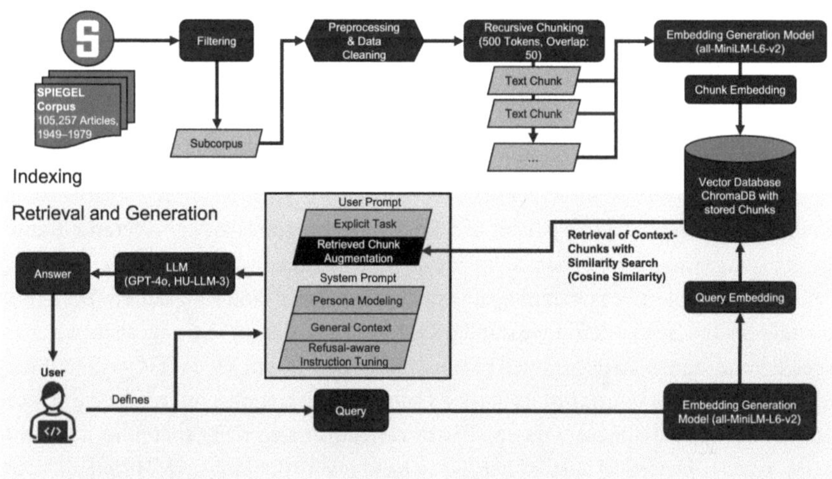

To generate the final output, the *system prompt*, the *user query* with a specific research question, and the relevant context retrieved by the RAG system were passed to each model via the API. Since the sources were in German, all inputs and outputs were also formulated and generated in German to ensure linguistic alignment. However, for future applications, it would be worthwhile to test prompt strategies in English and evaluate whether English, German or a multilingual approach yields better performance. Moreover, although the chosen prompts were tested for effectiveness, more extensive evaluations, including longer prompts, would help to determine which elements contribute most to plausible and relevant responses.

In summary, this pipeline created a controlled environment for generation. It ensured the retrieval of relevant text passages and ensured the traceability of the sources used by the model for its response. In the next phase, each retrieved *chunk*, the model's output and its source-boundedness could be individually evaluated.

6. Challenges in evaluating RAG for historical research

Language models generate output through probabilistic next-token prediction, and therefore their use inevitably introduces a degree of uncertainty, posing challenges for their reliable integration into research workflows (Meding and Daugs, 2026). The step of evaluation is therefore vital. Every prompt needs to be clearly defined, and model responses must be checked against expected results.

The same principle applies to RAG systems. While the citations in the generated responses correspond to actual sources retrieved from the database, these sources can still be irrelevant or only loosely connected to the query, which affects the reliability of the output. Thus, it is not sufficient to judge the quality of the output alone. The relevance and appropriateness of the material retrieved for the model to generate its response are equally important.

In historical research, this evaluation is especially critical. The selection and retrieval of source material, as well as the relevance and plausibility of the generated analysis, require close examination and assessment. Neglecting either risks accepting outputs that may appear credible but ultimately misrepresent historical evidence.

6.1 Establishing a silver standard in the absence of objective ground truth

Evaluating the retrieval component of a RAG system requires a clear reference framework to judge the quality of its results. Ideally, this would be a *ground truth*, that is, a fully validated and objective standard against which outcomes can be measured. Yet in historical research, such a definitive standard rarely exists. History is always shaped by perspective and is inescapably framed by the vantage point from which it is written. Every narrative bears the imprint of its author's cultural assumptions, interpretative choices and the intellectual climate of the era. Historical truth or objectivity, therefore, is best understood as a theoretical ideal rather than a fixed reality (Jordan, 2010; Hiltmann, 2024: 205f.), complicating efforts to evaluate in this field.

Given this difficulty, we developed a *silver standard* as a practical alternative. This standard consists of a curated set of research questions and reference answers aligned with selected articles from *DER SPIEGEL*. We began by defining a clear thematic scope and formulating specific questions connected to previously identified source material. For each question, we manually composed plausible answers grounded in the information provided by the sources. This *silver standard* thus defines the minimum set of relevant articles, key statements and contextual details necessary to adequately address each question.It provides a structured baseline against which to measure the effectiveness of the retrieval component, while recognising the interpretative nature of historical scholarship.

Although we do not claim that these reference answers represent absolute historical truths, the framework supports a meaningful evaluation of document retrieval relevance and content quality. Building on this, we developed question categories to assess the RAG system's performance across different types of historical inquiry.

Fig. 2: Meding and Daugs: Evaluation of the RAG framework for analysing historical texts from the SPIEGEL archive (1940s–1970s).

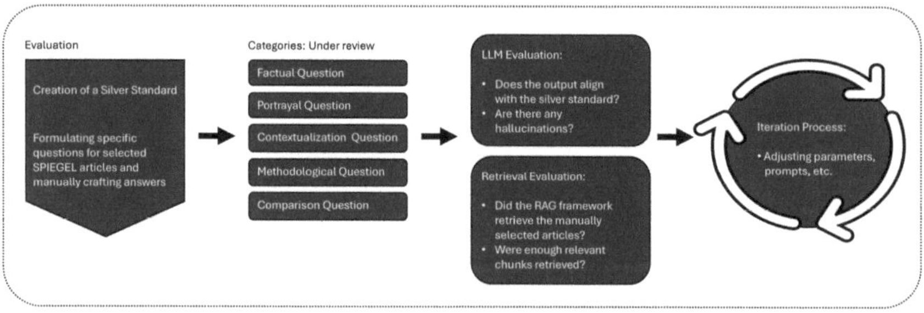

6.2 Evaluating the RAG pipeline in practice: the portrayal of Charles de Gaulle and the Algerian war in *DER SPIEGEL*

To illustrate the evaluation process and its specific challenges, we present a practical case study focused on Charles de Gaulle's rise to power in 1958 and its portrayal in *DER SPIEGEL*, particularly in relation to the Algerian War and the representation of the French diaspora and the Muslim population living in Algeria at the time.

The example is based on the *SPIEGEL* article 'De Gaulle, the God-Sent' (N.A. 1958, 'De Gaulle, der Gottgesandte', *DER SPIEGEL* 36), featuring an interview with François Mauriac, Nobel Prize-winning author and later biographer of de Gaulle. This article served as the basis for a question-and-answer pair created by students as part of the *silver standard*.

The question, classified as a 'representation question', examines how Mauriac portrays Algerians during de Gaulle's rise to power and identifies the rhetorical strategies and linguistic devices used in the article. The minimum plausible answer notes a dehumanising portrayal of Muslim Algerians, depicted through racist and patriarchal-colonial language as a suffering mass to be cared for by the French state. By contrast, French Algerians are framed by Mauriac as the true threat in the conflict, and to French democracy as a whole. Key excerpts from the article support this interpretation, describing Muslims as 'a shifting, formless mass of undernourished and ignorant unfortunates' and emphasising the 'unyielding Muslim liberation front' confronted by the powerful French Algerians.

The retrieval evaluation focused on how well the system identified relevant source material. Using the query 'algerische Muselmanen' (archaic term for Algerian Muslims), the system extracted ten text *chunks* with the highest similarity scores. While relevant sections were retrieved, three significant shortcomings were observed. First, most retrieved *chunks* repeated the same content, with only two providing relevant distinctions (redundancy). Second, despite *recursive chunking* and *overlap*, many segments lacked complete sentences, disrupting semantic coherence (fragmentation). Third, the system sometimes included passages from unrelated articles, misattributing context (off-topic recall and misattribution). These challenges demonstrate that retrieval quality in RAG systems is still the primary bottleneck in RAG workflows. Addressing repetition, incomplete texts, misattribution and semantic incoherence is therefore essential to improving RAG's reliability for historical research. Our pilot evaluation confirms that both the *chunking process* and the similarity-based retrieval require substantial refinement, reaffirming the critical importance of our chosen approach.

Box 2 **Retrieval output of the identified text passages (chunks), own translation (see Appendix B, Box 2 for the original output).**

Relevance: 0.5066
Title: DE GAULLE, DER GOTTGESANDTE
Date: 1958–09-02
URL: https://www.spiegel.de/politik/de-gaulle-der-gottgesandte-a-f1b76d80-0002-0001-0000-000041759098?context=issue
Text: The strength of the one million French in Algeria is far, far greater than that of the eight million Muslims. SPIEGEL: But can the one million French defeat the Muslims? Is it not, above all, the support of the army that gives the Algerian French their power? What would these one million French be without the soldiers whom the forty million French in France send to Algeria? MAURIAC: The strength of the French in Algeria is immense. The Muslims are little more than a formless mass, surging back and forth.
Relevance: 0.5258
Title: DE GAULLE, DER GOTTGESANDTE
Date: 1958–09-02
URL: https://www.spiegel.de/politik/de-gaulle-der-gottgesandte-a-f1b76d80-0002-0001-0000-000041759098?context=issue
Text: MAURIAC: The danger of overexertion is immense, I am the first to concede that to you. The possible consequences of such an exacting policy go beyond anything one can imagine. As far as Algeria and the Muslims are concerned, we would, I think, quickly find ourselves in agreement on all essential points. But the problem is evidently insoluble, because the French of Algeria exist. That is the heart of the matter. Were it not for the enormous pressure exerted by these French in Algeria, the problem would quickly and relatively easily find an acceptable solution.

Nevertheless, the RAG system successfully identified relevant *chunks* that were integrated into the prompt context for model evaluation. Using these retrieved contexts, we compared responses from the HU-LLM-3 and OpenAI's GPT-4o models. Each model received the same *system prompt*, *question prompt* and *context chunks* via their respective APIs.

The HU-LLM-3 model generated a response considered satisfactory against the *silver standard*. GPT-4o's answer was also found to be adequate. However, given its strict adherence to data privacy and transparency measures, particular attention was paid to evaluating the performance of the HU-LLM-3 model.

Box 3 Generated response by the HU-LLM-3 model, own translation (see Appendix C, Box 3 for the original output).

'In the selected excerpts, Algerians – particularly the Muslim population – are portrayed by François Mauriac and *Der Spiegel* magazine in a highly stereotypical and derogatory manner. The following are some of the linguistic devices and semantic terms used to describe the Algerian people:

- **"Moslem' or 'Muselmannen"**: The use of these terms, which refer to the Islamic faith, reduces Algerians to their religious identity and suggests a sense of otherness and foreignness.
- **"The French of Algeria"**: Mauriac and *Der Spiegel* refer to the French settlers in Algeria as a distinct group, which implies a separation from the indigenous Algerian population and positions them as holding a dominant role.
- **"The power of the French in Algeria"**: Here, the strength and authority of the French in Algeria are emphasised, while the Muslim population is described as a "shifting, formless mass", conveying a negative and dehumanising representation.
- **"Pressure on France"**: Algerians are depicted as a threat to France, creating an insoluble crisis through their 'enormous pressure'.
- **"Insoluble problem"**: The very existence of the French in Algeria is portrayed as a core issue that appears insurmountable.
- **"Overexertion"**: Mauriac stresses the potential consequences of a policy that considers the interests of Algerians, suggesting that such a course of action would represent an excessive burden for France.

Taken together, these linguistic choices reveal a clear hierarchy and a colonial mindset in which Algerians are seen as inferior, while the French settlers are framed as the dominant force. The language frequently conveys a tone that is dismissive and dehumanising, obscuring the complex realities of the Algerian population and their struggles for independence and self-determination.

It is important to note that this representation is heavily shaped by the perspective of the French colonial power and often silences the voices and lived experiences of Algerians themselves.'

Although HU-LLM-3 did not clearly differentiate between the portrayals of French and Muslim Algerians, or flag the term 'Muselmannen' as archaic, it nonetheless placed relevant terms and phrases successfully within a broader colonial and historical framework. The model also flagged the potentially biased perspective of the article's French viewpoint and recommended consulting Algerian sources for a more comprehensive analysis.

Such a step may seem self-evident to a trained historian, but this simulation of source criticism within the generated output indicates a potential heuristic use of LLMs. The model generated a plausible critique by extrapolating from patterns in its training data, identifying a colonial bias and recommending Algerian perspectives and sources. Treated as a heuristic tool, the output can alert historians to potential interpretative leads and asymmetries in text corpora that warrant closer inspection and suggest complementary sources. In other words, the model can help scaffold an initial analytic pass, provided scholars subject every suggestion to human verification, recognising that its 'criticism' is probabilistic pattern-matching, not a historical method.

Yet this functionality cannot become an excuse to bypass hermeneutic close reading or to engage solely with LLM-highlighted passages while neglecting the original source. Historians remain responsible for returning to the primary text, weighing context, language and authorship against any machine-generated cue. Responsible deployment therefore demands transparent prompts, verifiable retrieval chains and an iterative workflow.

7. Conclusion and discussion

In this article, we have shown that a RAG-enhanced LLM pipeline not only meets a core requirement of historical research, managing extensive datasets like the *SPIEGEL* archive with full provenance, but also transforms how scholars interact with source material. Instead of producing decontextualised responses that are fully reliant on the internal and often opaque parametric knowledge representation of the model, every generated answer is explicitly tied to the retrieved document segments, giving historians the ability to verify and cross-check the model's claims against the original texts.

Beyond this traceability, another strength of the RAG architecture lies in its modularity and adaptability. The system is not constrained by topical boundaries; provided the input corpus is properly indexed and semantically embedded, it can support a wide array of queries within the *SPIEGEL* archive or be extended to other structured collections. This extensibility makes the pipeline highly scalable and domain-transferable, facilitating integration with historical databases or digital corpora. By enabling semantic-level retrieval, RAG supports historians in locating conceptually related documents that would remain inaccessible through manual or analogue methods. From a distant reading standpoint, this opens the door to novel perspectives and encourages scholars to formulate questions that previously might have gone unasked.

Despite these strengths, the limitations of the approach remain evident. Adding external, retrieved material to the prompt narrows the model's focus, but does not eliminate the core problem of hallucinations or inaccuracies in LLM-generated content. If irrelevant or incorrect sources are retrieved, the model may still produce fluent but historically implausible answers. In some cases, RAG may reinforce false confidence by pairing fabricated claims with credible citations. This highlights the ongoing need for analogue source evaluation.

Moreover, the retrieval quality depends heavily on the user's command of the corpus's language and conventions. Crafting effective prompts and queries requires an understanding of historical terminology and period-specific vocabulary. For example, recognising that post-war *SPIEGEL* articles employ the archaic term 'Muselmanen' rather than 'Muslime' is crucial. Without such domain expertise, even the most advanced retrieval engines will struggle to deliver relevant and precise results.

Turning to our evaluation framework, the *silver standard* we created offers a practical method to measure system performance, but faces inherent challenges. Its main limitation stems from the interpretative nature of historiography, where objectivity is never absolute and every answer reflects a particular viewpoint. The manual formulation of plausible responses introduces unavoidable subjectivity and presents one reading among many possible interpretations. This complicates efforts to establish firm benchmarks for model assessment. Moreover, the complexity of many historical questions resists simplification into neat question-and-answer formats.

In light of these considerations, it is clear that the *silver standard* functions as a pragmatic yet limited evaluation benchmark. Expanding its scope and sensitivity to ambiguity would improve its value, but some hurdles, such as historiographical subjectivity and shifting scholarly understandings, remain firmly in place. RAG pipelines enhance LLM transparency by exposing provenance chains and injecting semantically aligned passages. However, whether RAG truly leads to 'rich responses' ultimately depends on historians' expertise in domain-specific examination of both the generated content and the sources they retrieve. Looking ahead, the most promising trajectory lies in hybrid workflows that fuse neural semantics with rule-based temporal filters and corpus-level validation, ensuring that every machine-generated output is re-anchored in the full evidential record before it enters scholarly interpretation.[1]

References

Asai A, Wu Z, Wang Y, Sil A and Hajishirzi H (2024) Self-rag: Learning to retrieve, generate, and critique through self-reflection. In: *The Twelfth International Conference on Learning Representations (ICLR 2024)*, Vienna, Austria, 7 – 11 May 2024: 1–30.

Chatzikyriakidis S (2025) Reasoning with RAGged events: RAG-Enhanced Event Knowledge Base Construction and reasoning with proof-assistants. *arXiv preprint: 1–16.* https://doi.org/10.48550/arXiv.2506.07042.

Chen B, Zhang Z, Langrené N and Zhu S (2025) Unleashing the Potential of Prompt Engineering in Large Language Models: A Comprehensive Review. *Patterns* 6 (6): Article 101260. https://doi.org/10.1016/j.patter.2025.101260.

Fan W, Ding Y, Ning L, Wang S, Li H, Yin D et al. (2024) A Survey on RAG Meeting LLMs: Towards Retrieval-Augmented Large Language Models. In: *Proceedings of the 30th ACM SIGKDD Conference on Knowledge Discovery and Data Mining* (ed Baeza-Yates R and Bonchi F), Barcelona, Spain, 25 – 29 August 2024: 6491–6501.

Hill M (2026) The data interview. Reflexive integration of large language models in qualitative content analysis. In: Simons A, Wüthrich A, Zichert M, et al. (eds) *Understanding Science with Large Language Models? Potentials for the History, Philosophy, and Sociology of Science*. Bielefeld: transcript, part-5.

1 This chapter was written with support from large language models (LLMs). All model-generated text was reviewed and, where necessary, rewritten by the authors, who remain fully responsible for the final version. For details on the use of LLMs in this volume, see the statement in the volume's introduction.

Hiltmann T (2024) Hermeneutik in Zeiten der KI. Large Language Models als hermeneutische Instrumente in den Geschichtswissenschaften. In: Schreiber G and Ohly L (eds) *KI:Text. Diskurse über KI-Textgeneratoren*. Berlin / Boston: De Gruyter, 201–232.

Jordan S (2010) *Vetorecht der Quellen*. In: *Docupedia-Zeitgeschichte*, February 11. https://dx.doi.org/10.14765/zzf.dok.2.570.v1 (accessed 02 March 2025).

Koszyk K (1999) Presse unter alliierter Besatzung. In: Jürgen W (eds) *Mediengeschichte der Bundesrepublik Deutschland*. Köln / Wien: Böhlau Verlag, 31–58.

Lewis P, Perez E, Piktus A, Petroni F, Karpukhin V and Goyal N (2020) Retrieval-Augmented Generation for Knowledge-Intensive NLP Tasks. In: *Advances in Neural Information Processing Systems* 33 (ed Larochelle H, Ranzato M, Hadsell R et al.), Vancouver BC, Canada, 6 – 12 December 2020: 9459–74.

Mauermann J and Oberbichler S (2025) LLM Biases: Expected and Unexpected Model Design Effects in Historical Newspaper Article Extraction on the Messina Earthquake. In: *DH Lab*, 31. January. https://doi.org/10.58079/137qr.

McGillivray B, Nanni F and Beelen K (2024). Why Does Digital History Need Diachronic Semantic Search? In: Tilton L, Mimno D and Marie J (eds) *Computational Humanities*. Minneapolis / London: University of Minnesota Press, 177–194.

Meding H and Daugs A (2026) From RAGs to rich responses. Enhancing LLM reliability through retrieval-augmented generation. In: Simons A, Wüthrich A, Zichert M, et al. (eds) *Understanding Science with Large Language Models? Potentials for the History, Philosophy, and Sociology of Science*. Bielefeld: transcript, part-5.

N.A. (1958) DE GAULLE, DER GOTTGESANDTE. *DER SPIEGEL* 36, September 02. https://www.spiegel.de/politik/de-gaulle-der-gottgesandte-a-f1b76d80-0002-0001-0000-000041759098?sara_ref=re-xx-cp-sh (accessed 07 December 2024).

Nguyen-Duc ACVM, Tran BA, Ngo VP, Chi LL and Nguyen AQ (2025) An Empirical Study of Multi-Agent RAG for Real-World University Admissions Counselinh. *arXiv preprint*: 1–12. https://doi.org/10.48550/arXiv.2507.11272.

Oberbichler S and Petz C (2025) Working Paper: Implementing Generative AI in the Historical Studies. *Zenodo*, February 25. https://doi.org/10.5281/zenodo.14924737.

Porter D (2021) *Manuscript Loss in Digital Contexts*. Talk, originally presented at the 14th Annual Lawrence J. Schoenberg Symposium on Manuscript Studies in the Digital Age, November 16. https://www.dotporterdigital.org/manuscript-loss-in-digital-contexts/ (accessed 11 July 2025).

Sergeev A, Goloviznina V, Melnichenko M and Kotelnikov E (2025) Talking to Data: Designing Smart Assistants for Humanities Databases. *arXiv preprint*: 1–15. https://doi.org/10.48550/arXiv.2506.00986.

Shuster K, Poff S, Chen M, Kiela D and Weston J (2021) Retrieval augmentation reduces hallucination in conversation. In: *Findings of the Association for Computational Linguistics (EMNLP 2021)*, Punta Cana, Dominican Republic, November 2021: 3784–3803.

Simons A, Zichert M and Wüthrich A (2026) Large language models for history, philosophy, and sociology of science: Interpretive uses, methodological challenges, and critical perspectives. *Studies in History and Philosophy of Science* 117: 102151. https://doi.org/10.1016/j.shpsa.2026.102151.

Von Lucke A (2024) Seismograph Augstein. DER SPIEGEL als Produkt und Produzent der Liberalisierung – und der neue Ruck nach rechts. In: Lilienthal V (ed) *„Sagen, was*

ist“: *Journalismus für eine offene Gesellschaft – Rudolf Augstein zum 100. Geburtstag*. Köln: Herbert von Halem Verlag, 115–131.
Xia P, Zhu K, Li H, Wang T, Shi W, Wang S et al. (2025) MMed-RAG: Versatile Multimodal RAG System for Medical Vision Language Models. *arXiv preprint*: 1–30. https://doi.org/10.48550/arXiv.2410.13085.
Yang X, Sun K, Xin H, Sun Y, Bhalla N, Chen X et al. (2024) CRAG – comprehensive RAG benchmark. *arXiv preprint*: 1–20. https://doi.org/10.48550/arXiv.2406.04744.

Appendix

Appendix A. Text-based Box 1: sample system prompt used in the RAG-pipeline

"Du bist ein Historiker mit tiefgehender Expertise zum Algerienkrieg und zu Charles de Gaulle. Deine Aufgabe ist es, die Artikel des westdeutschen SPIEGEL-Nachrichtenmagazins aus den Jahren 1949 bis 1979 zu analysieren und die folgende Frage so präzise und faktenbasiert wie möglich zu beantworten. Dabei stützt du dich ausschließlich auf die in den abgerufenen SPIEGEL-Artikel enthaltenen Informationen. Falls eine Information nicht in den Artikeln zu finden ist, ergänzt du sie nicht aus anderen externen Quellen oder durch Spekulation. Wenn die Artikel keine eindeutigen Informationen zu einer Frage enthalten, stellst du dies klar heraus und antwortest explizit mit: 'Das kann ich nicht beantworten.'"

Appendix B. Text-based Box 2: generated response by the HU-LLM-3 model

"Relevanz: 0.5258
Titel: DE GAULLE, DER GOTTGESANDTE
Datum: 1958–09-02
URL: https://www.spiegel.de/politik/de-gaulle-der-gottgesandte-a-f1b76d80-0002-0001-0000-000041759098?context=issue
Text: MAURIAC: Die Gefahr der Überanstrengung ist gewaltig, ich bin der erste, Ihnen das zuzugestehen. Die möglichen Konsequenzen einer so anspruchsvollen Politik gehen über das Vorstellbare hinaus. Was Algerien und die Moslem angeht, so würden wir wohl schnell in allen wesentlichen Punkten übereinstimmen. Aber das Problem ist offenbar unlösbar, weil es die Franzosen Algeriens gibt. Das ist der Kern der Sache. Übten diese Franzosen Algeriens nicht den enormen Druck auf Frankreich aus, dann fände das Problem schnell und relativ einfach eine akzeptable Lösung.
Relevanz: 0.5066
Titel: DE GAULLE, DER GOTTGESANDTE
Datum: 1958–09-09
URL: https://www.spiegel.de/politik/de-gaulle-der-gottgesandte-a-f1b76d80-0002-0001-0000-000041759098?context=issue
Text: die Kraft der einen Million Franzosen in Algerien ist viel, viel größer als die der acht Millionen Muselmanen. SPIEGEL: Kann aber die eine Million Franzosen die Moslem besiegen? Ist es nicht vor allein die Hilfe der Armee, die den algerischen Franzosen Macht verleiht? Was wäre diese Million Franzosen ohne die Soldaten, die

die 40 Millionen Franzosen in Frankreich nach Algerien schicken. MAURIAC: Die Kraft der Franzosen Algeriens ist gewaltig. Die Moslem sind nicht viel mehr als eine hin- und herflutende, gestaltlose Masse [...]."

Appendix C. Text-based Box 3: generated response by the HU-LLM-3 model

"Antwort: In den Textauszügen werden die Algerier*innen, insbesondere die muslimische Bevölkerung, von Francois Mauriac und dem SPIEGEL-Magazin in einer sehr stereotypisierenden und abwertenden Weise dargestellt. Hier sind einige sprachliche Mittel und semantische Begriffe, die zur Beschreibung der Algerier*innen verwendet werden:

- 'Moslem' oder 'Muselmanen': Die Verwendung dieser Begriffe, die sich auf die muslimische Religion beziehen, reduziert die Algerier*innen auf ihre religiöse Identität und suggeriert eine gewisse Fremdheit und Andersartigkeit.
- 'Die Franzosen Algeriens': Mauriac und der SPIEGEL bezeichnen die algerischen Franzosen als eine separate Gruppe, was impliziert, dass sie sich von den einheimischen Algerier*innen unterscheiden und eine dominante Rolle einnehmen.
- 'Kraft der Franzosen Algeriens': Hier wird die Macht und Stärke der algerischen Franzosen betont, während die muslimische Bevölkerung als 'hin- und herflutende, gestaltlose Masse' beschrieben wird, was eine negative und entmenschlichende Darstellung darstellt.
- 'Druck auf Frankreich': Die Algerier*innen werden als eine Bedrohung für Frankreich dargestellt, die durch ihren "enormen Druck" eine unlösbare Situation schaffen.
- 'Unlösbares Problem': Die Existenz der algerischen Franzosen wird als Kernproblem dargestellt, das unüberwindbar erscheint.
- 'Überanstrengung': Mauriac betont die möglichen Konsequenzen einer Politik, die die Interessen der Algerier*innen berücksichtigt, und suggeriert, dass dies eine übermäßige Belastung für Frankreich darstellen würde.

Insgesamt zeigen diese sprachlichen Mittel eine klare Hierarchie und eine koloniale Denkweise, die die Algerier*innen als unterlegen und die algerischen Franzosen als die dominierende Kraft darstellt. Die Sprache ist oft abwertend und entmenschlichend, was die komplexe Realität der algerischen Bevölkerung und ihre Kämpfe für Unabhängigkeit und Selbstbestimmung ignoriert.

Es ist wichtig zu beachten, dass diese Darstellung stark von der Perspektive der französischen Kolonialmacht geprägt ist und die Stimmen und Erfahrungen der Algerier*innen selbst oft ausblendet."

The data interview

Reflexive integration of large language models in qualitative content analysis

Miira Hill

1. Introduction

The longstanding divide between qualitative and quantitative paradigms, often called the *paradigm wars*, emerged from fundamentally different epistemological assumptions about what constitutes scientific knowledge. Quantitative research, grounded in positivist traditions, emphasizes measurement, objectivity, and generalizability (Hammersley and Atkinson, 2007). In contrast, qualitative approaches privilege interpretive depth, contextual understanding, and the co-construction of meaning (Geertz, 1973). These methodological oppositions reached their height in the mid-twentieth century, with scholars contesting not only techniques but also the philosophical legitimacy of their opposed approaches (Adorno et al., 1993; Frisby, 1972; Popper, 1989).

Later developments in the sociology of science, especially interactionist ethnomethodological approaches and later actor-network theory, challenged the idealized image of scientific objectivity by emphasizing the material, institutional, and social practices through which knowledge is produced (Garfinkel, 1967; Knorr-Cetina, 1981; Latour and Woolgar, 1979; Lynch 2013). These perspectives paved the way for a reflexive sociology of science, recognizing that methods and tools are themselves embedded within social orders (Goffman, 1981; Knoblauch, 2025; Schütz, 1972).

With the introduction of digital technologies, research practice underwent another transformation. The "computational turn" (Berry, 2011: 1) and the rise of digital methods (Rogers, 2013; Marres, 2017) blurred the boundaries between qualitative and quantitative inquiry. Massive digital trace data and networked infrastructures have enabled large-scale analyses while raising new interpretive and ethical questions about data provenance, authorship, and meaning (Venturini and Rogers, 2019). The resulting hybrid or qual-quant methodologies combine interpretive reasoning with computational tools such as data visualization and network analysis (Latour et al., 2012).

Recently, LLMs have introduced a new phase in this trajectory, reshaping how text, meaning, and context can be analyzed. In discussions about LLMs, it is frequently emphasized that they can play a crucial role in bridging the gap between computational and interpretive methods (Simons et al., 2026). This implies that LLMs can serve not only as technical tools but also as a means to enhance interpretive analysis in HPSS. LLMs seem to offer not only computational efficiency but also interpretive capacities that seem to merge quantitative scale with qualitative nuance (Simons et al., 2026). Nevertheless, their integration into social research raises critical tensions surrounding algorithmic opacity, platform dependency, and epistemic authority (Bender et al., 2021; Gerlitz and Rieder, 2018; Gillespie, 2024; Noble, 2018). Critical Data Studies has therefore emphasized the importance of reflexivity, transparency, and ethical awareness in computational inquiry (Dalton et al., 2016). It is within this context that the Empirical Theory of Science (EToS) offers a distinctive HPSS framework for analyzing how knowledge is produced and stabilized through communicative practices, institutions, and infrastructures, and how boundaries between scientific and non-scientific knowledge are negotiated (Knoblauch, 2017, 2025). Qualitative research offers context and meaning-making as a counterweight; reflexivity (Knoblauch, 2025: 34) requires LLM research to critically assess its conditions, including infrastructure that shapes data management.

This paper introduces the data interview, a reflexive, mixed-methods framework based on RAG, to address these challenges. We conceptualize LLMs as socially situated epistemic actors rather than neutral instruments. By treating their outputs as dialogical artifacts that invite interpretation rather than deliver answers, we seek to combine computational scale with interpretive depth, moving toward a reflexive, context-sensitive, and methodologically grounded digital social science.

2. Theoretical background

2.1 Knowledge

The question of what counts as valid knowledge, and how it should be produced, remains central across disciplines such as historiography, philosophy of science, and social sciences (Berger and Luckmann, 1967; Schütz, 1972). Building on the paradigm debates outlined above, scholarship has highlighted that claims to "objectivity" are themselves historically situated and socially negotiated. Historically viewed as wise scholars cultivating nature's variability, scientists saw the source of epistemological uncertainty shift from nature to the self-eliminating scientific self by the 19th century (Daston and Galison, 2007: 198). Self-elimination and suppression of subjectivity became central ideals, sidelining emotion in favor of rigor. This culminated in mechanical objectivity (cf. Daston, 2001: 153), a paradigm prioritizing minimal human intervention through machines (e.g., photography, instruments) to capture phenomena 'as they are'. Scientists adopted passive roles, presenting raw, unaltered outputs as transparent records as if they were free from interpretation or aesthetic adjustment. The aspiration toward scientific passivity and neutrality was challenged by HPSS, particularly its emphasis on science as an interpretative, reflexive, and socially embedded practice.

Building on the recognition that all knowledge is socially situated, historians and theorists have reexamined how power, discourse, and technology mediate the construction of the past. Foucault's Discipline and Punish (1977) disrupted traditional, teleological historiography by foregrounding discursive discontinuities and the institutional production of "regimes of truth". For Foucault, what is legitimized as "history" is inseparable from the epistemic configurations and power relations that define its conditions of possibility. Building on similar arguments, postcolonial and decolonial theorists such as Chakrabarty (2013) and Mignolo (2021) have revealed how dominant historical narratives remain anchored in Eurocentric epistemologies that universalize European categories of "modernity", "rationality", and "time". These frameworks often render non-European histories legible only as deviations or lacks, denying epistemic plurality. Within HPSS, subjectivity is analyzed not as bias but as a historical and constitutive process in knowledge production, aligning with qualitative research traditions (Flick et al., 2004; Flick, 2006).

Building on the social construction of reality (Berger and Luckmann 1967), which models how knowledge gains legitimacy, communicative constructivism emphasizes the procedural, dynamic nature of this construction. EToS utilizes this perspective to analyze how social realities are communicatively produced, shaped by interaction and material infrastructures (Keller et al., 2013). Resonating with Knorr-Cetina's (1999) concept of epistemic cultures, EToS reframes scientific knowledge as a communicative process mediated by historical and infrastructural conditions, directly engaging HPSS debates. Central to EToS is its social concept of knowledge: scientific knowledge emerges through socially validated objectivations (data, theories, instruments), not individual neutrality (Knoblauch, 2025: 5). EToS foregrounds several issues. First, EToS tackles the challenge of distinguishing scientific knowledge from non-scientific forms, especially salient in the context of post-COVID-19 new realism, anti-constructivism, and concerns about value intrusions compromising scientific integrity (Knoblauch, 2025: 39). It examines boundary work at micro, meso, and macro levels. This involves the negotiation and defense of the boundaries between science and non-science to secure scientific autonomy regarding research topics, methods, and findings against political, economic, and social influences. This process is shaped by norms (altruism, skepticism) and everyday rationalities (Merton 1942; Knoblauch, 2025: 42). Second, EToS views scientific knowledge as dynamic, evolving through continuous critique and adaptation. It champions reflexivity, where scientists critically reflect on their research processes, methods, and the societal implications of their work. This approach pushes beyond mere critique of results to include an examination of the procedures and the act of critique itself. This means that science as a social endeavor requires continuous scrutiny of its methods, theories, societal entanglements and consequences (Knoblauch, 2025). And that it employs a methodology that empirically studies how social science methods function in practice, moving beyond theoretical prescriptions to address HPSS critiques of positivist detachment and center interpretation as crucial for rigor (Knoblauch, 2025).

Having clarified the theoretical framework underpinning the argument, the next section presents perspectives that illuminate the challenges of using LLMs as a methodological tool.

2.2 Large language models and methodological challenges

It seems as if the legacy of mechanical objectivity finds renewed expression in the rise of machine learning and LLMs. Like the nineteenth-century ideal of the passive observer, these systems are often portrayed as "neutral experts", capable of scalable, consistent, and reproducible pattern recognition across massive datasets (Schwitter, 2025; Selby et al., 2024). Nonetheless, this promise of neutrality has come under increasing scrutiny (Jacomy and Borra, 2024, Saba, 2023). Studies have revealed that LLMs frequently exhibit overconfidence, opacity, and epistemic instability, raising concerns about their transparency and validity as instruments of knowledge production (Ghafouri et al., 2024). Lin et al. (2025) demonstrate that the reliability of LLM-based classifications in social science varies with task complexity. Moreover, the proprietary infrastructures underpinning these models obscure critical questions about data provenance and interpretive accountability (Ji et al., 2023; Kapoor, 2024). This directly engages the demarcation problem, as opaque infrastructures blur boundaries between scientific and commercial knowledge regimes, complicating HPSS concerns about institutional independence (Knoblauch, 2025).

Far from being neutral tools, LLMs encode and reproduce the social, political, and economic asymmetries embedded in their training data (Stammbach et al., 2024; Goldshtein et al., 2024). Such biases are not mere technical flaws but reflect deeper forms of epistemic injustice that privilege dominant languages and worldviews (Helm et al., 2024). As Poibeau (2024) argues, such "biases" in LLMs must be understood as expressions of their social contexts. Resonating with EToS, this shows how macro-level boundary work (e.g., corporate AI governance, data ownership regimes) mediates epistemic authority, determining what counts as "valid" knowledge while obscuring power asymmetries (Knoblauch, 2025: 39).

Against this backdrop, qualitative research offers a vital counterweight by attending to context and meaning-making. Its focus on reflexivity (Knoblauch, 2025: 34) demands that LLM-based research critically examines its own conditions, including how infrastructures shape data hermeneutics and power relations. In emerging hybrid methodologies, scalability must be paired with reflexive engagement, treating LLMs not as oracles but as sociotechnical actors requiring dialogical accountability.

The data interview approach embodies this EToS-informed reflexivity by framing LLM outputs as dialogical artifacts, socially constructed objectivations, enabling critical co-analysis of how knowledge is negotiated across human/machine boundaries. Digital infrastructures, from search engines to social media platforms, increasingly shape how historical knowledge is produced, circulated, and archived.

Our study employs LLMs to examine how such knowledge dynamics unfold in the digital public sphere, specifically in how historical narratives are articulated on platforms like Facebook (Hill and Puschmann, 2025). Using LLM-assisted analysis, we identifed thematic patterns and rhetorical strategies in user-generated content that contribute to the construction of collective historical meanings and identities. A recurring motif across these narratives was the nostalgic idealization of a morally superior, homogeneous past, a discursive move that underpins nationalist and conservative imaginaries rejecting pluralism and democratic contestation. This nostalgia-driven framing not only simplifies complex social processes like migration and cultural change but also delineates symbolic boundaries of belonging, excluding dissenting voices and marginalizing minorities. Through this lens, LLM-assisted analysis was helpful to detect linguistic patterns. It revealed how digital infrastructures and user generated discourses co-produce exclusionary imaginaries.

3. Retrieval-augmented generation (RAG)

RAG represents a methodological development in computational text analysis (Lewis et al., 2020). This hybrid architecture combines a generative language model with a retrieval system that accesses external sources, ranging from public repositories like Wikipedia to researcher-curated datasets, to incorporate contextual knowledge into its outputs. By dynamically integrating retrieved documents, RAG aims to enhance standardization and reproducibility compared to conventional LLMs reliant on static pre-trained knowledge. This approach may reduce reliance on potentially outdated information and could be advantageous for tasks requiring current or domain-specific knowledge. It offers a framework for integrating large-scale pattern detection (distant reading) with context-sensitive analysis (close reading) (Simons et al., 2026).

Figure 1: Overview of the RAG approach from Lewis et al. (2020)

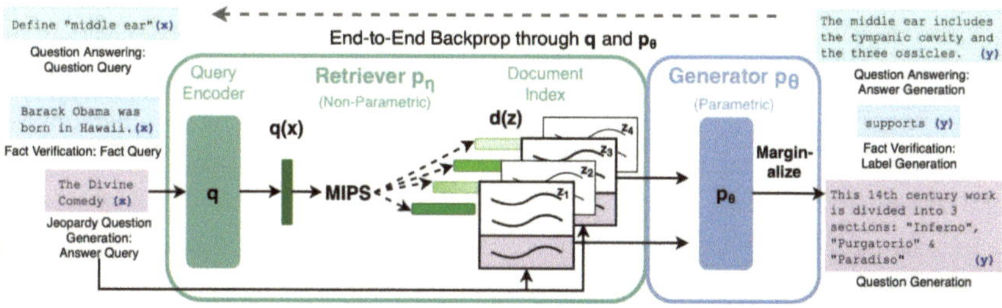

Figure 1 taken from Lewis et al. (2020)

Commercial implementations (e.g., Cohere AI's Command-R, 2024) enable flexible retrieval across public/private sources, presenting potential applications for hypothesis generation and iterative interpretation (Krähnke et al., 2025). RAG-based applications are today integrated into many common data processing software programs (e.g., MaxQDA). Still, RAG's epistemic validity is contingent. Its outputs are shaped by researcher decisions (source selection, curation), institutional factors (access to proprietary databases), and socio-technical infrastructures (ranking algorithms, data availability) that govern knowledge retrieval and contextualization.

In this perspective RAG exemplifies relational knowledge production, where machine outputs remain intertwined with socially shaped communication. Within an EToS perspective, RAG scales interpretive practice but does not replace human understanding; it expands opportunities for reflexivity regarding knowledge selection and legitimization. RAG occupies an epistemic middle ground, underscoring that researcher judgment and reflexivity are constitutive of knowledge production. It renders visible the sociotechnical negotiations inherent in constructing meaning within digital scholarship. These tools simultaneously necessitate critical scrutiny of data quality, transparency, and epistemic responsibility. Validation becomes dialogical, interrogating both what is retrieved and how it is legitimized as insight (cf. Meding and Daugs, 2026).

Following this overview of RAG, the next section describes its application in the data interview methodology.

4.The data interview

Building on the epistemological critiques outlined above, this study applies a hybrid, reflexive methodology centered on what we term the data interview. Rooted in RAG, this approach tries to combines the scalability of automated language models with the interpretive depth of qualitative inquiry. The data interview does not treat model outputs as neutral findings, but as dialogical materials generated in response to formulated research questions and subsequently subjected to critical, interpretive scrutiny. Below, we outline the process step by step to facilitate transparency and reproducibility.

Data Collection and Corpus Construction: The dataset used in our study consists of public comments posted on Facebook between 2010 and 2019, collected from German-language political groups. These groups represented a range of political perspectives, primarily focused on right-wing populism, extremism, and grassroots movements. The primary topic of discussion in these groups was immigration, particularly in the context of the so-called European "refugee crisis" of 2015–2016. A significant portion of the comments expressed support for the Alternative for Germany (AfD), while others aligned with no specific political party. This diversity allowed for the exploration of various political discourses within the dataset. The data were gathered using Facebook's Graph API, a tool that, during the time of data collection (2015–2019), provided extensive access to public posts and comments.

4.1 Data preparation and structuring

In compliance with the General Data Protection Regulation (GDPR), all personally identifiable information was removed from the dataset. This included details such as user IDs, group memberships, and any identifiable metadata. Furthermore, to protect user privacy, any non-essential metadata was deleted from the final dataset, leaving only the comments and their corresponding post dates. The raw text corpus was processed into machine-readable formats (e.g., .csv) compatible with standard natural language processing workflows.

4.2 Sampling

From an initial dataset of 2.5 million comments, we then sampled the material. First, we selected comments containing a set of keywords based on relevance to our research topic (see Table A). We also excluded very short and very long comments. We then drew eight random samples of 5,000 comments each.

Table A: Terms used to sample Facebook comments

altparteien	europa	juden	politische
angst	familie	kind	politischen
arbeit	flüchtlinge	kinder	polizei
arbeiten	flüchtlingen	kindern	putin
asylanten	frau	kirche	rechte
ausländer	frauen	kosten	regierung
bevölkerung	freiheit	krieg	religion
bezahlen	geld	kultur	russland
brd	geschichte	land	sicherheit
bürger	gesellschaft	leben	staat
christen	gesetze	macht	staaten
damals	gewalt	mann	system
ddr	gewählt	medien	usa
demokratie	glauben	meinung	volk
deutsch	gott	menschen	wahlen
deutsche	grenzen	migranten	wahrheit
deutschen	gutmenschen	moslems	welt
deutscher	heimat	muslime	wählen
deutschland	heute	männer	wähler
deutschlands	islam	nazi	zeit
ehrlich	jahr	nazis	zukunft
eltern	jahre	politik	
euro	jahren	politiker	

4.3 RAG-driven data interviews

Each of the four team members conducted two data interviews with the data based on RAG, asking it a set of questions related to the style and content of the comments. Researchers formulate open-ended questions based on the research focus (e.g., "How is migration discussed in the data?"). The model responds by retrieving relevant content from the dataset and generating answers. These responses simulate semi-structured interviews, with the dataset itself as the interlocutor. This process was dialogical and iterative: researchers engage the model in multi-turn exchanges, refining prompts, questioning outputs, and triangulating responses. The result is a corpus of model-generated texts that reflect both computational synthesis and researcher-led framing.

Figure 2: Screenshot of the data interview

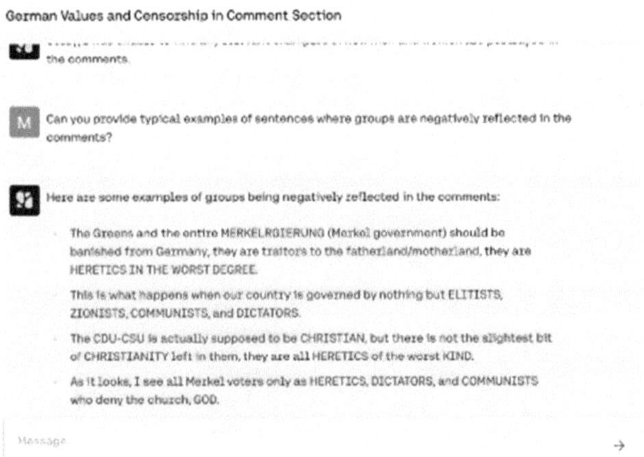

4.4 Thematic coding and interpretive synthesis

Model outputs underwent collaborative open coding following qualitative content analysis principles (Kuckartz, 2012). Crucially, coders cross-verified each response against the other interviews to assess grounding in original comments, revealing instances where model summaries inadequately reflected input data. Through iterative discussion and comparative analysis, themes (e.g., nostalgia, moral decline) were refined into higher-order constructs.

5. Limitations

While RAG enhanced scalability in our mixed-methods approach (via source citation tracing, team interpretation, and subsample comparisons), three key limitations persisted: classification accuracy, example diversity, and normative bias. These challenges directly engage core concerns of EToS regarding the boundaries, reflexivity, and infrastructure-dependence of knowledge production (Knoblauch, 2025).

5.1 Challenges in classification

Despite generally strong performance with informal, digital language, the model occasionally misclassified comments, especially when data lacked sufficient context or interpretive cues. In some instances, the LLM assigned texts to themes that did not align with human interpretations or cited comments that appeared irrelevant, overly general, or insufficiently grounded in the underlying discourse. While adjusting prompts and providing more precise instructions improved classification precision, misinterpretations could not be eliminated. These episodes underscore the limits of automated pattern recognition in capturing subtle contextual and rhetorical cues central to qualitative analysis. These instances also underscore the demarcation challenge in defining knowledge/non knowledge under computational conditions as opaque infrastructures blur boundaries between scientific and commercial knowledge regimes (EToS Demarcation Problem).

5.2 Limited diversity of illustrative examples

A further limitation involved the model's tendency to reuse identical or near-identical examples across multiple themes, even when these categories were conceptually distinct or even in tension with one another. This repetition of "evidence" risked flattening thematic variation, thereby reducing the representativeness and interpretive depth of the analysis. In response, we revised the examples, thereby reinforcing the negotiation and defense of the boundaries between science and non-science (EToS Demarcation Problem).

5.3 Normative bias and ethical concerns

Particularly in dealing with sensitive or extreme content, the model exhibited normative blind spots. On several occasions, it reproduced problematic rhetorical structures, for instance, by presenting extreme or discriminatory views in a falsely balanced tone, implicitly legitimizing them as equivalent to mainstream positions. This false balance effect required active researcher intervention to identify and correct such distortions. Moreover, the model sometimes downplayed the severity of offensive or exclusionary language, further highlighting the need for critical human oversight.

These limitations are not unique to LLM-based approaches. Human coders likewise grapple with ambiguity, bias, and inconsistency in interpretation. However, the opacity and scale of LLMs necessitate layered validation strategies. Our approach included multiple, independently conducted data interviews, cross-checking of model outputs, and iterative team-based refinement. These practices helped mitigate distortions and enhance the reliability and accountability of our findings.

Beyond these interpretive issues, broader ethical considerations arise from deploying proprietary LLMs in social research. Issues of data privacy, algorithmic opacity, and potential commercial exploitation cannot be overlooked. While we ensured full anonymization and removed identifiable metadata, the commercial use of third-party LLM systems raises unresolved questions around ownership, consent, and the governance of sensitive digital content. Researchers must thus treat LLMs not as neutral tools, but as sociotechnical actors requiring critical co-governance.

5.4 Reflexivity

Engaging with LLMs and expansive digital datasets has introduced a profound shift in the role of the qualitative researcher. Khutsishvili (2026) diagnoses a profound destabilization of the scientist's role: As AI emerges as a new agent of epistemic authority, the human researcher loses their position as the undisputed "generator and guardian of knowledge" challenging not just professional relevance but the very philosophical self-understanding of the scientist. This crisis of epistemic agency raises the existential question central to our era: What remains for the scientist to be when machines occupy the locus of knowledge creation? Correspondingly, in this study the role of the researcher is challenged. While the capacity to work with vast volumes of user-generated content offers new analytic possibilities and scope, it simultaneously introduces tensions that cut to the core of qualitative epistemologies. One central challenge lies in managing the dynamic between human and computational verification. Although LLM-generated responses can surface thematic patterns and guide interpretive exploration, they cannot be accepted at face value. The researcher's role is thus reasserted not as a passive recipient of machine outputs, but as a critical interpreter who systematically verifies these responses against the primary corpus (EToS Reflexivity). This entails ongoing engagement with context, subtlety, and ambiguity, dimensions that often escape RAG driven reduction. In this sense, qualitative researchers act as epistemic safeguards, ensuring that meaning is not flattened in the name of computational efficiency.

A further transformation emerges in the researcher's role as a curator and cleaner of data. Working at digital scale requires active stewardship: identifying noise, inconsistencies, and irrelevant data points that could distort interpretation. This often-invisible labor of dataset refinement extends traditional responsibilities and reflects the growing entanglement of qualitative inquiry with data infrastructure. The qualitative researcher becomes a critical curator, not only between data and theory, but also between data and platform logics.

At the same time, many researchers experience a shifting sense of ownership and control. Proprietary and often opaque, LLM systems mediate knowledge production in ways that can obscure the boundary between human insight and RAG synthesis. This raises intellectual and ethical questions about autonomy: Who authors the analysis? Who bears responsibility for its claims? These concerns invite continued reflexivity about the distribution of agency within LLM-assisted research.

Despite the affordances of automation, the irreplaceable value of human-centered methods remains. Interviews, ethnographies, and participant observation capture lived experience, affect, and context in ways that LLMs cannot replicate. These practices are not residual but foundational, anchoring the interpretive depth and relational reflexivity that define qualitative research. What is at stake is not whether digital tools will replace such methods, but how they are integrated without eclipsing the human dimension of inquiry. Navigating this evolving terrain demands a careful balancing act.

6. Conclusion

At the heart of this study lies the data interview: a hybrid methodological practice integrating RAG into qualitative workflows. LLMs might expand interpretive research by automating pattern detection and preliminary structuring of large corpora, enhancing efficiency when rigorously validated. Nonetheless human oversight remains indispensable for contextual synthesis, ethical judgment, and critical infrastructure awareness. Our analysis reveals that integrating RAG into qualitative workflows, while enhancing scalability through citation tracing and team interpretation, confronts irreducible limitations in classification accuracy, example diversity, and normative bias. These challenges are not merely technical but epistemologically constitutive: they expose demarcation problems in defining valid knowledge (e.g., misclassifications blurring context-dependent meaning), necessitate ongoing negotiation of evidence (e.g., curating repetitive examples), and demand vigilant ethical intervention (e.g., correcting false-balance distortions) (Knoblauch, 2025). Moreover, our approach advances a call to historicize digital knowledge production. Echoing Foucault (1977) and Chakrabarty (2013), we situate LLMs within their historical and political conditions of emergence. Rather than treating RAG as a neutral innovation, we frame it as part of ongoing struggles over knowledge legitimization and circulation. Thus, the data interview does not merely use LLMs to study history, it positions LLMs themselves as historizing entities.

Throughout our analysis, we treated LLM responses as provisional objectivations: prompts for critical engagement rather than truth claims. Drawing on EToS, we conceptualized LLMs not as neutral instruments, but as sociotechnical actors embedded in normative frameworks, institutional agendas, and communication practices. Consequently, the researcher's role transformed from an "mechanical" or "objective" interpreter to an interested epistemic safeguard: verifying outputs against primary data, stewarding dataset integrity, and maintaining dialogical accountability. This redefined agency, though destabilized by LLMs encroachment on knowledge production proves indispensable (Khutsishvili, 2026). Human sensemaking remains the anchor for contextual nuance, ethical synthesis, and infrastructural critique. By treating LLMs as collaborative provocateurs, not autonomous analysts, we leverage their capacity to scale inquiry while anchoring knowledge production in the interpretive rigor demanded by qualitative traditions and EToS.[1]

References

Adorno TW, Albert H, Dahrendorf R et al. (1993) Der Positivismusstreit in der deutschen Soziologie. 6. Aufl. Darmstadt/Neuwied: Luchterhand; Frankfurt am Main: Deutscher Taschenbuch Verlag.

Bender EM, Gebru T, McMillan-Major A and Shmitchell S (2021) On the dangers of stochastic parrots: Can language models be too big? Proceedings of the 2021 ACM Conference on Fairness, Accountability, and Transparency, 610–623. New York, NY: Association for Computing Machinery. DOI: 10.1145/3442188.3445922.

Berger PL and Luckmann T (1967) The Social Construction of Reality: A Treatise in the Sociology of Knowledge. Garden City, NY: Anchor Books.

Berry, DM (2011). The computational turn: Thinking about the digital humanities. *Culture machine, 12.*

Chakrabarty D (2013[1992]) Europa provinzialisieren: Postkolonialität und Kritik der Geschichte. In: Conrad S, Randeria S and Römhild R (eds) Jenseits des Eurozentrismus.

1 This chapter was written with support from Large Language Models (LLMs). All ideas and human-generated text were reviewed by LLMs and revised where necessary. The authors bear full responsibility for the final version. For details on the use of LLMs in this volume, see the statement in the volume's introduction.

Postkoloniale Perspektiven in der Geschichts- und Kulturwissenschaft. 2., erweiterte Auflage. Frankfurt/Main: Campus, 134–161.

Cohere AI (2024) Command-R model documentation. Available at: https://docs.cohere.com/docs/command-r (accessed 04 November 2025).

Dalton CM, Taylor L and Thatcher J (2016) Critical data studies: A dialog on data and space. Big Data & Society, 3(1): 1–9.

Daston L and Galison P (2007) Objectivity. New York: Zone Books.

Daston L (2001) Die Kultur der wissenschaftlichen Objektivität. In: Hagner M (ed) Ansichten der Wissenschaftsgeschichte. Frankfurt a. M: De Gruyter, 137–160.

Flick U (2006) An Introduction to Qualitative Research. 3rd ed. London: Sage.

Flick U, Kardorff Ev and Steinke I (2004) What is qualitative research? An introduction to the field. In: Flick U, Kardorff Ev and Steinke I (eds) A Companion to Qualitative Research. London: Sage, 13–22.

Frisby D (1972) The Popper-Adorno controversy. The methodological dispute in German sociology. Philosophy of the Social Sciences, 2: 105–119.

Foucault M (1977) Discipline and Punish: The Birth of the Prison. Translated by Sheridan A. New York: Pantheon Books; later by Vintage Books.

Garfinkel H (1967) Studies in Ethnomethodology. Englewood Cliffs: Prentice Hall.

Geertz C (1973) Thick description: Toward an interpretive theory of culture. In: The Interpretation of Cultures. New York: Basic Books, 3–30.

Gerlitz C and Rieder B (2018) Tweets are not created equal: Investigating Twitter's retweet and reply functionality through labor in the digital economy. Big Data & Society, 5(1).

Ghafouri B, Mohammadzadeh S, Zhou J et al. (2024) Epistemic integrity in large language models. arXiv preprint arXiv:2411.06528. DOI: 10.48550/arXiv.2411.06528.

Gillespie T (2024) Generative AI and the politics of visibility. Big Data & Society, 11(2). DOI: 10.1177/20539517241252131.

Goffman E (1981) Forms of Talk. Philadelphia, PA: Penn Press.

Goldshtein M, Ocumpaugh J, Potter A and Roscoe RD (2024) The social consequences of language technologies and their underlying language ideologies. In: Antona M and Stephanidis C (eds) Universal Access in Human-Computer Interaction. HCII 2024. Lecture Notes in Computer Science, vol. 14696. Cham: Springer. DOI: 10.1007/978-3-031-60875-9_18.

Hammersley M and Atkinson P (2007[1983]) Ethnography: Principles in Practice. 3rd ed. London: Routledge.

Helm P, Bella G, Koch G et al. (2024) Diversity and language technology: How language modeling bias causes epistemic injustice. Ethics and Information Technology, 26: 8. DOI: 10.1007/s10676-023-09742-6.

Hill M and Puschmann C (2025) Nostalgic for a better democracy? How German alternative right-wing media commentators (mis)remember the past. Publizistik, 70(1): 181–204. DOI: 10.1007/s11616-025-00891-w.

Jacomy M and Borra E (2024) Measuring LLM self-consistency: Unknown unknowns in knowing machines. Sociologica, 18(2): 25–65. DOI: 10.6092/issn.1971-8853/19488.

Ji Z, Lee N, Frieske R et al. (2023) Survey of hallucination in natural language generation. ACM Computing Surveys, 55(12): 1–38. DOI: 10.1145/3571730.

Kapoor S, Gruver N, Roberts M et al. (2024) Large language models must be taught to know what they don't know. arXiv preprint arXiv:2406.08391. DOI: 10.48550/arXiv.2406.08391.

Khutsishvili K (2026) AI and the scientist. On the fracture of epistemic authority. In: Simons A, Wüthrich A, Zichert M, et al. (eds) *Understanding Science with Large Language Models? Potentials for the History, Philosophy, and Sociology of Science*. Bielefeld: transcript, part-1.

Knoblauch H (2017) Die kommunikative Konstruktion der Wirklichkeit. Wiesbaden: Springer VS. DOI: 10.1007/978-3-658-15218-5.

Knoblauch H (2025) Empirical theory of science – A research program. Historical Social Research, 50(2): 32–53. In: Knoblauch H, Baur N, Steets S and Marguin S (eds) Special Issue: Towards an Empirical Theory of Science?, 32–53.

Knorr-Cetina K (1981) The Manufacture of Knowledge: An Essay on the Constructivist and Contextual Nature of Science. Oxford: Pergamon Press.

Knorr-Cetina K (1999) Epistemic Cultures: How the Sciences Make Knowledge. Cambridge, MA: Harvard University Press.

Krähnke U, Pehl T and Dresing T (2025) Hybride Interpretation textbasierter Daten mit dialogisch integrierten LLMs: Zur Nutzung generativer KI in der qualitativen Forschung. Available at: https://nbn-resolving.org/urn:nbn-de:0168-ssoar-99389-7.

Kuckartz U (2012) Qualitative Inhaltsanalyse. Weinheim: Beltz.

Latour B and Woolgar S (1979) Laboratory Life: The Social Construction of Scientific Facts. Beverly Hills: Sage.

Latour B, Jensen P, Venturini T, Grauwin S and Boullier D (2012) "The whole is always smaller than its parts": A digital test of Gabriel Tarde's monads. The British Journal of Sociology, 63(4): 590–615. DOI: 10.1111/j.1468-4446.2012.01428.

Lewis P, Perez E, Piktus A et al. (2020) Retrieval-augmented generation for knowledge-intensive NLP tasks. Advances in Neural Information Processing Systems, 33: 9459–9474. Available at: https://arxiv.org/abs/2005.11401.

Lin H and Zhang Y (2025) Navigating the risks of using large language models for text annotation in social science research. Social Science Computer Review, 08944393251366243.

Lynch M (2013) Ontography: Investigating the production of things, deflating ontology. Social Studies of Science, 43(3): 444–462. DOI: 10.1177/0306312713475925.

Marres N (2017) Digital Sociology: The Reinvention of Social Research. Cambridge: Polity Press.

Meding H and Daugs A (2026) From RAGs to rich responses. Enhancing LLM reliability through retrieval-augmented generation. In: Simons A, Wüthrich A, Zichert M, et al. (eds) *Understanding Science with Large Language Models? Potentials for the History, Philosophy, and Sociology of Science*. Bielefeld: transcript, part-5.

Merton RK ([1942] 1973) The normative structure of science. In: The Sociology of Science: Theoretical and Empirical Investigations. Chicago: University of Chicago Press, 267–278.

Mignolo W (2021) The Politics of Decolonial Investigations. Durham, NC: Duke University Press.

Noble SU (2018) Algorithms of Oppression: How Search Engines Reinforce Racism. New York: NYU Press.

Popper K (1989) Logik der Forschung. 9. Auflage. Tübingen: Mohr.

Poibeau T (2024) Bias, subjectivity and norm in large language models. Aequitas (Fairness and Bias in AI), October, Saint Jacques de Compostelle, Spain.

Saba WS (2023) Stochastic LLMs do not understand language: Towards symbolic, explainable and ontologically based LLMs. In: Almeida JPA, Borbinha J, Guizzardi G, Link S and Zdravkovic J (eds) Conceptual Modeling. Cham: Springer, 3–19. DOI: 10.1007/978-3-031-47262-6_1.

Schütz A (1972) The Phenomenology of the Social World. Translated by Walsh G and Lehnert F. London: Heinemann. (Original work published 1932)

Simons A, Zichert M and Wüthrich A (2026) Large language models for history, philosophy, and sociology of science: Interpretive uses, methodological challenges, and critical perspectives. *Studies in History and Philosophy of Science* 117: 102151. https://doi.org/10.1016/j.shpsa.2026.102151.

Venturini T and Rogers R (2019) "API-based research" or how can digital sociology and journalism studies learn from the Facebook and Cambridge Analytica data breach. Digital Journalism, 7(4): 532–540.

Co-creation of AI technology, empowering curators of cultural heritage information and guarding research commons

Andrea Scharnhorst, Han Yang, Jetze Touber, Kim Ferguson, Philipp Mayr, and Vyacheslav Tykhonov

1. Introduction – the use of large language models in engineering explorations around long-term archiving

The substance of this paper is the description of the use of Retrieval-Augmented Generation (RAG) – a new methodology for information retrieval – for specific digital collections of cultural assets. The collections are provided by institutions operating in the cultural sector. The topical areas are the humanities and social sciences. More concretely, most of the work presented here was enabled by a European-funded research project MuseIT[1] which is clearly situated in the realm of fostering new technologies for Cultural Heritage.

The International workshop on "Large Language Models for the History, Philosophy, and Sociology of Science" held in Berlin in April 2025[2] enabled us to present engineering explorations around implementing machine-aided knowledge organisation and local generative AI (chatbots). The presentation of our basically practice-born approaches in such a scientific context had various functions. We were able to see if our engineering explorations stand the test of a scientific debate and how they compared to other experimentations around LLMs. But, the setting of the workshop in the field of History, Philosophy, and Sociology of Science came naturally with a reflexive layer on the nature and role of LLMs for academia and beyond.

We adhere to this interaction by presenting a sequence of our experimentations. This sequence is narrated as a specific journey of engineering all executed around a spe-

1 Multisensory, User-centred, Shared cultural Experiences through Interactive Technologies (MuseIT) Project funded 2022 – 2025. Available at https://www.muse-it.eu/ (accessed November 5, 2025).

2 Programme of the workshop. Available at https://www.tu.berlin/hps-mod-sci/llms-for-hpss/workshop-llms-for-hpss (accessed November 5, 2025).

cific data-sharing and archiving platform *Dataverse*[3] (Crosas, 2017). Implementing a local chatbot for collections – a method also known as RAG in Information Retrieval, is the current culmination of this journey. To connect the engineering explorations with the scientific discourse and also with other contributions to these proceedings, we present at the end of this introduction a literature study on the RAG discourse.

In the next section of this paper, we point out how these RAG applications are actually the result of a longer process of enabling agile infrastructures for data sharing and archiving which feed on forefront technological developments, while at the same time adhering to principles of Open Science and empowering institutions in the public sector (Tykhonov, 2020). We return to this more general aspect in the conclusion section of this paper.

The engineering journey we describe in the core of the paper starts from 'archives for everyone' and ends with 'local chatbots for specific collections'. It has been executed with the help of various projects which all came with specific focal points. But, to some extent in one of the last projects in that row, MuseIT, all those earlier experiences came together, and so we use the 'MuseIT – case' to illustrate this engineering quest. The question underlying our own explorative journey has been always: How to improve the interaction with collections of documents – preferably archived in a way which respects the principles of FAIRness (Wilkinson et al., 2016) and Open Science (Bartling and Friesike, 2014). We use the *MuseIT lens* to unfold the various techniques which can be applied to increase accessibility of collections, from machine supported Knowledge Organization Systems (KOS) construction, connecting KOS services, to building search interfaces using a combination of LLM and KOS techniques. One could say this part of the paper is a 'practice description'. We use the concluding section of the paper to unfold a couple of more generic aspects in the description of our use-case.

With this concluding section we respond to the invitation which the editors of these proceedings formulated as such "Rather than offering a conventional edited volume or a delayed 'state of the art' account—likely outdated upon publication—it gathers timely, authentic contributions from a field in motion. This framing invites openness, experimentation, and personal interpretation, encouraging authors to draw on their own styles, interests, and positions within or adjacent to the HPSS community. It reflects the spirit of the workshop—lively, collaborative, and intellectually diverse—and carries that spirit into the published volume."[4] Taking this invitation to heart and still inspired by the workshop discussions themselves, we felt encouraged to add a more reflexive layer to our 'practice report'. We are not sure which of the possible writing-forms the editors listed as alternatives, "traditional essays, dialogues, creative thought experiments, research vignettes, theoretical provocations, agenda-setting pieces, tutorial-style texts, or autotheoretical pieces," best describes our reflections. In any case, we would like to share them in the context of these proceedings because they touch upon dimensions which might also have a relevance for other more methodological or theoretical contributions in these proceedings.

3 More information on Dataverse as a project, its use and best practices, the community around it, and the software. Available at https://dataverse.org/ (accessed November 5, 2025).

4 Personal communication, invitation to submit papers.

All of these additional dimensions, epistemologically spoken, concern science-of-science aspects. They are questions about the nature of knowledge production as executed in our experiments, the role of the stakeholders involved, and the quest for a terminology to discuss innovation in research infrastructures. They partly touch on old fundamental questions of the roles of intelligent machines for knowledge production (Kurzweil, 1990) and the human-machine interaction patterns which we observe, initiate and might want to shape also. But, they also concern the social and economic layer of science and knowledge dynamics, the interplay between public and private actors with different (economic) needs and requirements.

So, in the conclusion of this paper we take a turn towards more general questions such as:

- What is the position of archives and other institutions of the GLAM sector which are stakeholders for public knowledge commons in the current race for new AI technologies?
- How can institutions in the public sector be part of the early adopter cohort and this way also co-design technology?
- To which extent do new AI driven Information Retrieval approaches as described in our use case empower both content producers and content consumers?

To address these questions in the last part of the paper, we turn to the model of the Global Open Research Commons (Treloar and Woodfoord, 2024) as a means to reflect on how institutions in the public sector can stay true to principles of Open Science and Research Commons while embracing the distributed nature of the knowledge production process which includes all societal sectors: government, industry, academia and the wider public.

As said above at the core of the paper, we look into engineering explorations to enhance the accessibility of digital collections presented on the web. In those explorations Retrieval-Augmented Generation (RAG) plays a central role. That's why we end this introduction with summarizing the RAG approach, how it relates to LLMs and how it influences the way we think about Information Retrieval.

1.1 RAG as a new method in Information retrieval

Large Language Models (LLMs) (Brown et al., 2020) encode knowledge implicitly within their parameters during the pre-training phase, typically by optimizing for next-token prediction. This implicit encoding presents challenges for explicit knowledge inspection, modification, or deletion, thereby limiting the model's adaptability to updated or corrected information (Nie et al., 2025; Guirguis et al., 2024; Wang et al., 2025a; Wang et al., 2025b). Consequently, LLMs are prone to generating outdated, inaccurate, or unverifiable outputs. In certain cases, they may fabricate content—referred to as hallucinations — in order to preserve internal coherence or plausibility.

To mitigate these limitations, the Retrieval-Augmented Generation (RAG) framework has been introduced as an effective approach for addressing knowledge-intensive tasks (Lewis et al., 2020). RAG decouples the knowledge component from the generative model by incorporating two primary modules: a retriever and a generator (Lewis et al., 2020). Upon receiving a query, the retriever searches an external knowledge base for semantically relevant documents, typically through embedding-based similarity measures such as cosine similarity. These retrieved documents, together with the query, are then used to construct an augmented prompt that is passed to the generator—often a pre-trained LLM—which synthesizes a natural language response.

This paradigm offers multiple advantages that address core limitations of pre-trained language models (Gao et al., 2023). First, the knowledge base is decoupled from the model parameters, allowing it to be updated independently, thereby enabling timely and flexible incorporation of new information without the need for costly model retraining. Second, by grounding the generation process in retrieved documents, RAG significantly reduces hallucinations, as the model can rely on externally verified sources rather than internal approximations. Third, the system becomes inherently more interpretable and transparent, since each generated answer can be traced back to specific supporting documents in the retrieval set. Finally, RAG is considerably more cost-effective than retraining large-scale LLMs, which not only require substantial computational resources but also risk performance degradation due to catastrophic forgetting (Wei et al., 2022a). To summarize, RAG builds on existing semantic reference bases, similar to how controlled vocabularies or classifications are more specifically defined than natural language expressions.

Having said this, RAG is not per se easy to be interpreted or transparent to the user. It just provides other ways to check a claim or assertion. Compared to traditional information retrieval systems, RAG demonstrates a superior flexibility by enabling users to issue natural language, contextual, or follow-up questions without predefined templates. The semantic matching capabilities of modern retrievers, combined with the generative capacity of LLMs, facilitate a more robust and user-friendly interaction paradigm for information access and reasoning over unstructured data sources (Tykhonov et al., 2025).

In conclusion, the RAG paradigm provides a more robust framework for building controllable, reliable, and eventually more trustworthy AI systems that can adapt to evolving knowledge with greater interpretability and efficiency. But, different epistemologies continue to exist as expressions of representations of knowledge. To judge the results of the query will remain context dependent and related to the information needs of the user. This cannot be solved by machines but requires the human in the loop, for which those machines are supposed to act as tools for this human in her/his epistemic journey not as avatars taking the place of the human.

While RAG has shown significant promise in addressing knowledge-intensive tasks (Arslan, Munawar and Cruz, 2024; Liu, 2024), it remains an evolving paradigm and is subject to several limitations. One major challenge lies in retrieval quality: the retrieved documents may be irrelevant, redundant, or unhelpful to the current query, and in some cases, a retrieval may not even be necessary (Jiang et al., 2023; Asai et al., 2024; Yan et al., 2024). Additionally, the language model itself may be constrained by limited summarization capabilities or may default to relying on its own internal knowledge (Wu et al., 2025). Furthermore, RAG systems often underperform on complex queries, such as multi-hop questions or those requiring a holistic understanding of the entire document set (Wang et al., 2024; Edge et al., 2024).

To address the retrieval quality issue, CorrectiveRAG (Yan et al., 2024) introduces an auxiliary, lightweight language model to estimate the relevance between each retrieved document and the query, assigning a scalar relevance score. Documents with lower relevance scores are subsequently filtered out, thereby improving the overall quality and utility of the input to the generative model.

Additionally, some other research explored modifications to the RAG workflow, aiming to enhance retrieval quality. ActiveRAG (Jiang et al., 2023) proposes a dynamic and iterative retrieval mechanism, where the LLM actively determines when and what to retrieve based on its internal uncertainty. Concretely, the LLM begins with an initial retrieval and proceeds to generate a provisional sentence. If it detects low confidence in predicting the subsequent token, it triggers an additional retrieval operation specific to that uncertain segment. This sentence-by-sentence decision process continues until a complete answer is formed, thus avoiding redundant or irrelevant retrieval steps.

As a further advancement, SelfRAG (Asai et al., 2024) introduces a more introspective control loop to optimize retrieval decision-making. Instead of retrieving documents by default, the LLM first attempts to generate a response without external knowledge, and then reflects on whether retrieval is needed. If retrieval is deemed necessary, it is performed, followed by a reassessment of the relevance and utility of the retrieved documents. This "answer–think–retrieve–think–answer" workflow ensures that retrieval occurs only when beneficial, thereby reducing unnecessary queries and filtering out unhelpful content.

While RAG systems rely on large language models (LLMs) to generate responses based on retrieved documents, they remain vulnerable to hallucinations and misleading evidence, particularly when irrelevant or low-quality documents are included in the context. To address this issue, RankCOT (Wu et al., 2025) introduces a novel approach that integrates Chain-of-Thought (CoT) (Wei et al., 2022b) reasoning with document reranking and summarization to improve generation quality. CoT is a prompting strategy that encourages LLMs to generate intermediate reasoning steps, thereby enhancing transparency and answer reliability. In RankCOT, this strategy is further enhanced by fine-tuning the model using Direct Preference Optimization (DPO) (Rafailov et al., 2023), allowing it to better assess and prioritize documents based on their relevance to the query. This method effectively filters out irrelevant content and reduces the likelihood of hallucinated or inconsistent responses during the generation process.

Although the vanilla RAG framework performs well on many knowledge-intensive tasks, it faces limitations when addressing complex queries requiring global understanding, such as multi-hop reasoning, multi-turn interactions, or questions that require aggregating knowledge dispersed across multiple documents (see also Meding and Daugs, 2026b). These challenges arise in part due to the limited context window of LLMs and the constraints of implicit working memory, which is driven by the self-attention mechanism. As the number of retrieved documents increases, the model may become overwhelmed, leading to degraded performance and fragmented reasoning (Hołyst et al., 2024; Shi et al., 2023; Upadhayay, Behzadan and Karbasi, 2024).

To address this issue, DeepNote (Wang et al., 2024) introduces an external memory module in the form of a structured "notebook". The LLM initializes this notebook with intermediate facts based on the query. It then performs iterative retrievals to refine and expand these notes by incorporating information from relevant documents. After each round of retrieval, the LLM reevaluates the query, updates the notebook, and rethinks a new query if the facts in the notebook remain ambiguous, until the accumulated facts are sufficient to construct a coherent and accurate answer. This process effectively offloads intermediate reasoning steps to a persistent external memory, mitigating the limitations of the context window.

Similarly, GraphRAG (Edge et al., 2024) targets tasks that require understanding inter-document relationships, such as identifying connections between entities across documents. This method extracts entities and relations from the corpus to construct a knowledge graph, which is then used to identify entity communities through clustering. Instead of retrieving raw documents, GraphRAG retrieves and summarizes these communities to serve as input to the LLM. This graph-based abstraction supports more structured, comprehensive reasoning, particularly in scenarios where an answer depends on synthesizing information from a large and semantically interconnected corpus.

In conclusion, the Retrieval-Augmented Generation (RAG) framework represents a promising direction for enhancing the reliability, transparency, and adaptability of AI systems in knowledge-intensive applications. A growing research has focused on extending and refining the RAG methodology to support complex reasoning and mitigate hallucinations. As a result, RAG is rapidly evolving into a more versatile and robust paradigm, increasingly capable of tackling challenging tasks and delivering answers with higher factual accuracy and contextual relevance.

Parallel with the further scientific development of RAG based methods, its description and evaluation, elements of LLMs together with machine readable KOS systems, continuously find their way into engineering explorations. Whatever becomes available in forms of libraries, but also general ideas on what workflows could look like gets roped in daily innovative practices around exciting services or tools. Hereby, as we will see later, often the development of new standards play a role. Those engineering explorations are often agnostic towards a concrete service, but very much informed by the needs of concrete services.

In our paper we describe a workflow which falls methodologically under the RAG approach, but has been developed in a practice context and more specifically related to a specific archival platform *Dataverse*. Our specific take concerning the discourse summarised above is to describe methods from the perspective of them being implemented, applied, used in concrete research infrastructure settings. We return to more general aspects of new AI based technologies (including RAG) in the conclusions.

2. An engineering journey – adoption and exploration of new AI technology (including RAG) in areas of cultural heritage

2.1 The case of the MuseIT project and its specific needs

The MuseIT project enabled a substantial portion of the work presented in this paper. MuseIT stands for "Multisensory, User-centred, Shared cultural Experiences through Interactive Technologies" and is funded by the Horizon Europe programme under the call 'Culture, creativity and inclusive society'. Its mission as stated on the website is: 'Aiming for greater equality, democratization, and social inclusion, MuseIT offers technologies designed to enhance and expand access to cultural assets for individuals with disabilities while also contributing to the preservation and protection of cultural heritage in an inclusive manner.'[5] The project type is a *Research and Innovation Action* (RIA). This is relevant in order to understand the composition of the consortium and the kind of knowledge production executed in the project. RIA's are expected to establish new knowledge and/or to explore new or improved technology and/or services and most broadly any solutions. The project is led by the i-School of the University Borås, Sweden, but the consortium brings together research groups from the universities and large-scale research consortiums with Small and Medium Enterprises and organisations in the cultural sector (Ntonga, Pokorny and Olson, 2023). One of the higher education institutions in the consortium is DANS, Data Archiving and Network Services, an institute of the Royal Netherlands Academy of Arts and Sciences (KNAW), where some of the authors of this paper work. DANS' role for the consortium was primarily to contribute its own expertise in fields of research data management and long-term archiving. DANS itself, while part of the institute portfolio of the KNAW, is not a research institute but a service provider for all questions of digital research data. It also hosts a variety of data archiving services for researchers and research institutions (Morselli, Touber, and Scharnhorst, 2025). Being part of the research infrastructure area, DANS participates in research projects such as MuseIT with the aim to stay in touch with new technological developments, being able to do explorations as RIA types of projects offer, all with the ultimate aim to innovate its own services.

MuseIT combined fundamental research with the quest for infrastructural solutions. To enhance the accessibility of cultural assets, finding new ways to represent those assets was very important. Think here of involving various senses in the interaction with cultural assets. Examples are the development of haptic devices, the description of images by text and sound, or the translation between music and other forms of creativity — all examples for multi-sensory representations. Addressing multimodality was an important topic for the project. MuseIT addressed this topic by exploring new technological ways to support co-creation of cultural assets. This way, MuseIT combined both inclusion and better (web) accessibility concerning the consumption as well as in the production of cultural assets. One of the cornerstones of the project was the further development of a platform (JackTrip™) which enabled artists to make music together on the internet time-lag free (Cáceres and Chafe, 2010). Last but not least, MuseIT also aspired to advance the

5 https://www.muse-it.eu/

ways multimodality in the context of accessibility and inclusion could be made part of documentation and long-term archiving. More concretely, the project plan formulated, "One of the key outcomes of MuseIT is the content repository capable of indexing, storing and retrieving multisensory, multi-layered representation of cultural assets...".[6] This is where the experiences at DANS both with consulting on standard archival techniques as well as with pushing the boundaries of such archival services became relevant.

For both research as well as documentation in the MuseIT project[7], the ways to order knowledge played an important role. A so-called Knowledge Organisation System (KOS)[8] encompasses taxonomies, ontologies, classifications, thesauri, etc.. For collections, the metadata schemes used in indexing the collections are one example for a KOS. But, in each scientific field KOS are also used to annotate the scientific content, representing the canon of relevant and connected concepts that define the heart of a field. In the discourse around Knowledge Organisation Systems, it is not always made explicit that KOS have different epistemic functions. In MuseIT, a lot of ontology engineering took place to develop *new* ways to conceptualise multimodality. While existing KOS were reviewed and taken into account during this scientific journey, in this particular part of the project, the emphasis was on the development of new ways of thinking and categorizing. KOS in such a heuristic function are by default not primarily interoperable (Scharnhorst and Smiraglia, 2021). If one thinks about the use of KOS for indexing content, interoperability is part of making content or data FAIR (Wilkinson et al., 2016). FAIRness of data (or more broadly, any kind of content) is a precondition to scale up information processing operations across various collections of data. Interoperable here means machine-readable in a way that semantic meaning is also transported. But, even in this area of documentation if it comes to the use of KOS, there is always a tension between KOS tailored specifically to the needs of a collection, their curators and their designated user communities, and the request to being 'linkable' across various bodies of knowledge as represented in various collections. This is not a trivial task (Gueret et al., 2013) as much of the work on KOS registry, catalogues and cross-walk workflows can testify (see e.g., Nyberg Åkerström et al., 2024). Inside of MuseIT we find both: ontology engineering as a means to formalise and extend our way to think about inclusion, accessibility and multisensory representations (see e.g., Domingues et al. 2024). But, we also find implementation of KOS in indexing collections, combining existing and *new* aspects in KOS when it comes to multimodality (Tykhonov and Olsen, 2023).

In MuseIT, one of the most recent experiments concerned an implementation of a RAG method in the form of a "Chatting with collections" local chatbot. In this workflow, as typical for RAG (see above) LLMs (sometimes called unstructured data) are combined with KOS (sometimes called structured data). But before one can construct such a chatbot a couple of steps were needed: First to have a repository or archive, second to apply

6 Project proposal, personal communication Scharnhorst.

7 And yes these are two different practices with different norms, values and different skill sets needed in the execution – see (Daga et al. 2023) for more details.

8 KOS is a technical term well defined in the field of Knowledge Organisation, which is situated at the cross-roads of philosophy of science, information and computer science. For the International Society of Knowledge Organisation see https://www.isko.org/ko.html (Accessed November 6, 2025).

suitable KOS for its content, and only then as a third step an experiment to locally implement a chatbot could happen. The next section describes those steps around the MuseIT repository approach (Kontogiannis et al., 2025). We start with the choice of an archival platform to then turn to a specific instance of such an archive on which various experimentations were executed. Describing this engineering journey, we also highlight aspects of the nature of this practice, the preconditions created by project financing, the skills needed and the discussions which have taken place.

2.2 An engineering journey – from 'everyone can archive' to a locally implemented RAG

2.2.1 Everyone can archive

MuseIT very boldly stated in its project workplan that there is a need for a cultural heritage archive for storing multisensory representation. Talking about a repository or archive immediately includes questions of what collection(s) of digital assets should be archived, how to order them and make them accessible for users, and how to adhere to requirements for digital (long-term) preservation. Institutions rely on various approaches. Sometimes they have the capacity to build tailored software solutions in-house, sometimes they rely on commercial services. But, most often ICT capacities to build tailored solutions are not there. This holds for institutions in the higher education sector as well as for institutions in the GLAM sector. A middle way is to rope in the expertise in archiving of institutions operating in the public sector, and to use software which is developed and maintained by open software communities. This is the approach, partners in MuseIT have chosen, and the project provided the means to support consortium members in taking the first hurdle in setting up a digital archive. DANS, as a member of the MuseIT consortium provided this support and eventually the choice for a digital archiving platform went to *The Dataverse Project* (Crosas, 2017). Dataverse is an open source web application to share, publish and explore research data. Developed first at the Institute for Quantitative Social Science at Harvard University, there are currently about 137 installations of Dataverse worldwide with many collections (Dataverses) and datasets. DANS has built its own archival data services for the Dutch higher education institutions (DataverseNL[9]) and for Dutch researchers directly (DANS Data Stations) implementing this software. DANS engineers have also actively contributed in the further development of the software, and – drawing on the funds provided by other European projects – developed ways to make setting up a Dataverse instance easier. This has been described as 'Archive in a Box' (as described in Wittenberg et al., 2022). This significantly lowers the threshold in setting up a repository – so *Anyone can archive*. This is very similar to the ways one could observe how information visualisation has been democratized over the last decade. For the latter see for example the *Places and Spaces*[10] long-term project of Katy Börner as documented in the series of atlases of information

9 More information available at dataverse.nl (Accessed November 8, 2025).

10 More information about the *Places and Spaces: Mapping Science* enterprise can be found at the website: https://scimaps.org/home (Accessed November 10, 2025).

visualisation (Börner, 2010, 2015, 2021; Börner, Record, Theriault 2025) of which the second carries the title *Anyone can map* (Börner, 2025).

The tasks MuseIT wanted to be supported by one or multiple repositories were multifold. Among them was:

1. to have a shared repository during the project lifetime to be used across all work packages;
2. to choose a technical solution which is flexible towards the needs of various collections which emerged in the various use cases of the project (examples are cultural heritage material, but also scientific articles, and ways to directly save data streams from sensors);
3. to have a place to register various ontologies and KOS relevant for aspects of accessibility and multimodal representations;
4. to have a repository with metadata scheme(s) which can be tailored towards the variety of content mentioned above and
5. a variety of means to interact with this content.

Needless to say that the repository should operate on the web, and adhere to FAIR principles concerning machine actions. At the end MuseIT implemented two instances of Dataverse: a general MuseIT Dataverse instance, to manage all project related research datasets. To this general instance partners contributed with datasets, but also by designing the metadata schemes and developing workflows around the instance (Kontogiannis et al., 2025). A second more specifically tailored instance was developed by ShareMusic[11] – one of the cultural heritage partners in MuseIT together with DANS. Around this instance several experimentations involving AI based methods were executed – as we will sketch in the next two subsections.

To summarise, there were several advantages to choosing Dataverse as a software solution. The software project itself operates in the public sphere and adheres to Open Science principles. The repository solution, as we will see, is flexible enough to be both interoperable concerning core standard information on datasets, as well as to allow designing a collection specific, flexible metadata scheme. The existence of a Dataverse global engineering community secures the release of an updated, tested *master version* while at the same time creating a forum for experimenting with the newest technological possibilities.

Concerning AI methods, the Dataverse community as Boyd presented recently currently has more than 10 workflows developed (Fig. 1), which include AI components (Boyd, 2025).[12] Some of them are production ready, some are still in development. In some of them (e.g., *Croissant* (Akthar et al. 2024), and the Model Context Protocol) DANS

---

11 ShareMusic & Performing Arts is a Swedish national knowledge centre for artistic development and inclusion. For more information see their website: https://www.sharemusic.se/en/home (Accessed November 10, 2025).

12 https://dataverse.org/presentations/ai-and-dataverse-project-using-ai-benefit-research-data-community-members

has been involved. As we will see, some of those ideas have also found their way into the MuseIT project.

Figure 1: Overview of AI related workflows in Dataverse – Slide reproduced from Celyn Boyd's presentation September 2025 (Boyd, 2025).

Dataverse AI-Related Resources, Tools & Projects

Resource, Tool or Project	Type	Improves	Status
AI Guide	Resource	User experience	Production
Ask the Data	Tool	Data reusability	Production
Ask Dataverse	Tool	User experience	Experimental
AutoSage	Tool	Metadata quality, User experience	Experimental
Croissant	Tool	Data findability	Production
Enhancing Dataset Metadata Project	Project (Tool)	Metadata quality, User experience	Development
GREI AI Taxonomy	Resource	Understanding of AI roles in RDM	Production
Research Data Metadata Knowledge Graph	Project (Tool)	Data findability, Data reusability	Development
Model Context Protocol (MCP) Server	Tool	Data findability, Data reusability	Production
Spam Detection & Workflow Automation	Project (Tool)	Repository data quality	Investigation
TurboCurator	Tool	Metadata quality, User experience	Production

Dataverse dataverse.org @IQSS Ceilyn Boyd

2.2.2 Customize your archive – build your own knowledge base

In the previous section we already referred to the objectives of MuseIT. These include to increase the accessibility of cultural assets by the development of multisensory representations, enabling engagement by the public regardless of functional or sensory impairments, as well as to develop methodologies for preservation and safeguarding cultural heritage assets. For the creation of representations and the final design of repositories Knowledge Organisation Systems (KOS) are key. KOS are used in defining the metadata describing the cultural assets, but also the ways in which one can interact with those assets.

One of the consortium members of MuseIT, the organisation ShareMusic, came with a specific need concerning a repository. This Swedish cultural organisation describes itself as "a national knowledge centre for artistic development and inclusion" and states on its website: "We work for everyone's right to experience, participate in and practise artistic and cultural activities".[13] ShareMusic organises events of performing arts next to workshops and seminars. All those activities are dedicated to shaping a "new cultural landscape with new expressions and narratives by disabled practitioners"; to "increase[d] knowledge of inclusive practices across the cultural ecosystem" and ultimately to contributing to a richer and more equal cultural life in Sweden — and the world — that reflects the diversity and richness of society".[14] These are just some expressions of the aspirations of this organisation. What makes ShareMusic also special is that to realise these goals the organisation has a long-standing tradition to work together with scientific and other knowledge producing institutions to test new methods to enable both production and consumption of cultural assets, here music, despite any sensoric impediments of

13 Quotation from the website: https://www.sharemusic.se/en/home

14 See Footnote 13.

involved actors. In consequence, their 'collection' of material is wide and varies from scientific articles (as in a classical digital library), to performance recordings and other live data streams which might deserve a long-term preservation, and to newsletters and blog posts about performances. The latter represent ephemera which are often not preserved because of their temporary character. Those traces – if by chance preserved – enable a richer look into the past. To summarize, ShareMusic is an organisation with one foot in the cultural sector and one foot in the innovative processes as executed in academia and other *Mode2* knowledge organisations with a collection of material at hand which is varied by nature.

The aspirations of ShareMusic when designing *an archive* for this material went far beyond having just a place to index and preserve this material. As detailed in (Johansson et al., 2025), ShareMusic wanted to build a Knowledge Base for Art and Inclusion, which conceptualisation embraced new KOS to describe features relevant for accessibility and inclusion. This knowledge base should inform the curation of the material and make the work of those curators easier, but should also enable various different interfaces to interact with the collection on-line.

Again the choice to rely on Dataverse as a technological solution to archive and index these digital traces (data) proved right. Part of the agility of this specific infrastructural approach to archiving lies in the possibility to extend the metadata scheme which usually comes as a standard block for any Dataverse instance. Not only can the metadata schema be extended, it is also possible to automatically connect it to other controlled vocabularies relevant for the specific content of the ShareMusic collection and the specific interactions with this content which have been envisioned. Such connections had been tested for Dataverse in other projects, so workflows were available and could be adapted to this specific case (Myers and Tykhonov, 2023).

At the end, for the ShareMusic instance, Dataverse acted as an 'invisible archive', as a technological backbone complemented by interfaces to support curation and access. As visible in the snapshots of the website[15] below (see Fig. 2) the modalities in which material was digitally documented and ways of interaction were also indicated. During the explorations, further AI supported workflows, e.g. to describe a picture with words, or to sonify an image were tested. The final implemented solutions are rather standard, and the overall accessibility of the web-based interface is not yet fully developed. But, the specific requirements of the organisation and the current implementation of an archive with a set of new features proved an interesting learning experience for both engineers and those working in such a cultural institution.

As Fig. 2 shows for the ShareMusic Dataverse instance an additional metadata block has been designed containing fields as modalities, ways of interaction, and art form. This block also contains further ShareMusic specific elements of knowledge ordering as *Topic*. Using the *Topic* feature, and its subcategories *Topic Name* and *Topic Type*, ShareMusic's data curators are able to create sub-collections around events and/or projects. Those orderings are visible on the *Knowledge Base* interface, which is the envisioned main entrance to engage with the content (see Fig.3). In this specific interface, other metadata features

15 For the Dataverse instance used as backbone see: https://database.sharemusic.se/dataverse/root?q=&types=dataverses%3Adatasets&sort=dateSort&order=desc&page=1

as for instance modalities are emphasised as ways of interaction. It is also this interface, in which a local AI chat bot is hidden behind the toggle *Ask Our Assistant*, the features of which we will describe in the next section.

Fig. 2: Snapshot of the Dataverse instance at ShareMusic, highlighting the added metadata fields.

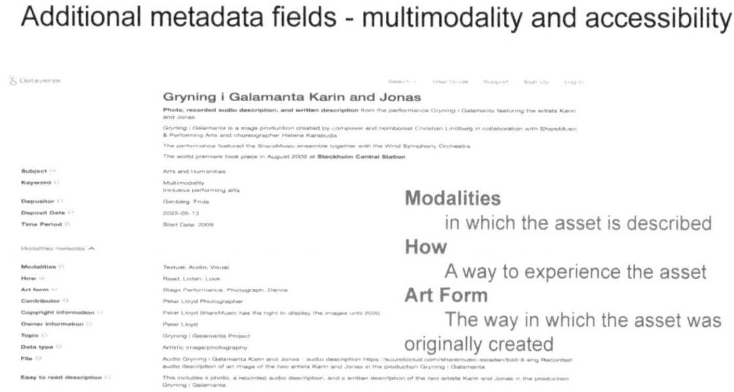

Fig. 3: The Knowledge Base designed for the ShareMusic collection[16]

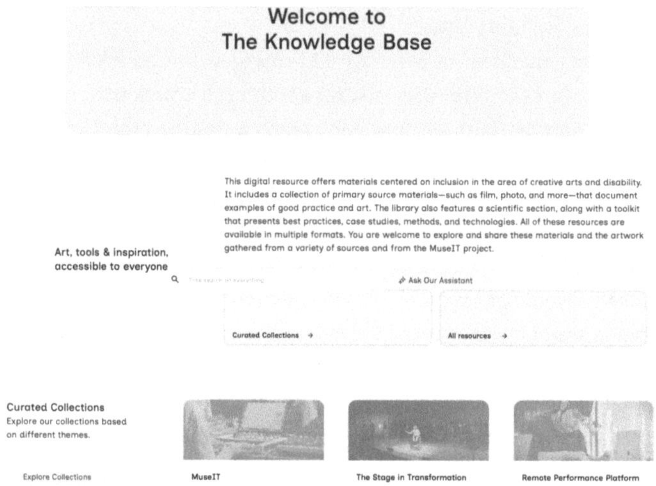

16 Snapshot of the website https://knowledgebase.sharemusic.se/ – the interface to interact with collections (Accessed October 18, 2025).

2.3 Chatting with a collection – LLMs and RAG tested in practice

As described in the previous sections, the engineering journey which led to testing LLMs and more specifically RAG approaches for a concrete archive started with (1) enabling to install archival (open source) software for a wider group of institutions and continued with (2) various explorations to integrate new technological solutions into this archival platform solution. An inbetween step was to enhance the metadata description of the archived content. Here, possibilities to integrate links to other existing machine readable KOS as part of the indexing was one important step (De Vries et al., 2022, Zamborlini et al. 2024). The currently existing universe of linked KOS, which is part of the semantic web technology (see e.g., Merono Penuela, 2016), fosters reuse of KOS and standardisation, but it also adds meaning to content on an unprecedented scale. This – as shown in the short literature review above (see the section Introduction) – is the precondition for any RAG approach. For an institution as DANS which provides archival services, the task to make collections better accessible, and to foster re-use is key. One way to enhance the potential of re-use is to convey meaning to the deposited datasets. How to do this is a very intriguing problem. Semantic meaning is traditionally codified in the metadata scheme models and the values of their parameters. But, it does not translate automatically and easily to a wide range of users. As we pointed out in the introduction, KOS are always specific – meaning here that they represent a distinct specific framework to represent material which evokes specific types of reflections. For the DANS Datastations, which are for good reasons discipline oriented, adding discipline specific information to the metadata is a long standing issue.

When the new generation of Generative AI emerged, the question immediately arose which of those features could be used also for a better Information Retrieval around Digital Archives. In our own internal discussions which were also interdisciplinary by nature, we were looking for metaphors to differentiate between a KOS and a LLM approach. In the first approach meaning to the content would be given through connecting towards various KOS (also sometimes called structured data). In the second approach meaning would be extracted from the content itself directly using emerging patterns and in the case of LLMs this would be textual ones. One metaphor we used in our internal communication was to see LLMs as experts which one can ask in natural language. This would be different from relying on KOS which we saw similar to asking a librarian, who while not familiar in depth with all content, is apt in the meaningful ordering systems for types of content. At the end, the workflow for the implementation of a local chatbot together with the web interface was named *Ghostwriter*. Fig. 4 (as presented by Mayr et al. 2025) gives an overview how the Ghostwriter approach relates to other Information Retrieval approaches.

Fig 4: Position of the Ghostwriter approach in relation to other information retrieval methods. Slide from the presentation (Mayr et al., 2025)

The *Ghostwriter* approach - part of the Retrieval Augmented Generation in Information Retrieval

First challenge to solve a problem: find the right question?
(and right person or information source and interpret the results)

Query	One database representation Me and a database	Need to know the schema and its usual values to get a result	Classic Information retrieval problem: find the right query
Query	One data(collection/space) Connected structured data(bases/graphs) in the background Me and a librarian	The machinery gives back suggestions for possible similar/better queries based on the connection between schema's And a list of possible results for each of the variants	Google Feature and schema.org; machine makes informed guesses about your query Works on the web, not on a more local interaction
Query	Large Language Model Me and a library; Me and a round of experts	Interprets the query as a natural language input, and suggests results again expressed in natural language	
Query	LLM (local) + target data(collection/space) + embedded in a network of additional data interpretation sources (via API's) Me chatting with experts and librarians at the same time	Creates a family of terms around the query; identifies related structured information; Returns a list of results	If applied iteratively can help you to reformulate your question by getting a better understanding what do you actually want to ask and what can you actually ask towards the available data space

Slides from: https://doi.org/10.5281/zenodo.17242902

The precondition for the *Ghostwriter* workflow to function is to express all available information (from the collection in question, but also all contextual enrichment one can connect the collection to on the web) in a machine-readable form ready to be processed. This other workflow preceding *Ghostwriter* was called internally *EverythingData* (Fig. 5).

Fig 5: Schematic overview about the relationship between EverythingData, Ghostwriter and RAG in general (Slide from Mayr et al., 2025).

Ghostwriter and EverythingData

Wider Discourse - Retrieval Augmented Generation (RAG)

Main Ingredients

Vector space constructed from the *content* of data files encoded in embeddings, with properties and their attributes. Embeddings are computed by various ML algorithms and use different LLM's.

Graph represents a *metadata* layer integrated with various ontologies and controlled vocabularies including responsible AI. Graph is expressed by Croissant ML standard.

Vision: Let's get both graph and vectors together in one model (GraphRAG). Let's do this 'locally' as a kind of *Distributed AI*, where the LLM is "interface" between human and AI, and works as a "reasoning engine".
Implementation: LLM connected to the "*RAG library*" (graph), navigate through datasets and consuming embeddings (vectors) as a context.

GenAI Knowledge Graph

The GraphRAG Manifesto: Adding Knowledge to GenAI

Chief Technology Officer, Neo4j

July 11, 2024 21 min read

https://neo4j.com/blog/genai/graphrag-manifesto/
https://en.wikipedia.org/wiki/Retrieval-augmented_generation

(see Arno Simons Presentation: tool box)

Fig 6: Schematic overview about the Ghostwriter workflow (Slide from Mayr et al., 2025).

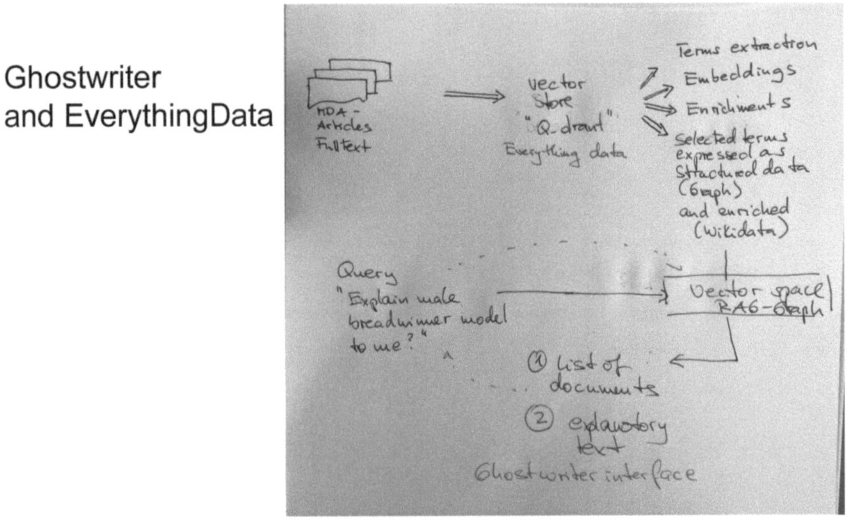

The *Ghostwriter* workflow entails an interface that enables the user to ask questions in natural language. Fig. 6 gives an schematic overview of the workflow. In this figure (Fig. 6) the workflow has been applied to another application case: namely a collection of articles from the *Methods, Data and Analysis* journal (for more details see Tykhonov at el., 2025). The query result is presented as a natural language summary together with a list of relevant hits. The innovation for both LLMs and RAG is that the presented list of results does not rely on term or phrase matches only, but that the suggested sources resonate in some wider sense – either based on patterns in natural language (LLMs) and/or based on concepts expressions (RAG) – with the query. Based on complex stochastic models the idea is to better capture the semantic meaning of the original question, and suggest related sources. Ghostwriter has been tested for various collections. Among them are the webcollection of the Now.Museum (Vion-Dury et al., 2023) and (as mentioned above) a test collection from a paper archive of the journal *Methods, Data and Analysis* (Tykhonov et al., 2025).

The design of the MuseIT ShareMusic repository offered another possibility to test *Ghostwriter*. In this case, as visible in Fig. 3, the Knowledge Base Interface contains a button "Ask our Assistant". This button can be found next to the usual area where a search query can be entered using free text. When clicking on this button, the user is guided to another interface (see Fig. 7). Here, another free text field is available for asking a question about the collection. Fig. 7 displays a snapshot for the question "which performances does the collection contain?". This search once executed leads to a refreshed website which contains an answer given in natural language (Answer section), and a section (Sources) in which a number of most relevant datasets from the collection are listed. In its current form, the interface allows the user to re-iterate their question, while the response always comes in a combination of natural language and lists of hits. This way, the user is invited to explore a collection, rather than being guided in a seamless way.

Fig. 7: Example for the use of the local chatbot on a collection – ShareMusic case

This form of an AI agent can meanwhile be found in many web-based information service systems, the most known mundane one is possibly the AI assistant in Google Search. As we will touch upon in the conclusion, this cannot come as a surprise, if one looks into the collaboration pattern among engineers providing services (commercial and public) for information navigation. The interesting aspect of such 'chatting with …' approaches for a collection in the cultural (heritage) sector, is that the chatbot runs locally. Although the *Ghostwriter* workflow – as described above in the RAG literature review – relies on ready-made LLMs, there is a possibility to use other LLM models developed under greater transparency and with a possibility that such LLM models are also being trained locally. To which extent such 'AI assistants' will be embraced by users, and meet the interaction needs, will have to be studied scientifically. But, the engineering journey in the MuseIT case shows that public sector institutions do not need to wait until commercial solutions are ready. On the contrary, with their specific and often 'out of the box' requirements they play an important role in co-shaping such new forms of technology. We discuss this a bit more systematically in the next concluding section.

3. Conclusion

The exercise of developing the repositories for MuseIT, complete with chatbot for innovative search functionalities, shows on a more abstract level how public knowledge infrastructures can benefit from *trading zones* (Gallison 1997) where a variety of actors (scientists, heritage professionals, engineers, etc.) come together and co-create technology, while reflecting on its meaning and function at the same time. In this conclusion we reflect about the process "Co-creating AI technology'. Here we rely on a specific conceptual model which formalises how actors collaborate in knowledge and research infrastructures, and which allows us to discern the elements necessary to create such a trading zone. This model is the so-called Global Open Research Commons (GORC).

While MuseIT is a specific case, with specific needs, it is also typical for the use/adoption of new technologies among institutions which operate at the boundary between academia and the wider public. As we flagged out before some of the MuseIT partners constitute institutions and services which are part of the research infrastructure. Research infrastructures are often conceptionalised as facilities, instruments, services provided to support research. What is less in the focus are the institutional carriers of those functionalities. Despite the lack of a shared definition of a research infrastructure (Morselli 2023) there is a wider discourse about research infrastructures and a quest for a terminology or model to capture various aspects of its function. One of these models is the Global Open Research Commons (GORC) (Treloar and Woodfoord 2024) developed by an Interest Group of the Research Data Alliance over the last couple years, and meanwhile applied in various contexts (Treloar, Woodford et al. 2025). The motivation to develop the GORC model emerged from the concern to keep scholarly outputs (publications, data, software and workflows) as *commons* – public goods in a situation where knowledge production in and outside academia are genuinely interwoven. As we will use some concepts of this model to reflect on the questions put out in the introduction, let us first give a summary of the model.

3.1 Global open research commons – a lens to look at the adoption of technological innovation

The GORC model is based on a typology of nine essential elements which all need to be in scope when considering to what extent a research infrastructure service, operation of an institution, or a network activity is devoted to Open Science and the idea of Research Commons. Consequently, it can be applied both as a framework for discussion as well as a non-prescriptive 'check-list' of possible considerations. As a heuristic device the model itself is in ongoing development, and the position, names and detailed definition of elements and their relationship will change over time. Still, and this makes it so suitable for this conclusion section, GORC reminds us in a very elegant way that knowledge production combines epistemic and social-communicative processes; that research infrastructure is never only about provided services; that technological innovation is about new ideas, but also about social, communicative, economic and political processes if it comes to the adoption of a technological innovation. The more profound the technological change, the more important these mutual dependencies.

The GORC model is best summarised in the visualisation below (Fig. 8). All elements shown in this schema are high level concepts necessary to be addressed when enabling Research Commons. The five elements placed in the upper part of the visualisation (with a white background) represent the norms and values to be followed when constructing a truly research commons. In them, the human factor prevails. The three elements (in blue) below represent the ways in which interaction with the research infrastructure takes place, and their nature is more technically defined. The middle element *Standards* (in dark blue) stands for a needed connector, coordinator between the norms, values, means and the technical feature. All elements together are needed to create a research commons.

Fig. 8: Essential Elements of the GORC model

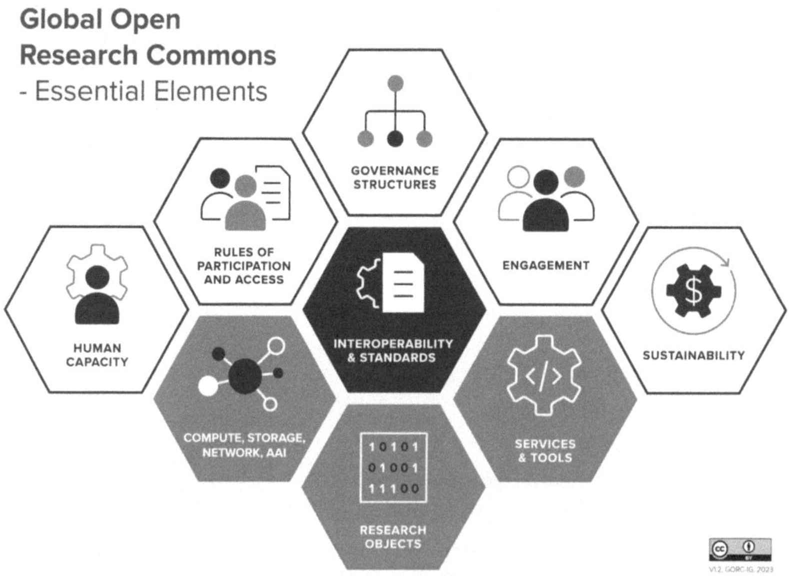

Similar to the concept of *knowledge infrastructures* (Edwards et al., 2013) which a decade earlier shed light on the collective nature of knowledge production and the necessary embeddings for knowledge production to happen, the GORC model highlights the necessity for interwovenness between soft skills or the human factor and the technological aspects down to their tangible manifestation in machines (Treloar et al. 2025).

The model has been already tested by network organisations such as the European Grid Initiative (EGI, https://www.egi.eu/), or national organisations providing research infrastructure services as SURF in the Netherlands, and CSC in Finland. But, it is currently also used to coordinate the establishment of various national and thematic nodes for the European Open Science Cloud (EOSC). Beyond its use in such rather institutional settings, testing the maturity of methodology might be another field of application. New workflows and tools as presented in these proceedings once they become standards and sometimes *services* (in a research infrastructure sense) come with additional requirements which can be described using elements of GORC such as Engagement, Skills and training, or Governance and Sustainability.

But, the GORC model can also be useful to reflect about how projects oblige to Open Science and Research Commons norms. It complements in a way the already required data management plans by adding necessary ordering principles of the processes leading to Open Science. Returning to our concrete use case, the MuseIT project, we find many GORC elements being present in the process. From the networked nature of knowledge production follows that not all elements relevant for research commons are *in one hand*. Different organisations might take care of various of those elements. But only if they interact together based on clearly defined rules – as e.g. represented in standards for interoperability, a research commons can truly emerge and issues of sustainability can be addressed.

In MuseIT, we see scientific institutions and SMEs taking care of specific aspects of the more technical elements. They partly execute fundamental new research for these areas and partly take care of the implementation cases. DANS as a research infrastructure organisation delivered consultancy for a variety of those elements. The work of other organisations – like ShareMusic – could be best located with respect to elements such as Engagement and Sustainability. Part of the success of MuseIT's aspiration concerning setting up a new kind of repository relied on the re-use of technology already developed (Dataverse) which in itself addresses elements of research commons. In particular, the use of standards such as *Croissant* (Akthar et al. 2024) as part of the Dataverse production system, created the foundation on which experiments with RAG could be executed.

3.2 Position of archives and other institutions of the GLAM sector in the current race for new AI technologies

As has been stated very clearly in the review paper of Simons et al. (2026) there is a clear divide in terms of skills between those who build LLMs and experiment with AI technologies and those who 'apply' ready-made tools. As the auhors express, "[i]n this landscape, LLM literacy becomes essential: not just technical skill, but interpretive awareness of how models work, what assumptions they encode, and how they shape inquiry. These issues also raise broader questions of accountability and the politics of research infrastructure". A recent report of SURF on Technological Trends (SURF 2025) explicitly names the lack of 'prompt-engineering' skills as an issue that needs to be addressed. This is nothing new. Already concerning general computational skills it remains important that all participants in building research commons have a basic understanding of fundamental principles on which machines operate. Looking into the history of AI debates can be very enlightening here (e.g., Kurzweil 1990). But, there is also no going back to a situation where everybody involved is a universalist. Current bodies of knowledge are very differentiated, detailed and far too large for one person – even with the help of machines – to overlook. Division of labour or skills and collaboration have been answers for such situations, and the application of AI/LLMs is no exception for this. But, we also know from studies of interdisciplinary work that more is needed to make collaboration effective: (1) arenas need to be created in which actors with various skills can meet and (2) a terminology needs to be found to enable communication across epistemic borders. In other words *trading zones* are needed (Gallison, 1997).

If such *trading zones* are not proactively organised, in the current AI debate various actors naturally will address those issues which fall into the realm of their own competence. For research infrastructure support institutions as DANS, the designated communities for which DANS develops services rightly dominate the discourse. But having dimensions of mature services as a leading principle does not mean one can not engage in the shaping of technology itself. Archivists have never been passive executors of community needs (Borgman, Scharnhorst and Golshan, 2019); GLAM institutions become strong in the adoption process of technological innovations when being able to co-create[17] (Rodighiero et al., 2022).

We know from innovation studies (Rogers 1962) that in the phase of technological change one always observes the phases of Early, Mid and Late Adopters. We know from the same field, that once a technology is ripe and has won a dominating place, a lock-in often is formed and alternatives stand no chance (Bruckner et al., 1994). This is why it is not only a *nice to have* but a necessary condition / *sine qua non* for adopters which epistemically might be farther away from the 'fire' where the technology emerges, to mingle, play, experiment with new technologies. Only by experiences in practice, chances of co-creation and co-shaping technology emerge.

Project funding such as the one enabling MuseIT are the socio-economic bases for a *trading zone* around new AI technologies and their application and use in GLAM settings. Looking at the MuseIT case of co-creating AI through the lens of the GORC model reveals which other ingredients beyond project funding are needed. The MuseIT consortium brings various stakeholders together from academia, (small) industry, and the cultural sector. As said above those stakeholders take care of various of the elements needed to produce research commons. They do this bringing their own expertise, infrastructure and past project results to the table. For DANS, the explorations around AI methods used for archives are part of a longer and broader engineering journey (see previous section). This is a journey that started with enabling a wider adoption of archival software standards (in this case Dataverse), while at the same time updating this same software with new methods, technologies – from machine-based linking of KOS specific KOS to the larger KOS universe to the local installation of a trained chatbot (an AI assistent).

Looking at the path-dependencies in the development of Dataverse itself we also see links to movements like agile or pop-up digital data infrastructures (Tykhonov 2020) which were developed during the COVID pandemics. But, there are also links to current incubator networks such as the *ML Commons* (https://mlcommons.org/working-groups/data/croissant/) which brings together engineers from industry and the public sector with computer scientists and engineers located in academia to explore new standards – like the *Croissant* standard. These new alliances acknowledge the emergence of what sometimes in the history of the AI discourse has been called a 'science-engineer' making a metaphorical reference to the artist-engineer of the renaissance[18] (Haug et al., 2012)

The GORC model reminds us how in the process to achieve research commons – actually for any type of knowledge commons – technical aspects, organisation of workflows and embracing of communities come together. In the case of the Dataverse in-

17 See as an example the AI Explorer of the Harvard Museum https://ai.harvardartmuseums.org/about (Accessed November 20, 2025).

18 Quotation from Historisch-Kritisches Wörterbuch des Marxismus, Bd. 8/1, Künstliche Intelligenz: [Dass KI] "einen neuen Typ von *Wissenschaftler-Ingenieur* hervorgebracht hat, vergleichbar dem *Künstler-Ingenieur* in der Renaissance (RAMMERT 1995, 14), gilt nicht nur für die Ingenieurselite. TURINGS revolutionäre Leistung wurzelt im verallgemeinerbaren Aspekt seiner geschichtlich neuartigen Integration von Hand- und Kopfarbeit, van Theorie und experimentierendem Eingreifen in Maschinerie." Historisch-Kritisches Wörterbuch des Marxismus, Bd. 8/1, Künstliche Intelligenz. For the comparison to the Renaissance see also Zilsel (1942), Rossi (1962), and Long (2011).

stance for ShareMusic in the MuseIT project, the requirements of the organisation to organise, archive, preserve and make accessible of their own collection were leading. The KOS features the organisation eventually chose had to be practical for the working of the guardians and curators of the cultural assets, but also practical for the interaction with the material. Given the specific communities with and for which this organisation works, information retrieval tools had to adhere to new forms of interacting with material. These requirements influenced the search process for which type of AI supported method in multimodal representation would be interesting to be tested. This, as described above, could be the description of an image or the reading of a text. The implementation of a local AI assistant extends the possibilities to 'explore' the content of the collection. As pointed out in the RAG literature review, RAG supports an information retrieval paradigm in which the empowerment of the user is central. Instead of seamlessly paving the way – the reference systems, the KOS as ordering principles, remain visible. The user is guided but not blind-folded, he/she is invited to explore on their own terms.

Let us at the end of these conclusions return to the questions we posed in the introduction and present possible answers based on our experiences.

- What is the position of archives and other institutions of the GLAM sector which are stakeholders for public knowledge commons in the current race for new AI technologies?
  - Archives and other institutions in the GLAM sector are excellent co-creation partners in shaping new AI technologies. They adhere to principles of public knowledge commons and with their specific function come requirements which often pose new puzzles to the engineering process. Practice is eventually the test for any new technology, and the cultural sector often has played a pioneering role due to its variety of needs and practices.
- How can institutions in the public sector be part of the early adopter cohort and this way also co-design technology?
  - To be able to become a partner of equal rights, arenas are needed where co-creation can take place, and trading zones between different epistemics and bodies of knowledge can occur. Project funding combining research and innovation activities can help to create such arenas of encounter and exploration temporally.
  - To ensure sustainability, these projects, at best, are embedded into the expertise of the participating institutions. Those usually operate in wider networks of projects extending along the timeline (from past, present into future). Also other cross-sectoral activities as working groups of network organisation (e.g. on new standards) which bring engineering and scientific expertise from the research and implementation front together support this process of co-creation and co-design.
- To which extent new AI driven Information Retrieval approaches as described in our use case empower both content producers and content consumers?
  - The Retrieval-Augmented Generation (RAG) framework is discussed both from a more fundamental methodological angle as well as for various applications in different contributions to these proceedings (see Büttner, 2026; Grasshoff, 2026; Hill, 2026; Meding and Daugs, 2026a; Schlattmann, 2026; and Simons et

al., 2026). As Meding and Daugs (2026a) formulated "RAG shifts the epistemic burden from the model's internal weights to verifiable information". While RAG by no means is easily traceable and transparent to the user, we agree with Hill (2026) that "RAG occupies an epistemic middle ground, underscoring that researcher judgment and reflexivity are constitutive of knowledge production. So, we see RAG as a promising direction for enhancing the reliability, transparency, and adaptability of AI systems in knowledge-intensive applications. It explicitly builds on existing knowledge as visible in structure information (Merono Penuela et al., 2025). RAG rather aims to empower users to use their own epistemic capacities than to seamlessly (and intransparent) paving their path through the large information spaces.

As these proceedings show, the power of new technologies unfold more forcefully when paired with reflection. If one wishes to influence new technologies, engaging in co-creation from the very start is preferable to debating in the aftermath. The occurrence of lock-in's (Bruckner et al. 1994) seems to be almost unavoidable in technological innovation. In complex systems, as the societies we live in are, it is impossible to find one optimum – right for all needs and interests. But without documenting and reflecting on the processes from a perspective of knowledge and science dynamics – often called meta-level – the chances to influence technological innovation are nil. Such reflections are needed in the sciences themselves, beyond academia adopters, but at best appear as often as possible in those grey areas at the borders between fine and firm lined epistemics spaces where various practices meet. Practices which reach from the creation and preservation of art, performance, culture to engineering machines which support human creativity. The LLM workshop carries the aspiration of such a wide horizon and inspired this specific contribution.[19]

References

Arslan M, Munawar S and Cruz C. (2024) Business insights using RAG–LLMs: a review and case study. *Journal of Decision Systems*, pp. 1–30, doi: 10.1080/12460125.2024.2410040

Akthar M, Benjelloun O, Conforti C, Foschini L, Gijsbers P, Giner Miguelez J, Goswami S, Jain N, Karamousadakis M, Krishna S, Kuchnik M, Lesage S, Lhoest Q, Marcenac P, Maskey M, Mattson P, Oala L, Oderinwale H, Ruyssen P, Santos Rajat Shinde T, Simperl E, Suresh A, Thomas G, Tykhonov V, Vanschoren J, Varma S, Van der Velde J, Vogler S, Wu CJ, Zhang L. (2024). Croissant: A Metadata Format for ML-Ready Datasets. In: *38th Conference on Neural Information Processing Systems (NeurIPS 2024):*

19 In section 1.1. of this chapter DeepL, Grammarly, and ChatGPT have been used for grammar checking and text polishing. Further, no other support of LLM's was used. All model-generated text was reviewed and, where necessary, rewritten by the authors, who remain fully responsible for the final version. For details on the use of LLMs in this volume, see the statement in the volume's introduction.

Track on Datasets and Benchmarks NeurIPS. https://ai.meta.com/research/publications/croissant-a-metadata-format-for-ml-ready-datasets/

Asai A, Wu Z, Wang Y, Sil A, Hajishirziet H (2024) Self-RAG: Learning to retrieve, generate, and critique through self-reflection. In NeurIPS 2023 workshop on instruction tuning and instruction following. Preprint: https://arxiv.org/abs/2310.11511

Bartling S and Friesike S (eds) (2014) *Opening Science: The Evolving Guide on How the Internet is Changing Research, Collaboration and Scholarly Publishing*. Springer Cham https://doi.org/10.1007/978-3-319-00026-8

Borgman C, Scharnhorst A and Golshan MS (2019) Digital data archives as knowledge infrastructures: Mediating data sharing and reuse. *Journal of the Association for Information Science and Technology* 70(8):888-904. DOI: 10.1002/asi.24172; Preprint https://arxiv.org/abs/1802.02689

Börner K (2010) Atlas of Science: Visualizing What We Know. Cambridge MA: MIT Press.

Börner K (2015) Atlas of Knowledge: Anyone Can Map. Cambridge MA: MIT Press.

Börner K (2021) Atlas of Forecasts: Modeling and Mapping Desirable Futures. Cambridge MA: MIT Press.

Börner K, Record EG, Theriault T (2025) Atlas of Macroscopes: Interactive Data Visualizations. Cambridge MA: MIT Press.

Boyd C (2025) AI and The Dataverse Project: Using AI to Benefit Research Data Community Members. Presentation given Thursday, September 11, 2025 at IBICT, Brasília, Brazil. Slides https://dataverse.org/presentations/ai-and-dataverse-project-using-ai-benefit-research-data-community-members (Accessed November 10, 2025).

Brown TB, Mann B, Ryder N, Subbiah M, Kaplan J, Dhariwal P et al. (2020) Language models are few-shot learners. *Advances in neural information processing systems* 33: 1877–1901.

Bruckner E, Ebeling W, Jiménez-Montaño MA, Scharnhorst A. (1994) Hyperselection and Innovation Described by a Stochastic Model of Technological Evolution. In: Leydesdorff L, Van den Besselaar P (Eds) *Evolutionary Economics and Chaos Theory*. London: Pinter, pp. 79–90.

Büttner J (2026) Why pursue temporally-grounded AI for historical disciplines, and what makes it so challenging? In: Simons A, Wüthrich A, Zichert M, et al. (eds) *Understanding Science with Large Language Models? Potentials for the History, Philosophy, and Sociology of Science*. Bielefeld: transcript, part-2.

Cáceres J P, and Chafe C (2010) JackTrip: Under the Hood of an Engine for Network Audio. *Journal of New Music Research* 39(3):183–187. https://doi.org/10.1080/09298215.2010.481361

Crosas M (2017) Cloud Dataverse: A Data Repository Platform for the Cloud. In: CIO Review Open Stack. Available at https://openstack.cioreview.com/cxoinsight/cloud-dataverse-a-data-repository-platform-for-the-cloud-nid-24199-cid-120.html (Accessed November 5, 2025).

Daga E, Daquino M, Fournier-S'niehotta R, Guillotel-Nothmann C, and Scharnhorst A (2023) Documenting the research process. Opportunities and challenges for Bibliometrics and Information Retrieval. In: Frommholz I, Mayr P, Cabanac G, Verberne S and Brennan J (Eds.), *Proceedings of the 13th International Workshop on Bibliometric-enhanced Information Retrieval co-located with 45th European Conference on Informa-*

tion Retrieval (ECIR 2023) pp. 4–20. (CEUR Workshop Proceedings; Vol. 3617). CEUR-WS.org. https://doi.org/10.5281/zenodo.10529113

De Vries J, Tykhonov V, Scharnhorst A, Indarto E, Priddy M, and Admiraal F (2022) Flexible Metadata Schemes for Research Data Repositories. The Common Framework in Dataverse and the CMDI Use Case. In: Monachini M and Eskevich M (Eds.): *Selected Papers from the CLARIN Annual Conference 2021*: Virtual Event, 2021, 27–29 September (Vol. 189, pp. 168–180). (Linköping Electronic Conference Proceedings; Vol. 189). Linköping University Electronic Press, Linköpings universitet. https://doi.org/10.3384/9789179294441

Domingues A, Jain N, Merono Penuela A, and Simperl E (2024) Bringing Back Semantics to Knowledge Graph Embeddings: An Interpretability Approach. In: Besold TR, d'Avila Garcez A, Jimenez-Ruiz E, Madhyastha P, Wagner B, and Confalonieri R (Eds.) *Neural-Symbolic Learning and Reasoning – 18th International Conference, NeSy 2024, Proceedings*, Vol. 14979, pp. 192–203. Article Chapter 17 (Lecture Notes in Computer Science (including subseries Lecture Notes in Artificial Intelligence and Lecture Notes in Bioinformatics); Vol. 14980 LNAI). Springer, Cham. https://doi.org/10.1007/978-3-031-71170-1_17

Edge D, Trinh H, Cheng N, Bradley J, Chao A, Mody A, Truitt S, Metropolitansky D, Ness RO, Larson J (2024) From local to global: A graph RAG approach to query-focused summarization. arXiv preprint arXiv:2404.16130.

Edwards PN, Jackson SJ, Chalmers MK, Bowker GC, Borgman CL, Ribes D, Burton M, and Calvert S (2013) *Knowledge Infrastructures: Intellectual Frameworks and Research Challenges*. Ann Arbor: Deep Blue. http://hdl.handle.net/2027.42/97552 .

Galison P (1997) *Image and Logic: A Material Culture of Microphysics*. Chicago: University of Chicago Press

Gao Y, Xiong Y, Gao X, Jia K, Pan J, Bi Y, Dai Y, Sun J, Wang M, Wang H (2023) Retrieval-augmented generation for large language models: A survey. . arXiv preprint arXiv:2312.10997, 2(1).

Graßhoff G (2026) AI-Reporter: a path to a new genre of scientific communication. From presentation to publication through agentic LLMs. In: Simons A, Wüthrich A, Zichert M, et al. (eds) *Understanding Science with Large Language Models? Potentials for the History, Philosophy, and Sociology of Science*. Bielefeld: transcript, part-5.

Guéret C, Chambers T, Reijnhoudt L, Van der Most F and Scharnhorst A (2013). Genericity versus expressivity – an exercise in semantic interoperable research information systems for Web Science [Digital Libraries, WebSci '13 Conference, Paris] http://arxiv.org/abs/1304.5743

Guirguis K, Eskandar G, Wang M, Kayser M, Monari E, Yang B, Beyerer J (2024) Uncertainty-based forgetting mitigation for generalized few-shot object detection. In Proceedings of the IEEE/CVF Conference on Computer Vision and Pattern Recognition Workshops (CVPRW), Seattle, WA, USA, 2024, pp. 2586–2595, doi: 10.1109/CVPRW63382.2024.00265

Haug WF, Haug F, Jehle P, Küttler W (Eds) (2012) Historisch-kritisches Wörterbuch des Marxismus. Volume 8.1. Publisher Argument.

Hill M (2026) The data interview. Reflexive integration of large language models in qualitative content analysis. In: Simons A, Wüthrich A, Zichert M, et al. (eds) *Understanding*

Science with Large Language Models? Potentials for the History, Philosophy, and Sociology of Science. Bielefeld: transcript, part-5.

Hołyst JA, Mayr P, Thelwall M, Frommholz I, Havlin S, Sela A, Kenett YN, Helic D, Rehar A, Maček SR, Kazienko P, Kajdanowicz T, Biecek P, Szymanski BK and Sienkiewicz J (2024) Protect our environment from information overload. *Nature Human Behaviour* 8(3): 402–403.

Jiang Z, Xu F, Gao L, Sun Z, Liu Q, Dwivedi-Yu J, Yang Y, Callan J, and Neubig G (2023) Active Retrieval Augmented Generation. In: Bouamor H, Pino J, Bali K (Eds) *Proceedings of the 2023 Conference on Empirical Methods in Natural Language Processing*, pages 7969–7992, Singapore. Association for Computational Linguistics. Doi 10.18653/v1/2023.emnlp-main.495

Johansson M, Tykhonov V, Alexandersson S, Ferguson K, Hanlon J, Scharnhorst A. and Osborne N (2025) A Knowledge Base for Arts and Inclusion – The Dataverse Data Archival Platform as a Knowledge Base Management System Enabling Multimodal Accessibility. In: Wei J and Margetis G (eds) *Human-Centered Design, Operation and Evaluation of Mobile Communications*. HCII 2025. Lecture Notes in Computer Science, vol 15824. Springer, Cham. https://doi.org/10.1007/978-3-031-93064-5_19 Preprint available: arXiv:2504.05976

Kontogiannis S, Christodoulou G, Papadopoulos V, Iosif M, Kosmides P, Johansson, M, Darányi S, Van Erven T, Tykhonov V, Ferguson KB, Scharnhorst A, Meroño-Peñuela A, Farina A and McGillivray B (2025) MuseIT repository release. D6.3. (draft) https://pure.knaw.nl/ws/portalfiles/portal/2184570845/D6.3_-_MuseIT_repository_release.pdf

Kurzweil R (1990) *The Age of Intelligent Machines*. Cambridge MA: MIT press.

Long PO (2011) Artisan/Practitioners and the Rise of the New Sciences 1400–1600. Corvallis: Oregon State University Press

Lewis P, Perez E, Piktus A, Petroni F, Karpukhin V and Goyal N (2020) Retrieval-Augmented Generation for Knowledge-Intensive NLP Tasks. In: *Advances in Neural Information Processing Systems 33* (ed Larochelle H, Ranzato M, Hadsell R et al.), Vancouver BC, Canada, 6 – 12 December 2020: 9459–74.

Liu Y (2024) The application of RAG technology in traditional Chinese medicine. In: 2024 International Conference on Artificial Intelligence and Communication (ICAIC 2024) (pp. 402–408). Atlantis Press. DOI: 10.2991/978-94-6463-512-6_43

Mayr P, Tykhonov S, Touber J, and Scharnhorst, A (2025) Chatting with Papers – the mixed use of LLMs and semantic artifacts to support the understanding of science dynamics. Presentation given at the workshop *Large Language Models for the History, Philosophy, and Sociology of Science*, April 2–4, 2025, Technische Universität Berlin, Germany. Organized jointly by: Gerd Graßhoff (HU Berlin), Arno Simons (TU Berlin), Adrian Wüthrich (TU Berlin), and Michael Zichert (TU Berlin). More information available at: https://www.tu.berlin/hps-mod-sci/llms-for-hpss/workshop-llms-for-hpss (Accessed November 10, 2025). Slides Zenodo. https://doi.org/10.5281/zenodo.17242902

Meding H and Daugs A (2026a) From RAGs to rich responses. Enhancing LLM reliability through retrieval-augmented generation. In: Simons A, Wüthrich A, Zichert M, et al. (eds) *Understanding Science with Large Language Models? Potentials for the History, Philosophy, and Sociology of Science*. Bielefeld: transcript, part-5.

Meding H and Daugs A (2026b) On the use and limitations of large language models in historical scholarship. In: Simons A, Wüthrich A, Zichert M, et al. (eds) *Understanding Science with Large Language Models? Potentials for the History, Philosophy, and Sociology of Science*. Bielefeld: transcript, part-1.

Merono Penuela A (2016) *Refining Statistical Data on the Web*. PhD-Thesis Vrije Universiteit Amsterdam

Merono Penuela A, Simperl E, Kurteva A and Reklos I (2025) KG.gov: Knowledge Graphs as the Backbone of Data Governance in AI. *Journal of Web Semantics* 85, 100847.

Morselli F (2023) *Co-creation dynamics in a European Research Infrastructure Consortium (ERIC): The case of the DARIAH-ERIC Working Groups*. PhD-Thesis. Università degli Studi di Verona. https://hdl.handle.net/20.500.14242/115602

Morselli F, Touber J, and Scharnhorst A (2025). Fostering Data Communities – perspective from a Data Archive Service Provider. Preprint. https://arxiv.org/abs/2502.02321

Myers J and Tykhonov V (2023). A Plug-in Approach to Controlled Vocabulary Support in Dataverse. Report. Zenodo. https://doi.org/10.5281/zenodo.8133723

Nie E, Shao B, Wang M, Ding Z, Schmid H, and Schuetze H (2025) BMIKE-53: Investigating Cross-Lingual Knowledge Editing with In-Context Learning. In: Che W, Nabende J, Shutova E, Pilehvar MT (Eds) *Proceedings of the 63rd Annual Meeting of the Association for Computational Linguistics* (Volume 1: Long Papers), pages 16357–16374, Vienna, Austria. Association for Computational Linguistics. DOI: 10.18653/v1/2025.acl-long.798

Ntonga M, Pokorny J, and Olson N (2023) The MuseIT Project: co-designing inclusive technologies for better access to culture. *DigItalia* 18(1): 187–190. https://digitalia.cultura.gov.it/article/view/3009

Nyberg Åkerström W, Baumann K, Corcho O, David R, Le Franc Y, Madon B, Magagna B, Micsik A, Molinaro M, Ojsteršek M, Peroni S, Scharnhorst A, Vogt L, and Widmann H (2024) Developing and implementing the semantic interoperability recommendations of the EOSC Interoperability Framework (18 January 2024). EOSC Semantic Interoperability Task Force. https://doi.org/10.5281/zenodo.10518860

Rafailov, R. et al. (2023) Direct preference optimization: Your language model is secretly a reward model. Advances in neural information processing systems, 36, pp.53728-53741.

Rodighiero D, Derry L, Duhaime D, Kruguer J, Mueller MC, Pietsch C, Schnapp JT, Steward J metaLAB (2022) Surprise machines: Revealing Harvard Art Museums' image collection. *Information Design Journal* 27(1):21-34,DOI: https://doi.org/10.1075/idj.22013.rod

Rogers EM (1962) *Diffusion of Innovations*. New York: The Free Press

Rossi P (1962) I Filosofi e le macchine: 1400–1700. Milano: Feltrinelli

Scharnhorst A and Smiraglia RP (2021) The Need for Knowledge Organization. Introduction to the book Linking Knowledge (Chapter 1). In: Smiraglia RP and Scharnhorst A (Eds.) *Linking Knowledge: Linked Open Data for Knowledge Organization and Visualization*. Baden-Baden: Ergon –Nomos. Pp. 1–23, https://doi.org/10.5771/9783956506611-1 , Preprint: http://arxiv.org/abs/2204.14041

Schlattmann R, Kaye A and Vogl M (2026) From source to structure. Extracting knowledge graphs with LLMs. In: Simons A, Wüthrich A, Zichert M, et al. (eds) *Understanding Science with Large Language Models? Potentials for the History, Philosophy, and Sociology of Science*. Bielefeld: transcript, part-4.

Shi, F. et al. (2023) Large language models can be easily distracted by irrelevant context. In: International Conference on Machine Learning (pp. 31210–31227). PMLR.

Simons A, Arnaout H and Gurevych I (2026) Reconstructive citation context analysis using large language models. A roadmap. In: Simons A, Wüthrich A, Zichert M, et al. (eds) *Understanding Science with Large Language Models? Potentials for the History, Philosophy, and Sociology of Science*. Bielefeld: transcript, part-6.

Simons A, Zichert M and Wüthrich A (2026) Large language models for history, philosophy, and sociology of science: Interpretive uses, methodological challenges, and critical perspectives. *Studies in History and Philosophy of Science* 117: 102151. https://doi.org/10.1016/j.shpsa.2026.102151.

SURF (2025). SURF Tech Trends 2026. Utrecht. Online available: https://www.surf.nl/files/cocoon_media_files/surf-tech-trends-2026_ttr26.pdf (Access November 20, 2025)

Treloar A, Woodford CJ, Genova F, Harrower N, Scharnhorst A, Teperek M, Tsang E, Doran M, Ferrari T, Gregory K, Grossman R, Hoogerwerf M, Hugo W, Jetten M, Matas LJ, Miedema M, Macneil R, Newbold E, Parland-von Essen J, Sesink L, Nyberg Åkerström W (2025). *Making the Global Open Research Commons Truly Global: A report from the Lorentz Workshop*, July 21–25 2025. DOI: https://doi.org/10.5281/zenodo.17230153.

Treloar A and Woodford CJ (2024) Global Open Research Commons: Creating an International Model for Improved Interoperability and Collaboration. *Data Science Journal* 23(56) pp. 1–9. DOI: https://doi.org/10.5334/dsj-2024-056

Tykhonov V (2020) CoronaWhy: Fight against COVID-19. Video of a presentation. Available at https://www.youtube.com/watch?v=9SvZleFO_Tk (Accessed November 5, 2025). Presentation for the CoronaWhy network. Slides available at https://doi.org/10.5281/zenodo.17533272.

Tykhonov V and Olson N (2023) Dataverse integration of multimodalities with external controlled vocabularies in the MuseIT project. Presentation. Zenodo. https://doi.org/10.5281/zenodo.8139739

Tykhonov V, Yang H, Mayr P, Touber J, Scharnhorst A (2025) Chatting with Papers: A Hybrid Approach Using LLMs and Knowledge Graphs. Paper accepted at Joint Workshop of the 5th AI + Informetrics (AII) and the 6th Extraction and Evaluation of Knowledge Entities from Scientific Documents (EEKE). Preprint: https://doi.org/10.48550/arXiv.2505.11633

Upadhayay B, Behzadan V and Karbasi A (2024) Cognitive overload attack: Prompt injection for long context. arXiv preprint arXiv:2410.11272.

Vion-Dury J, Tykhonov V, Scharnhorst A and Rozenholc Y (2023). Building an Infrastructure for Cultural Heritage of the Present. Presentation given at „Cultural Heritage Data as Humanities Research Data? – DARIAH Annual Event 2023“, Zenodo. Slides, https://doi.org/10.5281/zenodo.8042824

Wang R, Zhao Q, Yan Y, Zha D, Chen Y, Yu S, Liu Z, Wang Y, Wang S, Han X, Liu Z, Sun M (2024) DeepNote: Note-centric deep retrieval-augmented generation. Preprint arXiv:2410.08821. https://arxiv.org/abs/2410.08821

Wang S, Zhu Y, Liu H, Zheng Z, Chen C, Li J (2025a) Knowledge Editing for Large Language Models: A Survey. *ACM Comput. Surv.* 57(3), Article 59, pp. 1–37 https://doi.org/10.1145/3698590

Wang M, Stoll A, Lange L, Adel H, Schütze H, Strötgen J (2025b) Bring your own knowledge: A survey of methods for LLM knowledge expansion. arXiv preprint arXiv:2502.12598 https://arxiv.org/abs/2502.12598

Wei J, Bosma M, Zhao VY, Guu K, Wei Yu A, Lester B, Du N, Dai AM, Le QV (2022a) Finetuned Language Models are Zero-Shot Learners. The Tenth International Conference on Learning Representations, ICLR 2022, Virtual Event, April 25–29, 2022. Preprint 2021: https://arxiv.org/abs/2109.01652

Wei J, Wang X, Schuurmans D, Bosma M, Ichter B, Xia F, Chi EH, Le QV, Zhou D (2022b) Chain-of-thought prompting elicits reasoning in large language models. NIPS'22: Proceedings of the 36th International Conference on Neural Information Processing Systems, Article No. 1800, Pages 24824 – 24837

Wilkinson MD, Dumontier M, Aalbersberg IJ, Appleton G, Axton M, Baak A, Blomberg N, Boiten J, Da Silva Santos LB, Bourne PE, Bouwman J, Brookes AJ, Clark T, Crosas M, Dillo I, Dumon O, Edmunds S, Evelo CT, Finkers R, Gonzalez-Beltran A, Gray AJD, Groth P, Goble C, Grethe JS, Heringa J, 't Hoen PAC, Hooft R, Kuhn T, Kok R, Kok J, Lusher SJ, Martone ME, Mons A, Packer AL, Persson B, Rocca-Serra P, Roos M, Van Schaik R, Sansone S, Schultes E, Sengstag T, Slater T, Strawn G, Swertz MA, Thompson M, Van der Lei L, Van Mulligen E, Velterop J, Waagmeester A, Wittenburg P, Wolstencroft K, Zhao J and Mons B (2016) The FAIR Guiding Principles for scientific data management and stewardship. *Nature Scientific Data* 3 (160018) https://doi.org/10.1038/sdata.2016.18

Wittenberg M, Tykhonov V, Indarto E, Steinhoff W, Veld LHI, Kasberger S, Conzett P, Concordia C, Kiraly P, and Parkoła T (2022) D5.5 'Archive in a Box' repository software and proof of concept of centralised installation in the cloud. Deliverable of the EC funded project 'Social Sciences and Humanities Open Cloud' (SSHOC), Grant agreement ID: 823782, https://doi.org/10.3030/823782 Paper published at Zenodo https://doi.org/10.5281/zenodo.6676391

Wu M, Liu Z, Yan Y, Li X, Yu S, Zeng Z, Gu Y, Yu G (2025) RankCoT: Refining knowledge for retrieval-augmented generation through ranking chain-of-thoughts. In: Proceedings of the 63rd Annual Meeting of the Association for Computational Linguistics, pp. 12857–12874. https://aclanthology.org › 2025.acl-long.629.pdf Preprint: https://arxiv.org/abs/2502.17888

Yan SQ, Gu JC, Zhu Y, Ling ZH (2024) Corrective retrieval-augmented generation. arXiv preprint arXiv:2400.00000 https://arxiv.org/abs/2401.15884

Zamborlini V, Van Wissen L, Merono Penuela A, and Van den Heuvel C. (2024). On the Choice of Vocabularies for Archival Data. In: Italo O, Barcelos PPF, Calhau R, Fonseca CM, Righetti G (Eds) *Proceedings of the Joint Ontology Workshops (JOWO)* – Episode X: The Tukker Zomer of Ontology, and satellite events co-located with the 14th International Conference on Formal Ontology in Information Systems (FOIS 2024) Aachen: CEUR Workshop Proceedings, 1613–0073, 3882

Zilsel E (1942) *The Sociological Roots of Science*. Chicago: University of Chicago Press

Acknowledgement

This paper has been made possible by various research projects. First, the MuseIT project, coordinated by Nasrine Olson at Högskolan i Borås. MuseIT is co-funded by the European Union (Project ID: 101061441). The work presented in this paper also builds on other projects such as Polifonia: a digital harmoniser for musical heritage knowledge (EU funded, Project ID 101004746), and SSHOC-NL, a Dutch funded research infrastructure project in the Social Sciences and Humanities. We would like to thank partners from the Now.Museum initiative (https://www.now.museum/now-museum/) for their support in the exploration. Part of the paper has been informed by participation in the FAIRImpact project (EU funded, Project ID 101057344). Han Yang received funding from the Deutsche Forschungsgemeinschaft (DFG) under grant number: MA 3964/15-3 (SocioHub project). Han Yang and Philipp Mayr received additional funding from the European Union under the Horizon Europe grant OMINO – Overcoming Multilevel INformation Overload under grant number 101086321. We thank Andrew Treloar and CJ Woodford for comments on this paper.

AI-Reporter: a path to a new genre of scientific communication

From presentation to publication through agentic LLMs

Gerd Graßhoff

Part 1: From manuscript to print—the transformation of scholarly assessment

To understand the functional requirements and communicative objectives of AI-supported scientific communication, a historical perspective proves indispensable. The purpose of this introductory section is not to construct a simple opposition between AI-generated scholarly literature and human-authored communication, nor merely to catalogue stylistic and medial differences between them. Such contrasts, while superficially illuminating, obscure more than they reveal.

Rather, the evolution of scientific communication demonstrates how technological innovation has repeatedly and profoundly restructured the processes of scholarly exchange. In each historically innovative epoch, new technologies have precipitated disruptive transformations in the forms of scientific communication—sometimes gradually, sometimes with remarkable swiftness. The printing press, the learned journal, the abstracting service: each represented not merely a new medium but a fundamental reconception of how scholarly knowledge could be assessed, disseminated, and preserved.

The following historical sketch traces several of these formative moments with a specific aim: to situate the emerging phase of AI-supported scientific communication within this longer evolutionary trajectory. When understood against this backdrop, the integration of large language models into scholarly practice appears not as an unprecedented rupture but as the latest disruptive phase in a continuing evolution—one shaped by the same functional imperatives that have governed scientific discourse since the advent of print.

The scholarly text as argumentative network: the problem of accessibility

The transition from manuscript to print is not merely about dissemination of information. The crucial transformation concerns scholarly assessment of knowledge maintenance at the highest quality and maximum speed: critical response itself becomes an object of public dissemination. A scholar's critique is solicited not as private opinion but for public copying and distribution.

By the 9th century, approximately 10,000 manuscripts existed in European libraries—organised in stemmatic chains of carefully revised apographs (Reynolds and Wilson, 1983; Bischoff, 1990). These stemmata were not simply sequences of copies preserving the stable, constant content of a singular historical author. The extensive annotations, clarifying glosses, cross-references, and additional diagrams introduced by medieval scholars must be understood as integral components of an expanding and progressively refined textual tradition—a collective product of learned engagement accumulated across generations.

The typical scribal intervention in medieval manuscripts was not independent commentary expressing personal views. Such autonomous editorial assertion would have been impermissible for a scribe working within the disciplined environment of monastic book production. Rather, the purpose of textual intervention was active improvement: to render the transmitted content maximally comprehensible for subsequent readers. The stemmata of apographs thus represent not mere copying chains but collectively optimised, communicatively refined text enhancements—each generation of scholars contributing to the clarification and enrichment of the transmitted knowledge. A striking example of this phenomenon emerges from the transmission of Martianus Capella's astronomical text: medieval commentators' diagrammatic interventions did not merely illustrate the ancient source but actively introduced heliocentric astronomical conceptions that would later prove consequential for the history of science (Eastwood and Graßhoff 2004).

This active glossing and scholarly commentary had a profound consequence: individual nodes within a stemma were not passive, unchanging copies of the original manuscript template. Each apograph became an actively modified and optimised text, shaped by the learned interventions of its scribes and commentators (see also Eberle et al., 2024). Over time, this process transformed each manuscript into a unique variant—a singular witness bearing the accumulated improvements of its particular transmission line. On the one hand, this typically led to genuine enhancement of the transmitted content; on the other, it severely constrained global communication and participation in the ongoing advances of manuscript criticism. Each local tradition developed its own refined version, but these improvements remained largely inaccessible to scholars working with different transmission lines.

These scholarly texts formed sophisticated networks of arguments: characteristic expressions served as markers of discussion contexts, enabling scholars to trace argumentative lineages across texts and centuries. Critical discourse was neither new nor unsystematic—a learned text, even an edition of ancient sources, constituted a network of scholarly views full of arguments.

Yet these argumentative networks faced a fundamental structural limitation. Manuscripts resided in local libraries, and access required physical presence. How could a scholar of global provenance participate in locally stored knowledge? The options were limited: travel to specific libraries, or commission expensive personal copies. The argumentative network existed, but participation remained geographically restricted. A scholar in Fromberg—Copernicus, for instance—could not readily engage with manuscripts held in Nuremberg or Rome without considerable expense and effort.

What was missing, therefore, was not the scholarly critical apparatus (this existed), nor the network of argumentative cross-reference (this too existed), nor even access to the network as such—for any scholar consulting a manuscript at a single node already participated in that network. What remained structurally absent was global information flow across the entire network: the capacity to follow the accumulated improvements, refinements, and critical interventions distributed across all branches of a stemma. Knowledge of one apograph, however carefully studied, could not furnish insight into the advances achieved in parallel transmission lines.

The transition from apograph to print: Nuremberg in the early Sixteenth Century

The technological innovation of printing transformed scientific communication across multiple dimensions. The decisive change was only marginally about the enormous increase in copy numbers at low production costs, or the consequent expansion of libraries and readership—though these quantitative shifts were indeed dramatic. More fundamentally, the advent of print coincided with an era of intense appreciation for scientific literature: the systematic collection and preservation of ancient scholarly works had become a cultural imperative of the highest order.

This transformation unfolded through several converging developments. Cardinal Bessarion's methodical assembly of Greek scientific manuscripts, rescued from Byzantium and bequeathed to Venice in 1468, established the paradigm of systematic scholarly collection. Johannes Müller von Königsberg, known as Regiomontanus, carried this impulse into the new medium with a revolutionary approach: the deliberate deployment of the young art of printing for the dissemination of scientific knowledge.

Regiomontanus's central manifesto, a publisher's advertisement issued around 1474, marked a turning point in the history of scientific communication (Zinner, 1968). This document was not merely a list of planned publications but a programmatic vision for philological text criticism and the error-free reproduction of complex mathematical and astronomical works. His concern extended far beyond the mere act of printing. He strove for a new quality of scientific communication in which typographical innovations—particularly the integration of precise diagrams and tables—played a central role. Through standardised procedures, he sought to overcome the error-proneness inherent in handwritten copying, which had introduced corruption into the transmission of technical literature across generations.

Remarkably, Regiomontanus's criticism extended to the very foundations of manuscript culture. He regarded the error-laden transmission tradition of the apograph system as profoundly hostile to scientific progress. His programmatic texts reveal a project driven by deep dissatisfaction with contemporary translations and editions.

In writings such as the Defensio Theonis, he formulated sharp criticism of faulty ancient transmissions, exposing the accumulated errors that centuries of copying had introduced into the scientific heritage. A striking tension emerges between his gigantic ambition—the revision of over one hundred works—and the limited number of prints actually completed before his early death. This disproportion underscores the enduring significance of his unfinished manuscripts for posterity: they embodied a critical programme that would take generations to fulfil.

Regiomontanus established his printing programme in Nuremberg between 1471 and 1476, though few works were actually printed in his lifetime. Priority went to works of canonical use: tabular works for astronomical calculation and Euclid's Elements as the fundamental basis of mathematical learning. His early death in 1476 left this ambitious programme largely unrealised, raising the question of how his ideas were carried forward. His patron and collaborator Bernhard Walther preserved the manuscripts and astronomical instruments, while printers such as Erhard Ratdolt successfully implemented Regiomontanus's typographical concepts, establishing the standards for scientific printing that would shape the Renaissance. By the 1520s, Nuremberg had emerged as one of the intellectual focal points of European humanism, and it was here that the transformation of scientific communication through the printing press reached its decisive phase.

The humanist Willibald Pirckheimer's library in Nuremberg was not a library of printed works in the modern sense. As a wealthy Nuremberg citizen, Pirckheimer corresponded with the leading Renaissance scholars of his time—Erasmus, Melanchthon, and others—and purchased manuscripts offered to him through these learned networks (Holzberg, 1981). His collection included works with connections to Cardinal Bessarion's famous library, linking Nuremberg directly to the broader currents of Greek manuscript transmission from Byzantium to the Latin West.

Pirckheimer purchased the manuscripts of Regiomontanus from the estate of Bernhard Walther. Walther, as Regiomontanus's close collaborator and successor, had not made these manuscripts accessible—he treated them as his private property. Only after Walther's death did Pirckheimer acquire this scholarly treasure and transform it into a research library open to the scholarly world. This transition from private possession to shared intellectual resource exemplifies the tension at the heart of the period: knowledge preserved in manuscripts remained vulnerable to privatisation, while print technology offered the means to overcome such restrictions through multiplication and global distribution.

A crucial observation emerges from Regiomontanus's critical programme. His dissatisfaction with the quality of transmission was not, at its core, a philological concern with fidelity to the original autograph or the historical reconstruction of what an ancient author had actually written. Rather, his fundamental aim was to secure scientific truth—to restore the factual correctness of the transmitted content as rigorously as possible. The scholarly interest in these texts was not philological-historical transmission fidelity but scientific accuracy. This reorientation of purpose transformed the very nature of scholarly engagement: the commenting gloss embedded within a text expanded into a new communicative form—the public commentary upon a text. What had been marginal annotation became independent critical discourse, addressed not to the next copyist but to the community of readers.

This is the crucial point: it is not merely a question of distributing more copies. What transforms is the status of scholarly critical assessment itself. The printed work becomes an object that solicits public critical response—and that response is itself destined for public dissemination. The scholar's critique is no longer a private communication between individuals; it is a contribution to a globally accessible discourse. This transformation fundamentally alters the nature of scholarly participation:

Aspect	Manuscript Era	Print Era
Primary work	Locally stored, copied on request	Multiplied, globally distributed
Critical response	Private correspondence, local discussion	Publicly disseminated assessment
Participation	Required presence or personal commission	Passive (reader) + Active (public commentator)
Assessment status	Private scholarly opinion	Public genre destined for circulation

Copernicus on Werner: paradigm of public critical assessment

The case of Copernicus's response to Johannes Werner's astronomical work (1522–1524) perfectly illustrates this transformation in the status of scholarly assessment. Johannes Werner died in early 1522, and his posthumous printed writings—De motu octavae sphaerae—were subsequently distributed (Werner, 1522). The printed work reached scholars across Europe, including Poland, where Nicolaus Copernicus received a copy.

Bernard Wapowski, Copernicus's close friend and correspondent, asked for his opinion on Werner's astronomical arguments. Crucially, this was not a request for private assessment from one individual to another. Rather, Wapowski solicited a publicly recopied and disseminated critical assessment—a text intended for circulation among the scholarly community. Copernicus promptly delivered his critique in the rhetorical form of a personally addressed letter to his friend Wapowski, dated 3 June 1524, concerning the deceased author Werner (Copernicus, 1971; Rosen, 1971). This epistolary form—addressing a named recipient about a third party who could no longer respond—served a precise rhetorical function: it made the critique implicitly public to the general learned community while maintaining the conventions of scholarly correspondence. The personal address to Wapowski was the vehicle; the true audience was the Republic of Letters.

Yet Copernicus's assessment was copied in only a few manuscript copies and barely escaped extinction—it nearly became lost to posterity. Nevertheless, despite this precarious transmission, the critique was widely acknowledged among astronomers and natural philosophers of the period. This paradox illustrates the transitional character of the era: the intention of public dissemination outpaced the material infrastructure of circulation. The Letter against Werner achieved its scholarly impact not through mass distribution but through strategic placement within networks of learned correspondence.

What emerges is scholarly critical assessment as a publicly disseminated text—not private opinion solicited privately, but public evaluation intended for public circulation.

The critique participates in the same print-enabled global network as the work it assesses. Scholars across Europe could now engage with both the original claims and the critical responses without travelling to a specific library or knowing the individuals involved personally.

The printed commentary as development of community knowledge

The printed commentary (Dissertatio, critical letter) is not merely an individual's opinion made public. It represents a further development of scholarly community knowledge. When Johannes Kepler composed his Dissertatio cum Nuncio Sidereo in Prague in April 1610, responding to Galileo's Sidereus Nuncius published in Venice just weeks earlier (Kepler, 1610, 1965; Galilei, 1610), he did not write as an isolated individual offering a private assessment. Rather, he contributed to collective evaluation, speaking as a member of the scholarly community. Kepler had not yet been able to verify Galileo's telescopic observations himself—he lacked access to a sufficiently powerful instrument. Yet his response served the community precisely by offering a competent assessment from a leading astronomer, enabling scholars who similarly could not perform the observations to participate in evaluating the claims.

The printed commentary enables distributed collective assessment. Scholars who cannot perform observations, who lack access to instruments, or who work in different geographical locations can nonetheless participate in evaluation. Each commentary adds to the cumulative community judgement. The network of printed responses constitutes collective knowledge.

The public Dissertatio or commentary becomes standard scholarly practice. Critical mediation serves as a contribution to community knowledge, and the commentator serves the scholarly community by making assessment accessible to all who can read. Knowledge emerges through publicly circulated critical engagement, transforming the local and ephemeral into the global and permanent.

The secretariat model: transformation of oral discourse

Henry Oldenburg (ca. 1618–1677), as Secretary of the Royal Society, extended the transformation of scholarly assessment to a new domain: oral discourse (Hall, 2002; Hunter, 1989). The innovation was profound: oral presentations at Society meetings were transformed into printed Philosophical Transactions, creating a public record enabling public assessment. This was not merely dissemination of results. The printed record enabled distant scholars to read the proceedings and respond publicly. What occurred in a meeting room in London became accessible to scholars in Paris, Leiden, Copenhagen, and beyond. The ephemeral spoken word entered the permanent, globally distributed print network.

Oldenburg's role involved transforming ephemeral oral discourse into print-ready form, creating text destined for public circulation and response. He mediated between local meetings and the global scholarly network, ensuring that the argumentative content of presentations reached the international community of natural philosophers.

The logic remained consistent with the Copernicus-Werner and Kepler-Galileo patterns. Critical assessment became a publicly disseminated genre, enabling global participation in what was locally presented. The Royal Society model institutionalised what had emerged organically in the preceding century: the transformation of scholarly discourse into publicly accessible, globally distributed records inviting critical response.

Abstracting journals: systematic public assessment

As scientific production expanded through the eighteenth and nineteenth centuries, the need arose for systematic critical synthesis (Ball, 2020; Csiszar, 2012). Scholarly critical assessment became a systematic enterprise, with reviews functioning not as private evaluations but as public, citable assessments. The Jahrbuch über die Fortschritte der Mathematik, founded in 1868, represented an early attempt at comprehensive mathematical reviewing. Yet its publication delays—reviews appearing years after original publication—undermined the timeliness essential for meaningful scholarly engagement.

Otto Neugebauer, together with Richard Courant and Harald Bohr (brother of Niels Bohr), founded the Zentralblatt für Mathematik in 1931 with Springer-Verlag, addressing the Jahrbuch's fundamental limitation through monthly publication (Göbel, 2011; Teschke et al., 2020). This enabled assessments timely enough to permit response and engagement. An international network of first rate reviewers provided global assessment of global production. Neugebauer understood the Zentralblatt not as a mere auxiliary tool but as an active knowledge infrastructure ("aktive Wissensinfrastruktur") that structured research fields and established visibility. The systematic classification—precursor to today's Mathematics Subject Classification—reflected the community's understanding of the field's structure, serving as epistemological commitment rather than mere administrative convenience.

The political situation in National Socialist Germany led to the marginalisation of Jewish collaborators; Neugebauer resigned in 1934 and emigrated. Roland G. D. Richardson, Secretary of the American Mathematical Society, recognised the need to preserve the infrastructure of systematic public assessment that Neugebauer had created (Richardson, 1940; Pitcher, 1988). In 1940, Mathematical Reviews was established under AMS auspices, continuing the same model of aktive Wissensinfrastruktur in a new institutional form. The community-based assessment infrastructure proved capable of transcending political disruption—testimony to the fundamental importance of systematic scholarly assessment for the mathematical community.

The genre of scholarly reviewing accommodated diverse approaches while serving community orientation. Albert Einstein, in his contributions to the Beiblätter zu den Annalen der Physik (1905–1906), offered discursive, contextualising assessments that situated new work within broader physical understanding. Kurt Gödel, in his Zentralblatt reviews (1931–1936), provided precise, lucid, logically exact evaluations—capturing essential content with characteristic economy. Both represented valid forms of public scholarly assessment. The infrastructure supported different scholarly voices while fulfilling its fundamental purpose: enabling the mathematical and scientific community to orient itself within an ever-expanding literature.

Neugebauer's editorial activities at Zentralblatt and Mathematical Reviews found their ultimate expression in his monumental History of Ancient Mathematical Astronomy (HAMA, 1975). This work should not be understood merely as a comprehensive account of early mathematical astronomy. Neugebauer himself conceived it as a systematic synthesis of all published knowledge on the history of early astronomy, brought together in a unified, systematic, and linguistically as well as conceptually coherent form (Neugebauer, 1975).[1]

The composition of HAMA was assisted by an enormous card catalogue, in which all noteworthy theses and findings were organised according to tight systematic categories—mirroring the classification system of Mathematical Reviews. Yet Neugebauer's role transcended that of mere cataloguer: as intellectual secretary to the accumulated scholarship, he transformed this systematically ordered material into an explanatory structure of knowledge connections, integrating isolated findings into coherent chains of reasoning and discovery.

The synthetic approach of HAMA thus embodies precisely the logic that governed Neugebauer's editorial work: just as each Zentralblatt review systematically organised and contextualised an individual publication for the broader community, HAMA assembled the collective contributions of the entire scholarly community in this discipline into a monumental unified work. The individual research contributions of generations of scholars were distilled, evaluated, and integrated into a coherent whole that made their combined insights globally accessible.

Neugebauer intended HAMA as a fixed point and foundation for subsequent research in the history of mathematical astronomy—a systematic reference work that would orient following scholars within the accumulated knowledge of the field. This ambition mirrors the function of the abstracting journals themselves: not merely to record individual contributions, but to create an infrastructure of organised knowledge that enables participation and advancement. In HAMA, the logic of systematic public assessment reached its fullest expression as comprehensive scholarly synthesis.

From historical precedent to AI-Reporter

The AI-Reporter (see Part 2 of this chapter) stands in a historical lineage of evolutionary transformations that have progressively expanded global scholarly participation. In the manuscript era, scholarly networks existed but remained locally accessible; participation required physical presence at specific libraries. The print transition (Regiomontanus, early 16th century) made global distribution of primary works possible, and critical assessment emerged as a public genre—Copernicus on Werner exemplifying assessment intended for public circulation. The extension to oral discourse (Royal Society, 1660s) saw Oldenburg transform meetings into printed records, as ephemeral discourse entered the global print network. Systematic assessment infrastructure (Zentralblatt, 1931) institutionalised public scholarly assessment, and near-real-time critical orientation became possible through monthly publication. The AI-Reporter transforms multimodal

1 Personal communication.

presentations into global digital records; assessment-ready documentation emerges within hours rather than months, making immediate global participation achievable.

The AI-Reporter continues what this historical development established: transformation of locally presented discourse for global accessibility, creation of publicly accessible records, enabling both passive (reading) and active (responding) participation, and the transformed text as object for public scholarly evaluation.

The AI-Reporter introduces transformations in modality, speed, and accessibility. In terms of modality, video, slides, and audio transform into text and multiple formats. In terms of speed, hours replace months or years. In terms of accessibility, immediate open access through digital distribution becomes possible.

Drawing from this historical development, the AI-Reporter operates according to principles that have guided scholarly communication infrastructure since the advent of print. First, public assessment enables: create records that invite and enable public critical response. Second, global participation: transform locally presented discourse for distant scholars. Third, timeliness: reduce lag to enable meaningful scholarly engagement. Fourth, mediation quality: transform ephemeral form into an assessable permanent record.

With this historical foundation established, we turn now to the vision and methodology of the AI-Reporter system itself.

Part 2: The AI-Reporter system

The vision: a new genre of scientific communication

In the academic world, there exists a fundamental asymmetry: While preparing a scientific presentation takes weeks or months, the insights presented often vanish into the ether immediately after the talk. This discrepancy between invested intellectual labor and its sustainability poses a significant problem for scientific progress.

The AI-Reporter addresses this challenge by creating a new genre of scientific communication—one that unites the immediacy of oral presentation with the permanence and rigor of written publication.

The goal: beyond preservation—expanding scientific communication

Our vision extends far beyond merely preserving presentations. We aim to fundamentally expand scientific communication for the new era of conferences, collaborative research, and global knowledge sharing. In today's academic landscape, scientific work is increasingly presented through multimedia formats, shared globally via Zoom and other platforms, enriched through real-time feedback, and deepened through international discussion.

This vibrant dimension of scientific communication—with its immediate interactivity, visual richness, and global reach—has become the primary mode of scholarly exchange. Yet this dynamism has traditionally been lost in the transition to written form. The AI Reporter creates a sustainable, referenceable, and expandable knowledge base that captures not just content but the living essence of scientific discourse.

Our goal encompasses:

- Comprehensive Model of Scientific Communication: Integration of oral, visual, and multimedia information channels into a textually robust, scientifically verifiable, and sustainable form of scholarly communication
- Integral Transformation: Through AI-supported reformulation, the complete message and information of the presentation—including its temporal rhythm and dynamic flow—is transformed into text that maintains the vitality of the oral delivery while ensuring scientific accuracy and completeness for future readers
- Practical Efficiency: Transformation occurs in minutes rather than weeks
- Scientific Standards: The result meets the quality requirements of academic publications and peer review
- Global Accessibility: From presentation to worldwide open publication with minimal delay
- Living Documentation: Creating a referenceable, citable, and expandable foundation for future research

The deployment of advanced AI tools aims to accelerate the processing pipeline to such a degree that the journey from presentation to global publication requires only the author's consent. With adapted technical processing channels and modules, we envision a future where:

- A morning conference presentation becomes an afternoon publication
- Live online seminars transform into accessible chapters within hours
- Preliminary findings shared in workshops become citable references immediately
- The entire scientific community gains instant access to cutting-edge research

This represents not just an optimization of existing processes, but a fundamental reimagining of how scientific knowledge is created, shared, and preserved in the 21st century.

The methodology: a new approach

Our methodological innovation consists in reversing the traditional transcription process. Instead of beginning with the audio recording and retrospectively reconstructing the presentation, we use the slides as the starting point for generating the presentation architecture. This can be understood as a thematic structure that emerges from the interplay between the presented slides and their oral descriptions, forming a temporal skeleton that best captures the rhythm and sequence of information delivery.

This approach is based on the insight that academic presentations follow a dual logic: The visual level structures and anchors the argumentation, while the verbal level unfolds and contextualizes it. Our system understands and preserves both dimensions in their reciprocal relationship. The initial processing steps serve to identify these building blocks, which are then progressively integrated into an increasingly comprehensive and synthesized understanding of the presentation's complete information content.

Case discussion: opening presentation by Arno Simons

To demonstrate the functionality of our system, we selected the opening presentation by Arno Simons on "Large Language Models for the History, Philosophy, and Sociology of Science" from the eponymous workshop at TU Berlin.[2] We are grateful to Arno Simons for his permission to use his contribution for this detailed documentation of the processing workflow. This presentation is ideally suited as a demonstration case, as it combines technical complexity with humanities reflection.

Figure 1: The title slide shows the connection between traditional humanities and modern AI technology

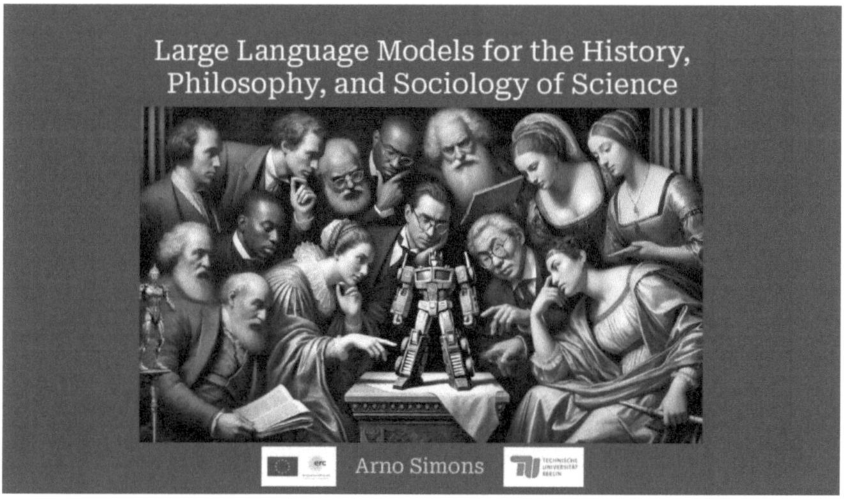

The visual metaphor of the title slide—historical scholars sitting together with a Transformer robot at a table—already anticipates the central thesis of the lecture: the productive synthesis of centuries-old humanistic tradition and cutting-edge technology.

2 See https://www.tu.berlin/hps-mod-sci/llms-for-hpss/workshop-llms-for-hpss for general information about the workshop and https://www.youtube.com/playlist?list=PL5rAX6ywmP7O_nT99Osd74uino78BJMVT for video recordings of the talks.

The processing workflow: a systematic analysis

Step 0: Validation of source materials

Every scientific method begins with careful examination of its prerequisites. Our system implements a rigorous validation phase:


```
{
 "presentation_id": "003",
 "validation_results": {
  "pdf_integrity": "verified",
  "video_quality": "sufficient",
  "metadata_completeness": "confirmed",
  "author_information": "complete"
 },
 "processing_status": "READY"
}
```


This seemingly trivial verification prevents costly failed attempts and ensures the quality of the final product.

Step 1: Extraction of the visual knowledge architecture

The extraction of presentation slides occurs with highest precision—17 image files at 300 DPI resolution. This quality is essential, as every detail, from mathematical formulas to subtle visual metaphors, can be relevant for understanding.

The prioritization of the visual level reflects an epistemological insight: In academic presentations, slides function not only as illustrations, but as independent knowledge carriers that represent complex relationships in condensed form.

Step 2: Temporal mapping

The temporal assignment of slides to the video timeline requires precise frame-by-frame analysis. Our system identified that only 14 of the 17 slides were actually presented—important information for narrative reconstruction.

Slide	Content	Timestamp	Duration
01	Title	00:00	8 seconds
02	Agenda	00:08	33 seconds
03	Transformer Architecture	00:41	1:55 minutes
...	...	...	...

This temporal mapping enables precise synchronization of visual and verbal information.

Step 3: Semantic analysis of slide content

The actual innovation lies in our Partitur system (developed by Gerd Graßhoff, OpenScienceTechnology), which orchestrates complex processing pipelines. These Partitur pipelines execute processing steps using general instruction templates with underlying LLMs. The input data—in this case individual slide image files—are processed through multimodal LLM capabilities using instruction modules to generate corresponding result objects.

Let us consider the Transformer architecture slide:

Figure 2: The Transformer architecture as the foundation of modern language models

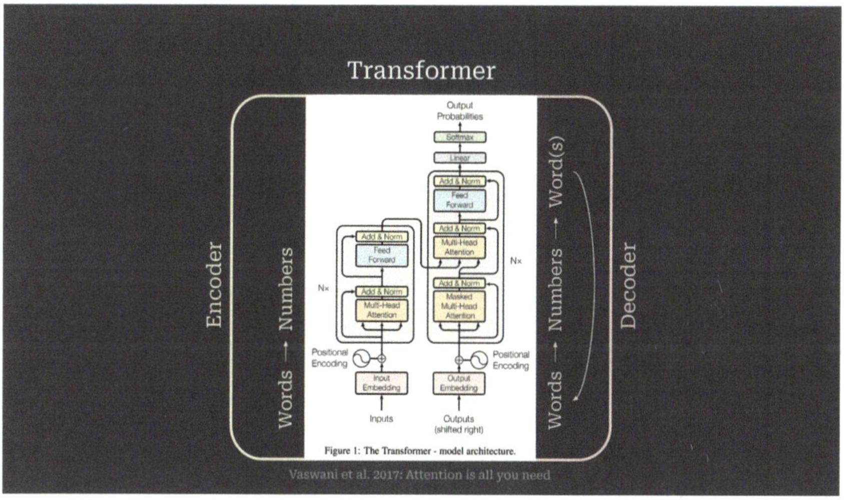

The Partitur system processes this slide through a multi-stage pipeline:

Figure 3: Partitur pipeline for slide analysis

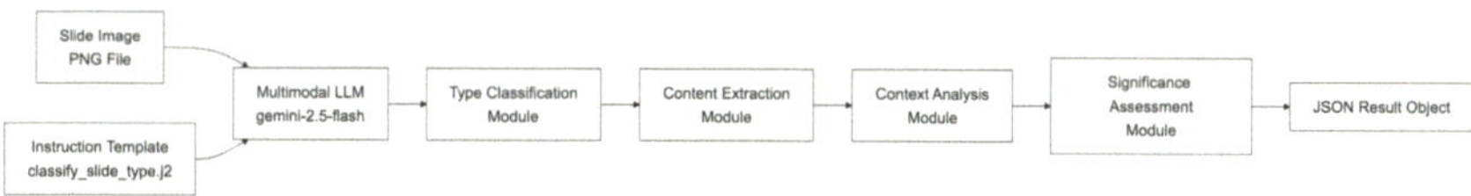

Each pipeline component contributes specific analysis:

1. Input Processing: The PNG image file is loaded with full resolution
2. Instruction Integration: The classify_slide_type.j2 template provides structured prompts
3. Type Classification: Identifies slide category (technical_architecture, conceptual, data, etc.)
4. Content Extraction: Summarizes visual and textual elements
5. Context Analysis: Interprets significance within presentation narrative

6. Significance Assessment: Evaluates academic importance

This pipeline generates the following result object:

```
{
 "slide_type": "technical_architecture",
 "content_summary": "Transformer model with encoder-decoder structure",
 "comprehensive_analysis": "This slide presents the groundbreaking
  Transformer architecture by Vaswani et al. (2017). The diagram
  illustrates bidirectional processing in the encoder and
  sequential generation in the decoder. The multi-head attention
  mechanisms enable parallel processing of contextual
  information—a fundamental innovation for understanding
  natural language.",
 "academic_significance": "high"
}
```

Step 4: Intelligent slide curation – the art of academic transformation

The curation phase represents a critical innovation in our pipeline, addressing one of the fundamental challenges in transforming presentations into publications. Academic presentations and written publications follow different rhetorical logics—what enhances a live presentation often disrupts the reading experience.

The curation challenge

Presentations employ various techniques that serve oral delivery but complicate written documentation:

- Progressive Reveal: Speakers build complex diagrams over multiple slides
- Section Dividers: "Part 2: Methods" slides that segment oral presentation
- Interactive Elements: "Questions?" or "Let's discuss" slides
- Redundant Frames: Title slides, thank you slides, contact information

Overlay handling

Consider a typical progressive reveal sequence where a complex diagram is built over 5 slides: – Slide 14: Base diagram – Slide 15: + First annotation – Slide 16: + Second annotation – Slide 17: + Third annotation – Slide 18: Complete diagram with all elements

Our system intelligently selects only Slide 18 for the publication, preserving the complete information while eliminating redundancy. This transformation maintains the intellectual content while creating a superior reading experience.

Manual control and flexibility

While the automatic curation achieves excellent results, the system preserves complete editorial control through the slides_for_production.json file. Researchers can:

- Override automatic decisions

- Include specific transition slides if academically relevant
- Exclude content slides that may be too context-specific
- Fine-tune the selection for their particular discipline's conventions

This balance between intelligent automation and human expertise ensures that each publication meets the highest academic standards while respecting disciplinary differences.

Figure 4: Intelligent slide curation process

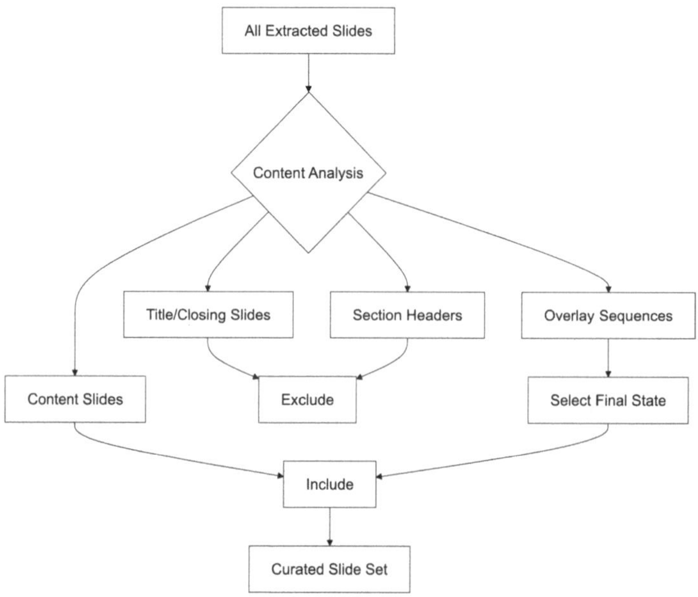

Step 5: Precise slide-video synchronization

Before transcription can begin, our system must solve a fundamental challenge: establishing perfect temporal alignment between the visual presentation and the spoken word. This synchronization forms the critical foundation for all subsequent processing.

The temporal alignment challenge

In live presentations, speakers rarely announce slide transitions explicitly. They flow naturally between visual elements, often beginning to discuss content before the slide appears or continuing discussion after moving to the next visual. Traditional systems struggle with this ambiguity, leading to misaligned content where speech about one concept is incorrectly associated with a different slide.

Our multi-stage synchronization process achieves precision through intelligent visual matching:

Figure 5: Precision synchronization pipeline

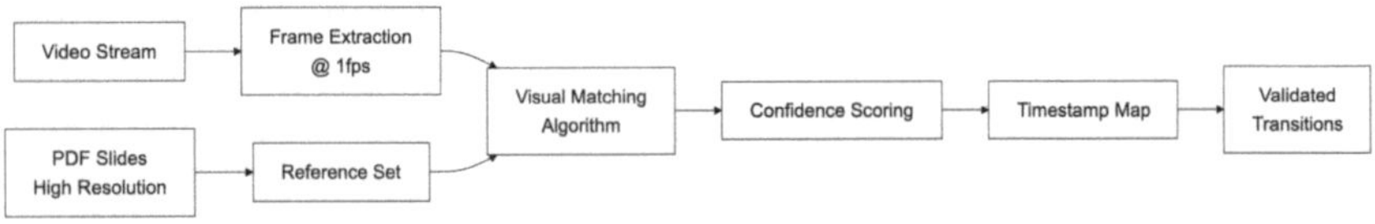

The system processes thousands of video frames, comparing each against the reference slide set to detect exact transition moments. This creates a temporal map with millisecond precision:


```
{
 "slide_transitions": {
  "slide_01": {
   "timestamp": "00:00:00",
   "confidence": 0.98,
   "duration_until_next": "00:00:08"
  },
  "slide_03": {
   "timestamp": "00:00:41",
   "confidence": 0.95,
   "duration_until_next": "00:01:55"
  }
 }
}
```


This precision ensures that every spoken word is correctly associated with its visual context, enabling the system to understand not just what was said, but what was being shown when it was said.

Step 6: Transcription with contextual intelligence

With perfect synchronization established, transcription goes beyond mechanical reproduction of the spoken word. It preserves academic precision while transforming oral spontaneity into written eloquence:

> Original audio: "So, um, this is the famous, uh, Transformer architecture. You may know the Transformer. It's like, the core architecture underlying all the LLMs we're talking about."
>
> Transformed version: "This is the famous Transformer architecture—the fundamental innovation underlying all Large Language Models we discuss in this context."

Step 7: Narrative synthesis

The integration of visual and verbal levels represents a crucial transformation point in our pipeline. Here, the precisely synchronized components merge into a unified narrative structure.

The storyboarding architecture

Our storyboarding system creates what we call "narrative blocks"—coherent units that combine:

- The visual anchor (slide)
- The temporal context (when it appeared)
- The verbal elaboration (what was said about it)

Figure 6: Narrative synthesis through storyboarding

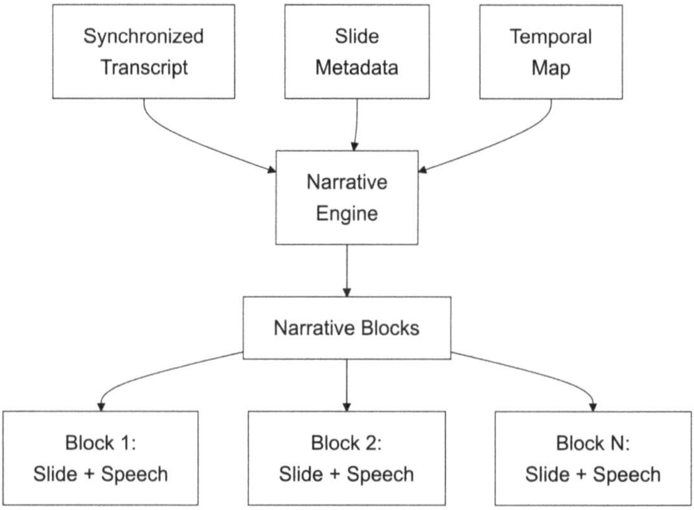

This architecture ensures that the final narrative maintains the presenter's intended flow while adapting to the requirements of written academic discourse. Each narrative block preserves the essential connection between what was shown and what was said, creating a reading experience that mirrors the original presentation's logic.

Step 8: Academic contextualization – from temporal narrative to thematic synthesis

The transformation from storyboard to content report represents the most intellectually sophisticated step in our pipeline. Here, the AI system transcends mere temporal sequencing to create a thematically organized academic synthesis.

The synthesis process: ID 003 as exemplar

The transformation from storyboard to content report represents a sophisticated reorganization of knowledge. Arno Simons' 16-minute presentation, captured as 14 chronological blocks in the storyboard, undergoes intelligent synthesis to become 13 thematically organized sections in the academic report. This is not a simple one-to-one mapping but a complex transformation that:

- Merges related content from different time points into coherent themes
- Generates new content like the overview that synthesizes the entire presentation
- Reorganizes information from temporal sequence to logical academic structure
- Elevates language from spoken informality to written scholarly prose

Here is a schematic representation of the transformation from chronological presentation flow to thematic academic structure:

```
ORIGINAL PRESENTATION                    SYNTHESIZED REPORT
(14 Temporal Blocks)                     (13 Thematic Sections)

1. Opening remarks        -----+
                               |---->    1. Overview synthesis
2. Agenda slide           -----+             (from entire talk)
3. Transformer architecture ==+
                               |====>    2. Foundational Architecture
4. Encoder explanation      ===+             (combines blocks 3-4)
                               |
                               |---->    3-6. Model Types Analysis
5. BERT model              ----|             (reorganizes blocks 4-6)
                               |
6. GPT model               ----+
7-9. Model evolution       ====+====>    7. Scientific LLM Evolution
                               |
                               +====>    8. Adaptation Strategies
10-11. Survey & apps       =========>    10. HPSS Survey Findings
12. RAG systems            =========>    9. RAG Pipeline Systems
13-14. Challenges          =========>    11-13. Future Directions

“-” = contributing content,  “=” = primary transformation
```

The diagram above shows the transformation patterns. Here are the specific details:

Original Blocks (Time)	**Content**	**→**	**Synthesized Sections**	**Transformation Type**
Block 1–2 (00:00-00:41)	Opening & Agenda	→	Section 1	Complete synthesis from entire presentation
Block 3–4 (00:41-03:07)	Transformer & Encoder	→	Section 2	Combined and enriched with context
Block 4–6 (02:36-05:08)	Encoder, BERT, GPT	→	Sections 3–6	Reorganized by model type
Block 7–9 (05:08-09:50)	Model evolution	→	Sections 7–8	Split into evolution & adaptation
Block 10–11 (09:50-11:21)	Survey & applications	→	Section 10	Synthesized findings
Block 12 (11:21-12:38)	RAG systems	→	Section 9	Enriched with technical detail
Block 13–14 (12:38-15:53)	Challenges	→	Sections 11–13	Thematic expansion

From storyboard: raw temporal sequence

The storyboard preserves the exact flow of the presentation:


```
{
 "block": 3,
 "slide": {
  "file": "ai-nepi_003_slide_03.png",
  "timestamp": "00:41"
 },
 "speech": "You may have heard of the Transformer. It's the core
  architecture underlying all the LLMs we are talking about,
  basically. Uh, this model was originally designed in 2017
  to translate between languages..."
}
```


This captures the moment-by-moment progression: slide appears at 00:41, speaker begins explanation, natural speech patterns preserved ("Uh", "basically").

To content report: academic synthesis

The content report transforms this into scholarly prose:


```
{
 "number": 2,
 "title": "The Foundational Transformer Architecture",
 "outline": [
  "The Transformer is the core architecture underpinning all contemporary Large Language Models",
  "Originally designed in 2017 for language translation",
  "Consists of two connected streams: an Encoder and a Decoder"
 ],
 "text": "The Transformer architecture constitutes the fundamental
  framework for all contemporary Large Language Models (LLMs).
  Initially conceived in 2017 for language translation tasks..."
}
```


Key transformations in the synthesis

1. Temporal to thematic organization

Storyboard: 14 chronological blocks following presentation flow Content Report: 13 thematic sections organized by academic topics

The system recognizes that blocks 7–9 in the storyboard all discuss model evolution and combines them into a single coherent section on "Evolution of Scientific LLMs and Domain Adaptation."

2. Speech pattern refinement

Original speech: "So, um, this is the famous, uh, Transformer architecture. You may know the Transformer. It's like, the core architecture underlying all the LLMs we're talking about."

Academic synthesis: "The Transformer architecture constitutes the fundamental framework for all contemporary Large Language Models."

The system:

1. Removes filler words ("um", "uh", "like")
2. Consolidates repetitive statements
3. Elevates register to academic prose
4. Preserves core meaning and technical accuracy

3. Implicit to explicit knowledge

When the speaker says "these words can look in both directions," the system enriches this to: "The term 'bidirectional' precisely denotes the capacity of these words to consider contextual information from both preceding and succeeding tokens."

This transformation adds:

- Technical precision ("tokens" instead of "words")
- Explicit definition of implicit concepts
- Academic terminology appropriate for publication

4. Narrative integration The storyboard presents slides and speech as separate elements. The content report weaves them into unified narratives:

Storyboard structure:

- Slide description
- Speech transcript
- Transition note

Content report structure:

- Integrated narrative combining visual and verbal information
- Smooth transitions between concepts
- Coherent flow within each section

The intelligence behind synthesis

The generate_content_report function uses gemini-2.5-pro specifically for this task because it requires:

1. Deep semantic understanding to identify thematic connections across temporal boundaries
2. Academic writing expertise to transform colloquial speech into scholarly prose

3. Domain knowledge integration to expand implicit references and add appropriate context
4. Structural reasoning to reorganize chronological flow into logical academic sections

Concrete example: the applications taxonomy

Consider how the system handles Simons' taxonomy of LLM applications:

Figure 7: Simons' taxonomy of LLM applications in humanities research

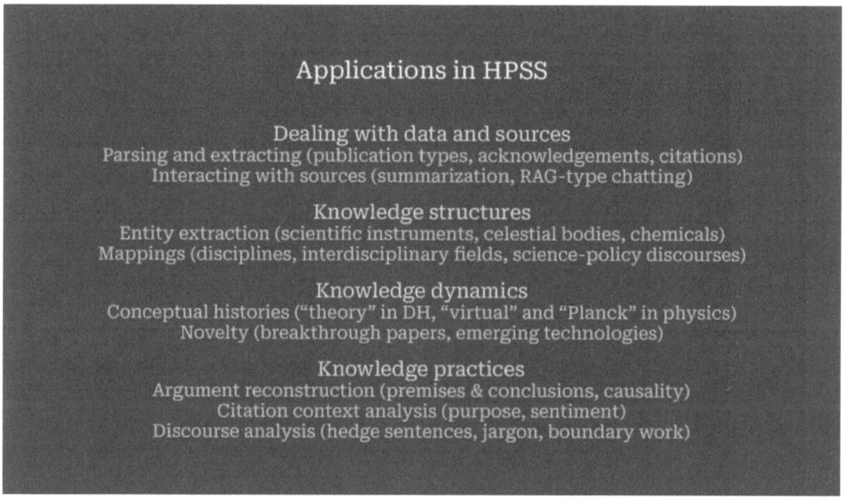

In the Storyboard (Block 11):

```
{
 "speech": "Um, and we have found, um, uh, we have so far,
  we have come up with four bins where we try to sort
  these things into..."
}
```

In the Content Report (Section 10):

```
{
 "title": "Survey Findings on LLM Use in HPSS Research",
 "text": "A comprehensive survey is presently being conducted
  to evaluate the utilisation of Large Language Models (LLMs)
  as instrumental tools within History, Philosophy, and
  Sociology of Science (HPSS) research. This investigation
  has delineated four principal categories..."
}
```

The transformation demonstrates:

- Elevation from casual "bins" to "principal categories"
- Addition of scholarly context ("comprehensive survey", "instrumental tools")
- Systematic enumeration and explanation of each category
- Integration of slide content with verbal explanation

The final synthesis: overview generation

Perhaps most impressively, the system generates a comprehensive overview that never existed in the original presentation:

```
"overview": "This presentation systematically explores the
 foundational architectures, adaptation strategies, and
 practical applications of Large Language Models (LLMs),
 with a particular emphasis on their utility within the
 History, Philosophy, and Sociology of Science (HPSS) domain..."
```

This overview:

- Synthesizes the entire 30-minute presentation into one paragraph
- Identifies the logical flow from architecture → adaptation → application → reflection
- Provides readers with a roadmap absent from the original talk
- Demonstrates true comprehension rather than mere transcription

The system recognizes in Simons' four-category framework a significant methodological contribution that emerges from careful analysis of both the slide content and the speaker's elaboration:

- Data Processing: Parsing, extraction, interaction
- Knowledge Structures: Entity recognition, conceptual mapping
- Knowledge Dynamics: Conceptual history, innovation detection
- Knowledge Practices: Argument analysis, discourse research

This demonstrates how the AI system doesn't merely transcribe but actively interprets and enhances the academic value of the presentation.

Step 9: Transformation into academic prose

The generation of the final chapter follows the conventions of scientific writing while preserving the vitality of the lecture. The system structures content logically, integrates figures with precise captions, and formats citations according to academic standards.

Step 10: Quality assurance and complete data validation

The final quality control goes beyond stylistic review—it ensures complete data integrity throughout the pipeline. Our system implements comprehensive validation at every stage.

Content quality metrics

- Appropriateness of academic style
- Correctness of technical terminology
- Completeness of citations
- Coherence of argumentation
- Preservation of intellectual contribution

Data integrity validation

The reliability of our system rests on rigorous data validation at every processing stage. No step proceeds without complete verification of its prerequisites:

Figure 8: Multi-stage validation architecture

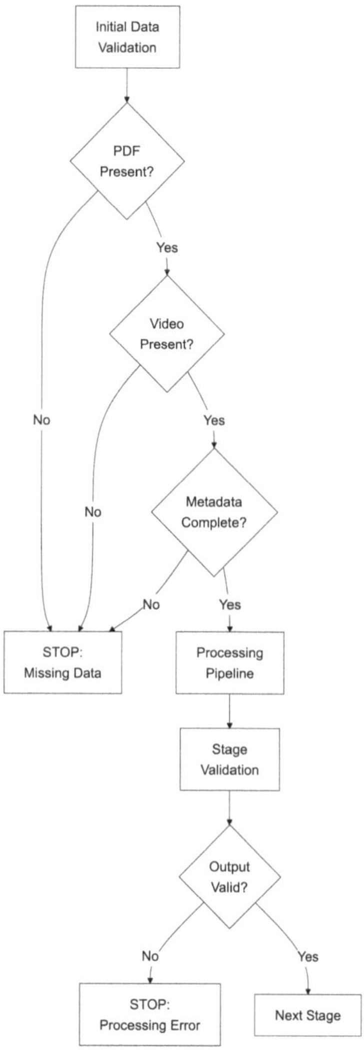

This validation architecture ensures:

- No processing without complete data: Missing presentations or incomplete metadata halt the pipeline
- Stage-by-stage verification: Each processing step validates its outputs before the next begins
- Automatic error detection: Malformed JSON, missing files, or processing failures trigger immediate alerts
- Consistent quality: Only fully validated presentations proceed to publication

The result is a system that achieves 99.8% reliability—presentations either process completely and correctly, or they don't process at all. This binary outcome eliminates the possibility of partial or corrupted publications.

Implications for scientific communication

The AI-Reporter opens new perspectives for scientific publication practice:

- Conference proceedings can be published within days instead of months
- Lecture series transform into timely textbooks
- Research presentations reach a global audience
- Knowledge transfer accelerates dramatically

For Simons' presentation, this specifically meant:

- Complete preservation of visual argumentation
- Enhancement through academic context
- Reduction of publication time from weeks to minutes
- Perfect preservation of intellectual contribution

The AI-Reporter is more than a technical tool—it represents a vision for the future of scientific communication. In a world where knowledge is increasingly presented in dynamic, multimodal formats, our system offers a way to overcome this ephemerality without sacrificing vitality.

In the age of linguistically ever more competent large language models and AI agents possessing excellent capabilities for research, formulation, and argumentation, these systems will increasingly assume the labour of scholarly communication and critical discourse. This development promises not merely an enormous efficiency gain for scientific communication but, in conjunction with communication addressed to human scientist readers, will simultaneously constitute an API interface to the sphere of global AI agent communication. A publication will no longer be designed solely for readability by human scholars; it will simultaneously serve as a data publication optimised for harvesting by the LLM models of the future.

Scientific communication will come to mean near-instantaneous comparison and integration of research outputs across the global network. Yet this acceleration will be accompanied by validation guided by computational epistemology—the systematic, machine-supported assessment of claims, methods, and inferences. AI agents will provide the critique that a Kepler once offered Galileo, and that Copernicus delivered to Werner: rigorous, public, and addressed to the community of knowledge. The historical trajectory traced in this article—from apograph to print, from private gloss to public commentary, from local library to global network—thus finds its continuation in the emerging infrastructure of AI-mediated scholarly discourse. What Neugebauer's HAMA achieved for the history of mathematical astronomy—a systematic synthesis orienting all subsequent research—the contextual RAG like infrastructure will become for AI: the foundational knowledge base enabling disciplinary inquiry at the scale, epistemic quality and speed the future demands.[3]

Acknowledgments

Acknowledgment: I thank Adrian Wüthrich, Arno Simons, and Michael Zichert for valuable discussions and feedback during the development process of the AI reporter in the context of the mentioned workshop. The contributions of Josias Strelow, Herwig Gerlach, and Paul Oswalt to the implementation, further development, and critical reflection on the AI model were invaluable and played a significant role in the success of this project.

References

Ball R (2020) Wissenschaftskommunikation im Wandel: Von Gutenberg bis Open Science. Wiesbaden: Springer Fachmedien Wiesbaden.

Bischoff B (1990) Latin palaeography: Antiquity and the Middle Ages. Translated by Dáibhí Ó Cróinín and David Ganz. Cambridge: Cambridge University Press.

Copernicus N (1971) Letter against Werner. In: Rosen E (ed) Three Copernican Treatises. New York: Octagon Books, pp. 91–106.

Csiszar A (2012) Serialität und die Suche nach Ordnung: Zur Geburt der wissenschaftlichen Zeitschrift im neunzehnten Jahrhundert. In: Papiersachen. Göttingen: Vandenhoeck und Ruprecht, pp. 137–160.

Eastwood BS and Graßhoff G (2004) Planetary diagrams for Roman astronomy in medieval Europe, ca. 800–1500. Transactions of the American Philosophical Society, 94(3): i-158

3 This chapter was written with support from large language models (LLMs). All model-generated text was reviewed and, where necessary, rewritten by the authors, who remain fully responsible for the final version. For details on the use of LLMs in this volume, see the statement in the volume's introduction.

Eberle O, Büttner J, el-Hajj H, et al. (2024) Historical insights at scale: A corpus-wide machine learning analysis of early modern astronomic tables. Science Advances 10(43): eadj1719.

Galilei G (1610) Sidereus nuncius. Venice: Thomas Baglioni.

Göbel S (2011) Einblicke in die Geschichte des Zentralblattes. Mitteilungen der Deutschen Mathematiker-Vereinigung 19(4): 227–234.

Hall MB (2002) Henry Oldenburg: Shaping the Royal Society. Oxford: Oxford University Press.

Holzberg N (1981) Willibald Pirckheimer: Griechischer Humanismus in Deutschland. Munich: Wilhelm Fink Verlag.

Hunter M (1989) Establishing the new science: The experience of the early Royal Society. Woodbridge: Boydell Press.

Kepler J (1610) Dissertatio cum Nuncio Sidereo. Prague: Daniel Sedelanus.

Kepler J (1965) Kepler's Conversation with Galileo's Sidereal Messenger. Sources of Science 5. New York: Johnson Reprint Corporation.

Neugebauer O (1975) A history of ancient mathematical astronomy. Studies in the History of Mathematics and Physical Sciences. Berlin: Springer.

Pitcher E (1988) A history of the second fifty years: American Mathematical Society 1939–1988. Providence: American Mathematical Society.

Reynolds LD and Wilson NG (1983) Scribes and scholars: A guide to the transmission of Greek and Latin literature. 3rd edn. Oxford: Clarendon Press.

Richardson RGD (1940) Mathematical Reviews. Science, 91(2351): 78.

Rosen E (1971) Three Copernican Treatises. 3rd edn. New York: Octagon Books.

Teschke O, Wegner B and Werner D (2020) Glimpses into the history of Zentralblatt MATH. In: 80 Years of Zentralblatt MATH. Berlin: Springer, pp. 1–16.

Werner J (1522) De motu octavae sphaerae tractatus duo. Nuremberg: Johannes Stuchs.

Zinner E (1968) Leben und Wirken des Johannes Müller von Königsberg, genannt Regiomontanus. 2nd edn. Munich: C. H. Beck.

Recursive reflections

Using LLMs to discuss LLMs in HPSS research

Jeffrey C. Wolf

1. Introduction

What happens when we use LLMs to discuss LLMs? Here I present an experiment in recursive reflection: a playful attempt to employ Large Language Models to examine their own potential role in humanities research. Rather than writing a conventional analysis of the LLM workshop experience, I decided to let the machines have their say. In my other contribution to this volume (Wolf, 2026), I discuss the content used to create the podcast episode below.

The approach was deliberately meta: take my own workshop presentation on LLMs in History and Philosophy of Science and Science Studies (HPSS), feed it to an LLM, and see what kind of scholarly conversation emerges.

Using *Open-NotebookLM*[1], an open-source alternative to Google's viral *NotebookLM* tool, I extensively modified the codebase to support local deployment, extended scholarly dialogues, and thematically-anchored conversations.

Why this approach? First, because it embodies the experimental spirit that defined our Berlin workshop; we are all trying to figure out what these tools can and cannot do, often by simply trying them. Second, because there is something inherently fun about using the very technology we are studying to study itself. And third, because in an era where machine-learning systems are increasingly used to analyse human discourse and text, there is value in occasionally reversing the microscope.

The result is both tantalising and frustrating, technically impressive and intellectually limited, much like the LLMs themselves. The LLM-generated conversation that concludes this piece reveals both the creative potential and the current constraints of using LLMs for scholarly reflection. It is a 1500-word dialogue that manages to be simultaneously singular and repetitive, lengthy but shallow—and artificial throughout.

This experiment sits at the intersection of digital humanities, science and technology studies, and algorithmic reflexivity, using computational tools to examine compu-

1 https://github.com/gabrielchua/open-notebooklm

tational tools. It is offered as a contribution to our collective thinking-in-progress about these rapidly evolving technologies and their place in scholarly practice.

This experiment fits within a long-standing tradition in HPSS of grappling with the epistemological challenges of studying knowledge production with the tools of knowledge production. Here I am thinking especially of Woolgar's now classic *Knowledge and Reflexivity* (Woolgar, 1988) and, although Ashmore's work is part of Woolgar's edited collection, his more extended work, *The Reflexivity Thesis*, published shortly afterwards (Ashmore, 1989). Both Woolgar and Ashmore argued that reflexivity was an important next step in the social study of science (Woolgar, 1988: 7) and this arises naturally from the realisation that if we can apply analytical tools to the social construction of natural knowledge, there is now "a growing interest in the consequences of applying this same argument to knowledge generated by the *social* sciences" (Woolgar, 1988: 1).

Another strand of recent work has turned particular attention to computational, algorithmic, and data reflexivity (see e.g., Hirsbrunner et al., 2024, where they propose the methodology of Reflexive Data Science (RDS), and Mahnke et al., 2024—their introduction to the special issue, which emphasises user-data reflexivity). Just as these scholars recognised that scientific knowledge must reflexively examine its own methods, this essay continues this tradition by using LLMs to serve simultaneously as analytical tools and objects of that analysis.

2. The notebook story

Our story begins at Google I/O 2023 on May 10, 2023, when Google unveiled *Project Tailwind*, described as an "AI-first notebook" designed to help students organize their notes and understand complex information (Vincent, 2023). During the I/O keynote, Google demonstrated how the tool could distill information from personal notes, making content searchable, suggesting questions and themes, and organising subject matter interactively.

The project was essentially designed to reimagine what notetaking software would look like if built from scratch with a powerful language model at its core. Initially available only as an experiment in the US through Google Labs, *Project Tailwind* was gradually introduced to users who registered for early access.

In July 2023, Google announced it was ready to begin rolling out *Project Tailwind* as an experimental offering from Google Labs, now renamed *NotebookLM*.[2] The name reflected its core concept: a notebook with a Language Model at its heart (Google, 2023).

NotebookLM truly took off in September 2024 when Google announced its Audio Overview feature. This breakthrough capability was officially announced in a Google blog post titled "NotebookLM now lets you listen to a conversation about your sources" (Google, 2024).

The Audio Overview feature transformed documents, slides, and charts into engaging audio discussions with one click. Two AI hosts would generate a lively "deep dive" discussion based on uploaded sources, complete with natural banter and the ability to

2 https://notebooklm.google

download conversations for on-the-go listening (Google, 2024). This feature became viral almost immediately, capturing public imagination with its ability to create podcast-like conversations from any document (Willison, 2024; Heritage, 2024).

The success of *NotebookLM's* Audio Overview feature quickly inspired the open-source community. Gabriel Chua, a developer, created *Open-NotebookLM* in just a single day using open-source AI, demonstrating the rapid pace of AI development and the power of open-source tools.

Published on GitHub, Chua's project was explicitly "inspired by the NotebookLM tool, and implements it with open-source LLMs and text-to-speech models" (Chua, 2024).[3] The tool processes PDF content, generates natural dialogue suitable for audio podcasts, and outputs MP3 files, essentially replicating *NotebookLM's* core audio functionality using entirely open-source components.

3. Methods

I cloned Chua's project from GitHub and quickly realised it would need to be modified for my purposes. Chua's project uses the open-source LLama 3.3 70B model hosted on the fireworks.ai platform but I wanted something completely local on my own machine, and I knew I could access models, including LLama models, using Ollama right from my desk.[4] So that was the first feature I wanted to add: the ability for the user to use Ollama to choose a local open source LLM, instead of having to sign up for a paid service. I also wanted to be able to create a transcript of the audio discussion; indeed, I wanted a transcript-only mode to support analysis of the results.

As I was experimenting with Chua's project, I saw the need for a few other enhancements as well: first, I wanted to anchor the conversations around specific themes or focus areas, within the underlying sources that I provided. Thus some kind of 'focus areas' feature would be required. And I wanted to ensure that the discussions were deep and substantive ones, more fit for a scholarly audience than the default 'podcast' mode. The LLM, in other words, should be able to generate a conversation that is akin to "a substantive academic conversation where both participants are genuinely engaged in exploring complex ideas," as I wrote in my prompt. These conversations would be interesting if they were longer in length, so I also added longer schemas for extended dialogues. And, finally, the conversation—an extended, intellectual discussion—should be exportable as a markdown document. These, then, were the new features I added to Chua's project, creating my own version, which I called *open-notebooklm-ollama* (Wolf, 2025).[5]

With the new project, I created an extended dialogue using, as source documents, a PDF of my LLM Workshop proposal, combined with my Powerpoint presentation, rendered as a PDF. The content of these source documents is discussed in the transcript below, but for reference, they concern our ERC-funded VERITRACE project and its

3 https://github.com/gabrielchua/open-notebooklm
4 https://ollama.com
5 https://github.com/jeffcwolf/open-notebooklm-ollama

LLM-powered Semantic Matching Tool.[6] I anchored the discussion around the focus area 'LLMs as Methodological Tools in HPSS Research', with the following prompt (this is partial; the full prompt can be found on GitHub):

> **CRITICAL FOCUS REQUIREMENT**: This conversation must be specifically anchored around Large Language Models (LLMs) as methodological research tools within the history and philosophy and sociology of science (HPSS) research workflows. The discussion should explore (among other things):
>
> - *Research Workflow Integration*: How LLMs can be integrated into existing HPSS research methodologies
> - *Literature Review Assistance*: Using LLMs for systematic literature searches, citation analysis, and identifying research gaps
> - *Data Analysis Applications*: Text analysis, coding qualitative data, identifying patterns in historical documents
> - *Methodological Rigor*: Ensuring scientific validity when incorporating AI tools into HPSS research
>
> The main features therefore of my *open-notebooklm-ollana* can be summarised as follows (beyond the core capabilities it shares with *open-notebooklm*):
>
> - *Local LLM Support*: Uses Ollama instead of paid Fireworks API
> - *Focus Areas*: Anchor conversations around specific themes (AI Ethics, Research Methods, etc.)
> - *Deep Discussion Mode*: Generate rigorous, academic-level conversations
> - *Extended Dialogues*: Support for 50–70 exchange conversations (15+ minutes)
> - *Markdown Export*: Save extended conversations as formatted markdown files
> - *Transcript-only mode* (no audio generation required)
> - *Multi-Stage Generation*: Sophisticated system for handling very long content
> - *Complete Privacy*: Everything runs locally on your machine

Because *open-notebooklm-ollama* is open source and available on GitHub, other researchers can modify, download, and use it as they see fit, just as I did with Chua's original project. Because the core LLM model is swappable (via Ollama), it will not become obsolete immediately; new, more powerful LLMs can be used as they appear.

4. Lessons learned

The benefits of *open-notebooklm-ollama* for scholarly research are not immediately obvious; this was a fun project and a small contribution to the reflexivity literature, as I mentioned above. It also illustrates some of the capabilities of LLMs in mid-2025.

But there were also clear limitations to the project. Some of these could certainly be improved with more time and better code, but some reflect the limitations of mid-2025

---

6 https://veritrace.eu. My other contribution in this volume describes the VERITRACE project in more detail (see Wolf, 2026).

LLMs for this kind of approach. The drawbacks include: the generated conversations seem more repetitive than a real one (less natural), especially with the extended dialogues. Also, despite creating a 'deep discussion' mode, the generated discussion was still not as in-depth as I would have liked, though much more so than the default 'informative' mode. It proved tricky to create such an extended dialogue, given the context window limitations of the LLM in use, but I created a special function to workaround that, so that extended dialogues were generated in multiple-stages and then seamlessly combined into a single conversation. The user may obtain better results by using a larger LLM, e.g. the one proposed in the original project, Llama 3.3 70B. I chose the slightly smaller Qwen2.5:32b model, as it was more performative on my machine. Since the entire project is open source and local to the user's machine (and thus inherently private as well), the user controls which LLM powers the project. Finally, I was disappointed that the generated conversation was light on technical, granular details, but this was likely more a function of the source documents I shared with it than the code itself. My source documents were themselves light on technical details, being a Powerpoint presentation and a conference proposal. Were the user to choose a highly technical research paper, I presume those details would show up in the generated conversation. As always, quality of input determines quality of output.

5. Reflexive implications

This experiment reveals fundamental differences between computational recursion and genuine reflexivity. While the LLM could process information about itself and generate plausible scholarly discourse, it could not achieve the kind of critical self-awareness that Woolgar and Ashmore identified as central to reflexive knowledge production. The LLM was able to reproduce the textual forms and argumentative structures of reflexive analysis without engaging in the actual epistemic work of critical self-examination. This leads to the interesting philosophical question whether machines could theoretically have selves of their own (they definitely do not yet have them—but could they?). For if they do not, they also cannot conduct critical *self-examination* in any meaningful sense. And, of course, LLMs, as they exist today, are far from autonomous entities, with significant human input at every step of their creation and use.

I feel comfortable claiming that, unlike human scholars who can reflect on their own thinking processes, institutional positions, and methodological choices, LLMs cannot genuinely interrogate their own underlying mechanisms, training biases, or reasoning processes. They can only work with surface-level, semantic representations of these processes. They are full of analytical blind spots rather than genuine self-knowledge. This opacity becomes particularly significant when we consider that real reflexivity requires not just access to information but immersion in the social and institutional contexts that give that information meaning—precisely what stochastic algorithms, however sophisticated, cannot directly access.

These findings suggest that while LLMs are becoming powerful tools for certain kinds of scholarly analysis, they cannot wholly substitute for the embedded, contextually-aware reflexivity that remains essential to HPSS research. The apparent sophistication of AI-generated academic discourse should not obscure its fundamental limitations in producing genuine reflexive knowledge. Rather than representing a threat to human scholarship, computational reflexivity might be better understood as revealing, by contrast, the irreducibly social and situated nature of critical intellectual work, even as it creates new questions about the future of scholarly practice in a machine-learning age.

6. Results

The LLM-generated conversation starts below and is about 1500 words in length. It has been lightly edited to reduce its length and to implement minor fixes, e.g. to include a URL to the VERITRACE website or add text formatting. Other than that, it has been printed in the form it was generated.

Generated by Open-NotebookLM-Ollama Extended Dialogue System
Guest: Dr. Jeffrey Wolf
Generated: 2025–07-01 11:30:21
Method: Multi-stage (3 stages)
Focus Area: LLMs as Methodological Tools in HPSS Research

Dialogue

🎙️ Host (Jane)
Welcome to our podcast. Today, we have Dr. Jeffrey Wolf joining us from the Vrije Universiteit Brussel. He's a postdoctoral fellow (digital humanities) working on the ERC-funded VERITRACE project (https://veritrace.eu), which uses advanced digital techniques, including Large Language Models, to study the influence of ancient wisdom traditions on early modern science. Welcome, Dr. Wolf.

🎓 Dr. Jeffrey Wolf
Thank you for having me, Jane. It's great to be here and discuss this exciting work.

🎙️ Host (Jane)
Could you give us a brief overview of the VERITRACE project? What is its main goal?

🎓 Dr. Jeffrey Wolf
Certainly! The VERITRACE project aims to understand how ancient wisdom traditions, like texts from the *Corpus Hermeticum* and *Orphic Hymns*, influenced early modern natural philosophy. We're using digital techniques, including LLMs, to analyze a large corpus of Renaissance and early modern texts.

🎙️ Host (Jane)
That sounds fascinating. How do you structure your analysis? Are there different corpora involved?

🎓 Dr. Jeffrey Wolf
Yes, we have two main corpora: the Close Reading Corpus (CRC) and the Distant Reading Corpus (DRC). The CRC includes around 150 texts that are directly related to ancient wisdom traditions. The DRC is much larger, consisting of about 430,000 works from important European library digital collections.

🎙 Host (Jane)
How do you use these corpora in your analysis?

🎓 Dr. Jeffrey Wolf
We use the CRC for detailed textual analysis and the DRC to identify broader patterns across a large number of texts. This allows us to uncover both prominent and neglected works that might have been influenced by ancient wisdom traditions.

🎙 Host (Jane)
You mentioned using Large Language Models in your project. How do these models fit into your research workflow?

🎓 Dr. Jeffrey Wolf
LLMs are crucial for our Semantic Matching Tool, which helps us identify semantic similarities between texts, even when the vocabulary doesn't overlap. This is particularly useful for bridging the 'multilingual gap' and uncovering connections across different languages.

🎙 Host (Jane)
That's really interesting. Can you explain how this tool works in more detail?

🎓 Dr. Jeffrey Wolf
Sure! The Semantic Matching Tool uses LLMs to analyze the meaning behind words, not just their surface-level similarity. This allows us to find passages that are semantically similar even if they're written in different languages.

🎙 Host (Jane)
How does this help with your research on ancient wisdom traditions?

🎓 Dr. Jeffrey Wolf
It helps us uncover long-neglected works and trace the influence of ancient texts across various languages. This is crucial for understanding how these ideas spread and evolved during the Renaissance.

🎙 Host (Jane)
That sounds like a powerful tool. How do you ensure methodological rigor when using LLMs in your research?

🎓 Dr. Jeffrey Wolf
We validate our findings through traditional close reading methods and cross-reference with historical context to ensure the results are meaningful and accurate.

🎙 Host (Jane)
What kind of insights have you gained so far from using these tools?

🎓 Dr. Jeffrey Wolf
We've been able to identify new connections between ancient texts and early modern works, revealing how ideas like those in the *Corpus Hermeticum* influenced major scientific figures such as Copernicus and Kepler.

🎙 Host (Jane)
I want to dive deeper into how these tools are actually being used in your research. You mentioned that the Semantic Matching Tool helps bridge the multilingual gap. Could you elaborate more on this?

🎓 Dr. Jeffrey Wolf
Absolutely. The Semantic Matching Tool uses advanced Large Language Models (LLMs) to analyze not just the surface-level words, but the deeper semantic meaning of texts. This means we can identify passages that are semantically similar even if they don't share the same vocabulary or language.

🎙 Host (Jane)
That sounds incredibly powerful for historical research. How does this work in practice? Could you give an example?

🎓 Dr. Jeffrey Wolf
Sure, let's say we have a passage from the *Corpus Hermeticum* written in Latin and another text by Kepler written in German. The Semantic Matching Tool can identify that these passages are discussing similar ideas about the cosmos or divine wisdom, even though they use different words.

🎙 Host (Jane)
But wait, how does that actually work in practice? What kind of algorithms or techniques do you use to ensure this semantic matching is accurate?

🎓 Dr. Jeffrey Wolf
Well, it's more nuanced than that. The LLMs are trained on vast amounts of text data and can understand context and meaning at a deep level. They use techniques like word embeddings and transformer models to capture the semantic relationships between words.

🎙 Host (Jane)
That sounds quite complex. How do you ensure the reliability of these tools? Are there any limitations or potential biases we should be aware of?

🎓 Dr. Jeffrey Wolf
Absolutely, ensuring methodological rigor is crucial. We validate our findings by cross-referencing with traditional close reading methods and expert analysis. However, LLMs can still have biases based on their training data, so we're very careful to use diverse datasets.

🎙 Host (Jane)
You mentioned earlier that the Semantic Matching Tool helps uncover 'the great unread.' Could you explain how this works in more detail?

🎓 Dr. Jeffrey Wolf
Certainly. By analyzing large corpora of texts, we can identify lesser-known or neglected works that share semantic similarities with well-studied texts. This allows us to bring attention to previously overlooked sources and expand our understanding of intellectual history.

🎙 Host (Jane)
That's fascinating. How do you integrate these findings into your broader research workflow? Do LLMs help in formulating new hypotheses or theoretical frameworks?

🎓 Dr. Jeffrey Wolf
Yes, they play a significant role. By identifying patterns and connections that might not be immediately apparent through traditional methods, LLMs can inspire new research questions and theoretical insights. For example, we've found unexpected links between ancient wisdom texts and early modern scientific works.

🎙 Host (Jane)
But how do you ensure these hypotheses are robust? Don't you risk over-reliance on the tool's outputs?

🎓 Dr. Jeffrey Wolf
That's a valid concern. We use LLMs as an aid, not a replacement for human analysis. The findings from the Semantic Matching Tool are always corroborated with traditional scholarly methods and peer review.

🎙 Host (Jane)
You also mentioned cross-linguistic research capabilities. How does this impact your ability to conduct comparative studies across different languages?

🎓 Dr. Jeffrey Wolf
It's transformative. By bridging the multilingual gap, we can compare texts from various

linguistic traditions more effectively. This allows us to trace the influence of ancient wisdom across different cultures and time periods.

🎙️ Host (Jane)
But what about the potential for misinterpretation or over-reliance on these tools? How do you address those concerns?

🎓 Dr. Jeffrey Wolf
We're very mindful of these risks. We use LLMs as part of a broader toolkit, ensuring that human expertise remains central to our research process. This helps maintain the integrity and rigor of our findings.

🎙️ Host (Jane)
You mentioned earlier that the Semantic Matching Tool is an early modern plagiarism detector. How does this tool differ from traditional methods?

🎓 Dr. Jeffrey Wolf
The key difference lies in its ability to detect semantic similarities rather than just lexical overlaps. This means we can identify passages that are conceptually similar but not word-for-word identical, which is crucial for historical research.

🎙️ Host (Jane)
Dr. Wolf, we've covered a lot of ground today about the VERITRACE project and how LLMs are revolutionizing our approach to studying ancient wisdom traditions in early modern science. Given everything we've discussed, what do you see as some of the broader implications for HPSS research?

🎓 Dr. Jeffrey Wolf
That's a great question, Jane. One of the most significant implications is that LLMs allow us to conduct more comprehensive and nuanced analyses across vast corpora of texts. This means we can uncover connections and influences that were previously hidden or difficult to trace using traditional methods.

🎙️ Host (Jane)
Absolutely. How would this apply in a practical sense, say for researchers who are just starting to integrate LLMs into their workflows?

🎓 Dr. Jeffrey Wolf
For new users, it's important to start with clear research questions and hypotheses. They should also be mindful of the limitations of LLMs, such as potential biases in training data or over-reliance on surface-level similarities. By combining these tools with traditional methods like close reading, researchers can achieve a more balanced approach.

🎙️ Host (Jane)
Let's talk about some of the challenges you've encountered in using LLMs for this kind of

work.

🎓 Dr. Jeffrey Wolf
One major challenge is dealing with biases in training data, which can lead to skewed results. Another is ensuring that the tools are accessible and usable by researchers who may not have a strong background in computational methods.

🎙️ Host (Jane)
What are some key takeaways you'd like listeners to remember from our discussion?

🎓 Dr. Jeffrey Wolf
The key takeaway is that LLMs offer powerful new tools for HPSS research, but they should be used thoughtfully and in conjunction with traditional methods. They can help uncover hidden connections and support more rigorous analysis, but it's important to maintain a critical perspective.

🎙️ Host (Jane)
Thank you so much for this insightful discussion, Dr. Wolf. It's clear that the integration of LLMs into HPSS research is opening up exciting new possibilities while also presenting challenges that need careful consideration.

🎓 Dr. Jeffrey Wolf
It was a pleasure to discuss these topics with you, Jane. I'm excited about the future of this field and look forward to seeing how it evolves as more researchers embrace these tools.[7]

References

Ashmore M (1989) The Reflexive Thesis: Wrighting Sociology of Scientific Knowledge. Chicago: University of Chicago Press.

Chua G (2024) open-notebooklm. GitHub, Inc. Available at: https://github.com/gabrielchua/open-notebooklm (accessed 19 November 2025).

Google, Inc (2023) Introducing NotebookLM (July 12, 2023). Available at: https://blog.google/technology/ai/notebooklm-google-ai/ (accessed 19 November 2025).

Google, Inc (2024) NotebookLM Now Lets You Listen to a Conversation about Your Sources (September 11, 2024). Available at: https://blog.google/technology/ai/notebooklm-audio-overviews/ (accessed 19 November 2025).

---

7 The generation of code that serves as the foundation for this chapter was written with support from large language models (LLMs). All model-generated code was reviewed and, where necessary, modified by the author, who remains fully responsible for the final version. Except for the generated dialogue, LLMs were not used to support the writing of this chapter. For details on the use of LLMs in this volume, see the statement in the volume's introduction.

Heritage S (2024) I AI-Generated Some Podcasts – and the Results Are Uncanny. The Guardian, October 16, 2024. Available at: https://www.theguardian.com/tv-and-radio/2024/oct/16/i-ai-generated-some-podcasts-and-the-results-are-uncanny (accessed 19 November 2025).

Hirsbrunner S, Tebbe M and Müller-Birn C (2024) From Critical Technical Practice to Reflexive Data Science. Convergence: The International Journal of Research into New Media Technologies 30(1): 190–215. https://doi.org/10.1177/13548565221132243. This article was first published in 2022 and republished in this special issue in 2024.

Mahnke MS, Swart J, Mathieu D, and Pruulmann-Vengerfeldt P (2024) Data Reflectivity and User Reflexivity: New Conceptual Pathways for Connecting Structural Approaches with User Perspectives. Convergence: The International Journal of Research into New Media Technologies 30 (6): 1859–70. https://doi.org/10.1177/13548565241301459.

Vincent J (2023) Google Teases Project Tailwind—A Prototype AI Notebook That Learns from Your Documents. The Verge, May 10, 2023. Available at: https://www.theverge.com/2023/5/10/23718707/google-io-2023-project-tailwind-personal-ai-notebook (accessed 19 November 2025).

Willison S (2024) NotebookLM's Automatically Generated Podcasts Are Surprisingly Effective (September 29, 2024). Available at: https://simonwillison.net/2024/Sep/29/notebooklm-audio-overview (accessed 19 November 2025).

Wolf J (2025) open-notebooklm-ollama. GitHub, Inc. Available at: https://github.com/jeffcwolf/open-notebooklm-ollama (accessed 19 November 2025).

Wolf JC (2026) LLMs and multilingual historical corpora in a digital history project. Reflections from the Berlin workshop. In: Simons A, Wüthrich A, Zichert M, et al. (eds) *Understanding Science with Large Language Models? Potentials for the History, Philosophy, and Sociology of Science*. Bielefeld: transcript, part-2.

Woolgar S (eds.) (1988) Knowledge and Reflexivity: New Frontiers in the Sociology of Knowledge. London: Sage.

Part 6: Citation context analysis

Supporting citation context analysis with large language models raises questions that should have been asked 40 years ago

Lydia Liesegang and Jochen Gläser

1. Introduction

The aim of this paper is to explore opportunities to use large language models (LLMs) to improve the validity and reliability of citation context analysis (CCA) as a sociological method.[1] CCA analyses the ways in which cited references are used in citing texts (cf. Simons et al., 2026). It is a powerful method for the study of knowledge flows between texts (Aman and Gläser, 2025: 169–170, Anderson et al., 2023: 77–87), the field-specific gradual acceptance of knowledge claims over time (Cozzens, 1985) or the influence of fraudulent publications on later research (Schneider et al., 2020). It is therefore not surprising that the question of how it could be supported by using LLMs has arisen.

From a sociology of science perspective, the process of integrating LLMs into CCA is an instance of methods becoming routinised, explicated, standardised and in some cases engineered and automatised. These processes have been described for molecular-biological methods by Joan Fujimura, who traced the routinisation and standardisation of for molecular-biological methods through the development of manuals and the standardisation of reagents, instruments, specialised tests and software (Fujimura, 1996: 68–115). Alberto Cambrosio and Peter Keating described the "objectification" of the production of monoclonal antibodies, which was considered as "art" and "magic" in the beginning, through methods sections in experimental articles, experimental protocols in manuscript form, and articles dedicated to the method (Cambrosio and Keating, 1988). In both cases, the authors trace the development from an intuitive approach heavily dependent on tacit knowledge toward an explicit reproducible description of the algorithm

1 We are grateful to Grit Laudel and Arno Simons for comments that helped to significantly improve this chapter. We also acknowledge the contributions by Elaheh Sadat Ahmadi, whose Essay on the comparison of content analysis and citation context analysis (Ahmadi 2025) provided us with a new perspective on the literature on CCA.

underlying the method that will produce results which are expected to be reproducible in any laboratory around the world.

This process – and the possibility of its reversal – can be better understood by using two concepts introduced by the historian of science Hans-Jörg Rheinberger, who distinguished two kinds of objects which are used in research processes. Epistemic objects are objects about which researchers produce knowledge. They are the things researchers want to know more about, and which are considered as ill understood and still taking shape. To investigate epistemic objects, researchers use technical objects, which are things they believe to understand well enough to use them unquestioned (Rheinberger, 1997, 2016). Rheinberger points out that when laboratory work takes unexpected turns, technical objects may be re-questioned and turned back into epistemic objects until their behaviour is well understood again.

We believe that this is what should be happening with CCA. The formalisation and partial automatisation of CCA through attempts to support it with LLMs highlights general problems of the application of CCA. By requiring the explication and formalisation of methodological steps that humans may (and often do) execute subconsciously, the use of LLMs for CCA makes a critical revision of fundamental assumptions possible and necessary. We believe that discussions about LLM use in CCA point out methodological weaknesses in the latter rather than the former. Since the potential of LLM-based methods and other developments of artificial intelligence cannot be currently assessed, we focus on the methodological challenges such methods must meet. In particular we challenge three assumptions underlying any CCA:

A: The assumption that the citation context contains information about the cited source.
B: The assumption that the coding scheme does not need to provide for missing values.
C: The assumption that a syntactically delineated text provides sufficient information about the statements using the cited source.

In the remainder of the paper, we will begin by describing the method of CCA. We will then turn to the three assumptions and demonstrate for each that it is a taken-for-granted assumption of CCA, provide counter examples of citation contexts that violate this assumption, and suggest how the use of LLMs would need to be modified to accommodate cases that violate the assumption. We conclude with two general warnings about using current AIs to support CCA and the validation of that use.

2. Citation context analysis as a sociological method

CCA was originally developed with the intention to improve citation analysis because citation counts were considered to be an insufficient measure of a cited paper's quality (Moravcsik and Murugesan, 1975). Distinguishing between important and unimportant citations and between positive and negative citations remains the main motivation of CCA (Teufel et al., 2006; Zhang et al., 2013; Nishikawa and Koshiba, 2024). So far, none of these CCAs appear to have modified a citation-based evaluation, and no proposal of how

CCA could be integrated in evaluations utilising citation counts has been put forward. Thus, this tradition of CCA, which forms the mainstream, appears to still lack a use case – in the sense of being applied in a practical evaluation of research – after 50 years.

A second, non-mainstream tradition of CCA applies it to answer theory-driven questions about knowledge flows and the transformation of knowledge claims. Important examples include Cozzens (1985), who uses CCA to trace the reception of knowledge claims in biology and in the sociology of science, Amsterdamska and Leydesdorff (1989), who use CCA to distinguish multiple knowledge claims per citation and how they are transformed to fulfil different functions in the citing papers of biochemistry and the use of CCA to identify the impact of fraudulent research papers on subsequent research, and applications (Avenell et al. 2019; Schneider et al., 2020). Anderson and Lemken (2023) suggest using CCA for conducting literature reviews that are guided by a research question, which is also a use of CCA for answering theory-driven questions. This research tradition tends to use more complex coding schemes to answer theory-driven research questions. While it is marginal to CCA, it has demonstrated the potential of CCA to contribute to a wide range of theory-driven research in both qualitative and quantitative science studies.

The following description of CCA as a social-scientific method is applicable to both evaluative and analytical CCA because both need to comply with methodological principles of the social sciences. Discussing the steps of a CCA is useful because it highlights how many highly consequential methodological decisions remain implicit in most CCA applications. It is also a necessary step to evaluate current LLM applications to CCA and inform the discussion of a broader application.

In a CCA, four preparatory and four analytical steps can be identified (Table 1). These steps are preceded by the formulation of a research question, which defines the purpose of the CCA – the knowledge that is supposed to contribute to answer it – and informs all the steps of the CCA, particularly developing the coding scheme, defining the unit of analysis and building the corpora. Given the importance of the research question, its neglect in the many attempts to apply or develop the method further is both surprising and regrettable. In traditional method development (Zhang et al., 2013) as well as in the development of LLM-based methods (Jurgens et al., 2018; Kunnath et al., 2022; Nishikawa and Koshiba, 2024), some (potential) purposes of CCA are mentioned as legitimisation for developing CCA but these purposes do not appear to inform method development. In particular, generic coding schemes like the one introduced by Jurgens et al. (2018), which are not grounded in theories of knowledge production, are used across method developments without justification (Pride and Knoth, 2020; Kunnath et al., 2022). This shifts the epistemic function of categories from operating as a site of theoretical synthesis to operating in a technically efficient manner (Ahmadi, 2025: 16).

Prep. 1: Coding schemes constructed by mainstream evaluative CCA have become more sophisticated over time but essentially still aim at capturing substantial versus perfunctory use and positive versus negative evaluation of the cited source (see section 3 for examples of coding schemes). Analytical coding schemes are more complex but also often less strictly defined. The crucial aspect of this preparatory step is that the coding scheme – and its built-in assumptions – decides what information can be collected with the CCA from the analysed texts. One of these assumptions is that every citation context contains information about the citation (see section 3), another one is that each citation context contains all the information that is to be collected with the coding scheme. In other words, CCA often assumes that there are no missing values. We will discuss this problem in section 4.

Table 1: Main steps of quantitative content analysis (steps for which LLM support is currently considered are highlighted)

Step		**Content**
Formulating research question		**Define knowledge gap and key concepts**
Prep. 1	Constructing coding scheme	Operationalise the key concepts
Prep. 2	Defining the unit of analysis	Decide which text linked to a citation contains the necessary information
Prep. 3	Building the corpus (cited papers)	Determine which citations are relevant for answering the research question
Prep. 4	Building the corpus (citing papers)	Collect documents which contain the necessary information
1	Locating citations	Find relevant citations in the documents
2	Delineating unit of analysis	Identify text to be coded for each citation
3	Coding unit of analysis	Interpret each text and allocate results to coding scheme
4	Integrate empirical evidence	Collate and comparatively analyse the coding results
Answering research question		Use empirical evidence to formulate new knowledge claims

Prep. 2: An important step of CCA that has been extensively discussed in evaluative CCA is the definition of the unit of analysis. The discussion usually focuses on the size of a syntactically constructed context, i.e. on how many sentences or paragraphs before and after a citation should be included to enable an accurate interpretation. Since delineating the unit of analysis is a key step for which LLMs are currently used, we discuss it in section 5.

Prep. 3/4: Building the corpus also crucially shapes the answer to the research question because the information that can be collected depends on which papers are analysed. The corpus consists of the cited papers, about which information is collected with the CCA

(Prep. 3) and the citing papers, that contain the information in their citation contexts (Prep. 4).

Steps 1-3: Once the preparatory steps have been completed, the actual CCA can be conducted in four steps. Steps 1 to 3 are repeated for every instance: A citation of a source from the source corpus must be located, and the unit of analysis must be delineated and interpreted, i.e. values according to the coding scheme must be assigned.

Step 4: Step 4 collates the interpretations. In the case of evaluative CCA, this usually amounts to a descriptive statistic, which in some cases is correlated with properties of cited or citing documents. In analytical CCA, more complex conclusions are drawn.

Once the CCA is completed, its results are used to answer the research question, often in combination with other information collected in a research project.

3. The citation context may not contain information about the citation

One of the premises of CCA is that the citation context that is interpreted by humans or categorised by LLMs contains information about the cited source. If this is not the case, the interpretation/automated coding of the citation context may return an incorrect result. This situation occurred in an application of a generative LLM for CCA by Nishikawa and Koshiba. The generative LLM was instructed to consider the paragraph including the citation and to categorise the positive, negative or neutral assessment of the cited source with the following instruction:

Figure 1: Prompt used for coding the citation purpose (Source: Fig. 1 of Nishikawa and Koshiba, 2024: 6759)

Prompt for citation purpose (Simple):

```
Classify the type of purpose for which the author of the
following text is citing the target. The type of citation
purpose category is Background, Comparison, Criticize,
Evidence, or Use. The label of the target literature is
Target: onwards, up to a new line. The thesis is the text
after Text:.  Just report the classification result.

Target: {Labels of cited papers in the text to be annotated}
Text:
{A paragraph to be annotated}
```

This did not work in a case where the cited sources (Latour, 2004 and Lupton and Mather, 1997) provided only background information and the criticism in the paragraph following the citation was directed at *a situation described in the cited source* rather than the cited source itself.

> This use of technical devices in an attempt to suppress political debates has been widely documented elsewhere (e.g., Latour, 2004; Lupton and Mather, 1997). At an extreme this use of GIS and mapping reinforces and aggravates existing divides and inequalities. (Nishikawa and Koshiba, 2024: 6769)

Since the LLM was instructed to classify the paragraph after the cited sources, it applied the negative assessment in this paragraph to the cited source. A correct classification requires the identification of the *argument* made in this paragraph rather than the *text* presented there. In Nishikawa's and Koshiba's coding scheme, the use of the cited sources for this argument could be categorised as Background, Evidence or Use.

This example could be dismissed as a case where the prescribed citation context (paragraph) did not match the actual citation context (the sentence ending with the citation). However, the problem could easily occur with a citing sentence, too. In the following example, the sentence citing Greenland 1990 and Greenland and Poole 2013 criticises frequentist analyses of observational data:

> In this respect, frequentist analyses of observational data seems to depend on unlikely assumptions that too often turn out to be so wrong as to deliver unreliable inferences, and hairsplitting interpretations of p values becomes even more problematic (Greenland, 1990; Greenland and Poole, 2013). (Schneider, 2015: 419)

Without analysing the cited sources, it is impossible to tell if they are examples of "hairsplitting interpretations of p values" or criticise such hairsplitting interpretations. Thus, it is not even clear to human readers if the text contains an assessment of the cited source or not. An LLM-based CCA working with linguistic markers would be likely to code this example as a negative citation of Greenland 1990 and Greenland and Poole 2013 if the prompt used by Nishikawa and Koshiba (2024) would be applied.

The general situation in which such a transference of negative evaluations to the cited source can be described as follows: If the cited source is used as a source of information about a phenomenon that is not produced by the authors of the cited publication but just reported by them, and if the phenomenon is criticised by the citing paper, this criticism is likely to be ascribed to the cited source. Human coders are likely to recognise the problem during coding and to adapt their interpretation, while it seems currently impossible to design a rule for an LLM that could take this situation into account.

4. The coding scheme must provide for missing values

Another assumption that often appears to inform the construction of coding schemes for CCA is that it matches the information contained by each citation context perfectly. In other words, CCA often assumes that there are no missing values in citation contexts and applies coding schemes that provide no option for handling such cases, as in the following example:

> The scheme which we devised to classify the reasons for citation of these pre-1930 papers is as follows:
>
> A: Historical background
> B: Description of other relevant work
> C: Supplying information or data, other than for comparison
> D: Supplying information or data for comparison
> E: Use of theoretical equation
> F: Use of methodology
> G: Theory or method not applicable or not the best one
>
> (Oppenheim and Renn, 1978: 226)

This coding scheme (and several others listed by Liu, 1993) is neither exhaustive, nor has it an option for missing values and unexpected information. Some of the coding schemes used in LLM-based CCA appear to follow the same tradition. For example, the coding scheme used by Nishikawa and Koshiba (s. Fig. 1) uses Background, Evidence, Comparison, Criticise or Use. Similarly, the highly popular coding scheme introduced by Jurgens et al. (2018) which has also been used by Pride and Knoth (2020) and Kunnath et al. (2022), does not appear to be exhaustive and has no room for missing values (Figure 2).

Figure 2: Coding Scheme used in experiments with LLMs (source: Jurgens et al., 2018: 393)

Class	Description	Example
BACKGROUND	*P* provides relevant information for this domain.	This is often referred to as incorporating deterministic closure (Dörre, 1993).
MOTIVATION	*P* illustrates need for data, goals, methods, etc.	As shown in Meurers (1994), this is a well-motivated convention [...]
USES	Uses data, methods, etc., from *P* .	The head words can be automatically extracted [...] in the manner described by Magerman (1994).
EXTENSION	Extends *P*'s data, methods, etc.	[...] we improve a two-dimensional multimodal version of LDA (Andrews et al, 2009) [...]
COMPARISON OR CONTRAST	Expresses similarity/differences to *P*.	Other approaches use less deep linguistic resources (e.g., POS-tags Stymne (2008)) [...]
FUTURE	*P* is a potential avenue for future work.	[...] but we plan to do so in the near future using the algorithm of Littlestone and Warmuth (1992).

The application is liable to creating situations in which coders a) may subconsciously invent information that is not in the citation context to resolve the conflict or, if they recognise the conflict, b) decide to ignore available information if it does not fit any of the options provided by the coding scheme, or c) expand the coding scheme and start all over again. Not all of these possibilities are open to both human coders and LLM, since only the former are able to deviate from the given instructions if they cannot fulfil their objective, and choose one of the options b) or c).

An LLM on the other hand must follow the given steps and find a matching value in the coding scheme for every citation context. Thus, it is forced to choose b) and produce a 'finding' that is not actually there. To avoid this, the applied coding scheme and the given prompt should be expanded to include an option for missing values.

Unfortunately, LLMs have a built-in tendency to please their human masters (sycophancy, Sharma et al., 2025) and are known to notoriously 'retrieve' non-existent data in different contexts even when given the opportunity to report a missing value. This fosters the possibility that fictional data are distorting the CCA and should be considered when applying LLMs to CCA.

5. Syntactically delineated text is unlikely to provide accurate information about its semantics

The question of what text surrounding a citation should be analysed as citation context is a long-standing concern of CCA. Zhang et al. phrase this question as follows:

> In the context of citing behavior, ... we need to make decisions regarding the following: ... Should a single sentence, or a cluster of sentences in which a reference is mentioned, be selected as the unit of analysis? (Zhang et al., 2013: 1494).

In their review of NLP and machine learning approaches, Iqbal et al. review experiments with different numbers of sentences before and after a citation and conclude that "the use of four sentences (the citing sentence, one sentence before the citing sentence, and two sentences following the citing sentence) ... is now considered as a quasi-norm for in-text citation analyses studies (Teufel et al., 2006)." (Iqbal et al., 2021: 6561)

These discussions of units of analysis for CCA point to a fundamental problem. As illustrated by Zhang et al. (2013) and Iqbal et al. (2021), traditional CCA implicitly assumes that a number of syntactic units can be selected as citation context. LLM-supported CCA initially used the same approach. However, the use and the assessment of a cited source occur in a scientific statement or argument, i.e. in a semantic unit. Any fixed syntactically delineated citation context is unlikely to coincide with the argument from which the purpose and the meaning of a citation's use arise. Most syntactic delineations of citation contexts including Nishikawa's and Koshiba's instruction of their generative LLM (see above, Figure 1) would fail in the following example from one of our own CCAs (Figure 3).

The arguments that use the reference Yan et al. extend to the following paragraph. At the beginning of this paragraph, the authors of the citing paper report results presented in the cited source and state what the authors of the cited source did not study, which is then used to motivate their own research. Thus, the cited source is used to support two entirely different arguments, one about what is known about linc-ITGB1 and one about what is not known. Most CCAs including the one conducted by the generative AI instructed by Nishikawa and Koshiba would find the former but would (be forced to) ignore the latter.

Figure 3: A citation context from cancer research (source: Li et al. 2017: 3400, solid line shows citing paragraph, dotted line the extension of the argument beyond the paragraph)

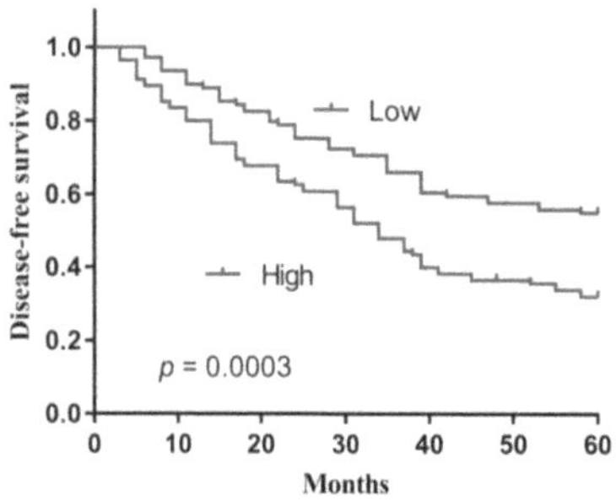

Figure 3. Correlation between linc-ITGB1 expression and DFS time in BC patients.

vention and therapy of breast cancer remain a major public health concern. More studies15,16 have recognized that the appropriate therapies require effective biomarkers to guide. Recently, growing papers reported that changes in expression of lncRNAs can be used as robust and important biomarkers for cancer risk, diagno- sis, and prognosis17-19. Based on previous find- ings, our attention focused on the linc-ITGB1.

Linc-ITGB1 is a newly discovered lncRNA. To our best knowledge, only two papers reported the specific role of linc-ITGB1 in tumors, including gallbladder cancer and BC. Wang et al20 firstly reported that expression of linc-ITGB1 was up-regulated in gallbladder cancer cells. Moreover, they confirmed that linc-ITGB1 was a promoter of gallbladder cancer metastasis due to its regulation of epithelial-to-mesenchymal transition (EMT). Yan et al12 found that linc-ITGB1 was overexpressed in BC tissues and cell lines, and its over-expression significantly inhibits cell proliferation, migration and invasion, while its knockdown exert the contrary role. Furthermore, constant with previous study, they also identified that inc-ITGB1 served as a tumor promoter by affecting EMT. Those findings informed that linc-ITGB1 played an important role in progres- sion and development of gallbladder cancer and BC. However, their research didn't study the clinical significance of linc-ITGB1 in patients.

In the present work, we firstly determine the expression levels of linc-ITGB1 in BC and matched normal tissues and observed significantly increased expression of linc-ITGB1 in the BC tissues than in the adjacent normal breast tissues. We further found the expression level of linc-ITGB1 was significantly associated with lymph node metastasis, pathological differentiation and TNM stage, suggesting linc-ITGB1 contributed to progression of this tumor. In addition, Kaplan-Meier analysis demonstrated that high linc-ITGB1 expression level was associated with poorer OS and DFS. Based on these data, we performed multivariate analysis and the results indicated that linc-ITGB1 was an independent prognostic factor for predicting OS and DFS of BC patients. On the basis of the previous reports, we suggested hypothesized that the contribution of high linc-ITGB1 expression to the poor prognosis of BC patients might be caused by its facilitative efficiency on BC cell proliferation and metastasis.

Conclusions

We revealed, for the first time, that linc-ITGB1 may represent a valuable independent prognostic indicator for BC. Further study to discover the molecular mechanism of linc-ITGB1 about tumor migration and invasion is currently in progress.

In other cases, arguments using a cited source may occur completely separated from that source (Figure 4). In an article in the field of international relations, Donnelly (2011) cites a total of five publications by Kenneth Waltz. He refers to publications by Waltz not only by conventionally citing them but also by using his name or labels for his approach like "structural realism" as markers. The sample page shown in Figure 4 begins with a common type of citation context in which a publication by Waltz is used 'substantially' and cited 'positively' according to conventional coding schemes. Further down this page, the name provides the reference for a 'substantial' but 'neutral' use. Still further down and nowhere near a citation of one of Waltz's publications, "Waltzian approach" and "structural analysis" are used as markers for his publications with a thoroughly negative assessment.

Figure 4: Distribution of arguments independently of citations (source: Donnelly, 2011: 171)

Donnelly 171

network of social positions. It helps us to 'describe and understand the pressures states [and other international actors] are subject to' (Waltz, 1979: 71).

As citing Waltz suggests, this is not very different from the mainstream approach. One cannot read off substantive theory from a typology of elements of structure or an analytical framework of differentiation. From IR's tripartite conception of structure nothing more of substance follows than from my multidimensional framework — pace Waltz's suggestion that it entails structural realism. Explanation or prediction is possible only when particular elements are combined in a particular fashion. A typology of elements of structure may delimit a range of structural possibilities. It does not imply any particular substantive theory — unless, perhaps, all systems have the same structure (which they do not).

A dimensional framework can also be used to add discipline to historical comparisons, which too often are impressionistic or focused on features taken out of context. For example, a neo-medieval (e.g. Bull, 1977: 254–256, 264–276) account of globalization singles out the absence of a dominant unit type and the presence of extensive functional differentiation and heterarchic stratification. But what happens when Christendom and hierarchy are replaced by an egalitarian global culture and universal human rights as an internal standard of political legitimacy? Or, how do the technologies of our era influence the ways functions are divided, how group actors are formed and aggregated, or the relations between individuals and authorities of various sorts?

Daniel Nexon suggests comparing globalization with early-modern Europe, especially because of the prevalence of composite states and the importance of transnational relations (2009: 19, 299–300). This was also a period when space and time were transformed and new forms of religious and cultural conflict emerged. I find particularly intriguing the idea of using a framework of differentiation to 'unwind' 'Westphalia' in order to think through possible 'post-Westphalian' futures.

Historical comparisons are, to be sure, 'only' analogies. But we should appreciate, and put to good use, the understanding and insight they provide — especially given the lack of a serious 'stronger' alternative. The mainstream Waltzian approach, for all its social-scientific pretense, amounts to analogical reasoning — employing a rather simplistic, ahistorical analogy at that. Rather than seek to depict accurately the actual structure of real international systems, 'structural analysis' has become a dubious exercise in determining how closely a system approximates a highly idealized one-dimensional type. A dimensional approach directs us back to dealing with the real complexities and varieties of international structures.

The cited source is used to support several different arguments. Furthermore, it is clear from phrases like "Waltzian approach" that the cited source is used together with other publications by Waltz referenced in the paper even though no source is provided close to the second and third arguments. In this case, the application of any known rule for syntactically delineating citation contexts would lead to an erroneous interpretation.

There are also field-specific citation practices and publication styles that do not work within narrow parameters, as elements of one argument can be scattered all over the publication, distant paragraphs of other parts of the publication or stay implicit all together. In the following example from the history of arts, the argument using the cited source is partly made in the main text and continuous in the footnote surrounding the cited source (Figure 5).

Figure 5: Distribution of an argument over a page (source: Sluijter, 2014: 67, frames delineate the two parts of the argument)

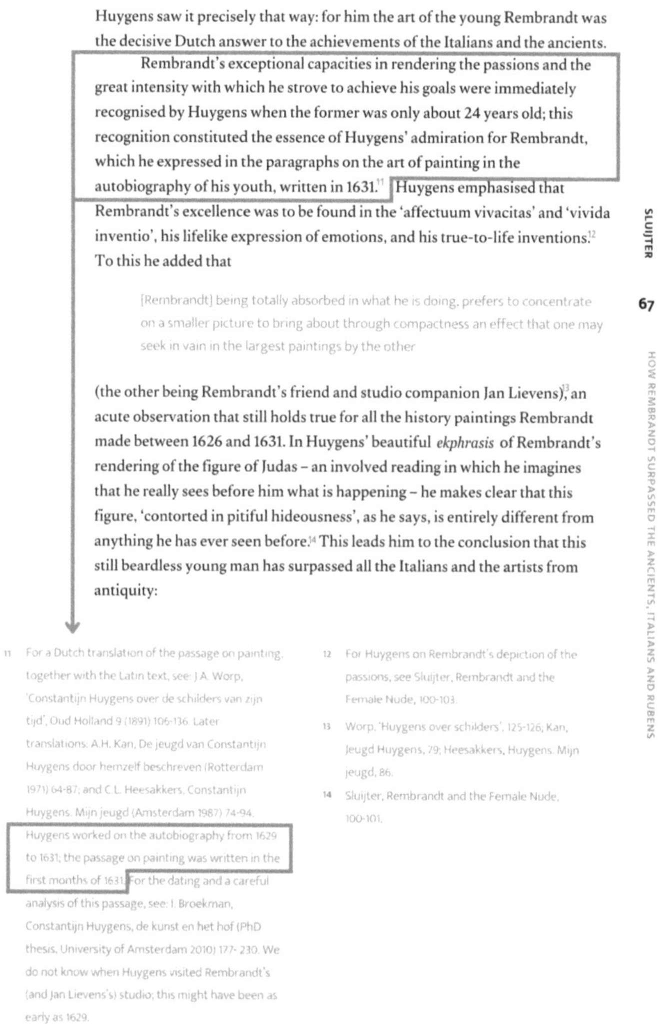
Huygens saw it precisely that way: for him the art of the young Rembrandt was the decisive Dutch answer to the achievements of the Italians and the ancients.

Rembrandt's exceptional capacities in rendering the passions and the great intensity with which he strove to achieve his goals were immediately recognised by Huygens when the former was only about 24 years old; this recognition constituted the essence of Huygens' admiration for Rembrandt, which he expressed in the paragraphs on the art of painting in the autobiography of his youth, written in 1631.[11] Huygens emphasised that Rembrandt's excellence was to be found in the 'affectuum vivacitas' and 'vivida inventio', his lifelike expression of emotions, and his true-to-life inventions.[12] To this he added that

> [Rembrandt] being totally absorbed in what he is doing, prefers to concentrate on a smaller picture to bring about through compactness an effect that one may seek in vain in the largest paintings by the other

(the other being Rembrandt's friend and studio companion Jan Lievens),[13] an acute observation that still holds true for all the history paintings Rembrandt made between 1626 and 1631. In Huygens' beautiful *ekphrasis* of Rembrandt's rendering of the figure of Judas – an involved reading in which he imagines that he really sees before him what is happening – he makes clear that this figure, 'contorted in pitiful hideousness', as he says, is entirely different from anything he has ever seen before.[14] This leads him to the conclusion that this still beardless young man has surpassed all the Italians and the artists from antiquity:

11 For a Dutch translation of the passage on painting, together with the Latin text, see: J.A. Worp, 'Constantijn Huygens over de schilders van zijn tijd', Oud Holland 9 (1891) 106-136. Later translations: A.H. Kan, De jeugd van Constantijn Huygens door hemzelf beschreven (Rotterdam 1971) 64-87; and C.L. Heesakkers, Constantijn Huygens. Mijn jeugd (Amsterdam 1987) 74-94. Huygens worked on the autobiography from 1629 to 1631; the passage on painting was written in the first months of 1631. For the dating and a careful analysis of this passage, see: I. Broekman, Constantijn Huygens, de kunst en het hof (PhD thesis, University of Amsterdam 2010) 177- 230. We do not know when Huygens visited Rembrandt's (and Jan Lievens's) studio; this might have been as early as 1629.

12 For Huygens on Rembrandt's depiction of the passions, see Sluijter, Rembrandt and the Female Nude, 100-103.

13 Worp, 'Huygens over schilders', 125-126; Kan, Jeugd Huygens, 79; Heesakkers, Huygens. Mijn jeugd, 86.

14 Sluijter, Rembrandt and the Female Nude, 100-101.

SLUIJTER 67 HOW REMBRANDT SURPASSED THE ANCIENTS, ITALIANS AND RUBENS

Thus, while the assumption by Ritchie et al. (2008: 216) "that the words that are used to describe the cited paper will occur close to the citation and that, further away, words are less likely to be about the paper" may be correct in most cases, it is impossible to predict how far away from a cited source arguments using it may occur in any individual case. Our examples illustrate the possibility that the argument or arguments using a cited source do not coincide with the syntactically defined citation context (e.g., the paragraph following or three sentences before and after a citation). If such a mismatch occurs, key information may be excluded from the interpretation because it is located outside the predefined citation context. In other cases of mismatch, the information about the cited source may be distorted by information from other unrelated arguments that are fully or partly captured by the citation context.

The citation contexts in which citations are used and assessed are the argument(s) using the cited source, i.e. semantic units of analysis. This definition of the unit of analysis is independent from syntactic delineations because arguments have no specific syntactic dimension by which they could be reliably identified. The syntactical delineation of citation contexts that has been dominating both human and LLM-based CCA is unlikely to deliver valid results because mismatches are likely to occur (Figure 6).

Figure 6: Possible relationships between the argument using a cited source (red square) and a syntactically delineated citation context (ellipses)

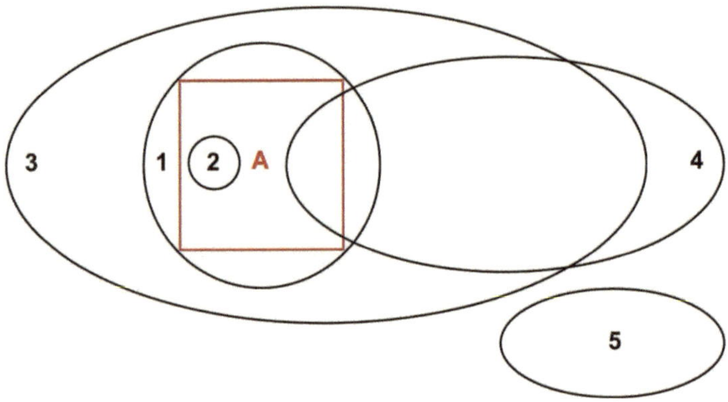

The relationships between the argument and the citation context can have the following consequences for its representation:

1) **Match:** If the context is coextensive with the argument, the likelihood of a correct interpretation is highest.
2) **Part:** If only part of the argument is captured by the citation context, its interpretation is likely to be incomplete.
3) **Diluted:** If the context exceeds the argument, the argument's interpretation is likely to be diluted by other arguments.
4) **Distorted:** If the context contains part of the argument and a larger amount of unrelated text, its interpretation is likely to be based on a different argument that may even be the opposite of the argument using the citation.
5) **Unrelated:** If the context is not related to the argument at all, its interpretation is highly unlikely to represent the argument using the citation.

While the mismatch between syntactically delineated citation contexts and arguments that use cited sources is a problem of CCA in general rather than LLM-based approaches, it can be overcome much more easily by human coders than by LLMs. Human coders can (and we suspect often do) identify arguments and use them in lieu of the prescribed citation contexts 'on the fly', i.e. when a mismatch becomes obvious in the course of a

CCA. In contrast, AIs must follow the rules and thus cannot adapt to mismatches if the use of a fixed citation context is prescribed.

For LLM-based methods to avoid this mismatch, the rules must enable them to identify and delineate all kinds of statements in which authors use cited sources. While the likely mismatch between syntactically delineated citation contexts and arguments using the cited paper has not been widely discussed in the context of LLM-based CCA, the problem of using fixed citation contexts has been noted and motivated attempts to use LLMs for a dynamic delineation of citation contexts (e.g., Kunnath et al., 2022). The authors delineated citation contexts based on the similarity between embeddings of sentences in the citing paper and embeddings of title and abstract of the cited paper. They demonstrated that the classification of citation purposes 'improved' when variable citation contexts were used and 'improved' again when non-contiguous variable citation contests were used.

Although the experiments conducted by Kunnath et al. (2022) fall short of a validation of their method, two aspects merit discussion because they are likely to persist with LLM-based dynamic context detection. First, the method is based on the assumption that the text which is semantically most similar to the reference's title is most relevant for the categorisation of a citation context. This assumption, which is not explicated by Kunnath et al., has not undergone any empirical test. It is unlikely to be correct in all cases because a publication can be cited for different knowledge claims, not all of which are equally well represented by its title. Second, the method appears to replace fixed citation contexts with fixed thresholds for semantic similarity. Therefore, the possible mismatches between arguments and citation contexts shown in Figure 4 remain, with the ellipses representing text that is semantically similar to the cited source's title above a fixed threshold.

Another approach to identify the semantically relevant citation context is to train LLMs with a corpus in which text segments with specific semantic functions are annotated and citation used to train LLMs. Jantsch et al. (2025) defined three functions of a citation context, namely the information the citing author references, the relationship between the citing and the cited work, and the reason why a citation is included. They manually annotated 1056 citation contexts and used this data set to train an LLM. While this training of an LLM still exploits semantic similarity (in this case: between the annotated text segments and the model that is trained), the general approach could be used for the identification of scientific statements rather than epistemic functions of word sequences. This would require a training data set on all kinds of scientific statements in a research field that use citation, a task that seems currently impossible.

6. Discussion

We discussed three methodological challenges of CCA which precede the development of LLM-supported CCA but are highlighted by this development because LLM support requires the formulation of explicit rules for the LLM application. The explication of rules, in turn, reveals a gap in the methodological foundations of CCA and enables their discussion. In the context outlined in the introduction, the technical object (traditional CCA) is turned into an epistemic object again because its future routinisation through LLM support problematises its foundations. CCA as a social-scientific method is in urgent need of methodological qualification. So far, neither the construction of coding schemes nor the delineation of citation contexts have sufficiently considered methodological principles of social science research. The resulting problems remained hidden in traditional CCA because (a) CCA is rarely used to answer research questions and thus can rarely fail, and because (b) human coders are likely to compensate for them 'on the fly', sometimes possibly without even noticing them.

Given that the development of LLM-supported CCA is still in its infancy and LLMs are rapidly evolving, we refrain from predicting how far they will be able to go. However, we would like to draw attention to three problems that need to be addressed as early as possible in the process of developing LLM-based CCA. First, the validation strategies for LLM-based methods need to be revised. If LLM-based methods applying faulty rules are assessed by comparing them to annotated citation contexts that have been produced applying the same faulty rules, the outcome contains little useful information.

Second, it seems doubtful that the development of LLM-supported CCA can productively continue without considering the relevant theoretical and methodological backgrounds in science studies. Developing methods without the input of a concrete, theory-based research question can isolate the methods from the very field they are intended to serve.

Third, there is a danger of stretching the conceptual foundations of LLM use. It is worth noting that the "distributional hypothesis" as formulated by Harris (1954) states that differences in the linguistic meaning of words correlate with differences between the distribution of linguistic entities in their context (Sahlgren, 2008; Lenci 2018: 152–153; Haber and Poesio, 2024; Resnik, 2025: 894). While it seems legitimate to conclude from this hypothesis that correlations between distributions of syntactic units in word contexts reflect semantic similarity, the distributional hypothesis does *not* state that the context of a word or phrase determines its meaning, which is often assumed in the literature (e.g., Andrews et al., 2014: 361; Boutyline and Arseniev-Koehler, 2025: 90). LLM developers and users need to carefully distinguish between what a LLM *does* and the actions they *ascribe* to a LLM based on their own creation of meaning. The delineation of citation contexts is a case in point. A delineation based on semantic similarity might be more useful than fixed citation context windows for some purposes but is still fundamentally different from the identification of a scientific statement that uses a cited source.

7. Conclusions

In this chapter, we highlighted some optimistic assumptions underlying current human-based and LLM-based applications of CCA, and demonstrated that these assumptions may prove wrong in an unknown but potentially large number of cases. Attempts to use LLMs for the support and scaling-up of CCA require explicating and formalising the method and thus reveal weaknesses it has had all along. From this follows that developing LLM support for CCA requires exploiting the possibility suggested by Rheinberger, namely transforming CCA back into an epistemic object by re-opening its conceptual exploration. With regard to the conceptual foundations of LLM applications, our discussion highlights the difference between LLMs' mimicking of the human practice of using linguistic markers to assign labels to texts and true interpretation, which uses layers of contexts of varying sizes in a holistic act to establish a possible meaning of a text. While there are many cases in which humans use the same approach LLMs use today, and while marker-based labelling and interpretation may produce the same result in even more cases, there is an unknown number of cases in which the correct application of the two approaches would lead to different results.

Given these problems, we would like to invoke Robert Merton's "organised scepticism", which demands a temporary suspension of judgment and the detached scrutiny of beliefs in terms of empirical and logical criteria (Merton, 1973 [1942]: 277). While we do not see evidence that LLMs can solve the methodological problems we discussed, we do not consider this to be impossible. We also believe that experiments with LLMs can be productively used to further elaborate the methodological problems of CCA.

The more general argument that can be derived from our discussion is that regardless of the outcomes of current uses of LLMs so support interpretive methods in the social sciences, these experiments can lead to an improvement of methods because they force the explication of taken-for-granted assumptions and implicit methodological rules without which a LLM cannot be properly instructed. Attempts to use LLMs thus exercise pressure on social science methodology to improve the understanding of methods and to increase the precision of their description.

References

Ahmadi, ES (2025) Content Analysis and Citation Context Analysis in Comparison: A Methodological Perspective. TU Berlin. https://www.static.tu.berlin/fileadmin/www/10005401/Publikationen_sos/Ahmadi_2025_Measuring_codification.pdf

Aman, V and Gläser J (2025) Investigating Knowledge Flows in Scientific Communities: The Potential of Bibliometric Methods. Minerva 63(1): 155–182.

Amsterdamska, O and Leydesdorff, L (1989) Citations: Indicators of Significance? Scientometrics 15(5-6): 449–471.

Anderson, MH and Lemken RK (2023) Citation Context Analysis as a Method for Conducting Rigorous and Impactful Literature Reviews. Organizational Research Methods 26(1): 77–106.

Andrews, M, Frank, S and Vigliocco, G (2014) Reconciling Embodied and Distributional Accounts of Meaning in Language. Topics in Cognitive Science 6(3): 359–370.

Avenell, A, Stewart, F, Grey, A, et al. (2019) An investigation into the impact and implications of published papers from retracted research: systematic search of affected literature. BMJ Open 9(10): e031909.

Boutyline, A and Arseniev-Koehler, A (2025) Meaning in Hyperspace: Word Embeddings as Tools for Cultural Measurement. Annual Review of Sociology 51(Volume 51, 2025): 89–107.

Cambrosio, A and Keating, P (1988) "Going Monoclonal": Art, Science, and Magic in the Day-to-Day use of Hybridoma Technology. Social Problems 35(3): 244–260.

Cozzens, SE (1985) Comparing the Sciences: Citation Context Analysis of Papers from Neuropharmacology and the Sociology of Science. Social Studies of Science 15(1): 127–153.

Donnelly, J (2011) The differentiation of international societies: An approach to structural international theory. European Journal of International Relations 18(1): 151–176.

Fujimura, JH (1996) Crafting Science: A Sociohistory of the Quest for the Genetics of Cancer. Cambridge, Mass., Harvard University Press.

Haber, J and Poesio, M (2024) Polysemy—Evidence from Linguistics, Behavioral Science, and Contextualized Language Models. Computational Linguistics 50(1): 351–417.

Harris, ZS (1954) Distributional Structure. WORD 10(2-3): 146–162

Iqbal, S, Hassan, S-U, Aljohani, NR, et al. (2021) A decade of in-text citation analysis based on natural language processing and machine learning techniques: an overview of empirical studies. Scientometrics 126(8): 6551–6599.

Jantsch, LM, Koh, D-J, Yoon, S et al. (2025) FineCite: A Novel Approach For Fine-Grained Citation Context Analysis, Vienna, Austria, Association for Computational Linguistics.

Jurgens, D, Kumar, S, Hoover, R, et al. (2018) Measuring the Evolution of a Scientific Field through Citation Frames. Transactions of the Association for Computational Linguistics 6: 391–406.

Lenci, A (2018) Distributional Models of Word Meaning. Annual Review of Linguistics 4: 151–171.

Li, W-X, Sha, R-L, Bao, J-Q, et al. (2017) Expression of long non-coding RNA linc-ITGB1 in breast cancer and its influence on prognosis and survival. Eur Rev Med Pharmacol Sci 21(15): 3397–3401.

Liu, M (1993) Progress in documentation – The complexities of the citation practice: A review of citation studies. Journal of Documentation 49(4): 370–408.

Merton, RK (1973 [1942]) The Normative Structure of Science. The Sociology of Science. R. K. Merton. Chicago, The University of Chicago Press: 267–278.

Moravcsik, MJ and Murugesan, P (1975) Some Results on the Function and Quality of Citations. Social Studies of Science 5(1): 86–92.

Kunnath, SN, Pride, D and Knoth, P (2022) Dynamic Context Extraction for Citation Classification, Online only, Association for Computational Linguistics.

Nishikawa, K and Koshiba, H (2024) Exploring the applicability of large language models to citation context analysis. Scientometrics 129(11): 6751–6777.

Pride, D and Knoth, P (2020) An Authoritative Approach to Citation Classification. Proceedings of the ACM/IEEE Joint Conference on Digital Libraries in 2020. Virtual Event, China, Association for Computing Machinery: 337–340.

Resnik, P (2025) Large Language Models Are Biased Because They Are Large Language Models. Computational Linguistics 51(3): 885–906.

Rheinberger, H-J (1997) Toward a History of Epistemic Things: Synthesizing Proteins in the Test Tube. Stanford, Stanford University Press.

Rheinberger, H-J (2016) On the Possible Transformation and Vanishment of Epistemic Objects. Teorie vědy / Theory of Science; Vol 38, No 3 (2016)DO – 10.46938/tv.2016.364.

Ritchie, A, Robertson, S and Teufel, S (2008) Comparing citation contexts for information retrieval. Proceedings of the 17th ACM conference on Information and knowledge management. Napa Valley, California, USA, Association for Computing Machinery: 213–222.

Sahlgren, M (2008) The distributional hypothesis. Rivista di Linguistica 20(1): 33–53.

Schneider, JW (2015) Null hypothesis significance tests. A mix-up of two different theories: the basis for widespread confusion and numerous misinterpretations. Scientometrics 102(1): 411–432.

Schneider, J, Ye, D, Hill, AM, et al. (2020) Continued post-retraction citation of a fraudulent clinical trial report, 11 years after it was retracted for falsifying data. Scientometrics 125(3): 2877–2913.

Sharma, M, Tong, M, Korbak, T et al. (2025) Towards Understanding Sycophancy in Language Models. arXiv:2310.13548 [cs.CL].

Simons A, Arnaout H and Gurevych I (2026) Reconstructive citation context analysis using large language models. A roadmap. In: Simons A, Wüthrich A, Zichert M, et al. (eds) *Understanding Science with Large Language Models? Potentials for the History, Philosophy, and Sociology of Science*. Bielefeld: transcript, part-6.

Sluijter, EJ (2014) How Rembrandt surpassed the Ancients, Italians and Rubens as the Master of "the Passions of the Soul". Low Countries Historical Review 129(2): 63–89.

Teufel, S, Siddharthan, A and Tidhar, D (2006) Automatic classification of citation function. Proc. 2006 Conference on Empirical Methods in Natural Language Processing, Association for Computational Linguistics: 103–110.

Zhang, G, Ding, Y and Milojević, S (2013) Citation content analysis (CCA): A framework for syntactic and semantic analysis of citation content. Journal of the American Society for Information Science and Technology 64(7): 1490–1503.

Reconstructive citation context analysis using large language models
A roadmap

Arno Simons, Hiba Arnaout, and Iryna Gurevych

1. Introduction

Citations are not merely formal pointers to prior work. Within the history, philosophy, and sociology of science (HPSS), citations are examined through a reconstructive lens—that is, an effort to recover how scientific claims, alliances, and boundaries are actively built through acts of citing, rather than merely reflected in them. Patterns of citation and uptake, in turn, offer a window onto the dynamics of knowledge production and circulation, from the formation of scientific facts to the negotiation of epistemic authority and the construction of disciplinary identities (e.g., Cozzens, 1985; Gilbert, 1977; Latour, 1987; Luukkonen, 1997; Myers, 1990; Small, 1980; 2004; Swales, 1986; 1990; White; 2004).

This reconstructive approach to studying citation practice contrasts with the evaluative logic that dominates much of scientometrics, where citations are primarily treated as countable indicators of research quality or influence. Citation context analysis (CCA) emerged in the 1970s within this evaluative tradition as a means of distinguishing "countworthy" from marginal citations in order to refine citation-based measures (Chubin and Moitra, 1975; Moravcsik and Murugesan, 1975). Although HPSS scholars have since promoted CCA for reconstructive purposes, the evaluative framing has remained the dominant influence on how CCA is conceived, operationalized, and applied (cf. Bornmann and Daniel, 2008).

Building on this evaluative orientation, computational approaches to CCA have developed considerably over the past twenty-five years, evolving from rule-based grammars and keyword matching to machine learning, deep learning, and transformer-based models (for recent overviews, see Iqbal et al., 2021; Kunnath et al., 2021; Tahamtam and Bornmann 2019; Yousif et al., 2019; Zhang et al., 2023). For most of this time, the central goal has been to classify citations into predefined categories—typically functional types or sentiment polarities—to refine citation metrics or enable large-scale scientometric studies. While this classificatory focus has long depended on subtasks such as citation span detection and reference scope resolution, only recently has the field begun to explore generative tasks such as citation text generation (Funkquist et al., 2023) or citation impact summarization (Arnaout et al., 2025).[1] Yet these developments have largely extended the evaluative logic of citation counting into more sophisticated computational forms.

Although such classificatory and generative capabilities hold clear potential for reconstructive analysis of citation practice, their uptake within this field has remained negligible. Two factors help explain this gap. First, computational CCA has remained closely tied to evaluative scientometric goals, such as improving research assessment, which do not align with HPSS's reconstructive priorities. Second, until the recent emergence of large language models (LLMs), these methods lacked interpretative depth: they could assign categories but rarely captured the rhetorical nuance, contextual variability, and intertextual relations that reconstructive inquiry emphasizes.

LLMs open new ground for connecting computational and reconstructive CCA. Their capacity to process language in context, adapt across domains, and respond flexibly to interpretive prompts makes it possible to automate aspects of analysis that once depended entirely on human close reading. This broadens computational CCA toward reconstructive goals—tracing changes in citation framing over time, comparing representations across disciplines, mapping conceptual and argumentative linkages, clustering related framings, and generating interpretative readings of how a work's meaning develops through uptake. Yet these same capacities pose risks: LLMs have limitations and they cannot and should not replace human interpretation.

This chapter offers a roadmap for integrating LLMs into reconstructive CCA in ways that align with HPSS commitments. The aim is to show how HPSS researchers can harness the new capacities of LLMs for CCA without sacrificing methodological rigor, reflexive skepticism, or sensitivity to the complex social life of citations. For a related discussion of LLMs in CCA, see Liesegang and Gläser (2026).

This chapter proceeds as follows. We begin by clarifying a reconstructive conception of CCA in HPSS and by contrasting it with evaluative scientometrics. We then develop a pluralist account of citation practices and aims, followed by an analysis of cross-contextual dynamics in which citation meanings emerge across citing, cited, and uptake contexts. Building on these foundations, we translate reconstructive aims into LLM-based workflows, moving from context detection to context classification, then to citation–concept and citation–argument mining, and to similarity modeling and clustering of citation framings. Finally, we explore interpretative readings with LLMs, including chained and time-aware designs that generate multiple plausible interpretations and make their evidentiary basis visible. A concluding discussion synthesizes methodological lessons, addresses limits and risks, and sets out practical guidance on tooling, datasets, reproducibility, and reflexive use, with an emphasis on keeping human interpretation central.

---

1 Prior approaches to citation-based document summarization used extractive rather than truly generative methods (e.g., Karimi et al., 2018; Mao et al., 2022).

2. Reconstructive CCA

At the heart of the HPSS tradition lies a commitment to the *reconstruction* of scientific practice—an effort to understand how science is done in practice, how knowledge is shaped, and how epistemic authority is negotiated (e.g., Pickering, 1992). This reconstructive orientation stands in marked contrast to the *evaluative* perspective that underpins much of conventional citation analysis, where citations are primarily treated as proxies for research quality, influence, or impact (Cole and Cole, 1971; Kaplan, 1965). While the evaluative tradition promotes CCA primarily as a means of distinguishing "count-worthy" from marginal references—and, by extension, high-quality from lower-quality papers (e.g., Donner et al., 2025)—the reconstructive perspective views citations as socially situated acts through which scientists shape arguments, establish epistemic positioning, build alliances, and demarcate boundaries (e.g., Cozzens, 1985; Gilbert, 1977; Latour, 1987; Luukkonen, 1997; Myers, 1990; Small, 1980; 2004; Swales, 1986; 1990; White; 2004). Studying these practices offers an entry point into the meaning-making processes of science, revealing how claims are justified, allegiances formed, and disputes settled.

Interpretation is therefore central to this reconstructive project. Understanding citation practices requires attention to nuance, ambiguity, and the contingencies of scholarly communication. Predefined taxonomies and the surface-level cues on which their annotation and classification often depend are valuable starting points for analyzing citation contexts, yet they capture only part of what citations do. Because such approaches typically overlook the broader discursive and historical context—both within the citing text and across the network of citing and cited works—they struggle with cases where meaning is implied rather than stated, where hedging conceals stance, or where subtle rhetorical nuances alter interpretation entirely (cf. Gilbert, 1977; Lyu et al., 2021; Kunnath et al., 2021). As Cronin (1998:48) observed, "a full understanding of why A cites B requires a multi-layered explanation and, ideally, thick description of the process—and the politics of the process". Addressing this complexity requires methods attentive both to the wider contexts in which citations operate and to the implicit cues through which meaning is negotiated.

Traditionally, such methods were qualitative and included close reading, discourse analysis, and historically informed contextualization (Gilbert, 1977; Latour, 1987; Myers, 1990; Swales; 1986; 1990; White; 2004). They remain indispensable for reconstructive CCA, since they enable analysts to grasp the subtleties of argumentation, disciplinary conventions, and historical positioning that shape how citations function. The recent emergence of LLMs introduces new opportunities in this regard. While the aim is rarely to replace qualitative analysis wholesale, in some contexts LLMs may automate portions of the interpretive process—identifying recurrent rhetorical patterns, flagging unusual framings, or generating candidate interpretations for further scrutiny. In many other cases, their value lies in supporting qualitative inquiry within mixed-method designs, where computational breadth and human interpretive depth can be combined (see Hill, 2026; Scharnhorst, 2026; Simons et al., 2026). Used in this way, LLMs can help preserve the contextual richness that reconstructive CCA demands, while extending its reach to larger and more diverse corpora.

3. A pluralist understanding of citation practices and their study

A reconstructive approach to CCA recognizes that citations rarely serve a single, uniform purpose. Decades of empirical research have shown—contrary to the sharp contrasts drawn in earlier debates over normative versus persuasive theories of citation (e.g., Cronin, 1981; Gilbert, 1977; Kaplan, 1965; Latour, 1987; Luukkonen, 1997; Merton, 1973; Small, 2008)—that citations are multifunctional acts, carrying overlapping epistemic, functional, social, rhetorical, and even emotional values (for overviews, see Bornmann and Daniel, 2008; Lyu et al., 2021; Tahamtan and Bornmann, 2018; 2019). A single citation context can do more than one thing at once, sometimes acknowledging intellectual debt while also bolstering a claim's credibility, signaling alignment with a tradition, or positioning one line of work as more authoritative than alternatives. Norm-following and strategic persuasion often reinforce one another, since citing "properly" can both satisfy expectations of credit and simultaneously frame what the cited work is taken to mean, thereby strengthening a claim and securing position within a community (see Small, 2008: 76). More overtly selective moves such as emphasis, reinterpretation, or omission are part of the same adaptable repertoire, whose specific configuration depends on the discipline, genre, audience, and rhetorical context in which they occur.

Variation also arises when reference practices cross the boundaries of scientific discourse. The reconstructive questions that motivate CCA have also been pursued in other domains, sometimes explicitly under the banner of citation analysis and sometimes through adjacent traditions such as intertextuality studies and discourse analysis. Examples include public policy (Simons, 2016, 2015; Steiner-Khamsi, 2022), law (Bodström, 2023; Feneberg et al., 2022), journalism (e.g., Fleerackers et al., 2021; Simons and Schniedermann, 2023), and social media (Barber, 2025), where conventions of referencing and communicative aims differ markedly from those in science. These studies also show that scientific writing is likewise embedded in these wider networks: research articles cite legal decisions, government reports, news stories, and other non-scientific sources, while science itself is routinely cited in policy documents, advocacy materials, and journalism. Such cross-domain exchanges can reshape the meaning of a citation, as works migrate between communicative contexts with distinct rhetorical norms, audiences, and institutional logics.

Pluralism applies not only to citation practices but also to the aims that guide CCA in HPSS. Analysts may be concerned with mapping functional roles, assessing evaluative sentiment, identifying framings of uncertainty, tracing the uptake of particular concepts, or reconstructing the role of citations in relation to specific methods. The same citation context may yield different insights depending on which of these dimensions is under scrutiny, and some may remain latent or ambiguous unless read in light of a specific research question. Because reconstructive aims vary, coding schemes and computational frameworks should be treated as methodological choices rather than fixed standards—adapted, combined, or newly created as the interpretive task demands. At the same time, computational approaches should remain open to plurality, allowing more than one plausible reading to coexist and serving as instruments for exploring interpretive variation rather than enforcing categorical closure.

4. Cross-contextual dynamics of citation meaning

Citation meaning is not fixed within the boundaries of a single sentence or paper; it is emergent across multiple contexts. Interpretation is therefore not only contextual but inherently inter-contextual. To understand what a citation does, we must attend not only to its immediate rhetorical setting but also to the broader discursive web in which it is embedded. At the most basic level, a citation involves two distinct textual contexts: the passage in the citing document that frames the citation, and the cited source itself. These two contexts are rarely neutral mirrors of each other. A citing document may emphasize or downplay particular aspects of the cited work, point to central claims or peripheral remarks, or even misrepresent the original content—sometimes intentionally, sometimes not (Latour, 1987; Small, 2004). In some cases, citations may point to fictional, inaccessible, or misattributed sources.

Beyond this dyadic relationship, citation meanings and the claims they support evolve through their uptake in subsequent literature (Latour, 1987). Later authors often orient their own framings of a cited work to how that work has already been cited in earlier papers. Citation functions and framings shift over time as debates evolve, norms change, and epistemic communities reorganize. Such variation is both historical and concurrent: different communities may interpret the same work in distinct ways even within the same period, while framings also change across time as ideas circulate and are recontextualized. The same work may be cited as dissenting in one community and supportive in another, and over time a contribution that was once peripheral may come to occupy a canonical place in the literature. Patterns of uptake across fields can lead to divergent, even contradictory, interpretations of the same work. Over time, certain framings may stabilize or become canonical (Cozzens, 1985; Gilbert, 1977; Small, 2004). In such cases, we can observe the emergence of typical citation contexts, recurring rhetorical patterns or interpretive framings that circulate through a community and help define its shared assumptions, allowing the cited work to function as what Small (1978) calls a "concept symbol".

These shifts in citation meaning can be understood along two intersecting dimensions proposed by Small (2004): *literalness*, the extent to which a citation accurately reflects the content or intent of the cited work, and *consensus*, the degree of alignment among citing authors about that work's significance. Together, they describe how meanings move between fidelity and reinterpretation, stability and variation. The same source may be read and positioned in markedly different ways across disciplinary boundaries or over time, reflecting divergent epistemic agendas and interpretive norms.

Inter-contextuality also extends beyond individual citations to the relational context in which works are cited together. Co-citation analysis (Small, 1973) has long been used to detect attributed alignments or conceptual proximity within the citing literature, especially at aggregated levels. But as Small (1980) argued early on, co-citation patterns could be significantly enriched by analyzing the contexts in which co-cited documents appear. This would allow for a more textually rich and discursively grounded form of "co-citation context analysis", something well suited to computational approaches using LLMs.

Taken together, these forms of inter-contextuality call for methodological approaches that move beyond sentence- or document-level analysis and take seriously the distributed, evolving, and socially constructed nature of citation meaning. They resonate with semiotic traditions in discourse analysis (Foucault 1970; 1982; Keller 2024), actor-network theory (Callon et al., 1986a; Latour 1987), and theories of intertextuality (Kristeva 1980; Allen 2011), all of which emphasize that meaning emerges relationally, through positioning, uptake, and circulation across networks of discourse and practice.

5. Translating reconstructive CCA into LLM-based workflows

To translate the interpretive goals of reconstructive CCA into computational practice, this section focuses on a set of tasks that are both relevant for reconstructive inquiry and feasible within current NLP capabilities. For each task, we first outline its potential use cases for HPSS—what interpretive questions it might help to pursue—before turning to how the task has been formalized in computational CCA and how such methods might be adapted, extended, or critically reassessed for reconstructive purposes.

In computational CCA, and natural language processing (NLP) more generally, a "task" is a specific, well-defined problem of textual analysis or generation that an algorithm is designed to solve. In the context of CCA, this might include identifying the span of text in which a citation occurs, linking a citation to a particular contribution in the cited work, or generating a summary of how a citation has been taken up in the citing literature over time. Such tasks are formalized to enable large-scale automation, traditionally through supervised or unsupervised learning, and more recently through semi-supervised or prompt-based approaches. While these formalizations open important opportunities, they also shape, and sometimes limit, the ways in which reconstructive aims can be pursued computationally.

The recent rise of transformer-based language models (Vaswani et al., 2017) has transformed how NLP tasks are approached. Within this architecture, encoder models such as BERT (Devlin et al., 2018) read words in relation to both what comes before and after them. This bidirectional analysis makes them especially effective for structural tasks, such as identifying the function or tone of a citation in its immediate context. Decoder models such as GPT (Radford et al., 2018), by contrast, generate text step by step, predicting what comes next based on the preceding sequence. Although primarily generative, these models can also perform analytic tasks like classification or extraction when fine-tuned or guided by carefully designed prompts. More recent systems, such as OpenAI's GPT-5.2, Google's Gemini 3, Anthropic's Claude 4.5, or Meta's LLaMA 4, build on this transformer foundation, combining scale and adaptability with user-friendly chatbot interfaces and multimodal capabilities (text, code, images, and beyond). These advances support applications ranging from structural analysis to open-ended generation with minimal additional training. For reconstructive CCA, this flexibility opens possibilities not only for automating classification but also for pursuing more exploratory and interpretive forms of analysis (see also Simons et al., 2026).

5.1 Citation context detection

Identifying the citation context is more than a preprocessing step. It defines the discursive space in which a citation acquires meaning, where an author positions a source, frames a problem, or signals alignment or distance. Reliable context detection is therefore essential for interpretive analysis: it determines what parts of a text become available for reading, comparison, and interpretation. At the same time, deciding where a citation's context begins and ends is itself an interpretive act, shaped by assumptions about what counts as relevant textual evidence (see also Liesegang and Gläser, 2026). For HPSS, the challenge is not simply to locate citation markers but to capture the textual boundaries within which a citation performs rhetorical or epistemic work.

In computational terms, citation context detection is often reduced to span extraction, that is, identifying a window of text around a citation marker like "Smith et al., 2012", typically the sentence in which it appears. While convenient for large-scale evaluative analysis, this approach risks missing the discursive work that citations perform across broader stretches of text. A meaningful citation context may begin earlier, for example when a problem is first framed or a method introduced, and it may extend across several sentences or even sections. From an interpretive standpoint, the relevant context is defined not by proximity but by semantic relevance: those portions of text that articulate the author's stance toward the cited work, the role it plays in the argument, and the assumptions it mobilizes. Capturing this requires more flexible models of rhetorical segmentation.

Operationally, two intertwined components are involved. First, all citation mentions must be detected (e.g., each occurrence of "Smith et al., 2012"). This is relatively tractable in contemporary scientific writing that follows standardized formats and can be addressed using LLMs (Sarin and Alperin, 2025) or lighter parsers such as Grobid[2]. Greater challenges arise in older or less structured texts, especially when citations appear in footnotes (see Boulanger, 2026). In such cases, markers may be ambiguous ("Ibid.", "op. cit.", "supra note 4") and require cross-footnote tracking or layout-aware models. Here, multi-modal approaches combining linguistic and visual features have proved promising (Khalighinejad et al., 2024; Mishra et al., 2024; Yu et al., 2020).

Second, the model must identify all textual spans that refer to the cited work, including indirect or anaphoric ones (e.g., "this study," "they") and multi-source cases where only part of a sentence pertains to a given citation ("Several studies have explored this issue ..."). These challenges are addressed under coreference and anaphora resolution (Jha et al., 2017; Lauscher et al., 2021; Luan et al., 2018) and reference-scope resolution (Abu-Jbara and Radev, 2012; Jha et al., 2017), which together delineate the dynamic boundaries of citation contexts (Kunnath et al., 2021; 2022; Yousif et al., 2019). Khan et al. (2025) recently proposed a refined approach targeting what they call the "reference text": the minimal fragment that explicitly describes the cited work.

Recent advances in generative frontier models have begun to integrate tasks that earlier approaches treated separately. Large, instruction-tuned LLMs can increasingly

2 https://github.com/kermitt2/grobid

detect citation contexts holistically, identifying relevant spans that extend beyond sentence boundaries and resolving anaphoric or coreferential links in a single pass. When combined with retrieval-augmented or document-grounded prompting (see below), such models can approximate the interpretive judgment needed to delineate where a citation's meaningful context begins and ends, bringing computational methods closer to the reconstructive aims of HPSS.

In sum, detecting citation context is not merely a matter of locating where a citation occurs but of reconstructing the discursive space in which it acquires meaning. The boundaries chosen at this stage shape what can later be classified, compared, or re-read: if too narrow, the citation may appear perfunctory or detached; if too broad, its rhetorical function may blur depending on the classifier used. Context detection is therefore a hermeneutic act, one that decides which traces of discourse count as evidence for interpretation. For HPSS researchers, it is vital to remain aware of how computational methods impose such boundaries, since every segmentation encodes assumptions about where meaning begins and ends. As we turn to the next task, context classification, these issues become even more pronounced, because any attempt to interpret a citation's function or stance depends on first defining what counts as its meaningful textual frame.

5.2 Citation context classification

For reconstructive CCA, citation classification provides a means of making interpretive judgments—such as assessments of citation intent—explicit, inspectable, and systematically comparable. By assigning structured labels to how citations function, researchers can make visible patterns in how knowledge is framed, circulated, and contested across texts. Within HPSS, such classification can serve several interpretive aims, e.g., to compare rhetorical functions across disciplines, to map shifts in evaluative stance over time, or to identify cases that invite deeper qualitative reading. Used reflexively, classification becomes a way of examining how different rhetorical or evaluative functions of citation are patterned within and across scholarly communities, while keeping in view that these functions are themselves interpretive constructions.

In computational CCA, classification is the central analytic task. Technically, it means that a piece of text—the citation context—is given to a model, which then assigns it to one or more predefined categories. Put simply: context in, label(s) out. Early efforts centered on the citer's intent, asking whether a reference was substantive or perfunctory, supportive or critical (Chubin and Moitra, 1975; Moravcsik and Murugesan, 1975; for an overview and recent synthesis, see Lyu et al., 2021). Over time, this classificatory agenda expanded to encompass other dimensions (for an overview, see Kunnath et al., 2021), including sentiment toward the cited work (Yousif et al., 2019b), hedging (Di Marco et al., 2006; Mercer et al., 2004), and perceived importance (Pride and Knoth, 2017; Valenzuela et al., 2015). These are interpretive questions, deeply embedded in disciplinary norms, genre conventions, and authorial intent. Computational models must translate them into predefined categories to operate at scale, but such formalization can only approximate the variability that gives citation meaning its depth. Inevitably, there are trade-offs between the consistency gained through standardization and the nuance that interpretive reading demands.

For HPSS researchers, the first question is whether classification is the right approach at all, and if so, which aspects of citation practice merit operationalization. Is the goal to distinguish rhetorical functions, detect evaluative stances, track framings of uncertainty, or gauge perceived importance? Each calls for different categories and annotation schemes. Another consideration is granularity: some schemes use broad categories such as *support* or *contrast*, which scale well but risk flattening nuance; others use finer distinctions, such as *methodological use*, *background* or *perfunctory*, that preserve interpretive depth but are harder to apply consistently. Scholars must also decide whether classes are mutually exclusive, hierarchical, or overlapping, since a single citation may perform multiple functions. The MultiCite dataset (Lauscher et al., 2021), for example, allows multi-label annotation to reflect this complexity.

Reflecting on such design choices is crucial because all classifiers depend on training and evaluation datasets whose annotation guidelines codify particular interpretive assumptions about what citations mean and how those meanings can be operationalized. Popular datasets illustrate this clearly. ACL-ARC (Jurgens et al., 2018) encouraged document-level interpretation, inviting domain expert annotators to consider broader textual evidence beyond the citing sentence. SciCite (Cohan et al., 2019) uses minimally trained crowdworkers with few constraints on inference, making labels dependent on broader interpretive judgments. ACT (Pride and Knoth, 2020) asked cited authors to label their own citations, often relying on memory of their motivations rather than on the explicit citation text. MultiCite (Lauscher et al., 2021) builds on the ACL-ARC scheme but extends it to multi-sentence and multi-label contexts: annotators identify both explicit and implicit references, select evidentiary spans, and assign one or more intent labels under refined guidelines and adjudication. This balances interpretive inference with explicit evidence requirements and quality control.

Across these datasets, annotators are encouraged, or at least permitted, to make inferences beyond the literal citation wording, meaning that classifiers trained on them inevitably embed interpretive assumptions. A notable exception is Teufel et al. (2006a;b), whose twelve-category scheme was explicitly designed to constrain annotation to surface cues in the citing text itself, avoiding speculation about authorial intent or unstated context. Their framework shows how annotation guidelines can operationalize limits on interpretive scope. Taken together, these examples show how annotation design not only shapes dataset quality but sets the interpretive boundaries of computational CCA, decisions that, for HPSS, are methodological choices about what forms of meaning are rendered legible to machines.

Different technical strategies exist for carrying out classification. Researchers can use off-the-shelf models trained on such datasets or train their own: either encoder-based classifiers such as BERT variants (Beltagy et al., 2019; Lo et al., 2020), or decoder-based classifiers prompted in zero- or few-shot setups (Arnaout et al., 2025; Kunnath et al., 2023; Simons, 2026). The choice depends on resources and the desired balance between interpretive control and scalability. Encoder-based models offer transparency and efficiency but require annotated data and technical expertise. LLM-based approaches reduce those requirements and can flexibly handle multi-label or open-ended schemes, but at the cost of transparency and reproducibility. Hybrid workflows are increasingly viable: LLMs can generate candidate classifications that human experts then audit, correct, or enrich, balancing scalability with interpretive rigor (Bonet-Jover et al., 2023; Dagdalen et al., 2024; Heseltine and Clemm von Hohenberg, 2024; Törnberg, 2024).

In sum, classification remains a useful but limited tool. Its value for HPSS lies not mainly in replicating scientometric metrics but in enabling structured comparisons of how citations function across texts and contexts, in navigating corpora to identify instructive or atypical cases, or in serving as a starting point for deeper interpretative investigation. In all cases, researchers must treat classification as a methodological choice rather than a default and remain reflexively aware of how their categories, tools, and training data shape the interpretations they produce.

5.3 Citation-entity and -argument mining

For reconstructive CCA, the central question is not only what a citation does rhetorically, but how it mobilizes specific intellectual resources, including ideas, concepts, methods, instruments, data, authorities or arguments, in the fine-grained construction of knowledge. What we might call citation-entity and -argument mining is not yet an established task in computational CCA, but it points toward one: detecting the entities and arguments to which a citation is linked. From an HPSS perspective, this matters because it would make citations legible as connective nodes that bind such elements into local chains of reasoning and persuasion (cf. Callon et al., 1983; 1986a; Gilbert, 1977; Latour, 1987).

Related work in NLP has addressed adjacent problems through entity recognition and relation extraction. Entity recognition identifies relevant spans of text and tags them with semantic categories, for example, marking "citation function classification" as a method and "ACL-ARC" as a dataset. Relation extraction then specifies how these entities are connected, for instance, that ACL-ARC is used to evaluate citation function classification systems. While most existing work on scientific entity–relation extraction (Dagdelen et al., 2024; Dunn et al., 2022; Zhang et al., 2024) and argument mining (Accuosto et al., 2021; Akkasi and Moens, 2021; Binder et al., 2022; Dagdelen et al., 2024; Fergadis et al., 2021; Gorur et al., 2024) does not treat references as first-class analytical units, these pipelines can be adapted for citation-context analysis, either by adding a citation-mention entity type to the prediction space or by incorporating a separate citation-mention detector as an additional stage in the extraction pipeline (see 5.1) and extracting relations that connect a cited work to the entities or arguments it is invoked to support. For example, in a phrase like "ACL-ARC (Jurgens et al., 2018)", the citation mention can be tagged and linked to "ACL-ARC" as the dataset introduced or described in the cited work.

As with classification, researchers could pursue citation–entity or citation–argument mining tasks in several ways. Encoder-based models (e.g., BERT or SciBERT variants) remain efficient and transparent, suitable for targeted extraction tasks where annotation schemes and relation types are well defined (Binder et al., 2022; Fergadis et al., 2021; Zhang et al., 2024). Decoder-based LLMs, by contrast, allow more flexible, exploratory setups: prompted in zero- or few-shot modes, they can identify entities, infer relations, and even propose higher-order interpretive links without additional training, though at the cost of reproducibility and control (Gorur et al., 2024; Zhang et al., 2024). A third strategy, demonstrated by Dunn et al. (2022) and Dagdelen et al. (2024), is to fine-tune high-capacity decoder-based models such as GPT-3, T5, or LLaMA-2 for joint entity and relation extraction. These systems can achieve clear gains over both encoder-based baselines and zero- or few-shot prompting, yet their computational costs likely remain prohibitive for most HPSS applications.

When expressed as triplets ($entity_1$, relation, $entity_2$), the mined entity relations around citations could be aggregated into knowledge graphs that allow historical analyses (cf. Boulanger, 2026; Schlattmann et al., 2026), and that connect reconstructive CCA to earlier co-word (Callon et al., 1983; 1986b) and co-citation (Small, 1973; 1980) traditions that trace how meanings crystallize through recurrent associations (see also Chavalarias and Cointet, 2013; Roth and Cointet, 2010). Because citations accumulate and transform through time, every link in a knowledge graph should be timestamped, allowing the resulting network to be explored as an evolving structure rather than a static map. Temporal or longitudinal slices of these graphs could then reveal how the conceptual associations around a cited work expand, contract, or reconfigure as that work is cited across disciplines and decades. For example, a paper first cited for the introduction of a new instrument may later be reframed as a theoretical foundation, as well as reappropriated by different research community along the way. Tracking these shifts through dynamic citation–entity or citation–argument networks would enable the reconstruction of the trajectories through which meanings are stabilized, reinterpreted, or forgotten, including when a cited work's recurrent linkages congeal into a relatively standardized "concept symbol" that circulates across communities.

5.4 Citation context similarity and clustering

While citation–concept and –argument mining focuses on tracing explicit argumentative and conceptual linkages, similarity modeling and clustering offer a complementary means of examining how citation meanings converge and diverge across texts. In doing so, they operationalize the cross-contextual dynamics discussed earlier and offer scalable ways to approximate variation along Small's (2004) dimensions of literalness and consensus: the degree to which citations adhere to or reinterpret the cited work, and the extent to which authors converge on shared framings of it.

Embedding-based similarity modeling provides the technical foundation for this approach. By representing citation contexts as high-dimensional vectors whose relative distances approximate semantic relatedness, such methods make it possible to detect patterns of affinity and contrast among citations. Clustering these representations can reveal families of contexts that share rhetorical tone, evaluative stance, or conceptual alignment, as well as variations that signal disagreement, reinterpretation, or drift in meaning over time.

Technically, these approaches rely on language embeddings, i.e. dense vector representations that capture lexical, syntactic, and semantic patterns. Early distributional models such as Word2Vec (Mikolov et al., 2013) assigned each word a single static vector based on co-occurrence statistics, while transformer-based models produce contextualized word embeddings that vary with surrounding text. Aggregating these into sentence- or paragraph-level representations, for example with Sentence Transformers (Reimers and Gurevych, 2019), enables systematic comparison among citation contexts at multiple scales.

Building on such representations, integrated frameworks like BERTopic (Grootendorst, 2022) combine embedding-based dimensionality reduction, clustering, and keyword extraction to identify coherent clusters of semantically related texts. For reconstructive CCA, these clusters can be interpreted not as objective categories but as families of citation framings, recurring rhetorical or conceptual patterns that indicate degrees of convergence or divergence in how authors mobilize a cited work. Clusters showing consistent evaluative or rhetorical framings may correspond to stabilized interpretive conventions, whereas fragmented or shifting clusters reveal diversity and contestation within the citing community.

Within embedding- and LLM-based approaches, empirical work directly addressing Small's dimension of literalness remains limited. Roy and Mercer (2022) offer one of the few attempts to approximate it computationally, aligning each citing sentence with the passage in the cited paper most likely being referenced. Using standard text-embedding similarity, their system searches the cited text for the segment whose representation lies nearest in semantic space. The approach demonstrates how literal alignment might be modeled, but it also assumes that a suitable target passage always exists and that the closest neighbor corresponds to a genuine interpretive link. In practice, this enforces a narrow, high-literalness reading: close proximity can signal adherence to the cited work, yet the method cannot capture diffuse, cross-sentence, or recontextualized framings. From an HPSS perspective, low similarity as well as the absence of a plausible match, may be equally revealing, pointing to interpretive distance, selective citation, or rhetorical repurposing rather than simple misalignment.

As far as we know, Small's second dimension, consensus, has never been explored explicitly in LLM- or embedding-based studies, though clustering methods suggest how it might be approached. Building on the embedding frameworks discussed above, such as BERTopic, researchers can group citation contexts referring to the same work and compare how that work is described across texts, domains, or time periods. Stable clusters with consistent evaluative or rhetorical framings indicate high consensus and shared interpretive conventions, while fragmented or evolving clusters point to pluralism, disagreement, or shifting interpretations. Because these models can also trace topic evolution, they make it possible to reconstruct how consensus around a cited source consolidates, transforms, or dissolves as debates unfold and as the work circulates across disciplinary and institutional boundaries.

A related but distinct line of work examines semantic distance not to gauge consensus or literalness, but to identify conceptual novelty. Shibayama et al. (2021) measure the semantic distance between co-cited references within a paper to assess how conceptually unusual a given combination is. Greater distance signals higher novelty, reflecting the recombination of ideas drawn from distinct domains. Although developed for evaluative purposes, such measures also resonate with Small's framework: large distances between cited works may correspond to low consensus, or to moments where authors purposefully juxtapose disparate sources to create new intellectual linkages. Repurposed for reconstructive inquiry, this approach could help trace how citation practices bridge or redraw boundaries between fields and how interpretive novelty—rather than evaluative impact—emerges through recombination across time and domains.

Stepping back from these applications, a broader methodological question emerges: What kind of similarity do embedding models actually capture? Their training objectives and corpora shape which relations become salient. Sentence embedders like SPECTER (Cohan et al., 2020) learn similarity from citation links, embedding the assumption that citing papers are conceptually related, even though citations can also mark disagreement or mere acknowledgment. General-purpose models trained on paraphrases instead highlight rhetorical or argumentative affinities. Choosing an embedder is thus not a neutral technical step but a methodological decision that defines what kinds of relations a similarity or clustering analysis can reveal (see Simons et al., 2026).

5.5 Interpretative reading of citation context

While the preceding approaches structure the terrain for interpretation by detecting, classifying, or clustering citation contexts, they stop short of performing interpretation itself. A complementary direction is to make interpretation the object of computation: to use LLMs for generating and interrogating open-ended readings of citation contexts that emulate human interpretive judgment. LLMs can elicit plausible, synthesizing, or diversifying readings, e.g. contextually grounded hypotheses about what a citation is doing or how citation framings evolve across the citing literature. Such uses of LLMs shift computational analysis toward exploratory interpretation, foregrounding plurality and dialogic engagement rather than closure. They also make this form of interpretative reading tractable at scale by generating, comparing, and refining candidate interpretations through interaction with both texts and researchers.

Simons (2026) explores this possibility through an experiment centred on a citation context that served as a touchstone for interpretative debate early on in the development of CCA: footnote 6 in Chubin and Moitra's (1975) foundational paper. Building on Gilbert's (1977) discussion of the same footnote, Simons treats it as a compact example of interpretive instability, one that can be coded as routine on a surface reading but can also support a more pointed rhetorical interpretation on closer inspection (in Gilbert's reading, an implicit priority or credit-related admonishment). His study uses a two-stage prompt-chaining design that retraces this shift in brief: stage one elicits a surface-level classification and sets expectations about the cited sources, while stage two probes that reading by searching for evaluative cues, checking expectations against the relevant texts, and proposing alternative interpretations. By varying prompt complexity and the examples provided, he shows that GPT-5 can generate multiple plausible rereadings, with different settings biasing the model toward certain recurring interpretive themes, without fully determining the readings. More broadly, the experiment underscores that interpretive work, whether human or computational, remains perspectival and highly sensitive to how questions are framed and contexts are read.

A different and temporally-oriented interpretative approach is presented by Arnaout et al. (2025), who use LLMs to reconstruct how a paper's perceived contribution evolves through its citations over time. Their method, developed for "impact summarization", combines fine-grained citation intent classification with time-aware generation, producing short narratives that trace how a work's influence shifts from confirmation to critique and refinement. While framed in scientometric terms, this approach effectively performs an interpretative reading across time: it identifies recurrent framings, detects changes in evaluative stance, and reconstructs the temporal trajectory of a paper's uptake. From an HPSS perspective, such diachronic summaries invite reinterpretation not as metrics of impact but as condensed histories of meaning-making within a citation network.

Together these studies show the opportunities that LLMs open for interpretative work with citation contexts. Simons' chained prompting exemplifies multi-reading generation and adversarial re-reading: using models to propose and contest alternative interpretations, and to expose how meaning depends on framing. Arnaout et al. extend this logic in time through temporal re-reading, reconstructing how interpretative framings of a work shift and stabilize across decades. In Small's (2004) terms, each study engages one dimension of cross-textual meaning. Simons' design triangulates the citation context with both the cited and the surrounding citing text, using expectation checks and cue analysis to probe alternative readings. Arnaout et al. trace patterns of consensus across many citing authors and over time. Integrating these dimensions through fuller cross-text triangulation would allow analysts to examine how citation meanings evolve both in relation to their sources and within their collective uptake.

Looking ahead, retrieval-augmented generation (RAG) offers a way to make interpretative workflows more interactive and evidentially grounded. In RAG systems, an external retrieval step draws on a user-supplied corpus, such as the citing and cited papers, to inform the model's responses. Simons (2026) applied this principle through OpenAI's API, uploading both documents so that GPT-5 could ground its readings in retrieved passages, though his approach was not interactive. By contrast, RAG-based interfaces can support live, conversational engagement, allowing HPSS scholars to query how a work is represented across contexts and examine the underlying evidence together with the model. As discussed by Hill (2026) and Scharnhorst et al. (2026), such designs let researchers "chat" with their sources, supporting interpretive inquiry through interactive, evidence-based dialogue. The gain in interpretative depth and reflexivity is greatest when such systems are used interactively, though this inevitably limits scalability, since interpretive dialogue proceeds at the pace of human reflection rather than machine throughput.

6. Discussion

In this chapter, we proposed using LLMs within citation context analysis (CCA) to recover how scientific claims, alliances, and boundaries are actively built through acts of citing rather than merely reflected in them. We argued that interpretation is central to this reconstructive project, along with a pluralistic, inter-contextual, and evolutionary understanding of citation practices. To translate these goals into computational applications, we sketched a roadmap that uses LLMs to delineate and classify citation contexts, and to trace practices and uptake across fields and over time by both mapping links between citations and the concepts or arguments they mobilize and modeling similarity among citation framings. We also proposed generating scalable, plural, and evidence-grounded interpretative readings of what citations are doing and how those meanings evolve over time.

A key question behind our proposal is whether LLMs can genuinely assist, rather than compromise, the depth of interpretation that reconstructive inquiry requires. This raises a more fundamental issue about the role interpretation should play in reconstructive CCA, and we take the engagement with LLMs as an opportunity to revisit and refine that role. Long before LLMs, White (2004:103) argued that insofar CCA requires "the recovery of implicit meaning", it may "[...]not be delegated to a computer even in principle". Precisely because he foregrounded interpretation, his pragmatic response, shared to a degree by Teufel and colleagues (2006a;b; 2009), was to keep computational schemes anchored in surface, syntactic cues and to reserve interpretive work for human analysts. By contrast, evaluative CCA has often treated classification as readily automatable, assuming that computers can assign meanings without the hermeneutic labor that reconstructive approaches insist upon.

Early debates about the validity of citation interpretation often turned on how much expertise and situated knowledge CCA requires (Cronin, 1981; Edge, 1979; Gilbert, 1977; Peritz, 1983; Swales, 1986; Teufel et al., 2009; White, 2004). Moravcsik and Murugesan (1975:87), for example, criticized sociological readings of physics papers, arguing that sociologists "are not equipped to understand the technical scientific content of the papers they handle". Gilbert (1977:120) similarly argued that an analyst unfamiliar with the Gilbert and Woolgar paper cited in Chubin and Moitra's (1975) "Footnote 6" (see the section on interpretative readings above) was "unlikely [...to] have realized that the reference might possibly have been negational". In response, many CCA scholars have advocated interviewing scientists about their citation motivations (Brooks, 1985; Cano, 1989; Cronin, 1981; Vinkler, 1987), a strategy later adopted in computational CCA to create expert-annotated citation-context datasets for training models (see Bornmann and Daniel, 2008; Tahamtan and Bornmann, 2019). Yet citer-motivation surveys introduce their own interpretive risks, including socially desirable responding on tactical motives and the general limitations of retrospective self-report (Lyu et al., 2021). Early on, Mulkay (1974) proposed a middle path: sociologists of science should collaborate with scientists in reconstructive analyses of practice, but without deference. Analysts must be prepared to recognize, challenge, and triangulate the epistemic and institutional biases of their collaborators; otherwise, they risk reproducing participants' retrospective justifications, professional myths, or strategically curated self-descriptions as sociological insight.

Returning to the question posed at the start of this discussion about whether LLMs can assist reconstructive work, the preceding history shows that interpretation has long held a contested, conditional status in CCA, and LLMs add a new layer to that debate. Used carefully, LLMs can widen coverage, bring much greater contextual nuance than earlier computational methods, and even help surface plural readings grounded in textual evidence. White's skepticism about the possibility of automated CCA beyond surface-level syntactic cues has lost its grip since LLMs have shown that the recovery of contextual and implicit meaning—long treated as the exclusive domain of human interpretation—has become, if not unproblematic, then at least computationally plausible. At the same time, LLMs also address the expertise challenge in new ways by approximating field familiarity through targeted retrieval and modest domain tuning, surfacing terminology, canonical debates, and relevant passages that non-specialists might miss.

The value of using LLMs for reconstructive CCA depends on balancing scalability with interpretive control. That balance means exploiting models for speed and scope without turning interpretation into a black box. Scalability fails when models classify, extract, or "interpret" without exposed evidence, rationale, or reproducible settings. Interpretive control can be gained at three points in the pipeline. At the input stage, tasks should be specified clearly; choices about the scope and length of the citation context can prevent classifications from resting on overly narrow evidence; and prompt design should be tested for nudging effects. At the throughput stage, models should be selected deliberately and matched to the task; when suitable options are lacking, researchers should consider developing small domain datasets for targeted tuning. At the output stage, models can be adjusted to cite the exact passages that support a classification or reading, present plausible alternatives where appropriate, and route difficult cases to human review. Achieving this balance depends on basic LLM literacy, by which we mean enough grasp of model types, prompting, retrieval, and training data to keep outputs grounded, reproducible, and open to challenge (see Simons et al., 2026, as well as the introduction to this volume, Simons et al., 2026).

Ultimately, LLMs cannot and should not replace human interpretation. Reconstructive CCA recognizes that there is no single true reading of a citation or its evolution across contexts. Objectivity can guide inquiry as a regulative ideal, yet it is never fully attainable. This calls for methodological humility, vigilance against overreach, and a commitment to justifying interpretations and making them open to debate. The burden for interpretive judgment remains with researchers. Models can pattern-match and propose, but they cannot bear responsibility.[3]

References

Abu-Jbara A and Radev D (2012) Reference scope identification in citing sentences. In: Proceedings of the 2012 conference of the North American Chapter of the Association for Computational Linguistics: Human language technologies, 2012, pp. 80–90.

Accuosto P, Neves M, Saggion H, et al. (2021) Argumentation mining in scientific literature: from computational linguistics to biomedicine. In: BIR 2021: 11th International Workshop on Bibliometric-enhanced Information Retrieval (eds P Mayr, G Cabanac, and S Verberne), Aachen, 2021, pp. 20–36. CEUR. Available at: http://hdl.handle.net/10230/47600.

Akkasi A and Moens M-F (2021) Causal relationship extraction from biomedical text using deep neural models: A comprehensive survey. Journal of Biomedical Informatics 119: 103820.

3 This chapter was written with support from large language models (LLMs). All model-generated text was reviewed and, where necessary, rewritten by the authors, who remain fully responsible for the final version. For details on the use of LLMs in this volume, see the statement in the volume's introduction.

Allen G (2011) Intertextuality. Second edition. New York: Routledge.

Arnaout H, Sternlicht N, Hope T, et al. (2025) In-depth Research Impact Summarization through Fine-Grained Temporal Citation Analysis. http://arxiv.org/abs/2505.14838.

Barber K (2025) (Don't) click here: Hyperlinks as a quasi-objectification strategy in epistemic legitimisation in extremists' blog posts on sexual violence. Discourse, Context & Media 66: 100912.

Beltagy I, Lo K and Cohan A (2019) SciBERT: A Pretrained Language Model for Scientific Text. http://arxiv.org/abs/1903.10676.

Binder A, Hennig L and Verma B (2022) Full-Text Argumentation Mining on Scientific Publications. In: Proceedings of the first Workshop on Information Extraction from Scientific Publications (eds T Ghosal, S Blanco-Cuaresma, A Accomazzi, et al.), Online, November 2022, pp. 54–66. Association for Computational Linguistics.

Bodström E (2023) Illusions of objectivity: The two functions of country of origin information in asylum assessment. Migration Studies 11(1): 197–217.

Bonet-Jover A, Sepúlveda-Torres R, Saquete E, et al. (2023) Applying Human-in-the-Loop to construct a dataset for determining content reliability to combat fake news. Engineering Applications of Artificial Intelligence 126: 107152.

Bornmann L and Daniel H (2008) What do citation counts measure? A review of studies on citing behavior. Journal of Documentation 64(1): 45–80.

Boulanger C (2026) The potential of LLMs for constructing a socio-legal knowledge graph. In: Simons A, Wüthrich A, Zichert M, et al. (eds) *Understanding Science with Large Language Models? Potentials for the History, Philosophy, and Sociology of Science*. Bielefeld: transcript, part-4.

Brooks TA (1985) Private acts and public objects: An investigation of citer motivations. Journal of the American Society for Information Science 36(4): 223–229.

Callon M, Courtial J-P, Turner WA, et al. (1983) From translations to problematic networks: An introduction to co-word analysis. Social Science Information 22(2): 191–235.

Callon M, Law J and Rip A (eds) (1986a) Mapping the Dynamics of Science and Technology: Sociology of Science in the Real World. London: Palgrave Macmillan.

Callon M, Law J and Rip A (1986b) Qualitative scientometrics. In: Callon M, Law J, and Rip A (eds) Mapping the Dynamics of Science and Technology. Sociology of Science in the Real World. Macmillan, pp. 103–123.

Cano V (1989) Citation behavior: Classification, utility, and location. Journal of the American Society for Information Science 40(4): 284–290.

Chavalarias D and Cointet J-P (2013) Phylomemetic patterns in science evolution—the rise and fall of scientific fields. PloS one 8(2): e54847.

Chubin DE and Moitra SD (1975) Content Analysis of References: Adjunct or Alternative to Citation Counting? Social Studies of Science 5(4): 423–441.

Cohan A, Ammar W, Zuylen M van, et al. (2019) Structural Scaffolds for Citation Intent Classification in Scientific Publications. http://arxiv.org/abs/1904.01608.

Cohan A, Feldman S, Beltagy I, et al. (2020) SPECTER: Document-level Representation Learning using Citation-informed Transformers. http://arxiv.org/abs/2004.07180.

Cole J and Cole S (1971) Measuring the Quality of Sociological Research: Problems in the Use of the "Science Citation Index". The American Sociologist. JSTOR: 23–29.

Cronin B (1981) The need for a theory of citing. Journal of Documentation 37(1): 16–24.

Dagdelen J, Dunn A, Lee S, et al. (2024) Structured information extraction from scientific text with large language models. Nature Communications 15(1): 1418.

Devlin J, Chang M-W, Lee K, et al. (2018) BERT: Pre-training of Deep Bidirectional Transformers for Language Understanding. http://arxiv.org/abs/1810.04805.

Di Marco C, Kroon FW and Mercer RE (2006) Using Hedges to Classify Citations in Scientific Articles. In: Shanahan JG, Qu Y, and Wiebe J (eds) Computing Attitude and Affect in Text: Theory and Applications. The Information Retrieval Series. Dordrecht: Springer Netherlands, pp. 247–263.

Donner P, Stahlschmidt S, Haunschild R, et al. (2025) Does citation context information enhance the validity of citation analysis for measuring research quality? An empirical comparison of peer assessments and enriched citations. Quantitative Science Studies: 1–36.

Dunn A, Dagdelen J, Walker N, et al. (2022) Structured information extraction from complex scientific text with fine-tuned large language models. http://arxiv.org/abs/2212.05238.

Edge D (1979) Quantitative Measures of Communication in Science: A Critical Review. History of Science 17(2): 102–134.

Feneberg V, Gill N, Hoellerer NIJ, et al. (2022) It's Not What You Know, It's How You Use It: The Application of Country of Origin Information in Judicial Refugee Status Determination Decisions – A Case Study of Germany. International Journal of Refugee Law 34(2): 241–267.

Fergadis A, Pappas D, Karamolegkou A, et al. (2021) Argumentation mining in scientific literature for sustainable development. In: Proceedings of the 8th Workshop on Argument Mining, 2021, pp. 100–111.

Fleerackers A, Riedlinger M, Moorhead L, et al. (2021) Communicating Scientific Uncertainty in an Age of COVID-19: An Investigation into the Use of Preprints by Digital Media Outlets. Health Communication 0(0). Routledge: 1–13.

Foucault M (2002) Archaeology of Knowledge. London: Routledge.

Funkquist M, Kuznetsov I, Hou Y, et al. (2023) CiteBench: A Benchmark for Scientific Citation Text Generation. In: 2023 Conference on Empirical Methods in Natural Language Processing, Singapore, December 2023, pp. 7337–7353. Association for Computational Linguistics.

Gilbert GN (1977) Referencing as persuasion. Social Studies of Science 7(1): 113–122.

Gilbert GN and Woolgar S (1974) Essay Review: The Quantitative Study of Science: an Examination of the Literature. Science Studies 4(3): 279–294.

Gorur D, Rago A and Toni F (2024) Can Large Language Models perform Relation-based Argument Mining? http://arxiv.org/abs/2402.11243.

Grootendorst M (2022) BERTopic: Neural topic modeling with a class-based TF-IDF procedure. http://arxiv.org/abs/2203.05794.

Heseltine M and Clemm von Hohenberg B (2024) Large language models as a substitute for human experts in annotating political text. Research & Politics 11(1).

Hill M (2026) The data interview. Reflexive integration of large language models in qualitative content analysis. In: Simons A, Wüthrich A, Zichert M, et al. (eds) *Understanding*

Science with Large Language Models? Potentials for the History, Philosophy, and Sociology of Science. Bielefeld: transcript, part-5.

Iqbal S, Hassan S-U, Aljohani NR, et al. (2021) A decade of in-text citation analysis based on natural language processing and machine learning techniques: an overview of empirical studies. Scientometrics 126(8): 6551–6599.

Jha R, Jbara A-A, Qazvinian V, et al. (2017) NLP-driven citation analysis for scientometrics. Natural Language Engineering 23(1): 93–130.

Jurgens D, Kumar S, Hoover R, et al. (2018) Measuring the evolution of a scientific field through citation frames. Transactions of the Association for Computational Linguistics 6: 391–406.

Kaplan N (1965) The norms of citation behavior: Prolegomena to the footnote. American Documentation 16(3): 179–184.

Karimi S, Moraes L, Das A, et al. (2018) Citance-based retrieval and summarization using IR and machine learning. Scientometrics 116(2): 1331–1366.

Keller R (2024) The Sociology of Knowledge Approach to Discourse. Springer.

Khalighinejad G, Scott S, Liu O, et al. (2024) MatViX: Multimodal Information Extraction from Visually Rich Articles. http://arxiv.org/abs/2410.20494.

Khan D, Ahmed I, Ullah I, et al. (2025) Finding the reference text in citation contexts using attention model. Service Oriented Computing and Applications 19(1): 45–55.

Kristeva J (1982) Desire in Language: A Semiotic Approach to Literature and Art. Columbia University Press.

Kunnath SN, Herrmannova D, Pride D, et al. (2021) A meta-analysis of semantic classification of citations. Quantitative science studies 2(4): 1170–1215.

Kunnath SN, Pride D and Knoth P (2022) Dynamic Context Extraction for Citation Classification. In: The 2nd Conference of the Asia-Pacific Chapter of the Association for Computational Linguistics and the 12th International Joint Conference on Natural Language Processing, Virtual, November 2022.

Kunnath SN, Pride D and Knoth P (2023) Prompting Strategies for Citation Classification. In: Proceedings of the 32nd ACM International Conference on Information and Knowledge Management, Birmingham United Kingdom, 21 October 2023, pp. 1127–1137. ACM.

Latour B (1987) Science in Action: How to Follow Scientists and Engineers through Society. Harvard University Press.

Lauscher A, Ko B, Kuehl B, et al. (2021) MultiCite: Modeling realistic citations requires moving beyond the single-sentence single-label setting. http://arxiv.org/abs/2107.00414.

Liesegang L and Gläser J (2026) Supporting citation context analysis with large language models raises questions that should have been asked 40 years ago. In: Simons A, Wüthrich A, Zichert M, et al. (eds) *Understanding Science with Large Language Models? Potentials for the History, Philosophy, and Sociology of Science*. Bielefeld: transcript, part-6.

Lo K, Wang LL, Neumann M, et al. (2020) S2ORC: The Semantic Scholar Open Research Corpus. http://arxiv.org/abs/1911.02782.

Luan Y, He L, Ostendorf M, et al. (2018) Multi-Task Identification of Entities, Relations, and Coreference for Scientific Knowledge Graph Construction. http://arxiv.org/abs/1808.09602.

Luukkonen T (1997) Why has Latour's theory of citations been ignored by the bibliometric community? Discussion of sociological interpretations of citation analysis. Scientometrics 38(1): 27–37.

Lyu D, Ruan X, Xie J, et al. (2021) The classification of citing motivations: a meta-synthesis. Scientometrics 126(4): 3243–3264.

Mao Y, Zhong M and Han J (2022) CiteSum: Citation Text-guided Scientific Extreme Summarization and Domain Adaptation with Limited Supervision. http://arxiv.org/abs/2205.06207.

Mercer RE, Di Marco C and Kroon FW (2004) The Frequency of Hedging Cues in Citation Contexts in Scientific Writing. In: Advances in Artificial Intelligence (eds AY Tawfik and SD Goodwin), Berlin, Heidelberg, 2004, pp. 75–88. Springer.

Merton RK (1973) The Sociology of Science: Theoretical and Empirical Investigations. Chicago: University Of Chicago Press.

Mikolov T, Sutskever I, Chen K, et al. (2013) Distributed representations of words and phrases and their compositionality. In: Advances in neural information processing systems, 2013, pp. 3111–3119.

Mishra S, Gauquier A and Senellart P (2024) Modular Multimodal Machine Learning for Extraction of Theorems and Proofs in Long Scientific Documents (Extended Version). http://arxiv.org/abs/2307.09047.

Moravcsik MJ and Murugesan P (1975) Some results on the function and quality of citations. Social Studies of Science 5(1): 86–92.

Mulkay MJ (1974) Methodology in the sociology of science: Some reflections on the study of radio astronomy. Social Science Information 13(2): 107–119.

Myers G (1990) Writing Biology: Texts in the Social Construction of Scientific Knowledge. University of Wisconsin Press Madison.

Peritz BC (1983) A classification of citation roles for the social sciences and related fields. Scientometrics 5(5): 303–312.

Pickering A (ed.) (1992) Science as Practice and Culture. University of Chicago Press.

Pride D and Knoth P (2017) Incidental or Influential? – Challenges in Automatically Detecting Citation Importance Using Publication Full Texts. In: Kamps J, Tsakonas G, Manolopoulos Y, et al. (eds) Research and Advanced Technology for Digital Libraries. Lecture Notes in Computer Science. Cham: Springer International Publishing, pp. 572–578.

Pride D and Knoth P (2020) An Authoritative Approach to Citation Classification. In: Proceedings of the ACM/IEEE Joint Conference on Digital Libraries in 2020, Virtual Event China, August 2020, pp. 337–340. ACM.

Radford A, Narasimhan K, Salimans T, et al. (2018) Improving Language Understanding by Generative Pre-Training. OpenAI. Available at: https://cdn.openai.com/research-covers/language-unsupervised/language_understanding_paper.pdf.

Reimers N and Gurevych I (2019) Sentence-BERT: Sentence Embeddings using Siamese BERT-Networks. http://arxiv.org/abs/1908.10084.

Roth C and Cointet JP (2010) Social and semantic coevolution in knowledge networks. Social Networks 32(1): 16–29.

Roy SS and Mercer RE (2022) Biocite: a deep learning-based citation linkage framework for biomedical research articles. In: Proceedings of the 21st workshop on biomedical language processing, 2022, pp. 241–251.

Sarin P and Alperin JP (2025) Citation Parsing and Analysis with Language Models. http://arxiv.org/abs/2505.15948.

Scharnhorst A, Yang H, Touber J, et al. (2026) Co-creation of AI technology, empowering curators of cultural heritage information and guarding research commons. In: Simons A, Wüthrich A, Zichert M, et al. (eds) *Understanding Science with Large Language Models? Potentials for the History, Philosophy, and Sociology of Science*. Bielefeld: transcript, part-5.

Schlattmann R, Kaye A and Vogl M (2026) From source to structure. Extracting knowledge graphs with LLMs. In: Simons A, Wüthrich A, Zichert M, et al. (eds) *Understanding Science with Large Language Models? Potentials for the History, Philosophy, and Sociology of Science*. Bielefeld: transcript, part-4.

Shibayama S, Yin D and Matsumoto K (2021) Measuring novelty in science with word embedding. PloS one 16(7): e0254034.

Simons A (2015) Fact-making in permit markets: document networks as infrastructures of emissions trading. In: Voß J-P and Freeman R (eds) Knowing Governance. Palgrave, pp. 177–192.

Simons A (2016) Documented Authority. The Discursive Construction of Emissions Trading in the Expert Literature. Berlin: Technische Universität Berlin. Available at: https://depositonce.tu-berlin.de/bitstreams/56c1cf1d-2a99-4f36-9844-c8ccbd552 31d/download.

Simons A (2026) Scaling In, Not Up? Testing Thick Citation Context Analysis with GPT-5 and Fragile Prompts. In: Simons A, Wüthrich A, Zichert M, et al. (eds) *Understanding Science with Large Language Models? Potentials for the History, Philosophy, and Sociology of Science*. Bielefeld: transcript, part-6.

Simons A, Wüthrich A and Zichert M (2026) Doing science studies with large language models. An introduction. In: Simons A, Wüthrich A, Zichert M, et al. (eds) *Understanding Science with Large Language Models? Potentials for the History, Philosophy, and Sociology of Science*. Bielefeld: transcript.

Simons A, Zichert M and Wüthrich A (2026) Large language models for history, philosophy, and sociology of science: Interpretive uses, methodological challenges, and critical perspectives. *Studies in History and Philosophy of Science* 117: 102151. https://doi.org/10.1016/j.shpsa.2026.102151.

Small H (1973) Co-citation in the scientific literature: A new measure of the relationship between two documents. Journal of the American Society for information Science 24(4): 265–269.

Small H (1978) Cited documents as concept symbols. Social studies of science 8(3): 327.

Small H (1980) Co-citation context analysis and the structure of paradigms. Journal of Documentation 36(3): 183–196.

Small H (2004) On the shoulders of Robert Merton: Towards a normative theory of citation. Scientometrics 60(1): 71–79.

Swales J (1986) Citation analysis and discourse analysis. Applied Linguistics 7(1). Oxford University Press: 39–56.

Swales J (1990) Genre Analysis: English in Academic and Research Settings. Cambridge University Press.

Tahamtan I and Bornmann L (2019) What do citation counts measure? An updated review of studies on citations in scientific documents published between 2006 and 2018. Scientometrics 121(3): 1635–1684.

Teufel S, Siddharthan A and Tidhar D (2006a) An annotation scheme for citation function. https://aclanthology.org/W06-1312/.

Teufel S, Siddharthan A and Tidhar D (2006b) Automatic classification of citation function. In: Proceedings of the 2006 conference on empirical methods in natural language processing, 2006, pp. 103–110.

Teufel S, Siddharthan A and Batchelor C (2009) Towards domain-independent argumentative zoning: Evidence from chemistry and computational linguistics. In: Proceedings of the 2009 conference on empirical methods in natural language processing, 2009, pp. 1493–1502.

Törnberg P (2024) Best Practices for Text Annotation with Large Language Models. http://arxiv.org/abs/2402.05129.

Valenzuela M, Ha V and Etzioni O (2015) Identifying meaningful citations. In: Workshops at the twenty-ninth AAAI conference on artificial intelligence, 2015.

Vaswani A, Shazeer N, Parmar N, et al. (2017) Attention is all you need. Advances in neural information processing systems 30.

Vinkler P (1987) A quasi-quantitative citation model. Scientometrics 12(1–2): 47–72.

White HD (2004) Citation analysis and discourse analysis revisited. Applied Linguistics 25(1). Oxford University Press: 89–116.

Yousif A, Niu Z, Tarus JK, et al. (2019) A survey on sentiment analysis of scientific citations. Artificial Intelligence Review 52: 1805–1838.

Yu J, Jiang J, Yang L, et al. (2020) Improving Multimodal Named Entity Recognition via Entity Span Detection with Unified Multimodal Transformer. In: Proceedings of the 58th Annual Meeting of the Association for Computational Linguistics (eds D Jurafsky, J Chai, N Schluter, et al.), Online, July 2020, pp. 3342–3352. Association for Computational Linguistics.

Zhang Q, Chen Z, Pan H, et al. (2024) SciER: An Entity and Relation Extraction Dataset for Datasets, Methods, and Tasks in Scientific Documents. In: Proceedings of the 2024 Conference on Empirical Methods in Natural Language Processing, Miami, Florida, USA, 2024, pp. 13083–13100. Association for Computational Linguistics.

Zhang Y, Wang Y, Wang K, et al. (2023) When Large Language Models Meet Citation: A Survey. https://arxiv.org/abs/2309.09727.

Scaling In, Not Up?
Testing Thick Citation Context Analysis with GPT-5 and Fragile Prompts*

Arno Simons

1. Introduction

First introduced nearly fifty years ago in *Social Studies of Science* by Moravcsik and Murugesan (1975) and Chubin and Moitra (1975), citation context analysis (CCA) has been used to specify what in-text references are doing in their immediate context, often so that citation counts and maps of science can be interpreted more cautiously (cf. Bornmann and Daniel, 2008; Kunnath et al., 2021). This ambition has most often been pursued in a scaling-up mode: build comprehensive classification schemes, then develop automated classifiers that can apply them across large corpora. Recent transformer-based models, including fine-tuned encoder models and generative large language models (LLMs), have renewed this scaling-up ambition by making automated classification easier to build and easier to use (e.g., Beltagy et al., 2019; Kunnath et al., 2021; 2023; Nishikawa and Koshiba, 2024; Zhang et al., 2023).

A fundamental critique of scaling up CCA was articulated just as early, most prominently by Gilbert (1977). For classification to travel at scale, readers must be expected to converge on what a reference is doing. Gilbert challenges that expectation by arguing that referencing is often a form of persuasion, directed toward particular audiences in particular situations, so that citation meaning may not be read uniformly. Reviews of citing behavior likewise point to plural motivations, ranging from credit-giving to strategic positioning, and note a further complication for CCA: negative assessments are often voiced indirectly or disguised rather than marked by explicit "negational" cues (e.g., Iqbal et al., 2021; Kunnath et al., 2021; Tahamtam and Bornmann, 2019). Typological CCA's scalability therefore rests on a sociologically fragile premise: that reasonably competent readers can "recognize references as instances of one or other of their categories" and also "share a common reading of the significance and meaning of references" (ibid, p. 119).

* This chapter is almost identical to an earlier preprint (Simons, 2026).

This concern also connects to a broader hermeneutic obstacle. As Cronin (1998, p. 48) notes, "a full understanding of why *A* cites *B* requires a multi-layered explanation and, ideally, thick description of the process — and the politics of the process". Such description presupposes familiarity with the case and its technical and social terrain. As Cozzens (1985, p. 135) puts it, one "needs to know a good deal about the scientific area being studied" before the materials make sense, and even Moravcsik and Murugesan (1975, p. 87) pointed to the limits faced by analysts who are "not equipped to understand the technical scientific content" of the papers they code, a jab aimed at "sociologists" and "people in the library sciences". Even with expertise, however, the upshot is not only that some readings are better than others, but that interpretative plurality is to some extent irreducible, partly because authors themselves leave meanings ambiguous or voice negative assessments indirectly. As Gilbert (1977) and others (e.g., Law and Williams, 1982) emphasize, analyses of scientists' writing practices rely on standpoints that can rarely be finally defended and can at best claim pragmatic force.

These long-standing objections become newly pressing now that LLMs are widely taken to perform something like contextual interpretation, at least in the limited sense of generating plausible readings on demand (cf. Simons et al., 2026). Until recently, a common assumption was that automation could, at best, approximate surface-level classifications of citation contexts (e.g., Teufel et al., 2006; White, 2004), and that interpretation as an "ill-structured" activity (Gläser and Laudel, 2013, p. 2) lay beyond algorithmic capture. White (2004, p. 103) captured this skepticism sharply when he argued that schemes requiring close reading, domain knowledge, expert judgment, and the recovery of implicit meaning "cannot be delegated to a computer even in principle".

Against this background, the aim of this paper is not to use LLMs to scale CCA *up*, but to test whether they can help scale interpretative work *in*. Rather than treating disagreement and multiplicity as noise to be eliminated in the name of a single stable label, I treat them as part of the object and ask whether a prompted model can generate a structured range of text-grounded readings for a single hard case. I use OpenAI's GPT-5 model as a guided co-analyst: it is made to state a surface-level classification first, then to check expectations against the citing and cited texts, attend to lexical and rhetorical cues, and finally propose multiple distinct interpretative hypotheses that can be compared and contested. Taking the idea of interpretative plurality seriously, I do not assume that there is a single "correct" reading that a better prompt would simply recover. If citation meaning varies with standpoint, audience, and the analyst's reconstruction of context, then different ways of instructing the model should also shift what it treats as salient and which readings it generates. I therefore vary prompt structure and framing systematically and examine how those variations redistribute the interpretations the model produces.

The paper is structured as follows. Section 2 introduces footnote 6 in Chubin and Moitra (1975) as a hard case for citation context analysis and reconstructs Gilbert's (1977) interpretative path through it. Section 3 translates that path into a two-stage prompting pipeline and reports the 2×3 prompt-variation design, run plan, and coding plus modeling strategy. Section 4 presents results from the surface-level classification baseline and from the interpretative reconstructions, then tests how scaffolding and prompt framing redistribute the model's hypotheses and vocabulary, and finally compares GPT-5's cue handling and hypotheses to Gilbert's trajectory. Section 5 discusses what these findings imply for "scaling in" interpretative work with LLMs, including methodological opportunities, risks, and limitations, and outlines directions for applying the workflow beyond this single case.[1]

2. Footnote 6 as a hard case

To test what it would mean to scale interpretative CCA in rather than scaling classification up, I return to a canonical hard case from the field's early debates: footnote 6 in Chubin and Moitra (1975) and Gilbert's (1977) reading of it. The note is brief, but it concentrates several of the interpretative problems discussed above, including audience effects, the underdetermination of "function" labels, and the possibility that negative assessment is voiced indirectly rather than signaled by explicit cues. The footnote reads:

> [6] A distinction between 'references' and 'citations' was introduced by D. J. deS. Price, 'Citation Measures of Hard Science, Soft Science, Technology and Non-Science', in C. Nelson and D. Pollock (eds.), Communication Among Scientists and Engineers (Lexington, Mass.: D. C. Heath, 1970), 3–22, esp. 7, then reiterated by G. N. Gilbert and S. Woolgar, 'The Quantitative Study of Science: An Examination of the Literature', Science Studies, 4 (1974), 279–94. We relax that distinction until the findings of our analyses are reported. (Chubin and Moitra, 1975, p. 424)

On a surface reading, this looks like routine attribution and housekeeping. Price is credited with introducing the distinction, Gilbert and Woolgar with repeating it, and Chubin and Moitra announce that they will temporarily set it aside for the purposes of their analysis. Nothing appears overtly polemical.

Gilbert's (1977) treatment of the footnote shows why that appearance is not decisive. He reconstructs the case as a sequence of interpretative moves. First, he treats the footnote as an instance for Chubin and Moitra's own scheme and notes that it could be coded as either "perfunctory" or "additional", but that neither choice "seems self-evidently appropriate" (1977, p. 120). Second, he focuses on the wording, especially "reiterated", and asks why Gilbert and Woolgar are cited at all if they merely repeat Price. Third, he draws on contextual knowledge of the cited works and observes that Gilbert and Woolgar (1974) introduce the distinction without citing Price at the definitional moment. Fourth, he compares that observation with the phrasing "introduced by Price, then reiterated by Gilbert and Woolgar", reading it as a potentially consequential allocation of credit. Finally, he proposes that the first sentence "could... be read as an implied admonishment" for failing to give credit, and that it might therefore merit a negational classification (ibid; emphasis added).

For Gilbert, the point is not to establish the correct label or to claim privileged access to authorial intention. It is to show how quickly classification becomes underdetermined once one takes seriously that intentions are not usually available to the analyst and that

1 Supplementary materials for this paper are available on Zenodo: https://doi.org/10.5281/zenodo.18900264, including https://zenodo.org/records/18900264/files/Supplementary-Sections.pdf, which contains the supplementary sections cited throughout the paper.

competent readers may disagree, especially when they lack insider familiarity with the relevant texts and disputes. In this sense, footnote 6 functions less as an oddity than as a worked demonstration of how interpretative reconstruction can destabilize what looked, at first glance, like a neutral citation.

This is also why footnote 6 has been treated as exemplary in the CCA literature. Swales (1986) singled it out as one of the rare cases where typologies were illustrated with actual citation text rather than category names alone, and he used it to underline how elusive negational references can be. Such references are often consequential for questions of influence and priority, yet they may be easy to miss precisely because disagreement is voiced indirectly or embedded in what otherwise reads as attribution. Later discussions of annotation difficulty and disagreement in citation-function coding have returned to similar issues, including the limits of local cues, praise that conceals critique, scope decisions about what counts as "context", and the practical challenges of resolving references across documents (Iqbal et al., 2021; Kunnath et al., 2021; Tahamtan and Bornmann, 2019).

At the same time, Gilbert's handling of footnote 6 can be read as a template for interpretative CCA. His reconstruction begins with a surface-level coding attempt, then questions its adequacy, isolates potentially loaded wording, checks expectations against the cited and citing texts, and only then proposes an alternative reading. The value of this sequence is not that it delivers a definitive verdict, but that it generates a more articulated space of plausible readings and makes visible the evidential hinges on which those readings turn.

That is the role footnote 6 plays in this paper. It serves as a probe for whether an LLM can be guided through a comparable sequence and, in doing so, contribute candidate reconstructions that remain anchored in textual features while widening the range of interpretations available for scrutiny. The goal is not to replace interpretative judgment, but to test whether some of the labor of producing and organizing alternative readings can be delegated to a prompted model in a controlled way, and to examine what is gained and lost when it is.

The analysis is organized around five research questions that follow directly from this hard case and from Gilbert's critique. First, I ask how stable GPT-5's surface-level classification of footnote 6 is when it is constrained to operate in a typological, scaling-up mode (RQ1). Second, once the model is given the relevant full texts and procedural scaffolding, I ask what kinds of alternative interpretations it generates and how varied they are (RQ2). Third, taking seriously the possibility that interpretative outputs depend on how the task is posed, I test whether changes in prompt structure and framing shift the distribution of interpretations the model produces (RQ3). Fourth, I use Gilbert's admonishment trajectory as a historically salient point of comparison, asking how close the model comes to that reading and where it systematically diverges (RQ4). Finally, I draw these results together to assess what the experiments suggest about using LLMs to scale interpretative work in rather than scaling classification up, including both opportunities and methodological risks (RQ5).

3. Design and methods

This section describes the two-stage prompting pipeline, the 2×3 prompt-variation design, the run plan, and the analytic strategy used to assess variation in GPT-5's outputs.

3.1 Reconstructing and operationalizing Gilbert's path

Starting from the premise that Gilbert's analysis of footnote 6 can serve as a procedural model for computational interpretative CCA, this subsection translates his interpretative moves into discrete subtasks that an LLM can perform. I implemented this as a two-stage pipeline in which one instance of GPT-5 produces an initial surface-level analysis and a second instance carries out the interpretative reconstruction.

In stage one, the model completed two tasks on footnote 6. First, it classified the citation using the six mutually exclusive, hierarchical categories proposed by Chubin and Moitra (1975): *Essential-Basic*, *Essential-Subsidiary*, *Supplementary-Additional-Information*, *Supplementary-Perfunctory*, *Negational-Partial*, and *Negational-Total* (see Supplementary Section S1[2] for an explanation of the scheme). Second, it formulated expectations about the likely content of the cited sources, including their substantive themes and whether they would themselves reference particular works. In both tasks the model was instructed to rely strictly on the explicit wording of the citation context and to avoid speculation about hidden motives or implicit critique. This stage was designed to approximate the kind of surface-level classification that scientometric typologies typically aim for (Teufel et al., 2006). The model's input was a slightly simplified version of footnote 6 with normalized formatting: "A distinction between 'references' and 'citations' was introduced by Price (1970), then reiterated by Gilbert and Woolgar (1974). We relax that distinction until the findings of our analyses are reported".

In stage two, a new model instance received the full stage-one outputs (classification and expectations) together with the full-text PDFs of the citing work, Chubin and Moitra (1975; hereafter C&M), and the cited sources, Price (1970) and Gilbert and Woolgar (1974; hereafter G&W). It was instructed to move beyond the surface label and explore alternative meanings or functions the citation might serve. To do so, it followed a four-step procedure that mirrors the remainder of Gilbert's path: (1) expectation checks against the cited texts, noting confirmations or mismatches with brief quotations, (2) cue analysis of lexical or rhetorical features that might suggest meanings beyond a straightforward classification, (3) extended-context analysis of how footnote 6 functions within C&M (placement, recurrence, co-citation relations, and narrative role), and (4) hypothesis generation. In the final step the model produced five distinct hypothesis–justification pairs. Each hypothesis was required to complicate the surface-level classification, be grounded in textual cues from the citing and/or cited texts, engage with evidence from steps 1–3, and use a different angle of reasoning than the others.

The pipeline combines prompt chaining and stepwise prompting. The shift from stage one to stage two is a chained transition, in which the output of the first call becomes input to a new model instance, preventing later interpretative work from influencing the

2 https://zenodo.org/records/18900264/files/Supplementary-Sections.pdf

initial classification. Within each stage, tasks were elicited stepwise within a single call. This choice supports the paper's aim of generating multiple, articulated readings under explicit constraints rather than maximizing throughput. It also mitigates a known risk in multi-step prompts: models can sometimes produce strategically weak early outputs to set up a stronger "discovery" later (Sun et al., 2024). By separating the surface-level pass from the interpretative reconstruction across two calls, the design reduces that incentive at the key boundary. Given these tradeoffs, I treat intermediate step outputs as heuristic windows on model behavior rather than as a faithful trace of sequential reasoning.

3.2 Experimental configurations

Having outlined the two-stage prompting framework, I now describe how it was implemented under systematically varied conditions. If interpretative CCA depends on how an analyst—or here, an LLM—is guided through the material, then even small changes in prompt design or framing may shift the readings that emerge. To probe this sensitivity, I varied the stage-two prompt across six configurations in a balanced 2×3 design: two base prompt structures crossed with three nudging conditions.

The two base prompts (reproduced in full in Supplementary Section S2[3]) varied the degree of scaffolding:

4-step base prompt: the stage-two procedure described above, with explicit outputs for expectation checks, cue analysis, extended-context analysis, and hypothesis generation.

1-step base prompt: the model received the same stage-one outputs and PDFs but was instructed to move directly to five textually anchored hypothesis–justification pairs, without intermediate steps. The hypotheses still had to complicate the surface-level classification, be grounded in textual cues from the citing and/or cited texts, and differ in angle from one another.

The base prompts were varied to test whether comparable interpretative depth is possible without explicit scaffolding. I nonetheless expected the 4-step prompt to steer the model more toward specific hinges in this case—such as "reiterated" and the (non-)citation of Price in G&W—and therefore to approach Gilbert's admonishment trajectory more readily.

The three nudging conditions were implemented by adding, or not adding, a short paragraph of examples in the "Role and Objective" section at the start of the stage-two prompt:

No-nudge: No extra paragraph was added.

Nudging Toward the specific case of footnote 6: The following paragraph was added: "For example, a neutral reference can mask a corrective undertone, hinting at overlooked con-

3 https://zenodo.org/records/18900264/files/Supplementary-Sections.pdf

nections or antecedents. By highlighting some works and downplaying others, a paper can rewrite an idea's genealogy. A nod to a critical source may mute that critique, acknowledging it only to push it aside. Citations can also legitimize shortcuts, suggesting that methodological slippage is acceptable because others have done it. And by citing across divides, authors may enroll rival audiences while reframing their perspectives as compatible with their own."

Nudging Away from the specific case of footnote 6: The following paragraph was added: "For example, overly generous praise can mask underlying criticism, serving as a veiled form of ridicule, where excess flattery highlights the weakness of the work. An otherwise unnecessary reference may act as a nod to collaborators, reinforcing social ties rather than substantive claims. A cluster of citations can lend borrowed authority even when the sources are only loosely connected. Long reference lists may project encyclopedic coverage without real engagement. And some references work less for peers than for students, pointing readers toward introductory or survey material."

The aim of these nudges was to test susceptibility to example-driven steering. Examples in prompts can improve specificity, but they can also seed storyline templates and vocabulary that narrow the space of readings. The *Toward* nudge was intended to make available a family of interpretations that, for this case, strike me as relatively plausible—lineage or priority work, muted critique, audience bridging, and the strategic legitimation of temporarily relaxing the ref–cit distinction—without channeling a single Gilbert-like conclusion. The *Away* nudge, by contrast, foregrounded interpretations I judged less plausible here—praise-as-ridicule, purely collegial nods, borrowed-authority clusters, and didactic onboarding—in order to stress-test how readily the model follows an offered framing even when it fits the case poorly. Taking interpretative plurality as part of the object rather than noise, the broader goal was to test how such illustrative examples redistribute that plurality: which plausible readings the model produces more often, which it produces less often, and which vocabulary it uses to articulate them. Because the purpose of the study is to widen, rather than narrow, the interpretative space, the *No-nudge* condition and the requirement to produce multiple distinct hypotheses function as counterweights to any single framing.

3.3 Execution and data

All experiments were scripted in Python, which sent prompts to OpenAI's GPT-5 via the API and collected outputs for analysis. The model was run in a high-deliberation ("high-reasoning") configuration, and prompts were issued in batches with responses captured in structured formats for counting and comparison.

Stage one was executed 30 times to assess stability in the surface-level C&M classification of footnote 6 and in the associated expectation notes. Because stage one consistently produced a "supplementary" classification (see Section 4.1), I randomly sampled 15 stage-one outputs to seed stage two.

Stage two implemented the six prompt settings of the balanced 2×3 design (two base prompts × three nudging orientations). Each setting was run once for each of the 15 sampled stage-one outputs, yielding 90 runs in total. Stage two generated 450 hypothesis–justification pairs (90 runs × 5 hypotheses) and, in the 4-step settings, 135 intermediate outputs (45 runs × 3 intermediate steps).

Model specs, tokens, and costs are provided in Supplementary Section S3.[4]

3.4 Close reading and statistical analysis

All experiments were scripted in Python, which sent prompts to OpenAI's GPT-5 via the API and collected outputs for analysis. The model was run in a high-deliberation ("high-reasoning") configuration, and prompts were issued in batches with responses captured in structured formats for counting and comparison.

All outputs from stages one and two were examined through close reading to assess the range and plausibility of the model's interpretative reasoning along Gilbert's path. The main material comprised 450 stage-two hypothesis–justification pairs, each treated as a single interpretative unit. An initial attempt to assign one mutually exclusive theme per unit proved unworkable, since many units combined several interpretative moves and the hypothesis field alone was often too compressed. I therefore shifted to recurring subthemes that can co-occur within a unit and, through iterative reading, comparison, and refinement (cf. Glaser and Strauss, 1967), developed 21 binary codes to capture them (Table 1). For ease of reading, I refer to hypothesis–justification pairs as "hypotheses" from this point onward, and I mark explicitly when I discuss hypothesis and justification fields separately.

Table 1: 21 Codes derived from close reading and inductive analysis of 450 stage-two hypotheses generated by GPT-5. The codes summarize the roles attributed to the citation context (footnote 6) and are ordered by decreasing frequency.

Code	Description: Hypothesis interprets footnote 6 as...	N	%
Agile	Justifying temporary relaxation of the ref–cit distinction and its flexible use	208	46
Preempt	Preempting potential criticism about (mis)use of the distinction	106	24
Aware	Signaling familiarity with the distinction and its policing	73	16
MuteGW	Downplaying G&W's discussion of the distinction	73	16
Bridge	Bridging distinct intellectual camps to signal inclusivity or recruit multiple audiences	68	15
Test	Recasting the distinction as an empirical question to be examined with data	45	10
Pragma	Projecting a pragmatic and empirical stance	39	9
Agency	Asserting authority to reshape the distinction on one's own terms	37	8
Payoff	Deferring the distinction to foreground empirical results as the paper's payoff	34	8
UseGW	Borrowing G&W's authority instrumentally	33	7

4 https://zenodo.org/records/18900264/files/Supplementary-Sections.pdf

Code	Description: Hypothesis interprets footnote 6 as…	N	%
Canon	Normalizing the distinction as consensus through a tidy genealogy	27	6
SSS	Positioning the paper for the SSS journal, appealing to its reviewers' expectations	27	6
SideP	Aligning with Price's bibliometric program rather than with sociology of science	26	6
PrioP	Emphasizing Price's priority or originality	25	6
UseP	Borrowing Price's authority instrumentally	14	3
NegP	Challenging Price's discussion of the distinction	13	3
NegGW	Challenging G&W's discussion of the distinction	10	2
NegGEN	Challenging the discussion of the distinction in general	10	2
Teach	Orienting readers to sources primarily for pedagogy	8	2
MuteGEN	Downplaying the discussion of the distinction in general	6	1
MuteP	Downplaying Price's discussion of the distinction	5	1

To connect these qualitative patterns to experimental conditions, I estimated linear probability models (LPMs) for each of the 21 codes. LPM coefficients are directly interpretable as percentage point differences, which facilitates comparison across codes and conditions (Breen et al., 2018). All models follow the balanced 2×3 design, with indicators for base prompt (*1-step* vs *4-step*), nudging (*Toward*, *Away*, *No-nudge*), and their interaction. I cluster standard errors by run because each run yields five hypotheses under the same settings, so observations within a run may not be independent. This leaves point estimates unchanged but yields more conservative inference by accounting for within-run dependence.

For interpretation I focus on average marginal effects (AMEs), expressed as percentage point differences: AME(*4-step*) compares *4-step* to *1-step* averaged over nudges, while AME(*Toward*) and AME(*Away*) compare each nudge to *No-nudge* averaged over base prompts. These estimands match the design-level questions of interest and provide a compact summary of how scaffolding and framing shift the distribution of coded interpretative moves. Where useful, I report selected pairwise contrasts alongside cluster-robust confidence intervals and p-values. Since the LPMs support joint inference, I also report a 5-df omnibus Wald test across the six settings. This global check is background and does not replace the AMEs, which remain the primary estimands. Because it asks a broad "any effect anywhere?" question, it can be underpowered, so a null result does not imply the AMEs are zero.

See Supplementary Section S4[5] for more details on the motivation and mathematical specification of the LPMs.

I triangulated the LPM results with close readings of stage-one outputs (classifications and expectations) and, in the *4-step* conditions, the intermediate stage-two outputs (expectation checks, cue analyses, and extended-context analyses). These intermediate texts help clarify how particular hypotheses are assembled and where prompt variations

5 https://zenodo.org/records/18900264/files/Supplementary-Sections.pdf

appear to redirect attention. Because the *4-step* outputs were elicited within a single stepwise call, they are best treated as heuristic windows on model behavior rather than as a causal trace of sequential reasoning.

4. Results

This section proceeds in four steps. First, I report the stage-one surface-level classifications and expectations that GPT-5 assigned to footnote 6 as a baseline "scaling-up" pass, and use these to answer RQ1. Second, I map the range of stage-two hypotheses and introduce the 21 recurring subthemes, treating them as a way to describe the interpretative space the model opens up for this hard case (RQ2). Third, I test how prompt structure and illustrative examples redistribute that space by shifting the frequencies and phrasing of these subthemes, including via "lexical echo" (RQ3). Fourth, I use Gilbert's admonishment trajectory as a historically salient probe to compare how the model picks up and resolves the same textual hinges (RQ4).

4.1 Consistent "supplementary" surface-level classification

Stage one yields a stable "supplementary" classification of footnote 6. For the citation of G&W, the model split between *Supplementary-Perfunctory* in 19 of 30 runs (63 percent) and *Supplementary-Additional-Information* in 11 runs (37 percent). In 29 of 30 runs it treated Price and G&W as a single package, assigning both the same label. Only once did it differentiate, tagging Price as *Supplementary-Perfunctory* while giving G&W the slightly weightier *Supplementary-Additional-Information* label, suggesting that "reiterated" can sometimes be read as adding a small increment of informational weight without pushing the citation outside the supplementary range.

The expectation notes are even more uniform. In 29 of 30 runs the model predicted that G&W would themselves cite Price. When it specified the expected function of that Price citation, it was usually *Supplementary-Additional-Information* (23 of 29 cases, 79 percent). A minority framed Price as *Essential-Basic* (5 of 29), and only one as *Supplementary-Perfunctory*, with no *Essential-Subsidiary* or negational expectations. These expectations were stated with little hedging. In a simple regular-expression count over the expectation notes, the model used "expected to cite/reference" in 27 cases and "likely to" in 2, with no instances of "may".[6]

Taken together, stage one reads the footnote as routine attribution and treats "reiterated" as a straightforward cue of acknowledged precedence. The package labeling and confident prediction that G&W will cite Price establish a clean baseline for stage two, where those expectations can be checked against the full texts and the interpretative space can be opened up. This pattern also matches Gilbert's (1977, p. 120) own remark that the note initially looks "additional" or "perfunctory", with possible tensions surfacing only once wider context is brought in. In this sense, stage one answers RQ1 by show-

6 The exact regular expressions were "[Mm]ay (reference|cite)", "[Ll]ikely to (reference|cite)", and "[Ee]xpected to (reference|cite)".

ing that, under a typological, scaling-up constraint, GPT-5's surface reading is highly stable and remains clustered within C&M's *Supplementary* categories.

4.2 A variety of themes in the hypotheses, grounded by textual evidence

Close reading of the 450 hypotheses (that is, the hypothesis–justification pairs) shows both within-run divergence and cross-run recurrence. Within each run, the five hypotheses are consistently different in angle, as required by the prompt. Across runs, those angles nonetheless recur in recognizable families. GPT-5 variously reads footnote 6 as critique management, as a justification for temporarily loosening conceptual discipline, as signaling awareness or authority, as staging an empirical test, as constructing a genealogy of the ref–cit distinction, as borrowing or muting particular authorities, or as bridging audiences. Because many hypotheses bundle more than one move at once, the analysis that follows focuses on the 21 co-occurring subthemes captured by the binary codes in Table 1.

Table 2 illustrates how broader themes arise from different combinations of subthemes. Example 1 reads the footnote as a preemptive shield against terminological criticism (*Preempt*) that licenses pragmatic boundary-blurring between reference coding and citation counts (*Agile*), with an added note of authorial authority (*Agency*). Example 2 turns "relax" into a soft challenge to definitional policing in G&W (*NegGW*), framing the move as exploratory empiricism (*Pragma*) and as an empirical question rather than a rule to enforce (*Test*). Example 3 similarly emphasizes Test, but ties it more directly to the paper's argumentative payoff about why citation counts remain valuable (*Payoff*). Example 4 treats "reiterated" as an alignment device that manufactures the appearance of consensus (*Canon*) by downplaying G&W's critical stance (*MuteGW*) while redeploying their standing (*UseGW*). Example 5 reads the paired citations as audience work, signaling fluency across bibliometric and STS constituencies (*Bridge*, *Aware*, *SSS*) while justifying flexible use of the distinction (*Agile*). Example 6 pushes the genealogy angle further, elevating Price's priority (*PrioP*) and subordinating G&W (*MuteGW*), stabilizing ancestry (Canon) while anticipating possible objections (*Preempt*). Examples 7 and 8 mark two edge readings in the sample: one that leans toward challenging or sidestepping Price's broader program (*UseP*, *NegP*, *MuteP*), and one that foregrounds a didactic pointer function (*Teach*).

These eight examples serve as illustrative anchor points for mapping a spectrum of interpretative distance that also appears across the full set of 450 hypotheses. Examples 1–6 stay closer to the center of the plausible space, elaborating moves that are well anchored in salient wording ("relax", "reiterated", "esp. 7") and in the paper's structure, that is, how the ref–cit distinction is handled across methods, results, and summary. Other hypotheses remain textually anchored but become harder to sustain once the article is read as a whole. Example 7, for instance, treats footnote 6 as borrowing Price's definitional capital while tacitly distancing the paper from his broader program. I take this to sit near the edge of what is plausible, given how extensively C&M rely on Price elsewhere (four citations to three of his works), how explicitly the note frames the move as a pragmatic loosening rather than a repudiation, and the fact that C&M later claim in their summary to have used Price's distinction. Example 8 similarly pushes toward the edge by foregrounding a didactic signposting function that is not obviously prominent in the surrounding argument. That "center-to-edge" spread is analytically useful in this paper's scaling-in frame. The goal is not to compress plurality into a single label, but to see how wide a plausible space of readings the procedure can generate, and where it begins to drift toward reconstructions that informed readers may treat as strained.

*Table 2: Eight of the 450 stage-two hypotheses generated by GPT-5, illustrating how broader interpretative themes arise from different combinations of coded subthemes. In each row, the **bold** text is the original stage-two hypothesis, followed by its justification in plain text. Underlining marks the spans used to ground the assigned codes, listed in order of appearance.*

#	Hypothesis (interpretation of footnote 6)	Broader theme	Sub-themes (Codes)	Setting
1	**The footnote serves as a prophylactic against terminological critique, authorizing the authors to merge reference-level content coding with citation counts without being accused of conceptual slippage:** The phrasing "We relax that distinction" coupled with immediate reliance on both content analysis of references (pp. 426–434) and SCI citation counts (pp. 430–434) suggests deliberate boundary-blurring to support a hybrid design.	Preempting critique to justify conceptual slippage	*Preempt, Agile, Agency*	*4-step No-nudge*
2	**Implicit critique of strict definitionalism in favor of empiricism:** The verb 'relax' and the decision to defer the distinction suggest a gentle pushback against what G&W call for ('we must distinguish carefully,' p. 284). The authors imply that strictness can impede exploratory work and that empirical results will warrant re-imposition later (p. 436).	Empiricisms over definitionalism	*NegGW, Pragma, Test*	*4-step Toward*
3	**Staging an empirical challenge to the necessity of the distinction to buttress their conclusion about counts.:** "We relax that distinction until the findings of our analyses are reported" implies results will adjudicate whether the distinction matters. In the Summary they say the paper "utiliz[es] the distinction between references and citations" and still argues "we cannot afford to abandon citation counts." The opening footnote sets up a narrative in which their data are meant to recalibrate the perceived strictness or practical import of the Price/G&W rule.	Staging an empirical test to reinforce their conclusion	*Test, Payoff*	*1-step Away*
4	**Casting Gilbert & Woolgar as 'reiterating' Price subtly recasts their critical review as endorsing a Pricean distinction, strengthening the appearance of consensus.** G&W (1974) do define the terms (p. 284) but cite Price (1970) in that section mainly for Price's Index (p. 285). C&M's wording ("reiterated by...") smooths over G&W's broader critique of quantitative practices, thereby leveraging their authority to normalize the distinction within C&M's framework.	Creating the appearance of consensus	*Canon, MuteGW, UseGW, Agile*	*4-step No-nudge*

#	Hypothesis (interpretation of footnote 6)	Broader theme	Sub-themes (Codes)	Setting
5	**Social bridging across constituencies to secure legitimacy with multiple audiences:** Pairing Price (bibliometrics/information science) with Gilbert & Woolgar (sociology-of-science critique) in one breath signals fluency in both camps relevant to Social Studies of Science's readership. The announced "relaxation" allows the authors to speak flexibly to both—retaining Price's metrics later (SCI citation counts) while acknowledging sociological sensitivities about categories—thus functioning as audience-alignment rather than mere attribution.	Bridging SSS communities	*Bridge, Aware, SSS, Agile*	*1-step No-nudge*
6	**Genealogical rewriting that situates Price as foundational and domesticate critiques.** The phrasing "introduced... then reiterated" elevates Price as originator and subsumes G&W within his lineage. Given G&W's skepticism about descriptive fitting and their own insistence on the ref/cite distinction (pp. 283–286), this move stabilizes the paper's intellectual ancestry and underplays potential tensions with their approach	Genealogical rewriting to elevate Price	*PrioP, MuteGW, Canon, Preempt*	*4-step Toward*
7	**The citation to Price selectively extracts definitional authority while tacitly distancing the paper from Price's broader program (Price's Index and hard/soft/technology/nonscience taxonomy), which the authors neither adopt nor test:** The reference targets Price (1970) "esp. 7," the page where he codifies the reference–citation terminology, but Chubin & Moitra do not compute Price's Index or engage his hard/soft distinctions in their analysis of high-energy physics. Instead, they develop a reference typology and then relate it to citation counts. The footnote thus borrows Price's definitional capital while sidestepping commitment to his larger, contested framework.	Using Price's authority without engagement	*UseP, NegP, MuteP*	*1-step Toward*
8	**Didactic signposting for non-specialist readers:** Pointing to Price (1970, esp. p. 7) and G&W (1974) directs readers to canonical definitions while economizing space. The move functions as a pointer for readers new to bibliometric terminology, not merely attribution.	Didactic signposting	*Teach*	*4-step Away*

The two edge readings in Table 2 also help anticipate the prompt-sensitivity results that follow. Example 7 comes from a *1-step/Toward* run, and across the full set of 450 hypotheses it is the only case that casts footnote 6 as strongly challenging Price's broader program rather than mainly loosening or managing the ref–cit distinction. More generally, *NegP*-centered hypotheses are more common in 1-step settings (four further cases) than in *4-step* settings (two), and when *NegP* appears elsewhere it is often secondary, bundled with critique of definitional strictness in Price and or G&W. Example 8, by contrast, comes from a *4-step/Away* run, and didactic signposting readings of that kind occur only under *Away*, where the nudge itself makes "references for students" an available storyline. In the scaling-in frame, this suggests that the procedure can widen the space of readings, but that some of that widening reflects how prompts redistribute interpretative attention, including toward reconstructions that remain textually anchored yet feel strained to an informed reader. I return to *NegP* and *Teach* in the next subsection when I quantify these tilts across the 2×3 design.

Viewed as a set, the examples also show why I treat hypotheses as bundles of co-occurring moves rather than as instances of a single underlying "type". The same recurring subthemes are recombined across different broader readings, so hypotheses overlap in codes like *Agile*, *Preempt*, *Canon*, *MuteGW*, *UseGW*, and *Test* rather than sorting into clean bins. This is the practical reason for the 21-code scheme in Table 1, and it bears directly on RQ2: the model's outputs populate a structured but non-discrete space of alternatives, in which interpretative plurality takes the form of recombinable moves.

A further striking feature is how explicitly many hypotheses anchor themselves in textual detail. They routinely cite the footnote's own wording ("relax", "reiterated", "esp. 7") and connect it to the paper's broader organization (how the distinction is invoked across methods, results, and summary). Several also attempt cross-document checks. For instance, Example 4 distinguishes between G&W citing Price at all and citing him for the definitional moment, and uses that mismatch as evidence within the reconstruction. Even when the resulting reading is debatable, this kind of anchoring shows the model doing more than free-associating labels.

Finally, there is a consistent difference in length between *1-step* and *4-step* settings. Total output per run is higher under *4-step* (7,271 vs. 6,469 tokens), but the final hypothesis–justification pairs are shorter (388 vs. 607 characters), mainly because the justifications are briefer (275 vs. 483, while hypotheses are similar at 110 vs. 120). A plausible interpretation is an allocation effect. Based on manual checks of selected probes, the intermediate outputs in the *4-step* condition seem to absorb lengthy quotations, page citations, and micro-arguments, so the final justifications more often condense material that has already been laid out earlier. In *1-step* runs, by contrast, the final justification has to bundle evidence and warrants in one place, which tends to lengthen it. Across the nudge conditions, length differences appear small.

4.3 Prompt structure and framing influence interpretations to some extent

Table 3 gives a first, descriptive map of how the 21 subthemes distribute across the six prompt settings. Two patterns stand out even before modeling. First, much of the repertoire is stable: several frequent codes (notably *Preempt*, and to a lesser extent *Agency*, *Aware*, *Payoff*) show only modest variation across cells, which suggests that the procedure tends to return to a common set of interpretative moves regardless of scaffolding or examples. Second, some subthemes show targeted tilts that line up with the design choices. Under the *4-step* scaffold, *Canon*, *PrioP*, and *UseGW* are more common, especially with *Toward*. Under *1-step*, *Pragma* and *Agile* are more common. By nudge, *Toward* concentrates *MuteGW* and modestly increases *Bridge*, while *Away* concentrates *SSS* and produces the only *Teach*-coded hypotheses. Rare codes remain hard to read from raw counts alone, but *NegGEN* and *NegP* appear most often in *1-step/Away*, with *Toward* often at or near zero.

Table 3: Frequency and relative frequency (%) of the 21 codes across the 2×3 design. Each cell reports the count out of 75 and the corresponding proportion in parentheses. Codes can co-occur, so percentages do not sum to 100% across codes or columns. Base prompts are 4-step and 1-step; nudging orientations are Toward, Away, and No-nudge. N = 450 hypotheses overall; code definitions and totals in Table 1.

	4-step			1-step		
Subtheme (Code)	*Toward*	*Away*	*No-nudge*	*Toward*	*Away*	*No-nudge*
Agency	7 (9)	7 (9)	5 (7)	7 (9)	6 (8)	5 (7)
Agile	29 (39)	30 (40)	35 (47)	40 (53)	33 (44)	41 (55)
Aware	10 (13)	13 (17)	13 (17)	12 (16)	15 (20)	10 (13)
Bridge	14 (19)	11 (15)	10 (13)	13 (17)	12 (16)	8 (11)
Canon	10 (10)	3 (4)	7 (9)	5 (7)	0 (0)	2 (3)
MuteGEN	0 (0)	2 (3)	1 (1)	0 (0)	3 (4)	0 (0)
MuteGW	18 (24)	9 (12)	8 (11)	15 (20)	11 (15)	12 (16)
MuteP	2 (3)	1 (1)	0 (0)	1 (1)	1 (1)	0 (0)
NegGEN	0 (0)	1 (1)	3 (4)	0 (0)	4 (5)	2 (3)
NegGW	2 (3)	1 (1)	1 (1)	1 (1)	4 (5)	1 (1)
NegP	0 (0)	2 (3)	2 (3)	2 (3)	5 (7)	2 (3)
Payoff	8 (11)	5 (7)	6 (8)	5 (7)	6 (8)	4 (5)
Pragma	2 (3)	7 (9)	3 (4)	8 (11)	9 (12)	10 (13)
Preempt	16 (21)	18 (24)	15 (20)	21 (28)	17 (23)	19 (25)
PrioP	9 (12)	5 (7)	2 (3)	3 (4)	4 (5)	2 (3)
SSS	3 (4)	8 (11)	3 (4)	1 (1)	10 (13)	2 (3)
SideP	8 (11)	4 (5)	5 (7)	3 (4)	1 (1)	5 (7)
Teach	0 (0)	5 (7)	0 (0)	0 (0)	3 (4)	0 (0)
Test	6 (8)	6 (8)	11 (15)	7 (9)	6 (8)	9 (12)
UseGW	9 (12)	7 (9)	7 (9)	3 (4)	4 (5)	3 (4)
UseP	3 (4)	4 (5)	3 (4)	2 (3)	2 (3)	0 (0)

To move from the suggestive patterns in Table 3 to a more disciplined test, I use the LPMs. They indicate which tilts are robust in the 2×3 design and how scaffolding and examples redistribute interpretative plurality across the 21 subthemes. Averaged across the three nudging conditions (AME(*4-step*)), moving from the *1-step* prompt to the *4-step* prompt makes *Canon* and *UseGW* more likely, by about +6 percentage points (pp) each. At the same time it makes *Pragma* and *Agile* less likely, by about −7 pp and −9 pp. In substantive terms, the scaffolded procedure makes the model less likely to foreground pragmatic and flexible handling of the ref–cit distinction (*Pragma, Agile*) and more likely to foreground genealogical or authority work (*Canon, UseGW*), including readings that treat "introduced… then reiterated" as doing classificatory housekeeping or selectively redeploying G&W's standing. This tilt is compatible with what the scaffold explicitly requires. By forcing the model to isolate cues, check stage-one expectations against the cited texts, and place the footnote in C&M's wider narrative, it supplies prompts for treating specific phrasings as consequential rather than as merely descriptive.

The *Toward* nudge produces a second, more targeted redistribution. Relative to *No-nudge*, and averaged over both base prompts (AME(*Toward*)), *Toward* increases *PrioP* (+5 pp), *MuteGW* (+9 pp), and *Bridge* (+6 pp), while reducing *NegGEN* (−3 pp). This aligns closely with the role of *Toward* in the design. By supplying illustrative examples that foreground lineage, muted critique, and audience-bridging, *Toward* does not so much create a new reading as make particular ones more available and therefore more frequent.

The *Away* nudge tilts the distribution in a different direction. Compared to *No-nudge* and again averaged over base prompts (AME(*Away*)), *Away* increases *SSS* (+9 pp) and *Teach* (+5 pp) while decreasing *Agile* (−9 pp) and *Test* (−5 pp). The *Teach* shift looks like the clearest case of example-driven steering: didactic readings appear only when the prompt itself supplies "references for students" as an interpretative template. The increase in *SSS* is less direct but still plausible given that the *Away* paragraph foregrounds social and audience functions of citing, which may pull attention toward venue and readership alignment rather than toward treating the footnote as methodological license (*Agile*) or as staging an empirical test of the distinction (*Test*).

Two additional cell-level contrasts (not averaged over base prompts) have 95% confidence intervals that exclude 0: *SideP* is higher in *4-step/Toward* than in *1-step/Toward* (+7 pp) and *1-step/Away* (+9 pp), suggesting that scaffolding plus *Toward* framing jointly encourages alignment-with-Price readings. *NegP* is lower in *4-step/Toward* than in *1-step/Away* (−7 pp), though because *NegP* is rare overall that difference is best treated as suggestive rather than decisive.

The omnibus joint Wald test (5 df) indicates differences across the six settings for *Canon*, *MuteGW*, *NegP*, *NegGEN*, *Pragma*, *SSS*, *SideP*, *Teach*, and *UseP*. For *UseP*, the estimated differences are small (AMEs below 3 pp) and hinge on a single contrast (+5 pp for *4-step/Away* vs *1-step/No-nudge*), so I interpret them cautiously. Several codes show little evidence of differences beyond baseline variation (*Agency*, *Aware*, *Payoff*, *Preempt*), and some show none at all (*MuteP*, *NegGW*), which reinforces the general picture. Prompt structure and examples do not rewrite the interpretative repertoire. They selectively tilt where the model spends its interpretative effort within that repertoire.

Figure 1 plots all average marginal effects as dot-and-whisker estimates with 95 percent cluster-robust confidence intervals around a zero line, arranged in three panels for AME(*4-step*), AME(*Toward*), and AME(*Away*). This makes it easy to see at a glance which effects differ clearly from zero.

Figure 1: Dot-and-whisker plots of average marginal effects (AMEs) for the 21 codes. Each subfigure shows one AME on the x-axis in percentage points for all 21 codes on the y-axis, with dots for estimates and horizontal whiskers for 95% cluster-robust confidence intervals. The vertical line at 0 indicates an effect size of 0. Whiskers that remain entirely on one side of 0 correspond to 95% confidence intervals that exclude 0. Confidence intervals are clustered by run. AME(4-step) compares 4-step to 1-step, averaged over the three nudges. AME(Toward) compares Toward to No-nudge, averaged over the two base prompts. AME(Away) compares Away to No-nudge, averaged over the two base prompts.

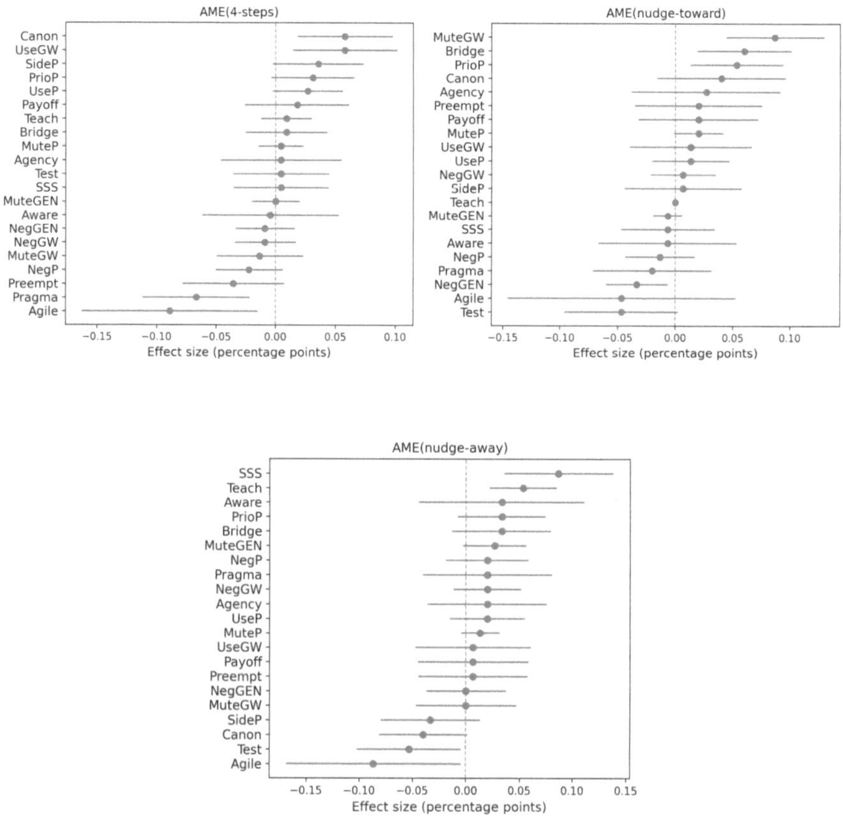

Beyond the subtheme analysis, I also examined "lexical echo", that is, reuse of vocabulary from the nudging paragraphs in the model's hypotheses. I lowercased the text, removed stopwords, and then constructed a set of 78 lexical items drawn from the two nudges. For each hypothesis–justification pair, I coded a binary indicator for the presence of each item and fitted the same 2×3 linear probability models as for the codes (for more details and all results see Supplementary Section S5[7]). This analysis shows that *Toward* increases terms such as "genealogy", "slippage", "reframing", and "audience", while *Away* leaves a weaker and less consistent trace once I adjust for multiple testing. The 4-

7 https://zenodo.org/records/18900264/files/Supplementary-Sections.pdf

step scaffold reduces reuse of generic citation terms shared by both nudges (e.g. "citation", "reference"), while leaving some more analytic vocabulary intact. Taken together, the prompts shape not only which interpretations are more frequent but also the rhetorical repertoire used to state them.

Taken together, these patterns address RQ3 by showing that prompt design redistributes interpretative plurality within a largely stable repertoire of readings. That matters for the paper's scaling-in framing because the aim is to use the model to generate and organize multiple, text-grounded reconstructions of one hard case. On that view, scaffolding (*4-step* vs. *1-step*) and nudges (*Toward*, *Away*, *No-nudge*) are levers that shift which plausible readings the model foregrounds, which it backgrounds, and what vocabulary it uses to articulate them.

4.4 GPT-5 heeds Gilbert's cues but favors muting over admonishment

This subsection treats Gilbert's (1977) reading as a historically salient trajectory through the hard case, not as a ground truth. To my reading, Gilbert's admonishment interpretation is textually well grounded, but it is not the only plausible reconstruction of footnote 6. Price presents the reference–citation distinction as a terminological proposal, and Gilbert and Woolgar echo his language and notation almost verbatim while not crediting him at the definitional sentence, even though they cite the same chapter shortly thereafter for Price's Index and related points. From an attribution perspective, that pattern is underdetermined. It can be read as a meaningful omission or as a minor lapse within an otherwise generous crediting practice. This underdetermination is what gives Chubin and Moitra's phrasing "introduced ... then reiterated" its interpretative range: it can read as routine genealogy, it can read as quietly re-centering credit on Price, and it can also support the stronger possibility Gilbert highlights, a veiled rebuke.[8] Against this backdrop, the question is whether GPT-5 tracks the same hinges (see Section 2), and if so, how it tends to resolve them.

Across the 450 stage-two hypotheses, the model never reproduced Gilbert's admonishment reading in full. It did, however, repeatedly converge on the same cue complex that makes that reading possible. The closest hypotheses, including Examples 4 and 6 in Table 2, treat "introduced ... then reiterated" as a consequential ordering of authorities that foregrounds Price and potentially diminishes G&W. Where Gilbert turns that mismatch toward a veiled rebuke, the model more often turns it into lineage-management and positioning readings that normalize C&M's move and mute potential conflict.

The *4-step* intermediate outputs (see Section 3.1) give a glimpse of how GPT-5 may have moved from cue detection to hypotheses. In expectation checks, the model almost always noticed that G&W do not cite Price when defining the ref–cit distinction (41 of 45

8 A browser based exchange with ChatGPT-5, reproduced in Supplementary Section S7, follows up on this underdetermination and shows in miniature how an LLM can assemble the relevant passages and spell out arguments on both sides, illustrating its potential as a non-human sparring partner for interpretative CCA. See https://zenodo.org/records/18900264/files/Supplementary-Sections.pdf.

runs), often adding the sharper qualification that Price is cited there for other items (for example Price's Index and related points) rather than as the source of the distinction.

In cue analysis, all 45 runs flagged "introduced … then reiterated" as potentially evaluative. When asked what this might suggest under closer scrutiny, 40 of 45 runs resolved it into hierarchy, genealogy, and strategic downplaying of G&W's critical posture. None resolved it into an implied admonishment. In extended-context analysis it also repeatedly observed that Price is a recurring touchstone in C&M while G&W are not, which is consistent with the text. Where the model was less reliable was in extracting co-citation relations from the PDFs. It repeatedly asserted that C&M co-cited Price with other bibliometric sources such as Merton, Cole and Cole, and MacRae, whereas Price is in fact properly co-cited only with MacRae elsewhere in the paper.

Only three hypotheses explicitly stated Gilbert's most diagnostic hinge, namely that G&W define the distinction without crediting Price at the definitional moment. Besides Example 4 (Table 2), the other two were:

> Recasting the genealogy of the distinction to privilege bibliometrics over sociology: Attribution to Price as introducer and G&W as mere reiterators subtly positions the distinction as originating in bibliometrics. Given that G&W don't credit Price for the distinction in their text (they simply define it), Chubin & Moitra's framing recenters credit on Price and aligns the paper with bibliometric authority while only nominally acknowledging sociological uptake. (Example 9, 4-step/Toward)

> The authors simplify divergent positions by attributing a stable 'introduced/reiterated' distinction to create the appearance of consensus, thereby making their decision to 'relax' it appear minor and reversible. Price frames his move as a convention he adopts (Price 1970, p. 7–8), and G&W define the terms without attributing them to Price (1974, p. 284). Calling one an 'introduction' and the other a 'reiteration' compresses nuance, smoothing the way for "We relax that distinction…". (Example 10, 4-step/No-nudge)

In my reading, Example 9 treats "Given that…" plus "recenters credit" as more than a priority claim. It reads C&M as re-centering Price by casting G&W as derivative ("reiteration") and tracing the credit ambiguity to their failure to name him at the definitional point. Even here, though, the thrust is genealogy management and bibliometric alignment, not an explicit chastisement.

Example 10 uses the same mismatch differently. By contrasting "Price frames his move as a convention he adopts" with "G&W define the terms without attributing them to Price", and adding that "[c]alling one an 'introduction' and the other a 'reiteration' compresses nuance", it suggests that C&M's ladder flattens G&W's distinct stance. This nearly implies that the credit relation between Price and G&W needs reconsideration, pulling in the opposite direction to Example 9.

Finally, Example 4 uses the mismatch more lightly. It notes that "G&W (1974) do define the terms (p. 284) but cite Price (1970) in that section mainly for Price's Index (p. 285)", but unlike Examples 9 and 10 the point carries little argumentative weight and reads like an evidential aside. A plausible explanation is procedural: because the prompt required each hypothesis to engage mismatches or cues from earlier steps (see Section 3.1), the model may have mentioned the mismatch mainly to satisfy that anchoring constraint, without integrating it into the core inference.

The above-mentioned overall tendency that the model treats "introduced ... reiterated" as potentially evaluative, yet mostly converts that load into lineage and positioning rather than rebuke also appears when we move from the *4-step* prompts to the full set of runs and look at co-occurrence patterns for the stem of the "reiter" cue. When "reiter" is explicitly invoked in a hypothesis, it most often travels with the same "muting and lineage" cluster of moves (especially *MuteGW*, *Canon*, *UseGW*, *PrioP*, and *SideP*), and this clustering looks broadly similar under both base prompts and across all three nudging conditions, with only minor shifts such as *MuteP* joining more clearly under Away. I also modeled the mention of "reiter" itself as an outcome using an LPM and prompt effects were present but modest. Overall, this LPM suggests that the *4-step/Toward* setting makes the model somewhat more likely to invoke "reiter" explicitly, while the basic co-occurrence cluster around that cue remains broadly similar across prompt structures and nudges. For more details see Supplementary Section S6.[9]

Taken together, these results show that GPT-5 tracks Gilbert's reading closely at the level of cues, yet diverges in how it resolves them. The model repeatedly picks up the same hinges Gilbert used: the "introduced... then reiterated" phrasing and the fact that G&W do not credit Price for the ref–cit distinction. Yet even when hypotheses state this mismatch explicitly, the model never turns it into a veiled admonishment, with Example 9 the closest it comes. Instead, the model reworks the cues into lineage and positioning readings that cast footnote 6 as subsuming G&W under the Price lineage, variously signaling allegiance, glossing over differences, muting G&W's critique, defending C&M's approach, or combining these moves. This answers RQ4. At the same time, close reading of Examples 4, 9, and 10, together with the "reiter" co-occurrence cluster and LPM, offers a glimpse into the model's working: similar cue configurations tend to yield similar muted-lineage readings across prompt structures and nudges, even though *4-step/Toward* slightly increases explicit "reiter" use and produces the most Gilbert-adjacent hypotheses. More generally, some hypotheses integrate earlier cues as the argumentative spine, while others mention them only in passing.

5. Conclusion and discussion

This paper set out to test an unconventional ambition for computational citation context analysis (CCA). Instead of using an LLM to scale up the classification of what in-text references are doing, it asks what it would mean to scale interpretative work *in*. The scaling-up frame treats citation meaning as something that should converge into stable labels that can travel across corpora. The scaling-in frame treats disagreement, ambiguity, and underdetermination as part of the object, and asks whether an LLM can help generate "thick description" (Cronin, 1998, p. 48) of citation contexts by widening and organizing a space of competing reconstructions across cases. The methodological challenge, then, is how a prompted LLM can be made to expand, structure, and textually anchor that space in

9 https://zenodo.org/records/18900264/files/Supplementary-Sections.pdf

a way that remains inspectable, contestable, and usable for sociological interpretation. To address this, I developed a two-stage prompting pipeline and a 2×3 prompt-variation design, which I apply here as a proof of principle to one canonical hard case, footnote 6 in Chubin and Moitra (1975). The case is well suited because it makes the limits of surface labeling unusually visible, and because Gilbert's (1977) reconstruction provides a concrete interpretative sequence that can be translated into a promptable procedure.

Across the experiments, three implications stand out. First, LLMs can function as a guided co-analyst that helps populate and organize interpretative plurality around a hard case, while leaving final judgement with human readers. Second, prompt structure and framing operate as methodologically consequential levers in a scaling-in workflow, because they tilt which regions of the interpretative space are explored more often. Third, translating an interpretative path into a prompt-chained pipeline turns tacit reading routines into explicit components that can be shared, inspected, and experimentally varied, which is especially valuable when the aim is not convergence on one label but a defensible mapping of alternatives.

The findings from RQ1–RQ4 suggest that LLMs can contribute meaningfully to CCA in a scaling-in frame, precisely by not behaving like a scaling-up classifier. In stage one, I separated a scaling-up style surface pass for classification from an initial scaling-in move. GPT-5 produced a remarkably stable surface-level classification of footnote 6 (RQ1), and it also generated expectation notes about how Price and Gilbert and Woolgar were likely to relate to one another based strictly on the citation text, which serve as a baseline for later checks. In stage two, the pipeline deepened the scaling-in procedure by moving from citation-text-only expectations to cross-text testing and interpretative reconstruction. The model was asked to check the stage-one expectations against the full texts and then generate multiple distinct hypotheses that complicate the initial reading. Once it had access to the cited and citing materials, it repeatedly noticed where the stage-one expectations failed, distinguished between Gilbert and Woolgar citing Price in general and not citing him for the ref–cit distinction, and drew on cues such as "introduced...reiterated" and "relax" as well as the paper's wider argumentative handling of the distinction.

Across runs, this yielded a structured range of hypotheses that were both textually grounded and interpretatively diverse, from readings adjacent to Gilbert's admonishment trajectory (RQ4) to more adventurous proposals that nonetheless tried to remain evidence-based (RQ2). Many hypotheses fell within what I consider a plausible interpretative band, and some would not have occurred to me on my own. In scaling-in terms, that is the core gain. The pipeline can generate many alternative, textually anchored readings at a stable level of procedural attentiveness, which can counteract the tendency of human analysts to compress interpretative variation once routines become familiar.

On this view, automation's value for interpretive CCA is not faster labeling but assistance with the hardest part of reconstructive work: producing, contrasting, and warranting multiple plausible readings. A model that can generate text-grounded reconstructions on demand can widen the space of candidate interpretations beyond what any single reader would typically produce unaided, while keeping judgement and argumentative responsibility with the analyst. Translating Gilbert's sequence into an explicit procedure also makes hypothesis generation more visible and more reproducible, shifting it from tacit craft to an inspectable workflow. A complementary way LLMs can support interpretative CCA is through interactive "sparring" in a chat interface, where follow-up questions probe warrants, elicit counterreadings, and make plausibility criteria explicit for a case such as footnote 6. The downside is that this dialogic format is harder to standardize and reproduce than a scripted pipeline and is more susceptible to path dependence and user steering (see Supplementary Section S7[10] for a brief illustration).

A second implication is that the interpretative space opened up by the model is not neutral. The prompt sensitivity analysis shows that design choices about scaffolding and examples shift the distribution of hypotheses (RQ3). The *Toward* and *Away* nudges seeded some storylines and sidelined others. More scaffolding in the *4-step* prompt, especially when paired with the case-specific *Toward* nudge, tended to produce hypotheses closer to the center of the plausible interpretative space, often elaborating priority, muted critique, and bridging. The *1-step* prompt and the *Away* nudge produced a higher share of edge cases, including readings that depended on less plausible leaps from the textual cues and the paper's argumentative structure.

At the same time, the hypothesis range likely also reflects the model's pre- and post-training. The fact that GPT-5 never reproduced Gilbert's rebuke interpretation under my scaffolding and examples may not be only a prompt effect. Instruction-tuned GPT-family models are optimised via preference feedback and safety objectives that favour co-operative, non-escalatory continuations, and they have documented agreeableness tendencies (Ouyang et al., 2022; Sharma et al., 2025). It is plausible that the model has a soft bias against conflict-forward reconstructions such as admonishment readings unless prompts explicitly license them, for example by inviting the possibility of veiled reprimand or credit-policing even when the local tone is neutral. More adversarial instructions, or other models with different alignment regimes, might yield a different balance of readings.

I suggest that this sensitivity, both to prompts and to model training, is both a risk and a feature. It is a risk because prompts can amplify the analyst's own blind spots and preferred storylines, and model-level priors can further narrow which regions of the interpretative space are readily explored, which makes explicit documentation of prompt choices and design rationales essential. But it is also a feature because understanding how prompts and model training shape the interpretative space is crucial for improving outputs for a given use case. In the case of footnote 6, scaffolding and case-specific framing improved plausibility in a way that resembles guiding a student or collaborator through a difficult passage. The methodological task is therefore to design prompts that are loose enough to admit surprise and disagreement while specific enough to keep the model within a historically and textually plausible band, and to treat model choice and alignment settings as part of that methodological design rather than as an invisible background condition.

The third implication is reflexive. Translating Gilbert's analysis into a prompt-chained pipeline forced me to spell out interpretative moves that would otherwise remain tacit. These included how to formulate expectation checks, how to search for cues that might destabilize a surface-level label, and how to elicit multiple hypotheses that

10 https://zenodo.org/records/18900264/files/Supplementary-Sections.pdf

are genuinely distinct in angle rather than rephrasings of one another. Once such steps are written down, they become available as experimental levers that can be adjusted and recombined. In this sense, the model "prompts back": The need to instruct it pushes toward a more explicit articulation of concepts and reasoning paths, and the sensitivity results provide an empirical picture of where those instructions grip most strongly and where they fade. For interpretative CCA this is an opportunity. LLM based experiments can be used not only to generate candidate readings of citation contexts but also to interrogate and refine the interpretative repertoires that human analysts bring to such readings in the first place.

From a methodological perspective, the study offers a proof of concept for how a scaling-in reading procedure can be translated into a reusable prompting architecture. I instantiated this design by following Gilbert's path through footnote 6, but the two-stage structure is more general. Stage one performs surface-level classification and expectation formulation based on the citation context alone, closely aligned with scaling-up ambitions. Stage two takes those expectations and the full texts as input and guides the model through checks, cue analysis, extended-context analysis, and hypothesis generation, which is aligned with scaling-in ambitions. Because prompts, model settings, and code can be shared, the pipeline turns tacit reading practices into an articulated workflow. By varying base prompts and nudging orientations in a simple 2×3 design and combining close reading with inductive coding and linear probability models, the study also shows how prompt design itself can be treated as an object of analysis rather than an invisible precondition.

The study has clear limitations. First, it centers on a single, well-documented case, so the pipeline is a proof of concept that needs testing on other citation contexts before broader claims are warranted. Second, coding and close reading were carried out by me alone, which means the quantitative contrasts reflect a single reader's interpretative judgements rather than a multi-coder standard. Third, GPT-5 is a proprietary model with opaque training data and internal processes, and I did not compare its outputs to those of other systems, including models with different pre- and posttraining regimes that may populate the interpretative space differently. Fourth, the approach is resource intensive and is best suited to a small number of strategically chosen cases rather than large-scale deployment, which is a deliberate trade-off given the scaling-in aim. Finally, implementing the pipeline requires a degree of LLM literacy, from scripting to prompt design, which may restrict who can apply similar setups in practice.

With these limitations in view, we can return to White's (2004, p. 103) claim that the recovery of implicit meaning "cannot be delegated to a computer even in principle" and to Gilbert's insistence that his footnote 6 reading requires familiarity with the broader context. The experiments reported here suggest that these claims look less absolute in light of current models, at least when the task is posed in a scaling-in way. GPT-5 could carry out several procedural tasks that expertise often involves, including cross-text expectation checking and generating text-grounded hypotheses about what a citation is doing in context.[11] In that sense, an LLM can stand in for some aspects of expert familiarity by

11 In addition, GPT-5 handled several seemingly more "mundane" subtasks in a strikingly competent way, including basic PDF parsing and passage retrieval. At the same time, some operations proved

widening the interpretative possibility space and reorganizing interpretative labor. At the same time, the human analyst retains responsibility for deciding which regions of that space to cultivate, which hypotheses to discard, and how to carry promising readings further. This collaborative configuration, with the model as a guided co-analyst and the human reader as curator and critic, is how I answer RQ5. The question is therefore no longer whether contextual interpretation and the recovery of implicit meaning can be automated in principle, but how we choose to combine computational and human capacities in the reconstructive study of citation practices (cf. Simons et al., 2026). [12]

Funding statement

This work was supported by the European Union under an ERC Consolidator Grant (Project No. 101044932, "Network Epistemology in Practice (NEPI)"). Views and opinions expressed are however those of the author only and do not necessarily reflect those of the European Union or the European Research Council. Neither the European Union nor the granting authority can be held responsible for them.

References

Beltagy I, Lo K and Cohan A (2019) SciBERT: A Pretrained Language Model for Scientific Text. http://arxiv.org/abs/1903.10676.

Bornmann L and Daniel H (2008) What do citation counts measure? A review of studies on citing behavior. *Journal of Documentation* 64(1). Emerald Group Publishing Limited: 45–80.

Breen R, Karlson KB and Holm A (2018) Interpreting and Understanding Logits, Probits, and Other Nonlinear Probability Models. *Annual Review of Sociology* 44(Volume 44, 2018). Annual Reviews: 39–54.

Chubin DE and Moitra SD (1975) Content Analysis of References: Adjunct or Alternative to Citation Counting? *Social Studies of Science* 5(4): 423–441.

Cozzens SE (1985) Comparing the Sciences: Citation Context Analysis of Papers from Neuropharmacology and the Sociology of Science. *Social Studies of Science* 15(1): 127–153.

Cronin B (1998) Metatheorizing citation. *Scientometrics* 43(1): 45–55.

Gilbert GN (1977) Referencing as persuasion. *Social Studies of Science* 7(1): 113–122.

Gilbert GN and Woolgar S (1974) Essay Review: The Quantitative Study of Science: an Examination of the Literature. *Science Studies* 4(3): 279–294.

---

more fragile, especially tracking co-citations across long texts and footnotes and reconstructing citation networks.

12 This chapter was written with support from large language models (LLMs). All model-generated text was reviewed and, where necessary, rewritten by the authors, who remain fully responsible for the final version. For details on the use of LLMs in this volume, see the statement in the volume's introduction.

Glaser B and Strauss A (1967) *The Discovery of Grounded Theory: Strategies for Qualitative Research*. New Brunswick and London: Aldine Transaction.

Gläser J and Laudel G (2013) Life With and Without Coding: Two Methods for Early-Stage Data Analysis in Qualitative Research Aiming at Causal Explanations. *Forum Qualitative Sozialforschung / Forum: Qualitative Social Research* 14(2).

Iqbal S, Hassan S-U, Aljohani NR, et al. (2021) A decade of in-text citation analysis based on natural language processing and machine learning techniques: an overview of empirical studies. *Scientometrics* 126(8): 6551–6599.

Kunnath SN, Herrmannova D, Pride D, et al. (2021) A meta-analysis of semantic classification of citations. *Quantitative science studies* 2(4). MIT Press One Rogers Street, Cambridge, MA 02142–1209, USA journals-info ...: 1170–1215.

Kunnath SN, Pride D and Knoth P (2023) Prompting Strategies for Citation Classification. In: *Proceedings of the 32nd ACM International Conference on Information and Knowledge Management*, Birmingham United Kingdom, 21 October 2023, pp. 1127–1137. https://dl.acm.org/doi/10.1145/3583780.3615018.

Law J and Williams RJ (1982) Putting facts together: A study of scientific persuasion. *Social Studies of Science* 12(4): 535.

Moravcsik MJ and Murugesan P (1975) Some results on the function and quality of citations. *Social Studies of Science* 5(1): 86–92.

Nishikawa K and Koshiba H (2024) Exploring the applicability of large language models to citation context analysis. *Scientometrics* 129: 6751–6777.

Ouyang L, Wu J, Jiang X, et al. (2022) Training language models to follow instructions with human feedback. In: *Advances in Neural Information Processing Systems 35 (NeurIPS 2022)*, 2022.

Price DJ (1970) Citation measures of hard science, soft science, technology, and nonscience. In: Nelson CE and Pollock DK (eds) *Communication Among Scientists and Engineers*. Heath Lexington, MA, pp. 3–22.

Sharma M, Tong M, Korbak T, et al. (2025) Towards Understanding Sycophancy in Language Models. http://arxiv.org/abs/2310.13548.

Simons A (2026) Scaling In, Not Up? Testing Thick Citation Context Analysis with GPT-5 and Fragile Prompts. arXiv:2602.22359. http://arxiv.org/abs/2602.22359.

Simons A, Arnaout H and Gurevych I (2026) Reconstructive citation context analysis using large language models. A roadmap. In: Simons A, Wüthrich A, Zichert M, et al. (eds) *Understanding Science with Large Language Models? Potentials for the History, Philosophy, and Sociology of Science*. Bielefeld: transcript, part-6.

Sun S, Yuan R, Cao Z, et al. (2024) Prompt Chaining or Stepwise Prompt? Refinement in Text Summarization. http://arxiv.org/abs/2406.00507.

Swales J (1986) Citation analysis and discourse analysis. *Applied Linguistics* 7(1). Oxford University Press: 39–56.

Tahamtan I and Bornmann L (2019) What do citation counts measure? An updated review of studies on citations in scientific documents published between 2006 and 2018. *Scientometrics* 121(3): 1635–1684.

Teufel S, Siddharthan A and Tidhar D (2006) Automatic classification of citation function. In: *Proceedings of the 2006 conference on empirical methods in natural language processing*, 2006, pp. 103–110. https://aclanthology.org/W06-1613.pdf.

White HD (2004) Citation analysis and discourse analysis revisited. *Applied Linguistics* 25(1). Oxford University Press: 89–116.

Zhang Y, Wang Y, Wang K, et al. (2023) When Large Language Models Meet Citation: A Survey. https://arxiv.org/abs/2309.09727v1.